Salvatore Greco, Ricardo Alberto Marques Pereira, Massimo Squillante, Ronald R. Yager, and Janusz Kacprzyk (Eds.)

Preferences and Decisions

Studies in Fuzziness and Soft Computing, Volume 257

Editor-in-Chief

Prof. Janusz Kacprzyk
Systems Research Institute
Polish Academy of Sciences
ul. Newelska 6
01-447 Warsaw
Poland
E-mail: kacprzyk@ibspan.waw.pl

Further volumes of this series can be found on our homepage: springer.com

Vol. 243. Rudolf Seising (Ed.)
Views on Fuzzy Sets and Systems from Different Perspectives, 2009
ISBN 978-3-540-93801-9

Vol. 244. Xiaodong Liu and Witold Pedrycz
Axiomatic Fuzzy Set Theory and Its Applications, 2009
ISBN 978-3-642-00401-8

Vol. 245. Xuzhu Wang, Da Ruan, Etienne E. Kerre
Mathematics of Fuzziness – Basic Issues, 2009
ISBN 978-3-540-78310-7

Vol. 246. Piedad Brox, Iluminada Castillo, Santiago Sánchez Solano
Fuzzy Logic-Based Algorithms for Video De-Interlacing, 2010
ISBN 978-3-642-10694-1

Vol. 247. Michael Glykas
Fuzzy Cognitive Maps, 2010
ISBN 978-3-642-03219-6

Vol. 248. Bing-Yuan Cao
Optimal Models and Methods with Fuzzy Quantities, 2010
ISBN 978-3-642-10710-8

Vol. 249. Bernadette Bouchon-Meunier, Luis Magdalena, Manuel Ojeda-Aciego, José-Luis Verdegay, Ronald R. Yager (Eds.)
Foundations of Reasoning under Uncertainty, 2010
ISBN 978-3-642-10726-9

Vol. 250. Xiaoxia Huang
Portfolio Analysis, 2010
ISBN 978-3-642-11213-3

Vol. 251. George A. Anastassiou
Fuzzy Mathematics: Approximation Theory, 2010
ISBN 978-3-642-11219-5

Vol. 252. Cengiz Kahraman, Mesut Yavuz (Eds.)
Production Engineering and Management under Fuzziness, 2010
ISBN 978-3-642-12051-0

Vol. 253. Badredine Arfi
Linguistic Fuzzy Logic Methods in Social Sciences, 2010
ISBN 978-3-642-13342-8

Vol. 254. Weldon A. Lodwick, Janusz Kacprzyk (Eds.)
Fuzzy Optimization, 2010
ISBN 978-3-642-13934-5

Vol. 255. Zongmin Ma, Li Yan (Eds.)
Soft Computing in XML Data Management, 2010
ISBN 978-3-642-14009-9

Vol. 256. Robert Jeansoulin, Odile Papini, Henri Prade, and Steven Schockaert (Eds.)
Methods for Handling Imperfect Spatial Information, 2010
ISBN 978-3-642-14754-8

Vol. 257. Salvatore Greco, Ricardo Alberto Marques Pereira, Massimo Squillante, Ronald R. Yager, and Janusz Kacprzyk (Eds.)
Preferences and Decisions, 2010
ISBN 978-3-642-15975-6

Salvatore Greco,
Ricardo Alberto Marques Pereira,
Massimo Squillante, Ronald R. Yager,
and Janusz Kacprzyk (Eds.)

Preferences and Decisions

Models and Applications

Springer

Editors

Prof. Salvatore Greco
Department of Economics and
Quantitative Methods
University of Catania,
Corso Italia 55
95129 Catania, Italy
E-mail: salgreco@unict.it

Prof. Ricardo Alberto Marques Pereira
Department of Computer and
Management Sciences
University of Trento
Via Inama 5
38122 Trento, Italy
E-mail: ricalb.marper@unitn.it

Prof. Massimo Squillante
Department of Economic and
Social Systems Analysis
University of Sannio at Benevento
Via delle Puglie
82100 Benevento, Italy
E-mail: squillan@unisannio.it

Prof. Ronald R. Yager
Machine Intelligence Institute
Iona College
New Rochelle, NY 10801, USA
E-mail: yager@panix.com

Prof. Janusz Kacprzyk
Systems Research Institute
Polish Academy of Sciences
ul. Newelska 6
01–447 Warsaw, Poland
E-mail: kacprzyk@ibspan.waw.pl

ISBN 978-3-642-15975-6 e-ISBN 978-3-642-15976-3

DOI 10.1007/978-3-642-15976-3

Studies in Fuzziness and Soft Computing ISSN 1434-9922

Library of Congress Control Number: 2010934863

Typeset & Cover Design: Scientific Publishing Services Pvt. Ltd., Chennai, India.

Printed on acid-free paper

9 8 7 6 5 4 3 2 1

springer.com

This volume is dedicated to Professor Mario Fedrizzi, our distinguished colleague and dear friend, on the occasion of his 60th anniversary which was celebrated by the scientific community during the TRENTO 2009 international workshop on Preferences and Decisions, held in Trento (Italy) on April 6-8, 2009.

Professor Mario Fedrizzi received his B.Sc. in Mathematics from the University of Padua and his M.Sc. in Operational Research from the University of Venice. Since then he has been associated with the University of Trento, where he founded the Applied Mathematics group at the Faculty of Economics around 25 years ago, remaining ever since the scientific and organizational leader of the group.

Moreover, Professor Mario Fedrizzi has over the years been appointed to prominent academic positions as Head of Department, Dean of Faculty, and Vice-Rector, and has also been entrusted with prestigious top management positions within the economic community of Trento and beyond.

The TRENTO 2009 international workshop was chosen to best celebrate Professor Mario Fedrizzi's anniversary by gathering a group of top scientists to present papers in areas in which he has long been active. The international workshop has also been an excellent opportunity to celebrate the close friendship and long term research collaboration between Trento and the Catania and Naples groups coordinated respectively by Professor Benedetto Matarazzo and Professor Aldo Ventre. For this reason, the organizing committee of TRENTO 2009 was extended to our colleagues from Benevento, Catania, and Naples.

As members of an informal seminal network of distinguished Italian scholars with strong international research connections, Professors Mario Fedrizzi, Benedetto Matarazzo, and Aldo Ventre have all played for years a leading role in the study and development of fuzzy set theory in Italy and worldwide, and our research community is greatly indebted for their devotion, support, advice, and encouragement.

Preface

Decision making is an omnipresent, most crucial activity of the human being, and also of virtually all artificial broadly perceived "intelligent" systems that try to mimic human behavior, reasoning and choice processes. It is quite obvious that such a relevance of decision making had triggered vast research effort on its very essence, and attempts to develop tools and techniques which would make it possible to somehow mimic human decision making related acts, even to automate decision making processes that had been so far reserved for the human beings. The roots of those attempts at a scientific analysis can be traced to the ancient times but – clearly – they have gained momentum in the recent 50 or 100 years following a general boom in science.

Depending on the field of science, decision making can be viewed in different ways. The most general view can be that decision making boils down to some cognitive, mental process(es) that lead to the selection of an option or a course of action among several alternatives. Then, looking in a deeper way, from a psychological perspective this process proceeds in the context of a set of needs, preferences, rational choice of an individual, a group of individuals, or even an organization. From a cognitive perspective, the decision making process proceeds in the context of various interactions with the environment. On the other hand, from a normative, formal perspective, the decision making process proceeds in the context of formal tools for the representation of sets of options (alternatives), preferences and utility functions, rationality, and mathematical tools that can be employed; that is, is concerned with a logic of decision making.

The perspective assumed in this volume is mainly within the formal approach to decision making. We will present some promising new developments that can help either get a deeper insight into the traditional formal models of decision making or show new conceptual tools that may lead to new models of a greater generality, an enhanced expressive power or a better computational efficiency. The authors deal with many types of decision making settings, notably with broadly perceived decision making under uncertainty, risk, imprecision (fuzziness), etc., and use a wide array of tools including probability theory, statistics, fuzzy logic, rough sets theory, etc.

A notable feature of contributions included in this volume is that they span a whole array of topics in the sense that they include – first – works dealing with mathematical tools which are indispensible for a meaningful analysis of virtually all realistic decision making models, exemplified by the modeling of various kinds of uncertainty and imprecision of information. Second, new developments in crucial underlying elements of decision making models, exemplified by preference modeling, are discussed. Third, multicriteria and multiperson decision making models are presented, notably those related to a crucial problem of consensus reaching. However, some papers are also included that present new mathematical tools that at present may only be viewed as a conceptually viable alternative to traditional mathematical tools. The above mentioned more basic works are complemented by a large group of papers which are concerned with the applications of various decision making models, notably in economics, finance, management, etc. Applications to problems related to new challenges related to, for instance, social networks are noteworthy.

Now, to give the reader a more detailed view of what is considered in this volume, we will present a brief description of the contents of the particular contributions in the order in which they appear. This may help the interested readers find the paper of interest.

Gianni Bosi and Romano Isler ("Continuous utility functions for nontotal preorders: a review of recent results") present some recent and significant results concerning the existence of a continuous utility function for a not necessarily total preorder on a topological space. First, the authors recall an appropriate continuity concept, a so-called weak continuity relative to a preorder on a topological space. Then, they provide a general characterization of the existence of a continuous utility function for a not necessarily total preorder on a topological space and show some relevant consequences, for the theory and applications.

Christer Carlsson and Robert Fullér ("Risk assessment of SLAs in grid computing with predictive probabilistic and possibilistic models") developed a hybrid probabilistic and possibilistic technique for assessing the risk of a service level agreement (SLA) for a computing task in a cluster/grid environment. The probability of success with the hybrid model is estimated higher than in the probabilistic model since the hybrid model takes into consideration the possibility distribution for the maximal number of failures derived from a resource provider's observations. The hybrid model shows that one can increase or decrease the granularity of the model in accordance to needs. One can reduce the estimate of the $P(S^*=1)$ by making a rougher, more conservative, estimate of the more unlikely events of $(M+1, N)$ node failures. The authors note that M is an estimate which is dependent on the history of the nodes being used and can be calibrated to "a few" or to "many" nodes.

Erio Castagnoli and Gino Favero ("From benchmarks to generalised expectations") are concerned with the case of the possibility of considering random variables as sets (hypo- or epigraphs), instead of mere functions which allows to treat random variables in the language and with the tools of measure theory, instead of the commonly adopted functional analysis. They show that, when looking at a random variable as a set, the concepts of the expectation and the expected utility (either "classical" or of the Choquet type) turn out to be slight variations of the same procedure of measuring a set (the truncated hypo- or epigraph corresponding to the given random variable) by means of a product measure (or capacity). They propose to extend this line of reasoning by using a generic ("non-product") measure or capacity to evaluate the set under examination, thus obtaining a broader concept of an expectation that includes dependence of the utility function on the state (or dependence of the probability on the amount). Basically, they justify the argument that the expectation of a random variable equals its own certainty equivalent, thus pointing out the equivalence between any random variable and a corresponding degenerate one. They also recover two different ways for defining the associative property of a generalized expectation.

Roy Cerqueti and Giulia Rotundo ("Memory property in heterogeneously populated markets") deal with the long memory of prices and returns of an asset traded on a financial market. They consider a microeconomic model of the market, and prove theoretical conditions on the parameters of the model that give rise to long memory. In particular, the long memory property is detected in an aggregation framework of agents under some distributional hypotheses on the market's parameters.

Giulianella Coletti and Barbara Vantaggi ("From comparative degrees of belief to conditional measures") are concerned with the "best" definition of conditional model for plausibilty functions and its subclass of possibility functions. They propose to use the framework of the theory of measurements by studying the comparative structure underlying different conditional models. This approach gives an estimate of the "goodness" and "effectiveness" of the model by pointing out the rules necessarily accepted by the user. Moreover, the results obtained by the authors that are related to the characterization of comparative degree of belief by means of conditional uncertainty measures are shown to be useful in decision theory. It is shown that they are in fact necessary when we need a model for a decision maker taking simultaneously into account different scenarios.

Salvador Cruz Rambaud and María José Muñoz Torrecillas ("Delay and interval effects with subadditive discounting functions") consider delay effect that appears as an anomaly of the traditional discounted utility model according to which a decrease of the discount rate is performed as waiting time increases. Since in this description it is not clear if the benchmark or the discounted amount availability is fixed or variable, and hence some authors use the terms like common difference effect, immediacy effect, interval effect, etc., the authors try to clarify the concepts of delay and interval effect and deduce some relationships between these concepts and certain subadditive discounting functions.

Bice Cavallo, Livia D'Apuzzo and Gabriella Marcarelli ("Pairwise comparison matrices: some issue on consistency and a new consistency index") consider multicriteria decision making with the pairwise comparisons of alternatives as an useful starting point for determining the ranking on the set of alternatives. The authors consider consistency conditions of the pairwise comparison matrix that allows to determine a weighted ranking that perfectly represents the expressed preferences. With reference to the new general unifying context proposed, the authors provide some results on a consistent matrix and a new measure of consistency that is easier to compute. Moreover, they provide an algorithm to check the consistency of a pairwise comparison matrix and an algorithm to build consistent matrices.

Fabio Baione, Paolo De Angelis and Riccardo Ottaviani ("On a decision model for a life insurance company rating") consider a rating system which is meant as a decision support tool for analysts, regulators and stakeholders in order to evaluate capital requirements of a firm under risky conditions. The authors define an actuarial model to measure the economic capital of a life insurance company basing the model on option pricing theory. In order to asses a life insurance company economic capital, they involve coherent risk measures already used in the assessment of banking Solvency Capital Requirements, according to Basel II standards. The authors show some results obtained by the application of the actuarial model to a portfolio of surrendable participating policies with minimum return guaranteed and option to annuitize.

Didier Dubois and Hélène Fargier ("Qualitative bipolar decision rules: toward more expressive settings") reconsider their previous approach to multicriteria decision-making whose idea is to choose between alternatives based on an analysis of the pros and cons, i.e. positive or negative arguments with various degrees of strength. Arguments correspond to criteria or affects of various levels of importance and ranging on a very crude value scale containing only three elements: good, neutral or bad. The basic decision rule considered in this setting is based on two ideas: focusing on the most important affects, and when comparing the merits of two alternatives considering that an argument against one alternative can be counted as an argument in favor of the other. It relies on a bipolar extension of comparative possibility ordering. Lexicographic refinements of this crude decision rule turn out to be cognitively plausible, and to generalize a well-known choice heuristics. It can also be viewed in terms of the cumulative prospect theory. The paper indicates several lines of future research, especially an alternative to the bicapacity approach to bipolar decision-making that subsumes both the cumulative prospect theory and our qualitative bipolar choice rule. Moreover, an extension of the latter to non-Boolean arguments is outlined.

Mario Fedrizzi, Michele Fedrizzi, Ricardo Alberto Marques Pereira and Matteo Brunelli ("The dynamics of consensus in group decision making: investigating the pairwise interactions between fuzzy preferences") present an overview of the soft consensus model in group decision making and investigate the dynamical patterns generated by the fundamental pairwise preference interactions on which the model is based. The dynamical mechanism of the soft consensus model discussed is driven by the minimization of a cost function combining a collective measure of dissensus with an individual mechanism of opinion changing aversion. The dissensus measure plays a key role in the model and induces a network of pairwise interactions between the individual preferences. The collective measure of dissensus is based on nonlinear scaling functions of the linguistic quantifier type and expresses the degree to which most of the decision makers disagree with respect to their preferences regarding the most relevant alternatives. In the extended formulation of the soft consensus model the extra degrees of freedom associated with the triangular fuzzy preferences, combined with non linear nature of the pairwise preference interactions, generate various interesting and suggestive dynamical patterns which are discussed.

János Fodor ("Fuzzy preference relations based on differences") introduces quaternary fuzzy relations in order to describe difference structures. He develops and discusses three models which are based on three different interpretations of an implication. Moreover, the author determines functional forms of the quaternary relation by solutions of functional equations of the same type.

Cesarino Bertini, Gianfranco Gambarelli and Angelo Uristani ("Indices of collusion among judges and an anti-collusion average") propose two indices of collusion among judges of objects or events in the context of subjective evaluation, and an average based on these indices. Their work may be viewed to have different aims, notably to serve as a reference point for appeals against the results of voting already undertaken, to improve the quality of scores summarized for awards by eliminating those that are less certain, and, indirectly, to provide an incentive for reliable evaluations. The authors present a computational algorithm and point out possible applications of their technique in various fields, from economics to finance, insurance, arts, artistic sports, etc.

José Luis García-Lapresta, Bonifacio Llamazares and Teresa Peña ("Scoring rules and consensus") consider that voters rank order a set of alternatives and a scoring rule is used for obtaining a set of winning alternatives using the scoring rule that is not previously fixed, but analyzing how to select one of them in such a way that the collective utility be maximized. In order to generate that collective utility, the authors ask voters for additional information in that agents declare which alternatives are good and their degree of optimism. With that information and a satisfaction function, for each scoring rule they generate individual utility functions such that the utility an alternative has for a voter should depend on whether this alternative is a winner for that scoring rule and on the position this alternative has in the individual ranking. Taking into account all these individual utilities, the authors aggregate them by means of an OWA operator and generate a collective utility for each scoring rule. By maximizing the collective utility, we obtain the set of scoring rules that maximizes consensus among the voters. Then, applying one of these scoring rules a collective weak order on the set of alternatives is obtained, that is, a set of winning alternatives.

Salvatore Greco, Benedetto Matarazzo and Roman Słowiński ("Dominance-based rough set approach to interactive evolutionary multiobjective optimization") present an application of the dominance-based rough set approach (DRSA) to interactive evolutionary multiobjective optimization (EMO). The preference information elicited by the decision maker in successive iterations consists in sorting some solutions of the current population as "good" or "bad", or in comparing some pairs of solutions. The "if ... then ..." decision rules are then induced from this preference information using the dominance-based rough set approach (DRSA). The rules are used within EMO to focus on populations of solutions satisfying the preferences of the decision maker. This makes possible to speed up convergence to the most preferred region of the Pareto front. The resulting interactive solution schemes, corresponding to the two types of preference information, are called DRSA-EMO and DRSA-EMO-PCT, respectively. Within the same methodology, the authors propose the DARWIN and DARWIN-PCT methods make it possible to take into account robustness issues in multiobjective optimization.

Janusz Kacprzyk and Sławomir Zadrożny ("Supporting consensus reaching processes under fuzzy preferences and a fuzzy majority via linguistic summaries") consider the classic approach to the evaluation of of degrees of consensus due to Kacprzyk and Fedrizzi in which a soft degree of consensus is a degree to which, for instance, "most of the important individuals agree as to almost all of the relevant options". The fuzzy majority, expressed as fuzzy linguistic quantifiers (most, almost all, ...) is handled via Zadeh's classic calculus of linguistically quantified propositions and Yager's OWA (ordered weighted average) operators. The soft degree of consensus is used for supporting the running of a moderated consensus reaching process along the lines of Fedrizzi, Kacprzyk and Zadrożny. Linguistic data summaries, in particular in its protoform based version proposed by Kacprzyk and Zadrożny are employed to indicate in a human consistent way some interesting relations between individuals and options to help the moderator identify crucial (pairs of) individuals and/options with whom/which there are difficulties with respect to consensus. An extension using ontologies representing both knowledge on the consensus reaching process and domain of the decision problem is discussed

Gabriella Marcarelli and Viviana Ventre ("Decision making in social actions") consider decision making in social action that involves both individual optimal choices and social choices. The theory of "perverse effects" by Boudon shows that the sum of rational individual choices can produce a very undesirable global effect. Then, decision making in social action must take into account the theory of cooperative games with many players in order to obtain the optimal strategies. Because of the semantic uncertainty in the definition of social actions, it is preferable assume that the issues are represented by fuzzy numbers. This is the basic idea proposed by the authors.

Antonio Maturo, Massimo Squillante and Aldo G.S. Ventre ("Coherence for fuzzy measures and applications to decision making") consider coherence, which is a central issue in probability, in a class of measures that are decomposable with respect to Archimedean *t*-conorms, in order to interpret the lack of coherence. Coherent fuzzy measures are utilized for the aggregations of scores in multiperson and multiobjective decision making. Furthermore, a geometrical representation of fuzzy and probabilistic uncertainty is considered in the framework of join spaces and, more generally, algebraic hyperstructures. The consider extensions of the coherence principle in nonadditive settings, exemplified by ambiguous or fuzzy settings, that is relevant for non-additive models in decision making, e. g., non-expected utility models.

Paola Modesti ("Measures for firms value in random scenarios") proposes a set of axioms in order to characterize appropriate measures of the (random) value of a company which provides a (sublinear) valuation functional consistent with the existence of a financial market. It makes it possible to give an upper and a lower bound to the value of a firm. The author considers also, in a random context, some classical valuation methods and test them with respect to the axioms.

Hannu Nurmi ("Thin rationality and representation of preferences with implications to spatial voting models") is concerned with some aspects of thin rationality that is of a primary concern in the current micro economic theory and formal political science. This concept refers to the behavioral principle stating that rational people act according to their preferences. Provided that the individual's preference is a binary, connected and transitive relation over alternative courses of action, one can define a utility function that represents the individual's preferences so that when acting rationally – i.e. in accordance with his/her preferences – he or she acts as if maximizing his/her utility. In the case of risky alternatives, i.e. probability mixtures of certain outcomes, a similar representation theorem states that the individual's preferences can be represented as a utility function with an expected utility property. These utility functions assign risky prospects utility values than coincide with weighted sums of the utility values of those outcomes that may materialize in the prospect. The weights, in turn, are identical with the probabilities of the corresponding outcomes. The author discusses also spatial models in which the individuals are identified as their ideal points in a space, and similarly the decision alternatives are represented as points in the space. The author approaches the spatial voting games from the angle of aggregation paradoxes, notably those of Ostrogorski, Simpson, the exam paradox, etc..

C.M. Sarris and A.N. Proto ("Quantum dynamics of non-commutative algebras: the SU(2) case") discuss the application of the maximum entropy formalism (MEP) which makes it possible to find the dynamics of Hamiltonians associated with non commutative Lie algebras. For the SU(2) case, it is easy to show that the Generalized Uncertainty Principle (GUP) is an invariant of motion. The temporal evolution of the system is confined to Bloch spheres whose radius lay on the interval (0;1). The GUP defines the fuzziness of these spheres inside domain for the SU(2) Lie algebra.

Rita A. Ribeiro, Tiago C. Pais and Luis F. Simões ("Benefits of full-reinforcement operators for spacecraft target landing") discuss the benefits of using full reinforcement operators for site selection in spacecraft landing on planets. Specifically, the authors discuss a modified uninorm operator for evaluating sites and a fimica operator to aggregate pixels for constructing regions that will act as sites to be selected at lower spacecraft altitude. An illustrative case study of spacecraft target landing is presented to clarify the details and usefulness of the proposed operators.

Giulia Rotundo ("Neural networks for non-independent lotteries") shows the density of the set von Neumann – Morgenstern utility functions on the set of utility functions that can represent arbitrarily well a given continuous but not independent preference relation over monetary lotteries. The main result obtained by the author is that without independence it is possible to approximate utility functions over monetary lotteries by the von Neumann – Morgenstern ones with arbitrary precision. The approach used is a constructive one. Neural networks are used because of their approximation properties in order to get the result, and their functional form provides both the von Neumann – Morgenstern representation and the necessary change of variables over the set of lotteries.

Romano Scozzafava ("Weak implication and fuzzy inclusion") defines a weak implication (H weakly implies E under P") through the relation $P(E|H)=1$, where P is a (coherent) conditional probability. In particular (as a by-product) the author obtains "inferential rules" that correspond to those of default logic, and discusses also connections between the weak implication and the fuzzy inclusion.

M. Socorro García-Cascales, M. Teresa Lamata and José Luís Verdegay ("The TOPSIS method and its application to linguistic variables") modify the known TOPSIS model to allow for the same linguistic values as the input and output of the process. The proposed method is applied to the process of quality assessment and accreditation of the industrial engineering schools within the Spanish university system.

Ronald R. Yager ("Information fusion with the power average operator") deals with the concept of a power average that provides an aggregation operator which allows argument values to support each other in the aggregation process, and describes the properties of this operator. Some formulations for the support function used in the power average are described. The author extends the facility of empowerment to a wider class of mean operators such as the OWA and generalized mean.

We wish to thank, first of all, all the authors for their excellent contributions and a great collaboration in this editorial project. Moreover, we wish to appreciate input and suggestion from the participants at long and inspiring discussions at TRENTO – 2009 The 5th International Workshop on Preferences and Decisions held in Trento, Italy on April 6 – 8, 2009 where the idea of publishing this volume has been conceived and then thoroughly discussed among the participants.

We wish to thanks Dr. Thomas Ditzinger and Ms. Heather King from Springer for their multifaceted support and help in the editorial process of this volume.

June 2010

Salvatore Greco
Ricardo Alberto Marques Pereira
Massimo Squillante
Ronald R. Yager
Janusz Kacprzyk

Contents

Continuous Utility Functions for Nontotal Preorders: A Review of Recent Results 1
Gianni Bosi, Romano Isler

Risk Assessment of SLAs in Grid Computing with Predictive Probabilistic and Possibilistic Models 11
Christer Carlsson, Robert Fullér

From Benchmarks to Generalised Expectations 31
Erio Castagnoli, Gino Favero

Memory Property in Heterogeneously Populated Markets 53
Roy Cerqueti, Giulia Rotundo

From Comparative Degrees of Belief to Conditional Measures 69
Giulianella Coletti, Barbara Vantaggi

Delay and Interval Effects with Subadditive Discounting Functions 85
Salvador Cruz Rambaud, María José Muñoz Torrecillas

Pairwise Comparison Matrices: Some Issue on Consistency and a New Consistency Index 111
Bice Cavallo, Livia D'Apuzzo, Gabriella Marcarelli

On a Decision Model for a Life Insurance Company Rating 123
Fabio Baione, Paolo De Angelis, Riccardo Ottaviani

Qualitative Bipolar Decision Rules: Toward More Expressive Settings 139
Didier Dubois, Hélène Fargier

The Dynamics of Consensus in Group Decision Making: Investigating the Pairwise Interactions between Fuzzy Preferences 159
Mario Fedrizzi, Michele Fedrizzi, Ricardo Alberto Marques Pereira, Matteo Brunelli

Fuzzy Preference Relations Based on Differences 183
János Fodor

Indices of Collusion among Judges and an Anti-collusion Average 195
Cesarino Bertini, Gianfranco Gambarelli, Angelo Uristani

Scoring Rules and Consensus 211
José Luis García-Lapresta, Bonifacio Llamazares, Teresa Peña

Dominance-Based Rough Set Approach to Interactive Evolutionary Multiobjective Optimization 225
Salvatore Greco, Benedetto Matarazzo, Roman Słowiński

Supporting Consensus Reaching Processes under Fuzzy Preferences and a Fuzzy Majority via Linguistic Summaries 261
Janusz Kacprzyk, Sławomir Zadrożny

Decision Making in Social Actions 281
Gabriella Marcarelli, Viviana Ventre

Coherence for Fuzzy Measures and Applications to Decision Making 291
Antonio Maturo, Massimo Squillante, Aldo G.S. Ventre

Measures for Firms Value in Random Scenarios 305
Paola Modesti

Thin Rationality and Representation of Preferences with Implications to Spatial Voting Models 321
Hannu Nurmi

Quantum Dynamics of Non Commutative Algebras: The SU(2) Case 339
C.M. Sarris, A.N. Proto

Benefits of Full-Reinforcement Operators for Spacecraft Target Landing 353
Rita A. Ribeiro, Tiago C. Pais, Luis F. Simões

Neural Networks for Non-independent Lotteries 369
Giulia Rotundo

Weak Implication and Fuzzy Inclusion 377
Romano Scozzafava

The TOPSIS Method and Its Application to Linguistic Variables 383
M. Socorro García-Cascales, M. Teresa Lamata, J. Luís Verdegay

Information Fusion with the Power Average Operator 397
Ronald R. Yager

Author Index 415

Continuous Utility Functions for Nontotal Preorders: A Review of Recent Results

Gianni Bosi and Romano Isler

Summary. We present some recent and significant results concerning the existence of a continuous utility function for a not necessarily total preorder on a topological space. We first recall an appropriate continuity concept (namely, *weak continuity*) relative to a preorder on a topological space. Then a general characterization of the existence of a continuous utility function for a not necessarily total preorder on a topological space is presented and some consequences of this basic result are produced.

1 Introduction

Since the seminal papers of Aumann [2] and Peleg [27], the existence of a continuous utility function for a not necessarily total preorder on a topological space has been deeply studied by several authors.

If $(X, \precsim, t)$ is a *topological preordered space*, then a function $u : (X, \precsim, t) \longrightarrow (\mathbb{R}, \leq, t_{nat})$ is said to be a *continuous utility function à la Richter-Peleg* (see Peleg [27] and Richter [29]) or simply a *utility function* if $u : (X, \precsim) \longrightarrow (\mathbb{R}, \leq)$ is an *order-preserving function* and $u : (X, t) \longrightarrow (\mathbb{R}, t_{nat})$ is continuous.

Despite of the fact that a utility function u for a nontotal preorder $\precsim$ does not allow to recover $\precsim$ (i.e., to characterize it), one may think that it provides enough information about $\precsim$. On the other hand, a representation of

Gianni Bosi
Dipartimento di Matematica Applicata "Bruno de Finetti", Università di Trieste, Piazzale Europa 1, 34127, Trieste, Italy
e-mail: giannibo@econ.units.it

Romano Isler
Dipartimento di Matematica Applicata "Bruno de Finetti", Università di Trieste, Piazzale Europa 1, 34127, Trieste, Italy
e-mail: romanoi@econ.units.it

S. Greco et al. (Eds.): Preferences and Decisions, STUDFUZZ 257, pp. 1–10.
springerlink.com

this kind is based upon the existence of only one real-valued function and therefore it is simpler than other representations of nontotal preorders (such as, for example, the *multi-utility representation* studied by Evren and Ok [16], according to which a preorder $\precsim$ on X is representable if there exists a family $\mathbf{U}$ of increasing real-valued functions u on $(X, \precsim)$ such that for all $x \in X$ and all $y \in X$ the inequalities $x \precsim y$ mean that for all $u \in \mathbf{U}$ the inequalities $u(x) \leq u(y)$ hold).

Herden [17, 18] characterized the existence of a continuous utility function for a not necessarily total preorder on a topological space by using the concept of a (*linear*) *separable system* in a preordered topological space and showed that the theorems of Debreu and Eilenberg can be obtained as corollaries of his main result. Herden continued Mehta's work (see e.g. Mehta [22, 24]), who followed the spirit of Nachbin [25] as regards the combination of the classical approach to mathematical utility theory with some of the most important results in elementary topology.

Then Herden and Pallack [21] introduced the concept of a *weakly continuous preorder* in order to generalize in the most appropriate way the notion of continuity to the case of a not necessarily total preorder on a topological space. According to Definition 2.3 in Herden and Pallack [21], a preorder $\precsim$ on a topological space (X, t) is said to be *weakly continuous* if for every pair $(x, y) \in \prec$ there exists a continuous increasing function $u_{xy} : (X, \precsim, t) \longrightarrow (\mathbb{R}, \leq, t_{nat})$ such that $u_{xy}(x) < u_{xy}(y)$.

Herden and Pallack then showed that the utility representation theorem of Debreu is generalizable to the case of a non total preorder while this is not the case of the utility representation theorem of Eilenberg (see Example 2.14 and Theorem 2.15 in the above mentioned paper of Herden and Pallack). This means that there always exists a continuous utility function for a weakly continuous preorder on a second countable topological space while there exist weakly continuous preorders on connected and separable topological spaces which are not continuously representable.

In a slightly more general framework, recently Bosi and Herden [5, 6] discussed the property according to which a topology t on a set X satisfies the *weakly continuous analogue of the Szpilrajn theorem* (i.e., every weakly continuous preorder on (X, t) can be *refined* or *extended* by a continuous linear (total) preorder, or equivalently for every weakly continuous preorder $\precsim$ on (X, t) there exists a continuous total preorder $\lesssim$ on (X, t) such that $\precsim \subset \lesssim$ and $\prec \subset <$). It is clear that the existence of a continuous utility function for a preorder $\precsim$ on a topological space (X, t) is equivalent to the existence of a continuous total preorder $\lesssim$ on (X, t) which extends $\precsim$ and is representable by a continuous utility function. Hence, if a topology t on a set X is such that every weakly continuous preorder on (X, t) admits a continuous utility representation, then t satisfies the weakly continuous analogue of the Szpilrajn theorem.

In this paper, we present a review of some of the most significant recent results concerning the existence of continuous utility functions for not necessarily total preorders. Despite of the fact that in this paper we are only concerned with the existence of continuous utility representations, we recall that in Bosi and Herden [8] the problem concerning the existence of semicontinuous utility functions for nontotal preorders is also treated.

2 Notation and Preliminaries

If we denote by $(X, \precsim)$ a *preordered set* (i.e., a set X endowed with a reflexive and transitive binary relation $\precsim$), then a function $u : (X, \precsim) \longrightarrow (\mathbb{R}, \leq)$ is referred to as a *utility function à la Richter-Peleg* (see Peleg [27] and Richter [29]) if the following two conditions are verified:

U1: For all $(x, y) \in X \times X$, $x \precsim y \Rightarrow u(x) \leq u(y)$.

U2: For all $(x, y) \in X \times X$, $x \prec y \Rightarrow u(x) < u(y)$.

Here $\prec$ is the *strict part* of $\precsim$ (namely, for all $x, y \in X$, $x \prec y$ if and only if $x \precsim y$ and not$(y \precsim x)$).

A utility function à la Richter-Peleg is more frequently called an *order-preserving function* or simply a *utility function*.

It is clear that if $\precsim$ is a *total* preorder on X (i.e., for all $x, y \in X$ either $x \precsim y$ or $y \precsim x$), then u is a utility function for $\precsim$ if and only if $x \precsim y$ is equivalent to $u(x) \leq u(y)$ for all $(x, y) \in X \times X$.

If a function $u : (X, \precsim) \longrightarrow (\mathbb{R}, \leq)$ only satisfies condition **U1**, then it is said to be *increasing*.

If $\precsim$ is a preorder on a set X then two elements $x, y \in X$ are said to be *indifferent* if we have that $x \precsim y$ and $y \precsim x$, while they are said to be *incomparable* if neither $x \precsim y$ nor $y \precsim x$.

Peleg [27] solved the problem raised by Aumann by providing for the first time sufficient conditions for the existence of a continuous utility function for a *partial order* on a topological space.

The reader may recall that a total preorder $\precsim$ on a topological space (X, t) is said to be *continuous* if it satisfies one of the following equivalent conditions:

C1: For all points $x \in X$ both sets $l(x) = \{z \in X \mid z \prec x\}$ and $r(x) = \{w \in X \mid x \prec w\}$ are open subsets of X.

C2: For all points $x \in X$ both sets $d(x) = \{z \in X \mid z \precsim x\}$ and $i(x) = \{w \in X \mid x \precsim w\}$ are closed subsets of X.

C3: For every pair $(x, y) \in \prec$ there exists some continuous increasing function $f_{xy} : (X, \precsim, t) \longrightarrow (\mathbb{R}, \leq, t_{nat})$ such that $f_{xy}(x) < f_{xy}(y)$.

Condition **C1** requires that the order topology $t^{\precsim}$ on X that is induced by $\precsim$ is coarser than t.

We recall that the utility representation theorems of Debreu and Eilenberg-Debreu (see Debreu [11, 12] and Eilenberg [13]) guarantee the existence of a continuous utility function for a total preorder on a second countable and respectively on a connected and separable topological space. Results of this kind are nice since the combination of continuity of the total preorder $\precsim$ on the topological space (X, t) and the topological conditions of second countability or respectively connectedness and separability of (X, t) allow us to get rid of suitable assumptions of *order separability* of $\precsim$, which must be invoked if one wants to characterize the existence of a continuous utility function u for a total preorder $\precsim$ on an arbitrary topological space (X, t) (see e.g. Definition 1.4.3 and Proposition 1.6.11 in the book of Bridges and Mehta [9]). The continuous utility representation problem in arbitrary concrete categories has been very recently discussed by Bosi and Herden [7].

In mathematical utility theory the above condition **C3** plays a very important role in order to determine sufficient conditions for the existence of a continuous utility function for a not necessarily total preorder on a topological space. Indeed, it is easy to show (see, for instance, Herden [17, Theorem 4.1]) that a preorder $\precsim$ on (X, t) can be represented by a continuous utility function $u : (X, \precsim, t) \longrightarrow (\mathbb{R}, \leq, t_{nat})$ if and only if there exists a countable family $\{u_k\}_{k \in K}$ of continuous increasing functions $u_k : (X, \precsim, t) \longrightarrow (\mathbb{R}, \leq, t_{nat})$ such that for every pair $(x, y) \in \prec$ there exists some $k \in K$ such that $u_k(x) < u_k(y)$.

3 Weakly Continuous Preorders

Herden and Pallack [21] first introduced the concept of a *weakly continuous preorder* in order to generalize in the most appropriate way the notion of continuity to the case of nontotal preorders. According to Definition 2.3 in Herden and Pallack [21], a preorder $\precsim$ on a topological space (X, t) is said to be *weakly continuous* if for every pair $(x, y) \in \prec$ there exists a continuous increasing function $u : (X, \precsim, t) \longrightarrow (\mathbb{R}, \leq, t_{nat})$ such that $u(x) < u(y)$.

Weak continuity is obviously a necessary condition for the existence of a continuous utility function for a not necessarily total preorder on a topological space. Such a notion coincide with the classical notion of continuity if one has to do with a total preorder.

Herden and Pallack then showed that the utility representation theorem of Debreu is generalizable to the case of a nontotal preorder while this is not the case of the utility representation theorem of Eilenberg-Debreu (see Example 2.14 and Theorem 2.15 in the above mentioned paper of Herden and Pallack).

We recall that a subset D of X is said to be *decreasing* if $x \in D$ and $z \precsim x$ imply that $z \in D$. Conversely, a subset I of X is said to be *increasing* if $x \in I$ and $x \precsim w$ imply that $w \in I$.

From Burgess and Fitzpatrick [10], if $(X, \precsim, t)$ is a topological space and $\mathbb{S}$ is a dense subset of $[0,1]$ such that $1 \in \mathbb{S}$, then a family $\{G_r\}_{r \in \mathbb{S}}$ of open decreasing subsets of X is said to be a *decreasing scale* in $(X, \precsim, t)$ if the following conditions hold:

S1: $G_1 = X$;

S2: $\overline{G_{r_1}} \subset G_{r_2}$ for every $r_1, r_2 \in \mathbb{S}$ such that $r_1 < r_2$.

A preorder $\precsim$ on a topological space (X, t) is weakly continuous if and only if for every $(x, y) \in \prec$ there exists a a decreasing scale $\{G_r\}_{r \in \mathbb{S}}$ that separates x from y (i.e., $x \in G_r$ and $y \in X \setminus G_r$ for every real number $r \in \mathbb{S} \setminus \{1\}$).

Decreasing scales have been recently invoked by Alcantud et al. [1] in order to prove the existence of a continuous utility function on a totally preordered topological space without using the well known *Debreu Open Gap Lemma.*

We recall that a preorder $\precsim$ on a topological space (X, t) is said to be *closed* if $\precsim$ is a closed subset of $X \times X$ with respect to the product topology on $X \times X$. In some relevant cases a closed preorder $\precsim$ on (X, t) even is weakly continuous. Indeed, if (X, t) is a compact (Hausdorff-)space or a locally compact second countable (Hausdorff-)space that is endowed with some closed preorder $\precsim$, then $\precsim$ is weakly continuous (see Herden and Pallack [21, Proposition 2.12]).

It is well known that for every closed preorder $\precsim$ on (X, t) both sets $d(x)$ and $i(x)$ are closed subsets of X. The reader may check that the converse holds for linear (total) preorders but not in general (see also Nachbin [25]).

Let $\precsim$ be a preorder on X. Then a pair $(x, y) \in \prec$ is said to be a *jump* of $\precsim$ if $]x, y[:= \{z \in X \mid x \prec z \prec y\} = \emptyset$. Further, $\precsim$ is said to be *dense* if it has no jumps.

With help of this notation we present the following proposition whose proof is found in Bosi and Herden [8]. The reader may observe that condition (jjj) below is just the property of *spaciousness* of the strict part $\prec$ of $\precsim$ according to the terminology introduced by Peleg [27].

Proposition 3.1. *Let $\precsim$ be a dense preorder on a topological space (X, t). Then the following assertions hold:*

(j) *For all pairs $(x, y) \in \prec$ there exist disjoint decreasing, respectively increasing, open subsets O_x, respectively U_y, of X such that $x \in O_x$ and $X \setminus l(y) \subset U_y$.*

(jj) *For all pairs $(x, y) \in \prec$ there exists an open decreasing subset O_{xy} of X such that $x \in O_{xy} \subset \overline{O_{xy}} \subset l(y)$.*

(jjj) *For all pairs $(x, y) \in \prec$ the set $l(y)$ is an open subset of X and $\overline{l(x)} \subset l(y)$.*

(jv) *For all pairs* $(x, y) \in \prec$ *the set* $\bigcap_{z \in r(x)} l(z)$ *is a closed and the set* $l(y)$ *an open subset of* X.

(ii) *Any of the equivalent conditions of assertion* (i) *implies that* $\precsim$ *is weakly continuous.*

The following proposition concerns the general case when the preorder is not necessarily dense.

Proposition 3.2. *Let* $(X, \precsim, t)$ *be a topological preordered space and assume that for every pair* $(x, y) \in \prec$ *the set* $l(y)$ *is an open and the set* $d(x)$ *is a closed subset of* X. *Then* $\precsim$ *is weakly continuous provided that one of the following conditions is verified for every jump* $(x, y) \in \prec$:

(i) $l(y)$ *is a closed subset of* X;

(ii) $r(x)$ *is an open subset and* $d(y)$ *and* $i(x)$ *are closed subsets of* X.

We recall that a preorder $\precsim$ on a set X is said to be *separable* if there exists a countable subset Z of X such that for any two points $x, y \in X$ such that $x \prec y$ there exists some point $z \in Z$ such that $x \prec z \prec y$.

From Proposition 3.1 we immediately obtain the following corollary by considering the already mentioned fact that the existence of a continuous utility function is equivalent to the existence of a countable family of continuous increasing functions separating points $x, y \in X$ such that $x \prec y$.

Corollary 3.3 *Let* (X, t) *be some topological space that is endowed with a dense preorder* $\precsim$ *which satisfies condition* (jjj) *of Proposition 3.1. Then in order for* $\precsim$ *to be representable by a continuous order preserving function* $u : (X, \precsim, t) \longrightarrow (\mathbb{R}, \leq, t_{nat})$ *it is necessary and sufficient that* $\precsim$ *is separable.*

4 Continuous Utility Functions

We recall that a family $\mathcal{N}$ of subsets of a topological space (X, t) is called a *network* for X if every non empty open subset of X is a union of elements of $\mathcal{N}$. It is well known that the existence of a countable network is equivalent to the existence of a countable basis in case that the topology is either metrizable or locally compact or else linearly ordered (se Engelking [14]).

Further, a topology t on a set X is a *hereditarily Lindelöf topology* if for every subset A of X and every open covering $\mathcal{C}$ of A there exists some countable subcovering $\mathcal{C}' \subset \mathcal{C}$ of A.

Herden and Pallack [21, Theorem 2.15] proved that every weakly continuous preorder on a second countable topological space has a continuous utility representation. Recently, Bosi, Caterino and Ceppitelli [3] proved the following theorem which appears as slightly more general.

Theorem 4.1. *Let $\precsim$ be a preorder on a topological space (X, t). Then the following conditions are equivalent:*

(i) *There exists a continuous utility function u on $(X, \precsim, t)$;*
(ii) *There exists a topology t' on X coarser than t such that $\precsim$ is weakly continuous on (X, t') and (X, t') is second countable;*
(iii) *There exists a topology t' on X coarser than t such that $\precsim$ is weakly continuous on (X, t') and there exists a countable network for (X, t');*
(iv) *There exists a topology t' on X coarser than t such that $\precsim$ is weakly continuous on (X, t') and the product topology $t' \times t'$ on $X \times X$ is hereditarily Lindelöf;*
(v) *There exists a topology t' on X coarser than t such that $\precsim$ is weakly continuous on (X, t') and the topology $(t' \times t')_{\prec}$ induced by the product topology $t' \times t'$ on $\prec$ is Lindelöf.*

Outline of the proof. (i) $\Rightarrow$ (ii). Let u be a continuous utility function on $(X, t, \precsim)$. Consider the total preorder $\lesssim$ on X defined by

$$x \lesssim y \Leftrightarrow u(x) \leq u(y) \quad (x, y \in X),$$

and let $t' = t^{\lesssim}$ be the order topology associated to $\lesssim$. From the Debreu Open Gap Lemma there exists a continuous utility function u' on the totally preordered topological space $(X, t', \lesssim)$. Since $\lesssim$ is (continuously) representable we have that t' is second countable. Further, t' is coarser than t and $\precsim$ is weakly continuous on (X, t').
(ii) $\Rightarrow$ (iii). Trivial.
(iii) $\Rightarrow$ (iv). Easy to show.
(iv) $\Rightarrow$ (v). Immediate.
(v) $\Rightarrow$ (i). For every pair $(x, y) \in X \times X$ such that $x \prec y$ consider a continuous increasing function u_{xy} on $(X, t', \precsim)$ with values in $[0, 1]$ such that $u_{xy}(x) < u_{xy}(y)$ and define

$$A_{u_{xy}}(x) := u_{xy}^{-1}([0, \frac{u_{xy}(x) + u_{xy}(y)}{2}[),$$

$$B_{u_{xy}}(y) := u_{xy}^{-1}(]\frac{u_{xy}(x) + u_{xy}(y)}{2}, 1]).$$

Then the family $C := \{A_{u_{xy}}(x) \times B_{u_{xy}}(y)\}_{(x,y) \in \prec}$ is an open cover of $\prec$ and since the topology $(t' \times t')_{\prec}$ is Lindelöf, there exists a countable subfamily C' of C which also covers $\prec$, and therefore there exists a countable family $\{u_n\}_{n \in \mathbb{N}}$ of continuous increasing functions on $(X, t', \precsim)$ with values in $[0, 1]$

such that for every $(x, y) \in X \times X$ with $x \prec y$ there exists some $n \in \mathbb{N}$ such that $u_n(x) < u_n(y)$. Hence, $u := \sum_{n=0}^{\infty} 2^{-n} u_n$ is a continuous utility function on $(X, t', \precsim)$ and therefore also on $(X, t, \precsim)$. □

We recall that from Herden [19] a topology t on a set X is said to be *useful* if every continuous total preorder $\precsim$ on the topological space (X, t) is representable by a continuous utility function $u : (X, t, \precsim) \longrightarrow (\mathbb{R}, t_{nat}, \leq)$ (see also Herden and Pallack [20]).

From Theorem 4.1 we immediately obtain the following corollary which provides some conditions under which a topology is useful.

Corollary 4.2 *A topology t on a set X is useful provided that the product topology $t \times t$ on $X \times X$ is hereditarily Lindelöf (in particular, in case that t has a countable net weight).*

We say that a topology t on a set X is *strongly useful* (see Bosi, Caterino and Ceppitelli [3]) if every weakly continuous preorder $\precsim$ on the topological space (X, t) is representable by a continuous utility function $u : (X, t, \precsim) \longrightarrow (\mathbb{R}, t_{nat}, \leq)$. It is clear that a strongly useful topology on a set X is also useful. The converse is not true. Indeed, in Bosi and Herden [8] (see also Example 3.2 in Bosi and Herden [7]) it is shown that the product topology t_{prod} on $[0, 1]^{\alpha}$ is not strongly useful when α is an uncountable ordinal number, since in this case the product ordering $\leq_{prod}$ that is induced by the canonical total preorder $\leq$ on $[0, 1]$ is weakly continuous but not continuously representable. On the other hand, we have that $([0, 1]^{\alpha}, t_{prod})$ is connected and separable for every ordinal number $\alpha \geq 1$ and therefore the product topology t_{prod} on $[0, 1]^{\alpha}$ is useful.

Bosi and Herden [8, Proposition 4.3] proved that a strongly useful topology which is in addition completely regular and Hausdorff must be normal.

We recall that a total preorder $\precsim$ on a topological space (X, t) is said to be *upper semicontinuous* if $l(x) = \{z \in X \mid z \prec x\}$ is an open subset of X for every $x \in X$.

From Bosi and Herden [4], a topology t on a set X is said to be *completely useful* if every upper semicontinuous total preorder $\precsim$ on the topological space (X, t) is representable by an upper semicontinuous utility function $u : (X, t, \precsim) \longrightarrow (\mathbb{R}, t_{nat}, \leq)$. Proposition 4.4 in Bosi and Herden [4] states that every completely useful topology is useful. On the other hand, there exist useful topologies which are not completely useful. Indeed, let $\aleph_1 := \Omega$ be the first uncountable ordinal and let t be the topology on $X := [0, \aleph_1[= [0, \Omega[$ that is induced by the sets $[0, \alpha]$, where α runs through all countable ordinals. Since $\overline{\{0\}} = \overline{[0, \alpha]} = X$ for every countable ordinal α, we have that t is a separable and connected topology on X. The natural order $\leq$ on X is an upper semicontinuous total preorder on X that, obviously, has no (upper semicontinuous) utility representation (see Bosi and Herden [4]).

The following corollary of Theorem 4.1 provides a characterization of strongly useful topologies in the metrizable case. The proof is based on the theorem in Estévez and Hervés [15], which states that a metrizable topology is useful if and only if it is separable (or equivalently second countable). Further, we use Corollary 4.5 in Bosi and Herden [4], according to which a metrizable topology is completely useful if and only if it is second countable.

Corollary 4.3 *Let t be a metrizable topology on a set X. Then the following conditions are equivalent:*

(i) *t is strongly useful;*

(ii) *t is completely useful;*

(ii) *t is useful;*

(iii) *t is separable.*

Since every completely useful topology is hereditarily Lindelöf (see Bosi and Herden [4, Lemma 4.1 and Proposition 4.2]), it is immediate to check that the following corollary of Theorem 4.1 holds.

Corollary 4.4 *Let t be a topology on a set X. Then t is strongly useful provided that the product topology $t \times t$ on $X \times X$ is completely useful.*

References

1. Alcantud, J.C.R., Bosi, G., Campión, M.J., Candeal, J.C., Induráin, E., Rodríguez-Palmero, C.: Continuous utility functions through scales. Theory and Decision 64, 479–494 (2008)
2. Aumann, R.: Utility theory without the completeness axiom. Econometrica 30, 445–462 (1962)
3. Bosi, G., Caterino, A., Ceppitelli, R.: Existence of continuous utility functions for arbitrary binary relations: some sufficient conditions (preprint)
4. Bosi, G., Herden, G.: On the structure of completely useful topologies. Applied General Topology 3, 145–167 (2002)
5. Bosi, G., Herden, G.: On a strong continuous analogue of the Szpilrajn theorem and its strengthening by Dushnik and Miller. Order 22, 329–342 (2005)
6. Bosi, G., Herden, G.: On a possible continuous analogue of the Szpilrajn theorem and its strengthening by Dushnik and Miller. Order 23, 271–296 (2006)
7. Bosi, G., Herden, G.: Continuous utility representations theorems in arbitrary concrete categories. Applied Categorical Structures 5, 629–651 (2008)
8. Bosi, G., Herden, G.: Utility representations and linear refinements of arbitrary binary relations (preprint)
9. Bridges, D.S., Mehta, G.B.: Representation of preference orderings. Lecture Notes in Economics and Mathematical Systems, vol. 422. Springer, Heidelberg (1995)
10. Burgess, D.C.J., Fitzpatrick, M.: On separation axioms for certain types of ordered topological space. Mathematical Proceedings of the Cambridge Philosophical Society 82, 59–65 (1977)

11. Debreu, G.: Representation of preference orderings by a numerical function. In: Thrall, R., Coombs, C.C., Davis, R. (eds.) Decision Processes, pp. 159–166. Wiley, New York (1954)
12. Debreu, G.: Continuity properties of a Paretian utility. International Economic Review 5, 285–293 (1964)
13. Eilenberg, S.: Ordered topological spaces. American Journal of Mathematics 63, 39–45 (1941)
14. Engelking, R.: General Topology. Heldermann Verlag, Berlin (1989)
15. Estévez, M., Hervés, C.: On the existence of continuous preference orderings without utility representations. Journal of Mathematical Economics 24, 305–309 (1995)
16. Evren, O., Ok, E.A.: On the multi-utility representation of preference relations. New York University (2007)
17. Herden, G.: On the existence of utility functions. Mathematical Social Sciences 17, 297–313 (1989)
18. Herden, G.: On the existence of utility functions II. Mathematical Social Sciences 18, 119–134 (1989)
19. Herden, G.: Topological spaces for which every continuous total preorder can be represented by a continuous utility function. Mathematical Social Sciences 22, 123–136 (1991)
20. Herden, G., Pallack, A.: Useful topologies and separable systems. Applied general topology 1, 61–82 (2000)
21. Herden, G., Pallack, A.: On the continuous analogue of the Szpilrajn Theorem I. Mathematical Social Sciences 43, 115–134 (2002)
22. Mehta, G.B.: Some general theorems on the existence of order-preserving functions. Mathematical Social Sciences 15, 135–146 (1988)
23. Mehta, G.B.: A remark on a utility representation theorem of Rader. Economic Theory 9, 367–370 (1997)
24. Mehta, G.: Preference and Utility. In: Barberá, S., Hammond, P., Seidl, C. (eds.) Handbook of Utility Theory, vol. 1, pp. 1–47. Kluwer Academic Publishers, Dordrecht (1998)
25. Nachbin, L.: Topology and order. Van Nostrand, Princeton (1965)
26. Ok, E.A.: Utility representation of an incomplete preference relation. Journal of Economic Theory 104, 429–449 (2002)
27. Peleg, B.: Utility functions for partially ordered topological spaces. Econometrica 38, 93–96 (1970)
28. Rader, T.: The existence of a utility function to represent preferences. Review of Economic Studies 30, 229–232 (1963)
29. Richter, M.: Revealed preference theory. Econometrica 34, 635–645 (1966)
30. Szpilrajn, E.: Sur l'extension de l'ordre partial. Fundamenta Mathematicae 16, 386–389 (1930)

Risk Assessment of SLAs in Grid Computing with Predictive Probabilistic and Possibilistic Models

Christer Carlsson and Robert Fullér

Abstract. We developed a hybrid probabilistic and possibilistic technique for assessing the risk of an SLA for a computing task in a cluster/grid environment. The probability of success with the hybrid model is estimated higher than in the probabilistic model since the hybrid model takes into consideration the possibility distribution for the maximal number of failures derived from a resource provider's observations. The hybrid model showed that we can increase or decrease the granularity of the model as needed; we can *reduce* the estimate of the P(S*=1) by making a rougher, more conservative, estimate of the more unlikely events of (M+1, N) node failures. We noted that M is an estimate which is dependent on the history of the nodes being used and can be calibrated to "a few" or to "many" nodes.

Keywords: risk assessment, predictive probabilities, predictive possibilities, grid computing.

1 Introduction

There is an increasing demand for computing power in scientific and engineering applications which has motivated the deployment of high performance computing (HPC) systems that deliver tera-scale performance. Current and future HPC systems that are capable of running large-scale parallel applications may span hundreds of thousands of nodes.

The current highest processor count is 131K nodes according to top500.org [16]. For parallel programs, the failure probability of nodes and computing tasks assigned to the nodes has been shown to increase significantly with the increase in number of nodes. Large-scale computing environments, such as the current grids CERN LCG, NorduGrid, TeraGrid and Grid'5000 gather (tens of) thousands of resources for the use of an ever-growing scientific community. Many of these Grids offer computing resources grouped in clusters, whose owners may share them only for limited periods of time and Grids often have the problems of any large-scale computing environment to which is added that their middleware is still

Christer Carlsson · Robert Fullér
Institute for Advanced Management Systems Research
Abo Akademi University, Åbo, Finland
e-mail: christer.carlsson@abo.fi

S. Greco et al. (Eds.): Preferences and Decisions, STUDFUZZ 257, pp. 11–29.
springerlink.com

relatively immature, which contributes to making Grids relatively unreliable computing platforms. Iosup et al [9] collected and present material from Grid'5000 which illustrates this contention. On average, resource availability (for a period of 548 days) in Grid'5000 on the grid level is 69% (±11.42), with a maximum of 92% and a minimum of 35%. The mean time between failures (MTBF) of the grid is of 744±2631 seconds (around 12 minutes); at the cluster level, resource availability varies from 39% up to 98% across the 15 clusters, the average MTBF for all clusters is 18404±13848 seconds (around 5 hours); at the node level on average a node fails 228 times (over 548 days), but some nodes fail only once or even never. Long et al [12] collected a dataset on node failures over 11 months by from1139 workstations on the Internet to determine their uptime intervals. Planck and Elwasif [14] collected a dataset on failure information for a collection of 16 DEC Alpha work-stations at Princeton University; the size of this network is smaller and is a typical local cluster of homogeneous processors; the failure data was collected for 7 months and shows similar characteristics as for the larger clusters.

Schroeder and Gibson [17] analyse failure data collected over 9 years at Los Alamos National Laboratory (LANL) and which includes 23000 failures recorded on more than 20 different systems, mostly large clusters of SMP and NUMA nodes. Their study includes root cause of failures, the mean time between failures, and the mean time to repair. They found that average failure rates differ wildly across systems, ranging from 20–1000 failures per year, mean repair time varies from less than an hour to more than a day. Most applications (about 70%) running in the LANL are short duration computing tasks of $\leq$ 1 hour, but there are also large-scale, long-running 3D scientific simulations. These applications perform long periods (often months) of CPU computation, interrupted every few hours by a few minutes of I/O for check-pointing. When node failures occur in the LANL, hardware was found to be the single largest cause (30-60%); software is the second largest contributor (with 5-24%), but in most systems the root cause remained undetermined for 20–30% of the failures (cf. [17]). They also found that the yearly failure rate varies widely across systems, ranging from 17 to an average of 1159 failures per year for several systems. The main reason for the differences was that the systems vary widely in size and that the nodes run different workloads.

Iosup et al [9] fit statistical distributions to the Grid'5000 data using maximum likelihood estimation (MLE) to find a best fit for each of the model parameters. They found that the best fits for the inter-arrival time between failures, the duration of a failure, and the number of nodes affected by a failure, are the Weibull, Log-Normal, and Weibull distributions, respectively. The results for inter-arrival time between consecutive failures indicate an increasing hazard rate function, i.e. the longer a computing node stays in the system, the higher the probability for the node to fail, which will prevent long jobs to finish. Iosup et al [9] also wanted to find out if they can decide where (on which nodes or in which cluster) a new failure could/should occur. Since the sites are located and administered separately, and the network between them has numerous redundant paths, they found no evidence for any other assumption than that there is no correlation between the occurrence of failures at different sites. For the LANL dataset Schroeder and Gibson [17] studied the sequence of failure events and the time between failures as stochastic processes. This includes two different views of the failure process: (i) the view as seen by an individual node; (ii) the view as seen by the whole system. They found that the distribution between failures for individual nodes is well modelled by a Weibull or a Gamma distribution; both distributions create an equally good visual fit and the same negative log-likelihood. For the system wide view of the failures the basic trend is similar to the per node view during the same time. The Weibull and Gamma distributions provide the best fit, while the lognormal and exponential fits are significantly worse.

A significant amount of the literature on grid computing addresses the problem of resource allocation on the grid (see, e.g., [Brandt [1], Czajkowski [7], Liu et al [11], Magana et al [13], and Tuecke [18]). The presence of disparate resources that are required to work in concert within a grid computing framework increases the complexity of the resource allocation problem. Jobs are assigned either through *scavenging*, where idle machines are identified and put to work, or through *reservation* in which jobs are matched and pre-assigned with the most appropriate systems in the grid for efficient workflow.

In grid computing a resource provider [RP] offers resources and services to other Grid users based on agreed service level agreements [SLAs]. The *research problem* we have addressed is formulated as follows:

- the RP is running a risk to be in violation of his SLA if one or more of the resources [nodes in a cluster or a Grid] he is offering to prospective customers will fail when carrying out the tasks
- the RP needs to work out methods for a systematic risk assessment [RA] in order to judge if he should offer the SLA or not if he wants to work with some acceptable risk profile

In the context we are going to consider (a generic grid computing environment) resource providers are of various types which mean that the resources they manage and the risks they have to deal with are also different; we have dealt with the following RP scenarios (but we will report only on extracts due to space):

- RP1 manages a cluster of n_1 nodes (where n_1 is ≤ 10) and handles a few (≤ 5) computing tasks for a *T*
- RP2 manages a cluster of n_2 nodes (where n_2 is ≤ 150) and handles numerous (~ 100) computing tasks for a *T*; RP2 typically uses risk models building on stochastic processes (Poisson-Gamma) and Bayes modelling to be able to assess the risks involved in offering SLAs
- RP3 manages a cluster of n_3 nodes (where n_3 is ≤ 10) and handles numerous (~ 100) computing tasks for a *T*; if the computing tasks are of short duration and/or the requests are handled online RP3 could use possibility models that will offer robust approximations for the risk assessments
- RP4 manages a cluster of n_4 nodes (where n_4 is ≤ 150) and handles numerous (~ 100) computing tasks for a *T*; typically RP4 could use risk models building on stochastic processes (Poisson-Gamma) and Bayes

modelling to assess the risks involved in offering SLAs; if the computing tasks are of short duration and/or the requests are handled online hybrid models which combine stochastic processes and Bayes modelling with possibility models could provide tools for handling this type of cases.

During the execution of a computing task the fulfilment of the SLA has the highest priority, which is why an RP often is using resource allocation models to safeguard against expected node failures. When spare resources at the RP's own site are not available outsourcing will be an adequate solution for avoiding SLA violations.

The risk assessment modelling for an SLA violation builds on the development of predictive probabilities and possibilities for possible node failures and combined with the availability of spare resources. The rest of the paper will be structured as follows: in section 2 we will work out the basic conceptual framework for risk assessment, in section 3 we will introduce the Bayesian predictive probabilities as they apply to the SLAs for RPs in grid computing, in section 4 we will work out the corresponding predictive possibilities and show the results of the validation work we carried out for some RP scenarios; in section 5 there is a summary and conclusions of the study.

2 Risk Assessment

There is no universally accepted definition of business risk but in the RP context we will understand risk to be a potential problem which can be avoided or mitigated (cf. [5] for eferences). The potential problem for an RP is that he has accepted an SLA and may not be able to deliver the necessary computing resources in order to fulfil a computing task within an accepted time frame *T*. *Risk assessment* is the process through which a resource provider tries to estimate the probability for the problem to occur within *T* and *risk management* the process through which a resource provider tries to avoid or mitigate the problem.

In classical decision theory risk is connected with the probability of an undesired event; usually the probability of the event and its expected harm is outlined with a scenario which covers the set of risk, regret and reward probabilities in an expected value for the outcome. The typical statistical model has the following structure,

$$R(\theta, \delta(x)) = \int L(\theta, \delta(x)) f(x|\theta)\, dx \tag{1}$$

where L is a loss function of some kind, x is an observable event (which may not have been observed) an $\delta(x)$ is an estimator for a parameter θ which has some influence on the occurrence of x. The risk is the expected value of the loss function. The statistical models are used frequently because of the very useful tools that have been developed to work with large datasets.

The statistical model is influenced by the modern capital markets theory where risk is seen as a probability measure related to the variance of returns. Markowitz [14] initiated the modern portfolio theory stating that investors should focus on selecting portfolios based on portfolios' risk-reward characteristics instead of compiling portfolios of assets that each individually has attractive risk-reward characteristics. The risk is classified in systematic ("market") risk and idiosyncratic ("company" or "individual") risk. The analogy would be that an RP – handling a large number of nodes and a large number of computing tasks – will reach a steady state in his operations so that there will be a stable systematic risk ("market risk") for defaulting on an SLA which he can build on as his basic assumption and then a ("small") idiosyncratic risk which is situation specific and which he should estimate with some statistical models.

We developed a hybrid probabilistic and possibilistic model to assess the success of computing tasks in a Grid. The model first gives simple predictive estimates of node failures in the next planning period when the underlying logic is the Bayes probabilistic model for observations on node failures. When we apply the possibilistic model to a dataset we start by selecting a sample of k observations on node failures. Then we find out how many of these observations are different and denote this number by l; we want to use the two datasets to predict what the $k + 1$: *st* observation on node failures is going to be. The possibility model is used to find out if that number is going to be 0, 1, 2, 3, 4, 5, … etc.; for this estimate the possibility model uses the "most usual" numbers in the larger dataset and makes an estimate which is "as close as possible" to this number. The estimate we use is a triangular fuzzy number, i.e. an interval with a possibility distribution. The possibility model turned out to be a faster and more robust estimate of the $k + 1$: *st* observation and to be useful for online and real-time risk assessments with relatively small samples of data.

3 Predictive Probabilities

In the following we will use node failures in a cluster (or a Grid) as the focus, i.e. we will work out a model to predict the probabilities that n nodes will fail in a period covered by an SLA ($n = 0, 1, 2, 3, \ldots$). In the interest of space we have to do this by sketches as we deal with standard Bayes theory and modelling (cf. [5] for references).

The first step is to determine a probability distribution for the number of node failures for a time interval $(t_1, t_2]$ by starting from some basic property of the process we need to describe. Typically we assume that the node failures represent a Poisson process which is non-homogenous in time and has a rate function $\lambda(t),\ t \geq 0$.

The second step is to determine a distribution for $\lambda(t)$ given a number of observations on node failures from r comparable segments in the interval $(t_1, t_2]$. This count normally follows a Gamma (α, β) distribution and the posterior distribution $p\ (\lambda_{t1,\ t2})$, given the count of node failures, is also a Gamma distribution according to the Bayes theory. Then, as we have been able to determine $\lambda_{t1,\ t2}$ we can calculate the predictive distribution for the number of node failures in the next time segment; Bayes theory shows that this will be a Poisson-Gamma distribution.

The third step is to realize that a computing task can be carried out successfully on a cluster (or a Grid) if all the needed nodes are available for the scheduled duration of the task. This has three components: (i) a predictive distribution on the number of nodes needed for a computing task covered by an SLA; (ii) a distribution showing the number of nodes available when an assigned set of nodes is reduced with the predicted number of node failures and an available number of reserve nodes is added (the number of reserve nodes is determined by the resource allocation policy of the RP); (iii) a probability distribution for the duration of the task.

The fourth step is to determine the probability of an SLA failure: p_1 (*n* nodes will fail for the scheduled duration of a task) x (*1*- p_2 (*m* reserve nodes are available for the scheduled duration of a task)) if we consider only the systematic risk. We need to use a multinomial distribution to work out the combinations.

Consider a Grid of *k* clusters, each of which contains n_c nodes, leading to the total number of nodes $n = Sum\ (n_c)$, where $c = 1, ..., k$, in the Grid. Let in the following $\lambda(t)$, $t \geq 0$, denote generally a time non-homogeneous rate function for a Poisson process $N(t)$. We will assume that we have the RP4 scenario as our context, i.e. we will have to deal with hundreds of nodes and hundreds of computing tasks with widely varying computational requirements over the planning period for which we are carrying out the risk assessment.

The predictive distribution of the number of events is a Poisson-Gamma-distribution, obtained by integrating the likelihood with respect to the posterior. Under the reference prior the predictive probability of having *x* events in the future on a comparable time interval equals

$$p(x|r, \frac{1}{2} + \sum_{i=1}^{r} x_i, 1) = \frac{\Gamma(\sum_{i=1}^{r} x_i + x + 1/2) r^{(\frac{1}{2}+\sum_{i=1}^{r} x_i)}}{\Gamma(1/2 + \sum_{i=1}^{r} x_i) x! (r+1)^{(\frac{1}{2}+\sum_{i=1}^{r} x_i)+x}} \qquad (2)$$

When a computing job begins execution in a cluster (or Grid) its successful completion will require a certain number of nodes to be available over a given period of time. To assess the uncertainty about the resource availability, we need to model both the distribution of the number of nodes and the length of the time required by the jobs.

Given observed data on the number of nodes required by computing tasks, the posterior distribution of the probabilities **p** is available in an analytical form under a Dirichlet prior, and its density function can be written as

$$p(\mathbf{p}|\mathbf{w}) = \frac{\Gamma(\sum_{m=1}^{u} \alpha_m + w_m)}{\prod_{m=1}^{u} \Gamma(\alpha_m + w_m)} \prod_{m=1}^{u} p_m^{\alpha_m + w_m - 1} \qquad (3)$$

where w_m corresponds to the number of observed tasks utilising *m* nodes, α_m is the *a priori* relative weight of the *m* th component in the vector **p**, and ***w*** is the vector (w_m), where $m = 1,..,u$. The corresponding predictive distribution of the number of nodes required by a generic computing task in the future equals the Multinomial-Dirichlet distribution, which is obtained by integrating out the uncertainty about the multinomial parameters with respect to the posterior distribution. The Multinomial-Dirichlet distribution is in our notation defined as,

$$p(M = m^*|\mathbf{w}) = \frac{\Gamma(\sum_{m=1}^{u} \alpha_m + w_m) \prod_{m=1}^{u} \Gamma(\alpha_m + w_m + I(m = m^*))}{\prod_{m=1}^{u} \Gamma(\alpha_m + w_m) \quad \Gamma(1 + \sum_{m=1}^{u} \alpha_m + w_m)}$$
$$= \frac{\Gamma(\sum_{m=1}^{u} \alpha_m + w_m)}{\Gamma(\alpha_{m^*} + w_{m^*})} \frac{\Gamma(\alpha_{m^*} + w_{m^*} + 1)}{\Gamma(1 + \sum_{m=1}^{u} \alpha_m + w_m)}. \tag{4}$$

By combining the above distributions, we will find the probability distribution for the number of nodes in use for computing tasks in a future time interval $(t_1, t_2]$, as the corresponding random variable equals the product *XM*.

To simplify the inference about the length of a cluster/Grid computing task affecting a number of nodes, we propose that the time used for the task follows a Gaussian distribution with expected value μ and variance σ^2. Obviously, it is motivated to have separate parameter sets for different types of tasks. Assuming the standard reference prior for the parameters, we obtain the predictive distribution for the expected time used for a future computing task, say *T*, in terms of the probability density for the expected time as follows,

$$p(t|\bar{t}, ((n-1)/(n+1))s^2, n-1) = \frac{\Gamma(n/2)}{\Gamma(\frac{n-1}{2})\Gamma(1/2)} \times$$
$$\left(\frac{1}{(n+1)s^2}\right)^{\frac{1}{2}} \left[1 + \frac{1}{n+1)s^2}(t-\bar{t})^2\right]^{-\frac{n}{2}} \tag{5}$$

The probability that a task requires more time than any given time t equals $P(T > t) = 1 - P(T \leq t)$, where $P(T \leq t)$ is the cumulative density function (CDF). The value of the CDF can be calculated numerically using existing functions. However, it should also be noted that for a moderate to large n, the predictive distribution is well approximated by the Gaussian distribution with the estimated mean t and the variance $s^2 = (n+1)/(n-3)$. Consequently, if the Gaussian approximation is used, the probability $P(T \leq t)$ can be calculated using the CDF of the Gaussian distribution.

We now consider the probability that a computing task will be successful. This happens as the sum *P* ("none of the nodes allocated to the task fails") + Sum ($m_{=1}$, m_{max}) P ("m of the nodes allocated to the task fail & at least m idle nodes are available as reserves"). Here m_{max} is an upper limit for the number of failures considered. Notice that we simplify the events below by considering the m failures to take place simultaneously. We then get

P (S=1) = 1 – P (S=0)

= *1* – ∑ *P* (*m* failures occur & less than *m* free nodes available) (*m* = 1, …, m_{max})

= *1* – ∑ *P* (*m* failures occur) *P*(less than *m* free nodes available) (*m* = 1, …, m_{max}) (6)

≥ *1* – ∑ *P* (*m* failures occur) *P*(less than *m* anytime) (*m* = 1, …, m_{max})

The probability $P(m$ failures occur) is directly determined by the failure rate model discussed above. The other term, the probability $P($less than m free nodes at any time point), is dependent on the resource allocation policy and the need of reserve nodes by the other tasks running simultaneously.

The predictive probabilities model has been extensively tested and verified with data from the LANL cluster (cf. Shroeder-Gibson [17]). Here we collected results for the RP1 scenario where the RP is using a cluster with only a few nodes; the test runs have been carried out also for the scenarios RP2-4 with some variations to the results.

In the following we have collected a number of test results on the predictive probabilities (cf. (2)) using LANL data (cf. figs. 1-4) and on the probability that m nodes will be available (cf. (4)) when the computing task has a duration of t (cf. (5)) using LANL data (cf. fig. 5-8).

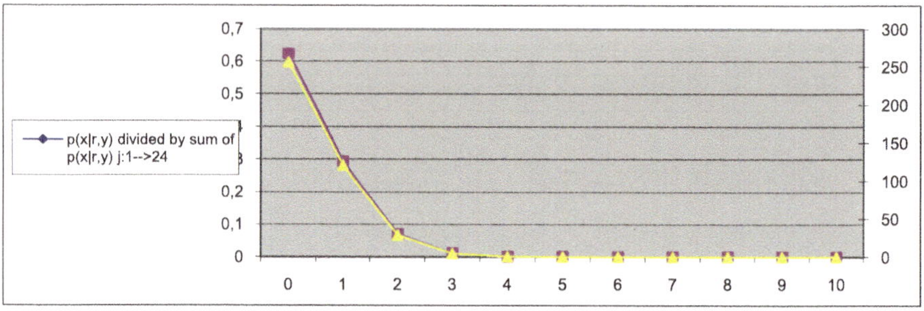

Fig. 1 Prediction of node failures [Prob. of Failure: 0.0644987] with 436 time slots of LANL cluster data for computing tasks running for 2 days, 19 hours

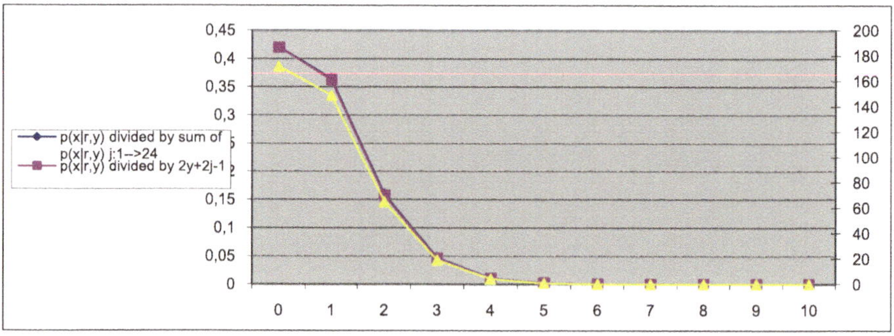

Fig. 2 Prediction of node failures [Prob. of Failure: 0.1040136] with 246 time slots of LANL cluster data for computing tasks running for 5 days, 4 hours

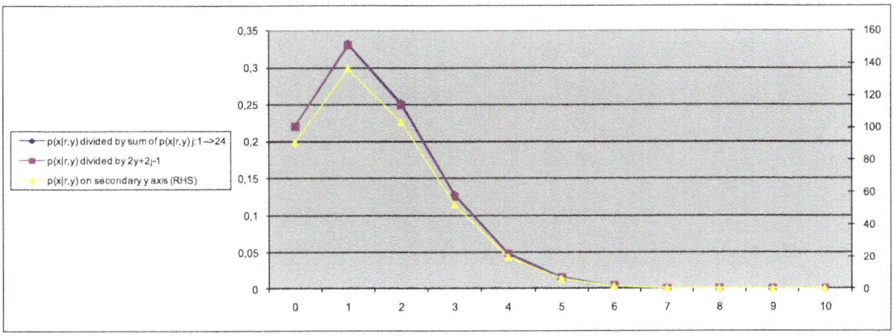

Fig. 3 Prediction of node failures [Prob. of Failure: 0.1509012] with 136 time slots of LANL cluster data for computing tasks running for 7 days, 0 hours

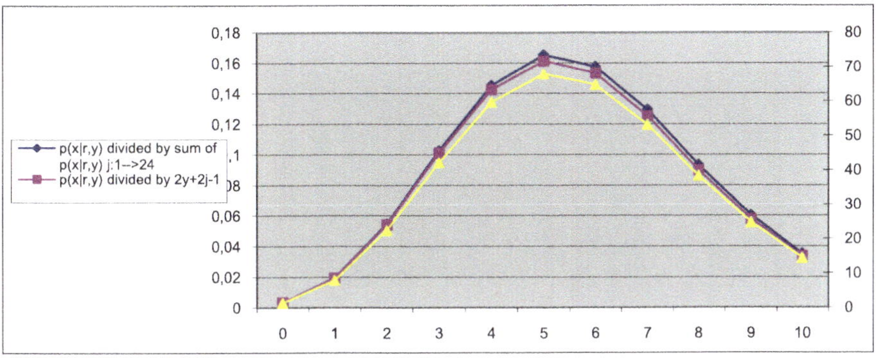

Fig. 4 Prediction of node failures [Prob. of Failure: 0.5132377] with 36 time slots of LANL cluster data for computing tasks running for 9 days, 0 hours

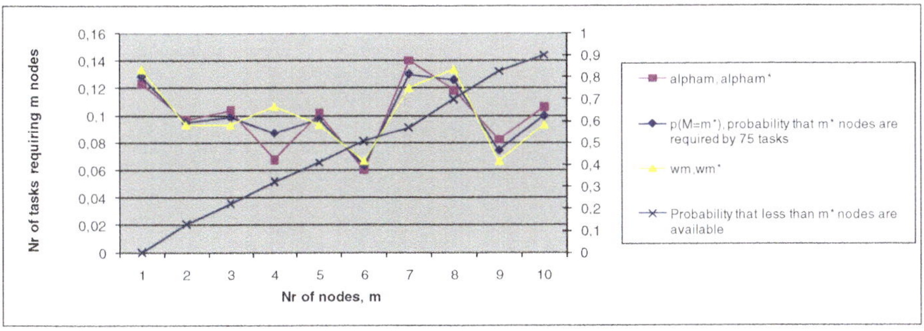

Fig. 5 Probability that less than m nodes will be available for computing tasks requiring ≤ 10 nodes; the number of nodes randomly selected; 75 tasks simulated over 236 time slots; LANL data

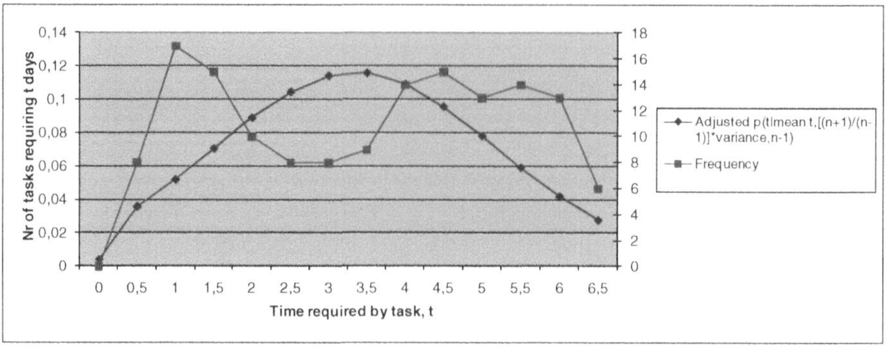

Fig. 6 Expected time needed for a computing task t; simulated 75 computing tasks requiring various t; data from LANL

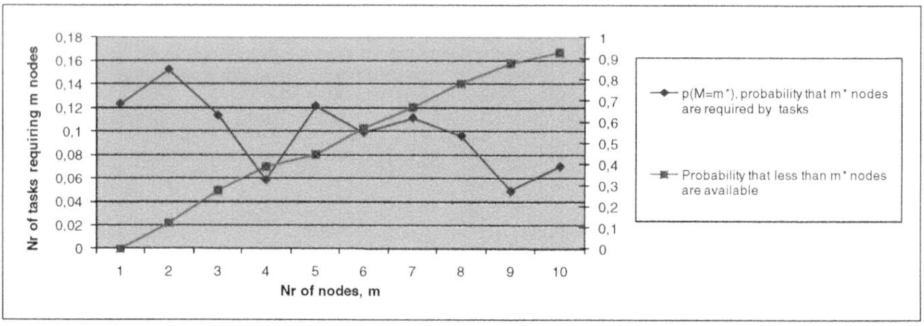

Fig. 7 Probability that less than m nodes will be available for computing tasks requiring $\leq$ 10 nodes; the number of nodes randomly selected; 100 tasks simulated over 236 time slots; LANL data

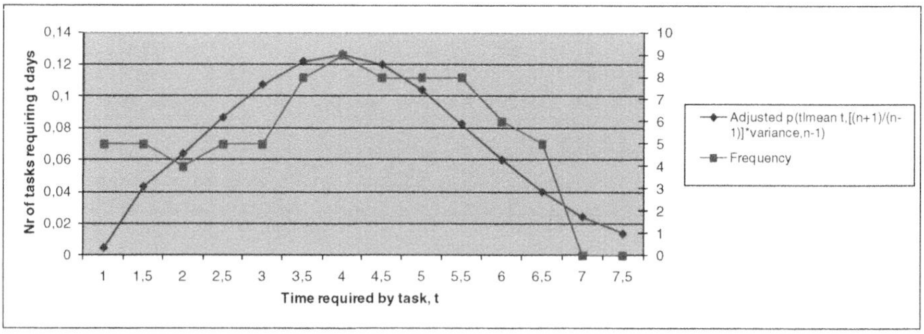

Fig. 8 Expected time needed for a computing task t; simulated 100 computing tasks requiring various t; data from LANL

4 Predictive Possibilities

In this section we will introduce a hybrid method for simple predictive estimates of node failures in the next planning period when the underlying logic is the Bayes models for observations on node failures that we used for the RA models in section 3. This is essentially a standard regression model with parameters represented by triangular fuzzy numbers – typically this means that the parameters are intervals and represent the fact that the information we have is imprecise or uncertain. We can only sketch the model here, for details cf. [5]; the model builds on some previous results in [2], [3], [4] and [6].

We will take a sample of a dataset (in our case, a sample from the LANL dataset) which covers inputs and fuzzy outputs according to the regression model; let this sample be x_i , Y_i , where $i = 1, 2, \ldots, n$. The main purpose with the fuzzy non-parametric regression model is to estimate $F(x)$ at any $x \varepsilon \mathbf{D}$. The membership function of the estimated Y is normally worked out to be as close as possible to the corresponding observations forming the fuzzy number, i.e. we should estimate $a(\mathrm{x})$, $\alpha(x)$, $\beta(x)$ for each $x \varepsilon \mathbf{D}$ so that we get a fit between the estimated Y and the observed Y' which is "a closest fit"; here we will use Diamond's distance measure (cf. [8]).

Let $A = (a, \alpha_1, \beta_1)$ and $B = (b, \alpha_2, \beta_2)$ be two triangular fuzzy numbers; then the squared distance between A and B is defined by,

$$d^2(A, B) = (\alpha_1 - \alpha_2)^2 + (a - b)^2 + (\beta_1 - \beta_2)^2 \tag{7}$$

Let us now assume that the observed (fuzzy) output is $Y_i = (a, \alpha_i, \beta_i)$, then with the Diamond distance measure and a local linear smoothing technique we need to solve a locally weighted least-squares problem in order to estimate $F(x_0)$, for a given kernel K and a smoothing parameter h, where

$$K_h(|x_i - x_0|) = K\left(\frac{|x_i - x_0|}{h}\right) \tag{8}$$

The kernel is a sequence of weights at x_0 to make sure that data that is close to x_0 will contribute more when we estimate the parameters at x_0 than those that are farther away, i.e. are relatively more distant in terms of the parameter h.

Let
$$F_{(i)}(x_i, h) = (\hat{a}_{(i)}(x_i, h), \hat{\alpha}_{(i)}(x_i, h), \hat{\beta}_{(i)}(x_i, h)) \tag{9}$$

be the predicted fuzzy regression function at input x_i. Compute $F_{(i)}(x_i; h)$ for each x_i and let

$$CV(h) = \frac{1}{l}\sum_{i=1}^{l} d^2(Y_i, F_{(i)}(x_i, h)) =$$

$$\frac{1}{l}\sum_{i=1}^{l}((\alpha_i - \hat{\alpha}_{(i)}(x_i, h))^2 + (a_i - \hat{a}_{(i)}(x_i, h))^2 + (beta_i - \hat{\beta}_{(i)}(x_i, h))^2) \tag{10}$$

We should select h_0 so that it is optimal in the following expression,

$$CV(h_0) = \min_{h>0} CV(h) \tag{11}$$

By solving the minimisation problem we can get an estimate of $F(x)$ at x_0 by,

$$\hat{F}(x_0) = (\hat{a}(x_0), \hat{\alpha}(x_0), \hat{\beta}(x_0)) \tag{12}$$

and the following estimate of $F(x)$ at x_0,

$$\begin{aligned}\hat{F}(x_0) &= (\hat{a}(x_0), \hat{\alpha}(x_0), \hat{\beta}(x_0)) \\ &= (\mathbf{e}_1^T \mathbf{H}(x_0, h)\mathbf{a}_Y, \mathbf{e}_1^T \mathbf{H}(x_0, h)\alpha_Y, \mathbf{e}_1^T \mathbf{H}(x_0, h)\beta_Y)\end{aligned} \tag{13}$$

We can use this model – which we have decided to call a *predictive possibility* model – to estimate a prediction of how many node failures we may have in the next planning period given that we have used statistics of the observed number of node failures to build a base of Poisson-Gamma distributions (for details cf. [6]).

We will use the Fréchet-Hoeffding bounds for copulas to show a lower limit for the probability of success of a computing task in a cluster (or a Grid). Let us recall that in (6) we have the notation *P(success)* as *P(S=1)* and *P(failure)* as *P(S=0)*; furthermore, let us use the abbreviation *less-than-m-anytime* fir the event *less than m free nodes available at any point of time*. Then we can rewrite (6) as,

$$\begin{aligned}
P(\text{success}) &= 1 - P(\text{failure}) = 1 - \sum_{m=1}^{m_{\max}} P(m \text{ failures occur \& less-than-m-anytime}) \\
&= 1 - \sum_{m=1}^{m_{\max}} P(m \text{ failures occur}) P(\text{less than } m \text{ free nodes available}) \\
&\geq 1 - \sum_{m=1}^{m_{\max}} P(m \text{ failures occur}) P(\text{less-than-m-anytime}) \\
&\geq 1 - \sum_{m=1}^{m_{\max}} P(m \text{ failures occur}) \\
&\times \left\{ \sum_{j=1}^{t^*} P(j - 1 \leq \text{the length of a task} < j, \text{less-than-m-anytime}) \right. \\
&\left. + P(\text{the length of a task} \geq t^*, \text{less-than-m-anytime}) \right\} \\
&\geq 1 - \sum_{m=1}^{m_{\max}} P(m \text{ failures occur}) \\
&\times \left\{ \sum_{j=1}^{t^*} P(\text{the length of a task} < j, \text{less-than-m-anytime}) \right. \\
&\left. + P(\text{the length of a task} \geq t^*, \text{less-than-m-anytime}) \right\}
\end{aligned} \tag{14}$$

Let us introduce the notations *G(m) = P(less-than-m-anytime)* and *F(t) = P (the duration of a task is less than t)*. Let t_* be chosen such that 1-F (t_*) ≥ 0.995. Furthermore, de4note the copula of F and G by H, where *H(t, m) = P(the duration of a task is less than t, less-than-m-anytime)*. Then using the Freéchet-Hoeffding upper bound for copulas we find that,

$$P(\text{success}) \geq 1 - \sum_{m=1}^{m_{\max}} P(m \text{ failures occur}) \times \left\{ \sum_{j=1}^{t^*} \min\left\{ \frac{\Gamma(b/2)}{\Gamma(\frac{b-1}{2})\Gamma(1/2)} \left(\frac{1}{(b+1)s^2}\right)^{\frac{1}{2}} \left[1 + \frac{1}{(b+1)s^2}(j-\bar{j})^2\right]^{-\frac{b}{2}} P(\text{less than } m \text{ free nodes at any time point}) \right\} + \underbrace{P(\text{the length of a task is } \geq t^*, \text{less-than-m-anytime})}_{\leq 0.005} \right\} \tag{15}$$

If we summarize these results we get,

$$P(\text{success}) \geq 1 - \sum_{m=1}^{m_{\max}} P(m \text{ failures occur}) \times \left\{ \sum_{j=1}^{t^*} \min\left\{ \frac{\Gamma(b/2)}{\Gamma(\frac{b-1}{2})\Gamma(1/2)} \left(\frac{1}{(b+1)s^2}\right)^{\frac{1}{2}} \left[1 + \frac{1}{(b+1)s^2}(j-\bar{j})^2\right]^{-\frac{b}{2}} P(\text{less-than-m-anytime}) \right\} \right\} \tag{16}$$

Now we can use the new model as an alternative for predicting the number of node failures and use it as part of the Bayes model for predictive probabilities. In this way we will have hybrid estimates of the expected number of node failures – both probabilistic and possibilistic estimates. An RP may use either one estimate for his risk assessment or use a combination of both.

We carried out a number of validation tests in order to find out (i) how well the predictive possibilistic models can be fitted to the LANL dataset, (ii) what differences can be found between the probabilistic and possibilistic predictions and (iii) if these differences can be given reasonable explanations. The testing was structured as follows:

- 5 time frames for the possibilistic predictive model with a smoothing parameter from the smoothing function: $h = 382.54 * \text{Nr of timeslots}^{-0.5325}$
- 5 feasibility adjustments from the hybrid possibilistic adjustment model to the probabilistic predictive model

In the testing we worked with short and long duration computing tasks scheduled on a varying number of nodes and the SLA probabilities of failure estimates remained reasonable throughout the testing. The smoothing parameter h for the possibilistic models should be determined for the actual cluster of nodes, and in such a way that we get a good fit between the probabilistic and the possibilistic models. The approach we used for the testing was to experiment with combinations of h and then to fit a distribution to the results; the distribution could then be used for interpolation.

We run the tests for all the RP1-RP4 scenarios (cf. [5] for details) but here we are showing only results for the RP1 scenario (cf. fig. 9-11) in the interest of space. The tests aimed at calibrating the possibilistic predictive model to generate estimates of the possible failure of n nodes that would be "close" or "somewhat similar" to the probabilistic predictions. This served the notion that the estimates should not deviate too much from each others but also the aim to find out under what conditions we could get estimates that are "close" or "somewhat similar". We run simulations with a varying number of time slots for 10 nodes and variations on the duration of computing tasks, from 8 days and down to 2 days. We found differences in the predictions for the longer durations of computing tasks but the estimates got "closer" for shorter durations; we also run simulations for computing tasks going down from 24 hours to 1 hour but these results are not shown because the differences in estimates became very small. All tests were based on the LANL data.

Then we next run a series of tests with the two models to find out how the estimates differ when we have 1, 2, 3 or 4 spare nodes dedicated to the computing task; the results are shown in figs. 12-15.

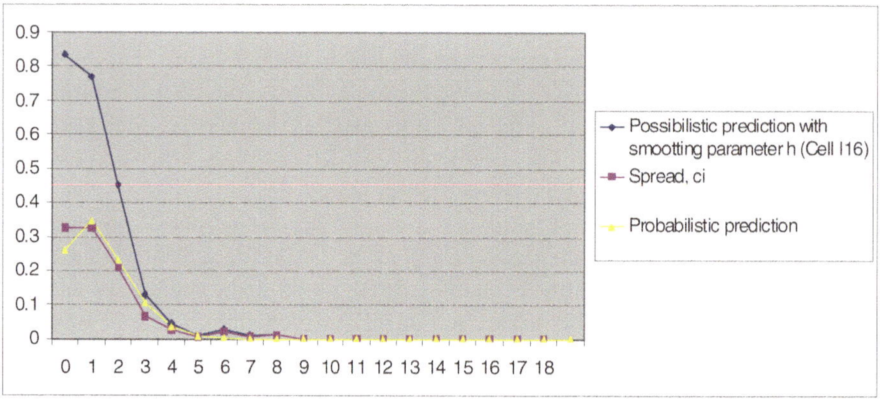

Fig. 9 Possibilistic and probabilistic prediction of n node failures for a computing task with a duration of 8 days on 10 nodes; 153 time slots simulated

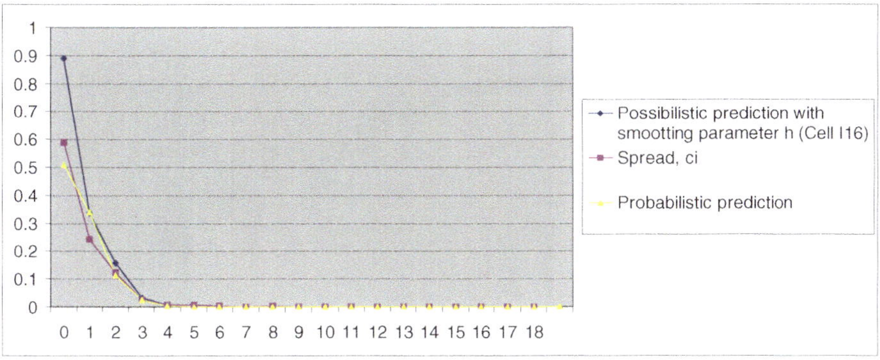

Fig. 10 Possibilistic and probabilistic prediction of n node failures for a computing task with a duration of 4 days on 10 nodes; 153 time slots simulated

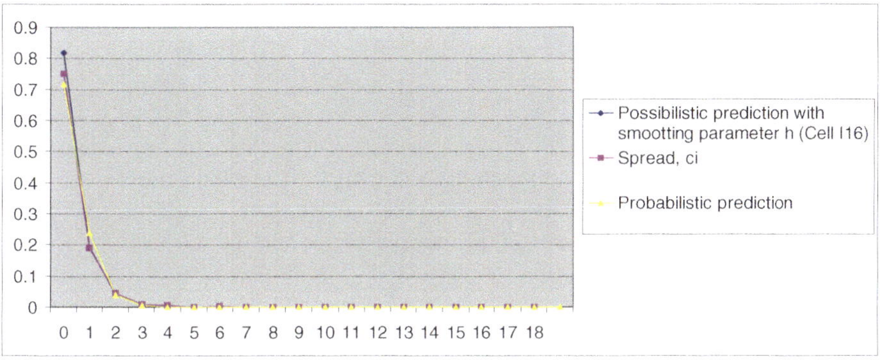

Fig. 11 Possibilistic and probabilistic prediction of n node failures for a computing task with a duration of 2 days on 10 nodes; 153 time slots simulated

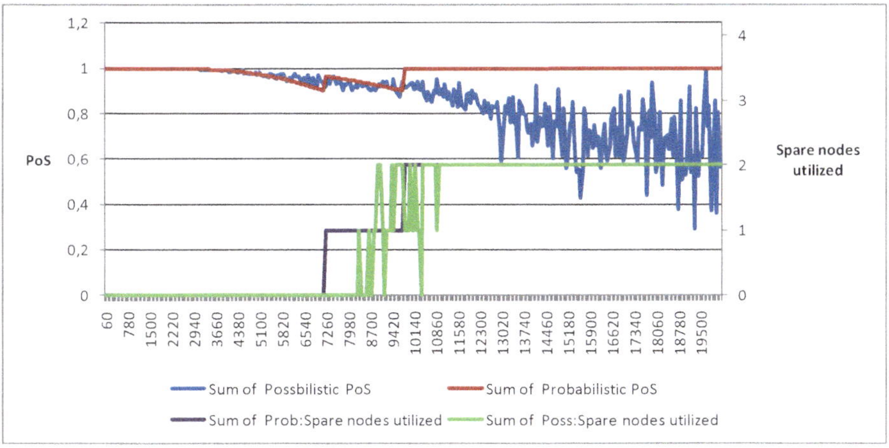

Fig. 12 Comparison of probabilistic and possibilistic success for an SLA for computing tasks on a 6 node cluster with two spare nodes; simulated for 60-19500 minutes.

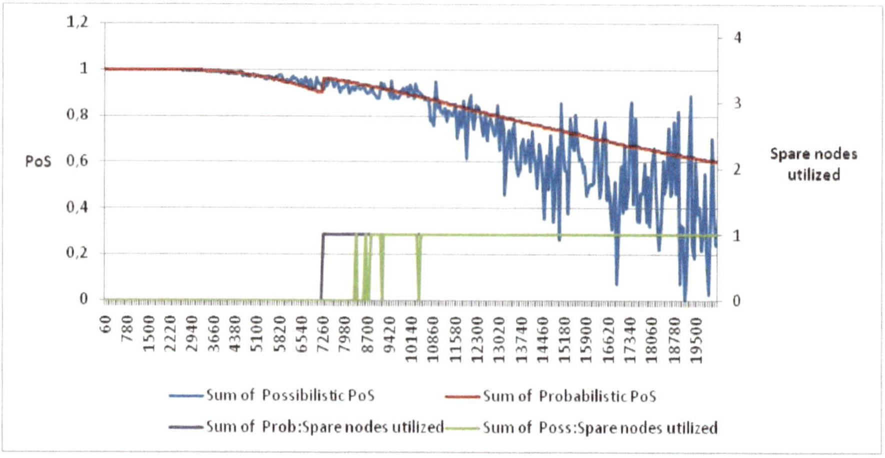

Fig. 13 Comparison of probabilistic and possibilistic success for an SLA for computing tasks on a 6 node cluster with one spare node; simulated for 60-19500 minutes.

The probability of success (PoS, red curve) for a computing task goes down (as expected) when the duration of the task increases. The PoS goes up momentarily when a spare node is reserved for the task (cf. fig. 13) but then starts to go down again; the possibility of success (PboS, blue curve) closely follows the PoS for computing tasks taking about 10 000 min but then starts to show strong fluctuations. The reason for this is that the predictive possibilities are calculated from a regression models taking samples from the LANL data where there are relatively few computing tasks with long durations (a majority of the computing tasks run for less than 360 min).

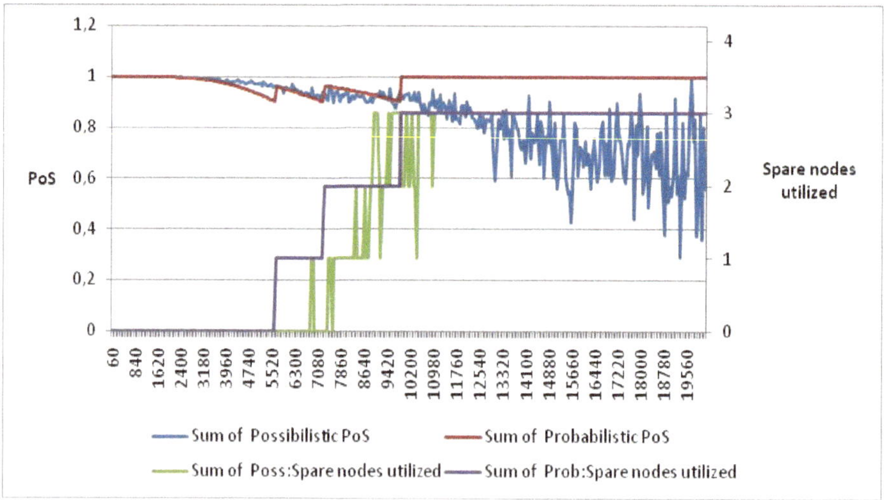

Fig. 14 Comparison of probabilistic and possibilistic success for an SLA for computing tasks on a 6 node cluster with three spare nodes; simulated for 60-19500 minutes

The same trends can be seen in fig.12 where two spare nodes are reserved for the computing tasks but here the PoS stays at 1.0 when the second spare node is added, i.e. the computing task on 6 nodes will probably always be completed despite taking up to 20 000 min (i.e. 13.88 days).

We then continued the testing by adding a third and a fourth spare node for the computing tasks, cf. figs. 14-15 and with the fourth spare node we reached a state where also the PBoS reaches 1.0.

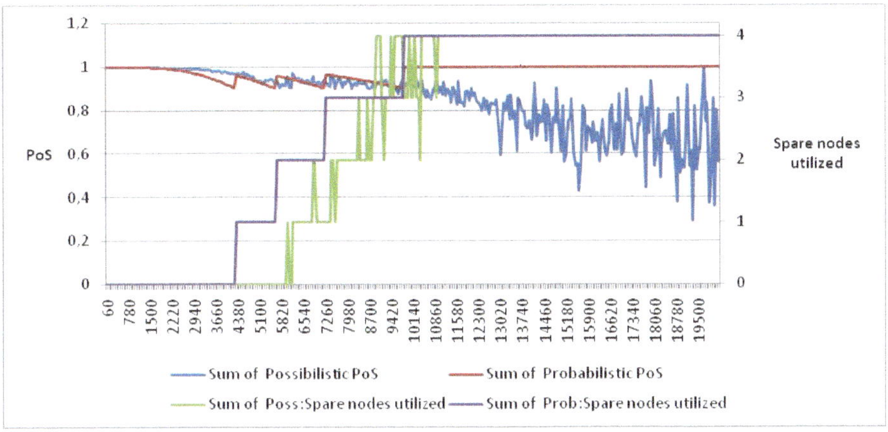

Fig. 15 Comparison of probabilistic and possibilistic success for an SLA for computing tasks on a 6 node cluster with four spare nodes; simulated for 60-19500 minutes

5 Summary and Conclusions

We developed a hybrid probabilistic and possibilistic technique for assessing the risk of an SLA for a computing task in a cluster/grid environment. The probability of success with a hybrid model is estimated higher than in the probabilistic model since the hybrid model takes into consideration the possibility distribution for the maximal number of failures derived from the RP's observations.

The hybrid model showed that we can increase or decrease the granularity of the model as needed; we can *reduce* the estimate of the $P(S^{*}=1)$ by making a rougher, more conservative, estimate of the more unlikely events of (M+1, N) node failures. We noted that M is an estimate which is dependent on the history of the nodes being used and can, of course, be calibrated to "a few" or to “many” nodes.

We have run series of validation experiments with the two series of models to find out how the theoretical models work and if they will give reasonable results. When we had verified that the models work as they should in terms of the theory we run series of explorative experiments to find out if we can get the PoS and PboS estimates to become “close” or “somewhat similar” and under what circumstances that would happen.

The probabilistic models scale from 10 nodes to 100 nodes and then on to any (reasonable) number of nodes; in the same fashion also the possibilistic models scale to 100 nodes and then on to any (reasonable) number of nodes.

The RP can use both the probabilistic and the possibilistic models to get two alternative risk assessments and then (i) choose the probabilistic RA, (ii) the possibilistic RA or (iii) use the hybrid model for a combination of both RAs – this is a cautious/conservative approach.

The risk assessment of SLAs in grid/cluster computing is an emerging field and we foresee continued work to refine the methods we now have developed.

References

1. Brandt, J.M., Gentile, A.C., Marzouk, Y.M., Pébay, P.P.: Meaningful Statistical Analysis of Large Computational Clusters (2005)
2. Carlsson, C., Fullér, R.: On Possibilistic Mean and Variance of Fuzzy Numbers. Fuzzy Sets and Systems 122, 315–326 (2001)
3. Carlsson, C., Fullér, R.: Fuzzy Reasoning in Decision Making and Optimization. Studies in Fuzziness and Soft Computing Series. Springer, Heidelberg (2002)
4. Carlsson, C., Fullér, R.: A Fuzzy Approach to Real Option Valuation. Fuzzy Sets and Systems 139, 297–312 (2003)
5. Carlsson, C., Weissman, O.: Advanced Risk Assessment, D4.1, The AssessGrid Project, IST-2005-031772, Berlin (2009)
6. Carlsson, C., Fullér, R., Mezei, J.: A lower limit for the probability of success of computing tasks in a grid. In: Proceedings of the Tenth International Symposium of Hungarian Researchers on Computational Intelligence and Informatic (CINTI 2009), Budapest, Hungary, pp. 717–722 (2009)
7. Czajkowski, K., Foster, I., Kesselman, C.: Agreement-Based Resource Management. Proceedings of the IEEE 93(3), 631–643 (2005)
8. Diamond, P.: Fuzzy least squares. Information Sciences 46, 141–157 (1988)
9. Iosup, A., Jan, M., Sonmez, O.O., Epema, D.H.J.: On the Dynamic Resource Availability in Grids. In: Proceedings of the 8th IEEE/ACM International Conference on Grid Computing (GRID 2007), pp. 26–33 (2007)
10. Lintner, J.: The Valuation of Risk Assets and the Selection of Risky Investments in Stock Portfolios and Capital Budgets. Review of Economics and Statistics 47, 13–37 (1965)
11. Liu, Y., Ngu, A.H.H., Zeng, L.: QoS Computation and Policing in Dynamic Web Service Selection. In: Proceedings of the 13th international World Wide Web (WWW 2004), pp. 66–73 (2004)
12. Long, D., Muir, R., Golding, R.: A Longitudinal Survey of Internet Host Reliability. In: 14th Symposium on Rel. Dist. Sys., pp. 2–9 (1995)
13. Magana, E., Serrat, J.: Distributed and Heuristic Policy-Based Resource Management System for Large-Scale Grids. In: Bandara, A.K., Burgess, M. (eds.) AIMS 2007. LNCS, vol. 4543, pp. 184–187. Springer, Heidelberg (2007)
14. Markowitz, H.M.: Portfolio Selection. Journal of Finance 7, 77–91 (1952)
15. Plank, J.S., Elwasif, W.R.: Experimental Assessment of Workstation Failures and Their Impact on Checkpointing Systems. In: 28th International Symposium on Fault-Tolerant Computing (1998)

16. Raju, N., Gottumukkala, N., Liu, Y., Leangsuksun, C.B.: Reliability Analysis in HPC Clusters. In: High Availability and Performance Computing Workshop (2006)
17. Schroeder, B., Gibson, G.A.: A Large-Scale Study of Failures in High-Performance Computing Systems. In: DSN 2006 Conference Proceedings, Philadelphia (2006)
18. Tuecke, S.: SNAP: A Protocol for Negotiating Service Level Agreements and Coordinating Resource Management in Distributed Systems. In: 8th Workshop on Job Scheduling Strategies for Parallel Processing, Edinburgh, Scotland (2002)

From Benchmarks to Generalised Expectations

Erio Castagnoli and Gino Favero

Abstract. A random variable can be equivalently regarded to as a function or as a "set", namely, that of the points lying below (when positive) and above (when negative) its graph. The second approach, proposed by Segal in 1989, is known as the *measure* (or *measurement*) *representation approach*. On a technical ground, it allows for using Measure theory tools instead of Functional analysis ones, thus making often possible to reach new and deeper conclusions. On an interpretative ground, it makes clear how expectation and expected utility, either classical or *à la* Choquet, are structurally analogous and, moreover, it allows for dealing with new and more general types of expectations including, e.g., state dependence.

Starting from the measurement approach and from a decision-theoretical result by Castagnoli and LiCalzi (2006), we present a new representation theorem in the same perspective and, finally, we propose a definition of generalised expectations and two different concepts of associativity that can be imposed to them.

1 Preliminaries

A very rough simplification leads to identifying two different philosophies in Decision Theory: a *maximising approach*, where some utility index is optimised, and a *satisficing approach*, where the aim is to reach (at least) a given target (see, e.g., Simon, 1955).

Erio Castagnoli
Università Commerciale "Luigi Bocconi", Dipartimento di Scienze delle Decisioni, via Röntgen 1, 20136 Milano, Italy; Accademia Nazionale Virgiliana, via Accademia 47, 46100 Mantova, Italy

Gino Favero
Università degli Studi di Parma, Dipartimento di Economia, Sezione di Matematica "Eugenio Levi", via Kennedy 6, 43125 Parma, Italy
e-mail: `gino.favero@unipr.it`

Both authors gratefully acknowledge financial support by MIUR, Parma University and Bocconi University.

S. Greco et al. (Eds.): Preferences and Decisions, STUDFUZZ 257, pp. 31–51.
springerlink.com

Castagnoli (1990), Castagnoli and LiCalzi (1993, 1994, 1996, 2006), Bordley and LiCalzi (2000), and Bordley (2002) have shown how to reconcile the two approaches while providing, moreover, a new insight in interpreting expected (and non expected) utility. Their basic idea is rather simple, and summarised hereafter.

Call $\mathscr{X}$ a set of random variables and denote them by means of capital letters (X, Y, ...). The expected utility functional is then

$$\mathrm{E}u(X) = \int_{\mathbb{R}} u(t)\mathrm{d}x(t) \qquad (X \in \mathscr{X}),$$

with $x(t) = \Pr\{X \leqslant t\}$ the cumulative distribution function of X.

We shall assume, here and in the rest of the paper, that u is bounded (as it is when, e.g., the random variables of $\mathscr{X}$ take their values in a bounded set as we assume later on). This way, u (or possibly its right-continuous version) can be supposed to take values in the interval $[0,1]$, so that it can be interpreted as the c.d.f. of some random variable U. Supposing such U to be independent of all $X \in \mathscr{X}$,

$$\mathrm{E}u(x) = \int_{\mathbb{R}} u(t)\mathrm{d}x(t) = \int_{\mathbb{R}} \Pr\{U \leqslant t\}\mathrm{d}x(t) = \Pr\{U \leqslant X\},$$

i.e., the expected utility of any $X \in \mathscr{X}$ turns out to be the probability that X "outperforms" a given (independent) benchmark U with c.d.f. u.

In other words, $\mathrm{E}u(x)$ and $\Pr\{U \leqslant X\}$ are simply two different ways of writing the same functional. Nevertheless, they are quite different from an interpretative point of view. In the first way the decision maker chooses according to the expectation of the "distorted" random variables $u(X)$. In the second way the decision maker acts *as if* he is choosing a random benchmark U (independent of all the random variables in $\mathscr{X}$), evaluating the probability to beat U for each $X \in \mathscr{X}$ and, finally, sorting $\mathscr{X}$ according to such probabilities.

Which interpretation to adopt is simply a matter of individual taste. Castagnoli and LiCalzi (1993, 1994, 1996) discuss several points, such as interpretation of the benchmark, risk aversion, stochastic dominance, and so on. They also discuss a couple of applications, showing that sometimes the second interpretation appears to be more proper and convincing than the first, and traditional, one. Cigola and Modesti (1996) and Modesti (2003) take into consideration SSB utility and lottery-dependent utility, respectively, from the point of view of stochastic benckmarking, and Beccacece and Cillo (2006) apply benchmarking to financial risk measurement.

It is time to enter more deeply into the problem.

Let Ω be the set of the fundamental states of the world, let $\mathscr{X}$ be the family of all of the random variables defined on Ω and taking values in a bounded interval $B \subseteq \mathbb{R}$, and let p be a probability on (a (σ-)algebra of events in) Ω. The c.d.f. function of any (measurable) random variable X is thus $x(t) = p\{\omega : X(\omega) \leqslant t\}$.

Take now a (possibly different) state space S, where we shall imagine the random benchmark U to be defined. Let S be endowed with a probability measure π, so that

$u(t) = \pi\{s : U(s) \leqslant t\}$ is the c.d.f. of U. The above equality can now be written, in greater detail,

$$\begin{aligned} \mathrm{E}_p u(X) &= \int_{\mathbb{R}} u(t)\mathrm{d}x(t) = \int_{\mathbb{R}} \pi\{U \leqslant t\}\mathrm{d}p\{X \leqslant t\} \\ &= P\big\{(\omega,s) : U(s) \leqslant X(\omega)\big\} = P\{U \leqslant X\}, \end{aligned}$$

where P is the product probability $p \otimes \pi$ on $\Omega \times S$, reflecting (or maybe clarifying) the assumed stochastic independence among U and all of the Xs.

We want to point out that it is quite natural to suppose U and the Xs to be defined on different state spaces. We reckon indeed to be extremely common that a decision maker has his own state space, but that he needs to assess preferences on random variables defined on another one. The introductory pages of Castagnoli and LiCalzi (2006) contain a detailed discussion about this feature; here, we limit ourselves to sketching a couple of ideas.

Think, for instance, of a decision maker who intends to invest his money in a financial portfolio in order to protect himself and his family against bad events. His state space S is likely to be formed by events related, e.g., to illnesses, family needs, accidents, and so on. On the other hand, the financial assets he is considering are defined on a state space Ω containing events related e.g., to economic growth, inflation rate, and so on.

Another example might be offered by an Insurance company aiming at investing in assets as a protection against the insured risks.

We argue that the setting described here can be of some importance in Experimental Economics, when the aim is to deal with a functional form which is sufficiently general so as to encompass the observed preferences of a decision maker.

Remark. Since U is the only random variable defined on the state space S, assigning a probability measure on S is undistinguishable from assigning a capacity, because the only events to be measured belong to the increasing chain $\{s : U(s) \leqslant t\}_{t\in\mathbb{R}}$. Any positive and monotonic function defined on this chain can be extended to the (σ-)algebra it generates on S in such a way to obtain either a capacity or a(n additive) measure, both of them being, in general, not unique.

Consequently, what precedes can be equivalently restated by saying that there exists a set function v defined on $\Omega \times S$ with the properties of being additive with respect to the events in Ω (for any fixed event in S) and just monotonic with respect to the events in S (for any fixed event in Ω).

2 The Result of Castagnoli and LiCalzi

If we allow for stochastic dependence among U and the Xs, the two functionals $\mathrm{E}u(X)$ and $P\{U \leqslant X\}$ are, of course, no longer equivalent. It is nevertheless possible to interpret the second one as a "state-dependent expected utility". Castagnoli and LiCalzi (2006) provide a full axiomatisation which guarantees a given preference preorder to be represented by the functional $P\{U \leqslant X\}$. Such an axiomatisation simply requires the weak (preference) preorder "$\succcurlyeq$" to be monotonic, continuous and satisfying Savage's *Sure Thing Principle* (Axiom P2 in his setting), as we are going to summarise.

Call $XAY := X\mathbb{I}_A + Y\mathbb{I}_{A^c}$ the random variable that coincides with X if the event A prevails and with Y elsewhere: the Sure Thing Principle amounts to asking that

$$YAZ \succcurlyeq XAZ \iff YAV \succcurlyeq XAV$$

for every $X,Y,Z,V \in \mathscr{X}$ and every event A. It is moreover necessary that there are at least three *essential events*, i.e., such that $XAY \succ Y$ for some X,Y. Under such assumptions, the following theorem holds.

Theorem 1 (Castagnoli – LiCalzi). *Let $\succcurlyeq$ be a monotonic and continuous (with respect to the* sup *norm) weak preorder among random variables defined on the same set Ω of states of the world. Then, the Sure Thing Principle holds if and only if there exist:*

(i) a second state space S, and a random variable U defined on it, and
(ii) a finitely additive, normalised set function (i.e., a probability charge) P on $\Omega \times S$

such that the functional $P\{\cdot \geqslant U\}$ represents $\succcurlyeq$, i.e., such that

$$Y \succcurlyeq X \iff P\{Y \geqslant U\} \geqslant P\{X \geqslant U\}.$$

Moreover, if $\succcurlyeq$ is pointwise continuous, then P is countably additive (i.e., a probability measure).

The proof given by Castagnoli and LiCalzi (2006) is based upon a very deep result given by Debreu (1960) for the case when the state space is finite. Indeed, in some sense, Theorem 1 might be considered an extension of Debreu Theorem for any (not necessarily finite) state space, although it features a different interpretation of the representing functional.

Remarks. 1. Whenever Ω is a finite set, both the sup norm and the pointwise continuity collapse into the classical, Euclidean one.

2. The above representation is a "cardinal" evaluation of an "ordinal" assessment ($X \geqslant U$). This looks quite appealing and natural in a decision problem, where an agent is required to perform comparisons: the preference $\succcurlyeq$ is founded upon ordinal evaluations, which can be effectively measured by a probability. In other words: in order to compare random variables, i.e., to express a preference preorder among them, the usual interpretation wants the decision maker to (cardinally) measure both events (by means of p) and outcomes (by u). Apparently, this is quite demanding: in order to express solely a relative and ordinal assessment, the decision maker is required to master two "absolute" and cardinal risk measures. The interpretation we propose simply wants the decision maker to compare all of the alternatives at hand with a chosen benchmark U and, on such a basis, to express a preference given by the likelihood that U is outperformed. As we pointed out in the final remark of Section 1, such a likelihood might be intended to be either cardinal (i.e., a probability) or simply ordinal (i.e., a capacity).

3. The set S of states of the world can *never* be elicited by preferences. It is indeed possible to observe the final choices of the decision maker, but there is no way to understand *how* they have been obtained. The c.d.f. of the random benchmark U is nevertheless observable, by taking into consideration the (degenerate) random variables $X = t\mathbb{I}_\Omega, t \in \mathbb{R}$. Therefore, we can always take $U(S) \subset \mathbb{R}$ as an observable "surrogate" for S, and hence the benchmark can be taken to be $U(s) = s$ (or, equivalently, any injective function $S \to U(S)$). We shall call *essential values* the elements of $U(S)$.

Example 1. A firm wants to hedge against a random loss by means of random variables defined on the set $\Omega = \{\omega_1, \omega_2\}$.

By observing its preferences (supposed to be continuous, monotonic and satisfying the Sure Thing Principle), we can elicit the values of its random benchmark U. Let us begin by noticing that if $\tau > t$ and $t\mathbb{I}_\Omega \sim \tau\mathbb{I}_\Omega$, then U takes no value in the interval $(t, \tau]$; assume that we have established in such a way that U takes the values 100, 400 and 1000 only. This way, S can be taken so as to contain just the three states of the world where U is valued 100, 400 and 1000: we can conveniently take precisely $S = \{s_1, s_2, s_3\} = \{100, 400, 1000\}$ and $U(s) = s$. We shall also refer to the "cumulative" states of the world in S, defined by $s'_1 := s_1 = \{U \leqslant 100\}$, $s'_2 := s_1 \cup s_2 = \{U \leqslant 400\}$, and $s'_3 = S = \{U \leqslant 1000\}$.

Let us further assume that the preference is:

$$\begin{aligned} 1000\mathbb{I}_\Omega &\sim 1000\mathbb{I}_{\omega_1} + 400\mathbb{I}_{\omega_2} \succ 1000\mathbb{I}_{\omega_1} + 100\mathbb{I}_{\omega_2} \succ 400\mathbb{I}_\Omega \succ \\ &\succ 400\mathbb{I}_{\omega_1} + 100\mathbb{I}_{\omega_2} \succ 1000\mathbb{I}_{\omega_1} \succ 100\mathbb{I}_{\omega_1} + 1000\mathbb{I}_{\omega_2} \sim 100\mathbb{I}_{\omega_1} + 400\mathbb{I}_{\omega_2} \succ \\ &\succ 100\mathbb{I}_\Omega \succ 400\mathbb{I}_{\omega_1} \succ 1000\mathbb{I}_{\omega_2} \sim 400\mathbb{I}_{\omega_2} \succ 100\mathbb{I}_{\omega_2} \succ 100\mathbb{I}_{\omega_1} . \end{aligned}$$

Note that the preference above is inconsistent with an expected utility: indeed, e.g., $100\mathbb{I}_{\omega_2} \succ 100\mathbb{I}_{\omega_1}$ would imply $p(\omega_2) > p(\omega_1)$, whereas $400\mathbb{I}_{\omega_1} \succ 1000\mathbb{I}_{\omega_2}$ would imply $p(\omega_1) > p(\omega_2)$. Call

$$\begin{aligned} P_1 &= P(\omega_1, s'_1) \ , \ P_2 = P(\omega_2, s'_1) \ , \\ P_3 &= P(\omega_1, s'_2) \ , \ P_4 = P(\omega_2, s'_2) \ , \\ P_5 &= P(\omega_1, s'_3) \ , \ P_6 = P(\omega_2, s'_3) \ : \end{aligned}$$

the above preferences entail that $P_5 > P_3 > P_1$, $P_6 = P_4 > P_2$, $P_5 + P_6 = 1$, and

$$P_5 + P_2 > P_3 + P_4 ; \ P_3 + P_2 > P_5 ; \ P_5 > P_1 + P_4 ; \ P_1 + P_2 > P_4 ; \ P_2 > P_1$$

which are consistent, for instance, with

$$(P_1, P_2, P_3, P_4, P_5, P_6) = (0.2, 0.3, 0.45, 0.35, 0.65, 0.35),$$

corresponding to the joint distribution

$$(p_1, p_2, p_3, p_4, p_5, p_6) = (0.2, 0.3, 0.25, 0.05, 0.2, 0).$$

on $\Omega \times S$ given by:

$$\begin{aligned} p_1 &= P(\omega_1, s_1) = P_1 \,, & p_2 &= P(\omega_2, s_1) = P_2 \,, \\ p_3 &= P(\omega_1, s_2) = P_3 - P_1 \,, & p_4 &= P(\omega_2, s_2) = P_4 - P_2 \,, \\ p_5 &= P(\omega_1, s_3) = P_5 - P_3 \,, & p_6 &= P(\omega_2, s_3) = P_6 - P_4 \,. \end{aligned}$$

This choice for P yields measures 0.5, 0.3 and 0.2 respectively for $\{s_1\}$, $\{s_2\}$ and $\{s_3\}$. Such measures can be intended as coming from a (unique) probability, as well as from infinitely many capacities v such that $v(\{s_3\}) = 0.2$, $v(\{s_2, s_3\}) = 0.5$ and $v(S) = 1$. Note that P is not unique: a different choice for it might be, e.g., $(p_1, p_2, p_3, p_4, p_5, p_6) = (0.14, 0.35, 0.3, 0.06, 0.15, 0)$.

The functional $P\{\cdot \geqslant U\}$ is extremely general (maybe even too much), and it can easily accommodate almost all of the classical paradoxes in Decision Theory. Indeed, besides the usual and obvious assumptions of monotonicity and continuity, it simply requires the weak form of "additivity" (with respect to the events) captured by the Sure Thing Principle: such a requirement is actually enough to ensure additivity of P, which turns then out to be a probability measure.

It is easy to see that, if the Sure Thing Principle is taken away (while maintaining monotonicity and continuity), Theorem 1 holds with a non additive P, i.e., the representing functional turns out to be $v\{\cdot \geqslant U\}$ with v a capacity on $\Omega \times S$.

Remark. The main feature of Theorem 1 does not lie in the fact that P is a probability, but rather in that it is a measure: it is actually clear that normalisation is unimportant for sorting purposes. Therefore, the functional $P\{\cdot \geqslant U\}$ can be equivalently written $m\{\cdot \geqslant U\}$, with m a bidimensional measure on $\Omega \times S$.

3 Measure of the Hypograph

In order to understand fully the implications of the two preceding sections, we need to address some remarks about the so-called *Measure* (or *Measurement*) *Representation Approach* as introduced in Segal (1993). Its major feature lies in identifying a random variable with a set, as explained hereafter.

To fix the ideas and focus effectively on the important details, let us consider a finite set $\Omega = \{\omega_1, \omega_2, \ldots, \omega_n\}$ of states of the world. Consider a random variable X and suppose, for the moment, that $X \geqq 0$: we denote its values by $x_i := X(\omega_i) \geqslant 0$. Such a random variable identifies the histogram, frequently called the (truncated) *hypograph*:

$$\text{hypo}(X) = \bigcup_{i=1}^{n} \left(\{\omega_i\} \times [0, x_i]\right) ;$$

the correspondence is one-to-one, as a histogram uniquely identifies a random variable. The expected value of X is then the measure of its hypograph according to the

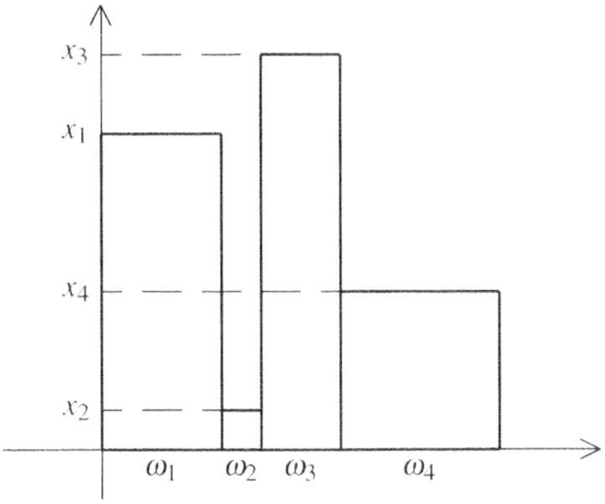

Fig. 1 Histogram of the values of a random variable

(bidimensional) measure $m = p \otimes \lambda$, the product of a probability measure p on S and the Lebesgue measure λ on $\mathbb{R}$:

$$m(\text{hypo}(X)) = m\Big(\bigcup_{i=1}^{n}[\{\omega_i\} \times [0,x_i]]\Big) = \sum_{i=1}^{n} m(\{\omega_i\} \times [0,x_i]) =$$
$$= \sum_{i=1}^{n} p(\{\omega_i\}) \cdot \lambda([0,x_i]) = \sum_{i=1}^{n} p_i \cdot x_i \, .$$

If, instead of the Lebesgue measure λ, we adopt the Lebesgue-Stiltjes measure μ generated by the increasing function $u\colon \mathbb{R}_+ \to \mathbb{R}_+$ with $u(0) = 0$, the measure of the hypograph becomes

$$m'(\text{hypo}(X)) = m'\Big(\bigcup_{i=1}^{n}[\{\omega_i\} \times [0,x_i]]\Big) = \sum_{i=1}^{n} m'(\{\omega_i\} \times [0,x_i]) =$$
$$= \sum_{i=1}^{n} p(\{\omega_i\}) \cdot \mu([0,x_i]) = \sum_{i=1}^{n} p_i \cdot [u(x_i) - u(0)] = \sum_{i=1}^{n} p_i \cdot u(x_i) \, ,$$

i.e., the expected utility of X.

Note that, in any case, we are just interested in measuring intervals of the type $[0,x_i]$, and that the measure of such intervals is of course a function, call it h, of the right endpoint only: in the two cases above, either $h(x_i) = x_i$ or $h(x_i) = u(x_i)$. Such functions can be extended to the remaining subsets of $\mathbb{R}$ by simply taking any measure or capacity which coincides with $h(x_i)$ on $[0,x_i]$. The results of Castagnoli and LiCalzi (2006) can be restated by saying that, under the stated assumptions, the functional F amounts to measuring the hypograph of a random variable by means of a bidimensional measure m (not necessarily a product measure).

In the two ways seen above, the hypograph of X has been measured by being "cut" into "vertical" slices, but an interesting approach is to make use of "horizontal" cuts instead. To such an extent, consider (any one of) the permutation(s) putting the values x_i in decreasing order, i.e., such that $x_{(1)} \geqslant x_{(2)} \geqslant \ldots \geqslant x_{(n)} \geqslant 0$[1]. This way,

$$\mathrm{hypo}(X) = \bigcup_{i=1}^{n} \left[\{\omega_{(1)}, \omega_{(2)}, \ldots, \omega_{(i)}\} \times (x_{(i+1)}, x_{(i)}] \right] ,$$

where we agree that $x_{(n+1)} = 0$ and that the last interval $(x_{(n+1)}, x_{(n)}]$ is taken closed, $[0, x_{(n)}]$. In other words, this amounts to interpret the random variable X as follows:

(*i*) X always gives at least the (minimum) value $x_{(n)}$;
(*ii*) if $\omega_{(n)}$ does not prevail, then X gives the extra amount $x_{(n-1)} - x_{(n)}$;
(*iii*) if neither $\omega_{(n)}$ nor $\omega_{(n-1)}$ prevail, then X gives the extra amount $x_{(n-2)} - x_{(n-1)}$ as well,

and so on. With such a decomposition, the above measures become, respectively,

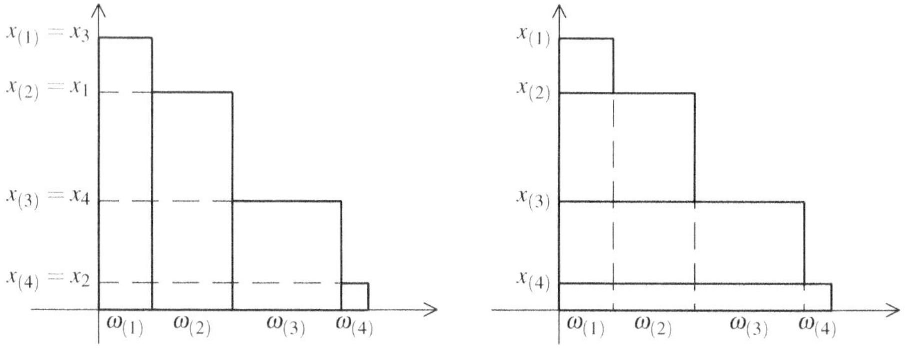

Fig. 2 The random variable in Figure 1 "rearranged"; both "vertical" and "horizontal" cuts are shown

$$\begin{aligned} m(\mathrm{hypo}(X)) &= m\Big(\bigcup_{i=1}^{n} \left[\{\omega_{(1)}, \omega_{(2)}, \ldots, \omega_{(i)}\} \times (x_{(i+1)}, x_{(i)}] \right]\Big) = \\ &= \sum_{i=1}^{n} m\big(\{\omega_{(1)}, \omega_{(2)}, \ldots, \omega_{(i)}\} \times (x_{(i+1)}, x_{(i)}]\big) = \\ &= \sum_{i=1}^{n} p\big(\{\omega_{(1)}, \omega_{(2)}, \ldots, \omega_{(i)}\}\big) \cdot \lambda\left((x_{(i+1)}, x_{(i)}]\right) = \\ &= \sum_{i=1}^{n} p\big(\{\omega_{(1)}, \omega_{(2)}, \ldots, \omega_{(i)}\}\big) \cdot (x_{(i)} - x_{(i+1)}) \end{aligned}$$

[1] This simply amounts to a basis change in $\mathbb{R}^n$: "vertical" cuts are based on the canonical basis $\mathbb{I}_{\omega_1}, \mathbb{I}_{\omega_2}, \ldots, \mathbb{I}_{\omega_n}$, whereas "horizontal" cuts refer to the basis $\mathbb{I}_{\omega_{(1)}}, \mathbb{I}_{\{\omega_{(1)}, \omega_{(2)}\}}, \ldots, \mathbb{I}_{\Omega}$.

and, in an analogous way,

$$m'(\text{hypo}(X)) = \sum_{i=1}^{n} p(\{\omega_{(1)}, \omega_{(2)}, \ldots, \omega_{(i)}\}) \cdot \mu((x_{(i+1)}, x_{(i)}]) =$$
$$= \sum_{i=1}^{n} p(\{\omega_{(1)}, \omega_{(2)}, \ldots, \omega_{(i)}\}) \cdot [u(x_{(i)}) - u(x_{(i+1)})],$$

depending on whether the Lebesgue measure or a Lebesgue-Stiltjes one is used. Note that, in the above way, $m(\text{hypo}X)$ and $m'(\text{hypo}X)$ yield the expectation or (respectively) the expected utility of any random variable X consistent (i.e., "comonotonic" in the sense of Nehring, 1999, that is, taking decreasing values) with the given permutation.

It is quite important to emphasise that the two expressions above only involve the increasing "chain" of events $\{\omega_{(1)}, \omega_{(2)}, \ldots, \omega_{(i)}\}$ and the corresponding probabilities, with respect to the permutation (or any one of them, in the case it is not unique) of the states of the world which sorts the values x_i in decreasing order.

In the above way, the expectation or the expected utility of any random variable consistent with the given permutation is recovered.

Now, if a capacity ν is adopted instead of a (probability) measure p, such formulae maintain their meaning, because $\nu(\{\omega_{(1)}, \omega_{(2)}, \ldots, \omega_{(i)}\})$ is still increasing on every "chain". Such a substitution amounts to "measuring" the hypograph with a set function being the product of a capacity ν and the Lebesgue (or a Lebesgue-Stiltjes) measure:

$$m(\text{hypo}(X)) = \sum_{i=1}^{n} \nu(\{\omega_{(1)}, \omega_{(2)}, \ldots, \omega_{(i)}\}) \cdot \lambda((x_{(i+1)}, x_{(i)}]) =$$
$$= \sum_{i=1}^{n} \nu(\{\omega_{(1)}, \omega_{(2)}, \ldots, \omega_{(i)}\}) \cdot (x_{(i)} - x_{(i+1)}),$$
$$m'(\text{hypo}(X)) = \sum_{i=1}^{n} \nu(\{\omega_{(1)}, \omega_{(2)}, \ldots, \omega_{(i)}\}) \cdot \mu((x_{(i+1)}, x_{(i)}]) =$$
$$= \sum_{i=1}^{n} \nu(\{\omega_{(1)}, \omega_{(2)}, \ldots, \omega_{(i)}\}) \cdot [u(x_{(i)}) - u(x_{(i+1)})],$$

i.e., the Choquet expectation and the Choquet expected utility of X, respectively.

The two settings ("vertical" and "horizontal" cuts) appear to be perfectly symmetric. Indeed, they amount to evaluating the bidimensional measure we use (be it m or m') on an entire (σ-)algebra of subsets in one of the two components, and simply on increasing chains of sets (either $[0, x_i]$ or $\{\omega_{(1)}, \omega_{(2)}, \ldots, \omega_{(i)}\}$) in the other one. On the side of this second component, the measure can be extended to the remaining subsets becoming either a measure or a capacity: since other types of subsets never enter into evaluating hypographs, the completion is immaterial.

It is straightforward to reconcile the functional $m\{X \geqslant U\}$ with the measure of the hypograph of X. Call $Q\colon \mathbb{R}_+ \to \mathbb{R}_+$ the "identity random variable", i.e., $Q(s) = s$

for every $s \in \mathbb{R}_+$ (but the same can be said with Q any injective function). The hypograph of X can be written as

$$\text{hypo}(X) = \bigcup_{i=1}^{n} (\{\omega_i\} \times [0,x_i]) = \{(\omega_i, s) : X(\omega_i) \geqslant Q(s)\} ,$$

because the pair (ω_i, s) belongs to hypo(X) if and only if $s \in [0,x_i]$, that is, if and only if $X(\omega_i) = x_i \geqslant s = Q(s)$. As a consequence, the functional $m(\text{hypo}(X))$ can be written as $m(\{(\omega_i,s) : X(\omega_i) \geqslant Q(s)\})$, or, briefly, $m\{X \geqslant Q\}$.

In the particular case when $m = m_1 \otimes m_2$ is a product measure, one gets

$$m(\{(\omega_i,s) : X(\omega_i) \geqslant Q(s)\}) = \sum_{t \in \mathbb{R}_+} m_1(\{\omega_i : X(\omega_i) = t\}) \cdot m_2(\{s : t \geqslant s\}) .$$

If, moreover, m_1 and m_2 are probability measures, $m_2(\{s : s \leqslant t\})$ is the c.d.f. $q(t)$ of Q, so that

$$m(\{(\omega_i,s) : X(\omega_i) \geqslant Q(s)\}) = \sum_{t \in \mathbb{R}_+} m_1(\{\omega_i : X(\omega_i) = t\}) \cdot q(t) = \mathrm{E}_{m_1} q(X) ,$$

and we are back to the expected utility.

Recall that we assumed that the random variables in $\mathscr{X}$ are taking values in a bounded interval $B \subseteq \mathbb{R}_+$; we can always suppose that $m(\Omega \times B) < +\infty$[2]. The *marginal measures* of m are naturally defined by

$$m_1(A) = m(A \times B) \qquad \text{with } A \subseteq \Omega \text{ any event} ,$$
$$m_2(I) = m(\Omega \times I) \qquad \text{with } I = [0,k] \subseteq B^3 .$$

Analogously, the *conditional marginal measures* are defined by

$$m_1(A|I) = \frac{m(A \times I)}{m_2(I)} \cdot m_2(B) , \qquad m_2(I|A) = \frac{m(A \times I)}{m_1(A)} \cdot m_1(\Omega) . \tag{1}$$

So far, we have decomposed the measure of hypo(X) into the sum of the measures of the "rectangles" $\{\omega_i\} \times [0,x_i]$ which build it up. It is noteworthy that they are in their turn hypographs of the (elementary) random variables $X \cdot \mathbb{I}_{\omega_i} = x_i \mathbb{I}_{\omega_i}$, and thus

$$m(\text{hypo}(X)) = \sum_{i=1}^{n} m(\{\omega_i \times [0,x_i]\}) = \sum_{i=1}^{n} m(\text{hypo}(X \cdot \mathbb{I}_{\omega_i})) .$$

Generally speaking, the terms $m(\text{hypo}(X \cdot \mathbb{I}_{\omega_i}))$ that evaluate the contribution of the different "rectangles" to the overall measure do not turn out to be the product of the

[2] The same conclusions hold, for instance, when the random variables are not bounded but $m(\Omega \times \mathbb{R}_+) < +\infty$; what matters is that the overall measure be finite.

[3] We only take intervals $[0,k]$ because, for other measurable subsets of B, the measure or capacity can be arbitrarily completed.

measures of their "basis" by their "height". Nevertheless, a standard trick allows to write them as if they were:

(1) taking any (strictly positive) probability p (in particular, p might be taken to be the marginal measure m_1), we can write

$$m(\{\omega_i\}\times[0,x_i]) = p\{\omega_i\}\cdot\frac{m(\{\omega_i\}\times[0,x_i])}{p\{\omega_i\}} = p\{\omega_i\}\cdot\tilde{u}(x_i,\omega_i)\,,$$

where $\tilde{u}(x_i,\omega_i)$ can be intended as a *state-dependent utility*, i.e., the evaluation attached to x_i in the case that it is realised in ω_i;

(2) taking any (strictly positive) utility function u (in particular, u might be taken to be the marginal measure m_2), we can write

$$m(\{\omega_i\}\times[0,x_i]) = u(x_i)\cdot\frac{m(\{\omega_i\}\times[0,x_i])}{u(x_i)} = \pi(x_i,\omega_i)\cdot u(x_i)\,,$$

where $\pi(x_i,\omega_i)$ can be indended as a *value-dependent probability*, that is, the probability assessed for ω_i in the case that x_i is the corresponding value.

The same conclusion can be drawn when considering the random variables $X\mathbb{I}_A$, with A any event in Ω. Such a setting can then accommodate the "inversion of preferences" when a decision maker compares random variables or their multiples. Think, e.g., of a guy with the necessity to have at least 100 in the case when the event A occurs. If the infimum of the values taken by $X\mathbb{I}_A$ and $Y\mathbb{I}_A$ are 80 and 110, respectively, he will consequently assess that $Y\mathbb{I}_A \succ X\mathbb{I}_A$, but the preference might be naturally reversed when comparing $2X\mathbb{I}_A$ and $2Y\mathbb{I}_A$. In such cases, we need to resort to a state-dependent utility (or, equivalently, to a value-dependent probability).

Write $f_A(k) := m\big(\text{hypo}(k\mathbb{I}_A)\big)$: it is immediate that such f_A is increasing with respect to k (and to A as well). We can therefore call $f_A^{-1}\big(m(\text{hypo}(X\mathbb{I}_A))\big)$ the *certainty equivalent of X conditional on A*. In particular, the (unconditional) certainty equivalent of X is $f_\Omega^{-1}\big(m(\text{hypo}(X))\big)$.

When dealing with random variables taking arbitrary values and not just positive ones, the above setting still holds, provided the following convention is adopted. If $x_i < 0$, the interval $[0,x_i]$ becomes the "reversed" interval $[x_i,0]$, and it is enough to define its measure to be the opposite of measure of the interval in the natural sense. Of course, this amounts to to setting anyway

$$\begin{aligned}\lambda\big([0,x_i]\big) &:= x_i \quad &&\text{(Lebesgue measure)},\\ \mu\big([0,x_i]\big) &:= u(x_i) &&\text{(Lebesgue-Stiltjes measure)},\end{aligned}$$

with $u\colon\mathbb{R}\to\mathbb{R}$ an increasing function such that $u(0)=0$. Note that this amounts to measuring the (truncated) *epi*graph of negative random variables. Explicitly, if $X \leqslant 0$, its (truncated) epigraph is:

(i) with vertical cuts, $\quad \text{epi}(X) = \bigcup_{i=1}^{n}\big(\{\omega_i\}\times[0,x_i]\big)\,;$

(*ii*) with horizontal cuts, being again $0 \geqslant x_{(1)} \geqslant x_{(2)} \geqslant \ldots \geqslant x_{(n)}$:

$$\begin{aligned}\mathrm{epi}(X) &= \bigcup_{i=1}^{n} \{\omega_{(i+1)}, \omega_{(i+2)}, \ldots, \omega_{(n)}\} \times [x_{(i)}, x_{(i+1)}) = \\ &= \bigcup_{i=1}^{n} \left[\Omega \setminus \{\omega_{(1)}, \omega_{(2)}, \ldots, \omega_{(i)}\}\right] \times [x_{(i)}, x_{(i+1)}).\end{aligned}$$

For a generic random variable X, the set to be measured is $\mathrm{hypo}(X^+) \cup \mathrm{epi}(X^-)$.

4 Another Interpretation of the Functional Proposed by Castagnoli and Li Calzi

The representing functional $P\{X \geqslant U\}$ can be obviously intended as the probability that X outperforms a given random benchmark U, but a different interpretation can be proposed.

As already mentioned in the remark at the end of Section 2, Theorem 1 can be rephrased by saying that the representing functional takes the form $m\{X \geqslant U\}$ with m a bidimensional measure on $\Omega \times S$; moreover, it can be equivalently written as the measure $m(\mathrm{hypo}(X))$ of the hypograph of X. Whenever X is a discrete random variable, we have that

$$m(\mathrm{hypo}(X)) = m\left(\bigcup_{i=1}^{n} (\{\omega_i\} \times [0, x_i])\right) = \sum_{i=1}^{n} m(\mathrm{hypo}(X \cdot \mathbb{I}_{\omega_i})) ,$$

and the single terms $m(\{\omega_i\} \times [0, x_i]) = m(\mathrm{hypo}(X \cdot \mathbb{I}_{\omega_i}))$, $i = 1, 2, \ldots, n$, evaluate the different "pairs" (ω_i, x_i) which additively give rise to the functional.

Under the stated assumptions (notably, under the Sure Thing Principle), it is therefore possible to elicit an evaluation for every single pair (ω_i, x_i), i.e., for every amount x_i in the case that it takes place in the event ω_i. In general, it is impossible to go on and to further split such an evaluation into the product of a probability and a utility: the "finest" elicitable evaluation is the one of the pair (ω_i, x_i) as a whole. Nevertheless, in the previous section we have seen that it is possible to take into consideration any (strictly positive) probability p and force the formula to decompose into

$$m(\{\omega_i\} \times [0, x_i]) = p\{\omega_i\} \cdot \frac{m(\{\omega_i\} \times [0, x_i])}{p\{\omega_i\}} = p\{\omega_i\} \cdot \tilde{u}(x_i, \omega_i) .$$

This allows to see the functional as a *state-dependent expected utility*, in the sense that the utility $\tilde{u}$ attached to a value x_i depends on the state of the world as well.

Note that the random variable U (and the set S) of Theorem 1 are far from being unique. Nonetheless, the pair (U, P) is "essentially" unique in the sense that follows. Take any injective function $h\colon \mathbb{R} \to \mathbb{R}$, and consider the functional $P\{h(X) \geqslant U\}$. We have

$$P\{h(X) \geqslant U\} = \sum_{i=1}^{n} m(\{\omega_i\} \times [0, h(x_i)]) = \sum_{i=1}^{n} m_h(\{\omega_i\} \times [0, x_i]) \,,$$

with m_h the measure such that $m_h(\{\omega_i\} \times [0,x_i]) = m(\{\omega_i\} \times [0,h(x_i)])$. Thus if, instead of the benchmark U, we take into consideration $g(U)$, with $g\colon \mathbb{R} \to \mathbb{R}$ an injective function, we get

$$P\{X \geqslant g(U)\} = P\{g^{-1}(X) \geqslant U\} = P_{g^{-1}}\{X \geqslant U\} \,,$$

with g^{-1} the left inverse of g. This shows that the two pairs (U,P) and $(g(U), P_{g^{-1}})$ are in some sense "interchangeable". Thus, whenever only the ordering among different random variables matters, we can use the functional $g^{-1}(P\{X \geqslant g(U)\})$, which can be equivalently written $g^{-1}(m\{X \geqslant g(U)\})$. In such a case, and assuming U to take values in B, the functional has the convenient property that $g^{-1}(m\{k \geqslant g(U)\}) = k$ for every $k \in \mathbb{R}$.

The most remarkable meaning of Theorem 1 is that, under the stated assumptions, the representing functional is a "measure" of the hypograph of the random variable X evaluated by a bidimensional measure m (not necessarily a probability) on the set $\Omega \times S$. Such a m turns out to be additive (i.e., a measure indeed) just because of the additivity property embedded in the Sure Thing Principle.

It is therefore quite intuitive that:

(*i*) if the Sure Thing Principle is removed, m is no longer additive, but it rather turns out to be a monotonic set function (because $\succcurlyeq$ is supposed to be monotonic), i.e., a capacity;

(*ii*) if the Sure Thing Principle is imposed on a restricted family $\mathscr{A}$ of events only, them m turns out to be (a capacity) additive on $\mathscr{A}$;

(*iii*) if even the monotonicity requirement is removed (i.e., $\succcurlyeq$ is just supposed to be continuous), then m turns out to be a general set function (which is additive or not depending on whether the Sure Thing Principle is assumed or not on some sets).

5 The Reversed Sure Thing Principle

The Sure Thing Principle amounts to cutting "vertically" (i.e., event-wise) the hypograph of a random variable. We want now to investigate shortly what happens when dealing with "horizontal" (i.e., amount-wise) cuts.

Let $X \in \mathscr{X}$. For every $k \in \mathbb{R}$, we define the *upper truncation* and the *lower truncation* of X at the amount k to be

$$X^k := X \cdot \mathbb{I}_{\{X>k\}} = \begin{cases} X & \text{if } X > k \\ 0 & \text{if } X \leqslant k \end{cases} \quad ; \quad X_k := X \cdot \mathbb{I}_{\{X\leqslant k\}} = \begin{cases} 0 & \text{if } X > k \\ X & \text{if } X \leqslant k \end{cases}$$

respectively. It is clear that $X^k + X_k = X$ for every X and every k.

The *Reversed Sure Thing Principle* requires that

$$Y^k + Z_k \succcurlyeq X^k + Z_k \iff Y^k + V_k \succcurlyeq X^k + V_k$$

for every $X, Y, Z, V \in \mathscr{X}$ and every $k \in \mathbb{R}$. (Note that this is equivalent to requiring that $Z^k + Y_k \succcurlyeq Z^k + X_k \Leftrightarrow V^k + Y_k \succcurlyeq V^k + X_k$ as soon as $\mathscr{X}$ has suitable symmetry properties. If, e.g., $\mathscr{X}$ is closed under opposites, the two properties clearly imply each other. If, as in the previous sections, the random variables take values in a bounded interval $[a,b]$, think that $X^k = (b+a-X)_{b+a-k}$.)

The interpretation is straightforward: if two random variables coincide below a certain threshold k (i.e., if their truncations X_k and Y_k are the same, called Z_k above), replacing the "common part" with a different one (i.e., V_k) does not affect the preference.

If the Reversed Sure Thing Principle is assumed, a "reversed" version of Theorem 1 can be obtained. We shall hereby state and prove the result in the case of a finite set Ω of states of the world, but we conjecture the same result to hold for more general sets. Provided that there are at least three essential values for k, we have:

Theorem 2. *Let $\succcurlyeq$ be a monotonic and continuous preorder. The Reversed Sure Thing Principle holds if and only if it is represented by the functional*

$$F(X) = \nu\{X \geqslant U\} = \nu\big(\text{hypo}(X)\big),$$

with ν a capacity, additive on S and monotonic on Ω.

Proof. Let $\Omega = \{\omega_1, \omega_2, \dots, \omega_n\}$ and suppose $B = [0,b]$. Given a permutation $\omega_{(1)}, \omega_{(2)}, \dots, \omega_{(n)}$ of the states of the world, write $A_i := \{\omega_{(1)}, \omega_{(2)}, \dots, \omega_{(i)}\}$ for every $i = 1, 2, \dots, n$ and consider the chain $\mathscr{C} := \{\emptyset, A_1, A_2, \dots, A_n = \Omega\}$. The preference among the random variables $b \cdot \mathbb{I}_{A_i}$ can be trivially represented by a one-to-one function $\alpha \colon \mathscr{C} \to [0,1]$ such that $\alpha(\emptyset) = 0$, $\alpha(\Omega) = 1$ and $\alpha(A_i) \leqslant \alpha(A_{i+1})$ for every i.

As in Section 3, any random variable X consistent ("comonotonic") with $\mathscr{C}$, i.e., such that $x_{(1)} \geqslant x_{(2)} \geqslant \dots \geqslant x_{(n)} \geqslant 0$, can be identified with the set

$$\bigcup_{i=1}^{n} \big(A_i \times (x_{(i+1)}, x_{(i)}]\big),$$

or, equivalently (α being one-to-one), with the function $\xi \colon [0,b] \to [0,1]$ such that $\xi(t) = \alpha(A_i)$ if $t \in (x_{(i+1)}, x_{(i)}]$. Theorem 1, exchanging the roles of Ω and B, guarantees the existence of (a probability or of) a measure $m_{\mathscr{C}}$ on $\mathscr{C} \times B$ such that the preference among all of the random variables consistent with $\mathscr{C}$ is represented by $m_{\mathscr{C}}\big(\text{hypo}(X)\big)$ or, equivalently, by $m_{\mathscr{C}}\{X \geqslant U\}$.

Consider now a "neighbour" permutation of the previous one in the sense of Nehring (1999): namely, it only differs by the exchange of the two "consecutive" states $\omega_{(j)}$ and $\omega_{(j+1)}$. Thus, such a permutation identifies the chain $\mathscr{C}' = \{\emptyset, A_1, \dots, A_{j-1}, A'_j, A_{j+1}, \dots, A_n = \Omega\}$ with $A'_j = \{\omega_{(1)}, \dots, \omega_{(j-1)}, \omega_{(j+1)}\}$. It is straightforward that α can be extended to represent the preference among the random variables $b \cdot \mathbb{I}_A$ with $A \in \mathscr{C} \cup \mathscr{C}' = \mathscr{C} \cup \{A'_j\}$. As above, we obtain a measure $m_{\mathscr{C}'}$ on $\mathscr{C}' \times B$ representing the preference among the random variables consistent with $\mathscr{C}'$, with the further property that $m_{\mathscr{C}'} = m_{\mathscr{C}}$ on all of the sets $I \times A_i$ with I a subinterval of B and $A_i \in \mathscr{C} \cap \mathscr{C}'$. This allows to define a capacity v on $(\mathscr{C} \cup \mathscr{C}') \times B$ which agrees with $m_{\mathscr{C}}$ on $\mathscr{C} \times B$ and with $m_{\mathscr{C}'}$ on $\mathscr{C}' \times B$, and thus represents the preferences among all of the random variables consistent either with $\mathscr{C}$ or with $\mathscr{C}'$.

Since any two permutations can be connected by a "path" of neighbour ones, we can repeat this argument to end up with a capacity v defined on the entire $2^{\Omega} \times B$ and completely representing the given preference.

Example 2.(1) Suppose that the two random variables:

		Y	X
A	ω_1	10	9
	ω_2	6	9
A^c	ω_3	8	8
	ω_4	4	4
	ω_5	1	1

are indifferent to a decision maker. Since they are identical on $A^c = \{\omega_3, \omega_4, \omega_5\}$, the Sure Thing Principle implies that the decision maker remains indifferent when the common part, i.e., the particular values taken by the two random variables on A^c, is changed. For instance,

	Y'	X'	Y''	X''	$Y\mathbb{I}_A$	$X\mathbb{I}_A$
ω_1	10	9	10	9	10	9
ω_2	6	9	6	9	6	9
ω_3	20	20	1	1	0	0
ω_4	30	30	10	10	0	0
ω_5	5	5	12	12	0	0

are pairwise indifferent. In particular, the fact that $Y\mathbb{I}_A$ and $X\mathbb{I}_A$ are indifferent reveals that the functional, call it F, representing the preference, has the property

$$F(Y) = F(Y\mathbb{I}_A) + F(Y\mathbb{I}_{A^c})$$

for every event A and every Y.

(2) Suppose now that the two random variables:

	Y	X
ω_1	12	7
ω_2	8	14
ω_3	6	6
ω_4	2	2
ω_5	1	1

are indifferent. Since they coincide below $k = 6$ (that is, again on $\{\omega_3, \omega_4, \omega_5\}$, the Reversed Sure Thing Principle implies that the decision maker remains indifferent when the common part is changed with values still below 6. For instance:

	Y'	X'	Y''	X''
ω_1	12	7	12	7
ω_2	8	14	8	14
ω_3	0	0	2	2
ω_4	6	6	4	4
ω_5	3	3	6	6

are pairwise indifferent. Let us now write Y and X from the point of view of "horizontal" cuts:

	ΔY
ω_1	4
$\{\omega_1, \omega_2\}$	2
$\{\omega_1, \omega_2, \omega_3\}$	4
$\{\omega_1, \omega_2, \omega_3, \omega_4\}$	1
Ω	1

	ΔX
ω_2	7
$\{\omega_1, \omega_2\}$	1
$\{\omega_1, \omega_2, \omega_3\}$	4
$\{\omega_1, \omega_2, \omega_3, \omega_4\}$	1
Ω	1

The two random variables have the same lower part (4, 1, 1), and the two upper parts (4, 2, and 7, 1) are indifferent. In particular,

	Y^6	X^6
ω_1	6	1
ω_2	2	8
ω_3	0	0
ω_4	0	0
ω_5	0	0

i.e.,

	ΔY^6
ω_1	4
$\{\omega_1, \omega_2\}$	2
Ω	0

	ΔX^6
ω_2	7
$\{\omega_1, \omega_2\}$	1
Ω	0

are indifferent. This amounts to saying that the functional, call it G, representing the preference, has the property

$$G(Y) = G(Y^k) + G(Y_k)$$

for every $k \in \mathbb{R}$ and every Y.

Note that the "classical" Sure Thing Principle does not ensure indifference between Y^6 and X^6, but rather between $Y\mathbb{I}_A$ and $X\mathbb{I}_A$ that assume, respectively, values 12 and 7 on ω_1 and values 8 and 14 on ω_2.

If both Sure Thing Principles are imposed, the measure m turns out to be a product one. Indeed, formula (1) holds for every event $A \subseteq \Omega$ and for every measurable set $I \subseteq B$ and thus it has to be

$$m(A \times I) = m_2(I|A) \cdot \frac{m_1(A)}{m_1(\Omega)} = m_1(A|I) \cdot \frac{m_2(I)}{m_2(B)},$$

that is,

$$m_1(A|I) = \frac{m_1(A)}{m_1(\Omega)} \quad \text{and} \quad m_2(I|A) = \frac{m_2(I)}{m_2(B)} \quad :$$

both the conditional measures, then, do not depend on the conditioning event.

6 Generalised Expectations

What comes above suggests to call a *generalised expectation* of a random variable a suitable measure of its hypograph. The most general concept appears to be the following one.

Definition 1. A *generalised expectation*, denoted with $\mathscr{E}$, of a random variable X defined on a set Ω is the real number

$$\mathscr{E}(X) := \nu\big(\text{hypo}(X)\big) ,$$

where ν is a bidimensional capacity on $\Omega \times \mathbb{R}$ such that

$$\nu(\Omega \times [0,k]) = k \qquad \text{for every } k \in \mathbb{R} .$$

The condition imposed on ν is nothing but the natural consistency requirement that $\mathscr{E}(k) = k$ for every $k \in \mathbb{R}$. This amounts to requiring that the marginal capacity ν_2 coincides with the Lebesgue measure on all of the intervals of the form $[0,k]$. Thus, the values $\frac{\nu(A\times[0,k])}{k} = \frac{\nu(\text{hypo}(k\mathbb{I}_A))}{k}$ are the "value-dependent probabilities" of the events of A (being, of course, not additive, but simply monotonic).

We can write as well

$$\mathscr{E}(X) = g^{-1}\big(\nu\{X \geqslant g(U)\}\big) ,$$

with $U : \mathbb{R} \to \mathbb{R}$ an injective function and $g : \mathbb{R} \to \mathbb{R}$ a strictly increasing one.

In order to obtain a (less general, but) more significant and useful concept of generalised expectation, we need to impose some associativity property that, roughly speaking, requires the "average" of partial expectations to yield back the overall one. We reckon that there are two natural ways to ask for such a property, each of them mimicking one of the two Sure Thing Principles commented above.

(1) If we take into consideration partial expectations with respect to events, it is natural to require that, for every event A (and every X),

$$\mathscr{E}(X) = \mathscr{E}(X \cdot \mathbb{I}_A) + \mathscr{E}(X \cdot \mathbb{I}_{A^c}) = \mathscr{E}(X|A)p(A) + \mathscr{E}(X|A^c)p(A^c), \qquad (2)$$

where p can be amount-dependent. This amounts to imposing the Sure Thing Principle, and thus the marginal capacity ν_1 turns out to be additive. Moreover, since the only sets to be measured on $\mathbb{R}$ are those of the form $[0,k]$, we can always take ν_2 to be the Lebesgue measure.

We can call a *generalised expectation of the first type* any generalised expectation satisfying (2), i.e., of the form:

$$\mathscr{E}(X) = m\big(\text{hypo}(X)\big)$$

with m a bidimensional capacity defined on $\Omega \times \mathbb{R}$ whose marginal m_1 is additive (and, as mentioned above, m_2 can always be taken to be the Lebesgue measure).

Example 3. The certainty equivalent of a random amount X is classically defined to be $u^{-1}\big(\mathrm{E}u(X)\big)$ and, as such, cannot incorporate state dependency, although the latter might be welcome for the problem into consideration. Note that $u^{-1}\big(\mathrm{E}u(X)\big)$ is an associative expectation.

As a simple example, consider the bet where 1000$ are won (respectively, lost) in the case when the Dollar to Euro ratio raises (respectively, falls) in the next month; consider as well the "opposite" bet yielding the same amounts on the complementary events. If the two events are supposed to be equiprobable, it is clear that the first bet is better than the second one, as it gives more valuable dollars when winning and forces to pay less valuable dollars when losing. Yet no utility function, and therefore no certainty equivalent, can incorporate such a preference, whereas the proposed functional $g^{-1}\big(v\{X \geqslant g(U)\}\big)$ can.

Example 4. As a second example, imagine two firms considering the possibilities to acquire a financial asset yielding 100 or -100 depending on whether the oil price will increase (U) or decrease (D). The first firm is a car producer and will make a profit of 1000 or 600 depending on whether sales of new cars increase (u) or decrease (d), while the second one produces car spare components and will make a profit of 500 in u or 800 in d.

Both firms evaluate the joint probability as follows:

	U	D	
u	0.1	0.3	0.4
d	0.4	0.2	0.6
	0.5	0.5	

The asset is more valuable to the first firm, as it better matches the risk of the final profit. Again, no associative mean can incorporate such an effect, while generalised means of the first type can.

(2) If we take into consideration partial expectations with respect to values, we want that

$$\mathscr{E}(X) = \mathscr{E}(X_k) + \mathscr{E}(X^k) \qquad \text{for every } k \in \mathbb{R} \text{ (and every } X\text{)}, \tag{3}$$

which amounts to reversing the Sure Thing Principle, that is, to refer to horizontal cuts instead of vertical ones.

We can call a *generalised expectation of the second type* any generalised expectation satisfying (3), i.e., of the form

$$\mathscr{E}(X) = m\big(\text{hypo}(X)\big)$$

with m a bidimensional capacity defined on $\Omega \times \mathbb{R}$ whose marginal m_2 is the Lebesgue measure.

Example 5. Consider the classical Ellsberg urn with 30 red balls and 60 black or green balls in unknown proportion.

A decision maker prudentially underestimates the probabilities according to the amount obtained in the corresponding events; such an underestimation is higher for ambiguous events. Take into consideration the four random variables

	R	B	G
X	1000	0	0
Y	0	1000	0
Z	1000	0	1000
V	0	1000	1000

Since $Z = X(R \cup B)1000$, $V = Y(R \cup B)1000$ (and, of course, $X = X(R \cup B)0$ and $Y = Y(R \cup B)0$), the Sure Thing Principle would entail that $X \succ Y \Rightarrow Z \succ V$.

Suppose now that, consistently with the Reversed Sure Thing Principle:

$$p(R, 1000) = 0.3\,, \qquad p(B, 1000) = 0.25\,,$$

and

$$p(R \cup G, 1000) = 0.55\,, \qquad p(B \cup G, 1000) = 0.65\,,$$

which allow for $X \succ Y$ and $V \succ Z$.

Example 6. The functional defined by:

$$F(X) = \max_{p \in \mathscr{P}} \mathrm{E}_p(X),$$

with $\mathscr{P}$ a closed and convex set of probabilities, violates the Sure Thing Principle but satisfies the reversed one. Indeed, $Y^k + Z_k$ and $X^k + Z_k$ determine two chains that coincide in all of the terminal events (i.e., the ones related to values $\leqslant k$). Therefore, $F(X)$ and $u^{-1}(F(u(X)))$ are generalised expectations of the second type but not of the first one. The same holds for $\min_{p \in \mathscr{P}} \mathrm{E}_p(X)$.

Imposing both versions of associativity amounts to requiring both the Sure Thing Principles. In such a case, m turns out to be a product measure (or, better, it can be completed to be as such), so that writing the functional as $g^{-1}(m\{X \geqslant g(U)\})$ yields back the classical formula of Nagumo, Kolmogorov and de Finetti for associative means.

7 Conclusions

The possibility of considering random variables as sets (hypo- or epigraphs), instead of mere functions, was originally proposed in the seminal papers of Segal (1989, 1993). Such a choice allows to treat random variables in the language and with the tools of Measure theory, instead of the commonly adopted ones of Functional analysis: the former being very powerful and very well developed, this allows one to obtaining results that would be quite tougher, or straightaway impossible, to reach by means of the latter. For instance, the possibility of extending a measure from a given family of subsets to a larger one was extensively deployed in Castagnoli and LiCalzi (2006) to axiomatise a preference preorder simply by means of continuity, monotonicity and the Savage's Sure Thing Principle.

In this paper we have shown that, when looking at a random variable as a set, the concepts of the expectation and the expected utility (either "classical" or *à la* Choquet) turn out to be slight variations of the same procedure of measuring a set (the truncated hypo- or epigraph corresponding to the given random variable) by means of a product measure (or capacity). We propose to push this line of reasoning further by using a generic ("non-product") measure or capacity to evaluate the set under examination, thus obtaining a broader concept of an expectation that includes dependence of the utility function on the state (or dependence of the probability on the amount).

We deem that any expectation, in order to represent a satisfactory summary of the random variable taken into consideration, should satisfy a reasonable requirement. Namely, it should yield the unique value that, when replaced to the different ones taken by the random variable, "leaves things unchanged", i.e., does not alter the expectation itself. This amounts to saying that the expectation of a random variable equals its own *certainty equivalent*, thus pointing out the equivalence between any random variable and a corresponding degenerate one. We have shown by means of some simple example that an approach as such can be quite fruitful in many case of practical interest.

Finally, we also recovered two different ways for defining the associative property of a generalised expectation.

References

Amarante, M.: Foundations of neo-bayesian statistics. J. Econ. Theory 144, 2146–2173 (2009)

Beccacece, F., Cillo, A.: Applying the benchmarking procedure: a decision criterion of choice under risk. Theory Dec. 61, 75–91 (2006)

Bordley, R.F.: Foundations of Target-based Decision Theory. In: Brachinger, H.W., Monney, P. (eds.) Decision Analysis, from Encyclopedia of Life Support Systems (EOLSS). Eolss Publishers, Oxford (2002), http://www.eolss.net

Bordley, R.F., LiCalzi, M.: Decision Analysis Using Targets instead of Utility Functions. Decis. Econ. Finance 23, 53–74 (2000)

Castagnoli, E.: Qualche riflessione sull'utilità attesa. Ratio Math. 1, 51–59 (1990)

Castagnoli, E., LiCalzi, M.: Expected Utility without Utility: Constant Risk Attitude. Rend. com. studi econ. XXX/XXXI, 145–160 (1993)

Castagnoli, E., LiCalzi, M.: Expected Utility without Utility: A Model of Portfolio Selection. In: D'Ecclesia, R.L., Zenios, S.A. (eds.) Operations Research Models in Quantitative Finance, pp. 95–111. Physica-Verlag, Heidelberg (1994)

Castagnoli, E., LiCalzi, M.: Expected Utility without Utility. Theory Dec. 41, 281–301 (1996)

Castagnoli, E., LiCalzi, M.: Benchmarking real-valued acts. Games Econ. Behav. 57, 236–253 (2006)

Castagnoli, E., Favero, G., Maccheroni, F.: Marchetto, il prezzatore perfetto (Elogio dell'internalità). Edizioni Čakuntala, Milano (still in progress) (2009)

Cigola, M., Modesti, P.: On SSB Utility Theory. In: Bertocchi, M., Cavalli, E., Komlosi, S. (eds.) Modelling Techniques for Financial Markets and Bank Management, pp. 272–284. Physica-Verlag, Heidelberg (1996)

Debreu, G.: Topological methods in cardinal utility theory. In: Arrow, K.J., Karlin, S., Suppes, P. (eds.) Mathematical Methods in the Social Sciences, 1959: Proceedings of the First Stanford Symposium. Stanford Mathematical Studies in the Social Sciences, vol. 4, pp. 16–26. Stanford University Press, Stanford (1960) (reprinted as Cowles Foundation paper 156)

Greco, G.H.: Integrale monotono. Rend. Sem. Mat. Univ. Padova 57, 149–166 (1977)

Klement, E.P., Mesiar, R., Pap, E.: A Universal Integral as Common Frame for Choquet and Sugeno integrals. Work in progress (2009)

Modesti, P.: Lottery-dependent Utility via Stochastic Benchmarking. Theory Dec. 55, 45–57 (2003)

Nehring, K.: Capacities and probabilistic beliefs: a precarious coexistence. Math. Soc. Sci. 38, 197–213 (1999)

Savage, L.J.: The Foundations of Statistics. Wiley, New York (1951) (revised edition, Dover, New York (1972))

Segal, U.: Anticipated Utility: A Measure Representation Approach. Ann. Oper. Res. 19, 359–373 (1989)

Segal, U.: The Measure Representation: A Correction. J. Risk Uncertainty 6, 99–107 (1993)

Simon, H.A.: A Behavioral Model of Rational Choice. Quart. J. Econ. 69, 99–118 (1955)

Memory Property in Heterogeneously Populated Markets

Roy Cerqueti and Giulia Rotundo

Abstract. This paper focuses on the long memory of prices and returns of an asset traded in a financial market. We consider a microeconomic model of the market, and we prove theoretical conditions on the parameters of the model that give rise to long memory. In particular, the long memory property is detected in an agents' aggregation framework under some distributional hypotheses on the market's parameters.

1 Introduction

During last years quantitative studies of financial time series have shown several interesting statistical properties common to many markets. Among the others, long memory is one of the most analyzed. This concept raised by time series empirical analysis in terms of the persistence of observed autocorrelations. The long memory property is fulfilled by a time series when the autocorrelation decays hyperbolically as the time lag increases. Therefore, this statistical feature is strongly related to the long run predictability of the future phenomenon's realizations.

Long memory models were introduced in the physical sciences since at least 1950, when some researches in applied statistics stated the presence of long memory within hydrologic and climatologic data. The earliest studies on this field are due to Hurst (1951, 1957), Mandelbrot and Wallis (1968), Mandelbrot (1972), and McLeod and Hipel (1978) among others.

Roy Cerqueti
University of Macerata, Faculty of Economics, Department of Economic and Financial Institutions, Via Crescimbeni, 20 - 62100 - Macerata, Italy
Tel.: +39 0733 2583246; Fax: +39 0733 2583205
e-mail: roy.cerqueti@unimc.it

Giulia Rotundo
University of Tuscia, Faculty of Economics, Via del Paradiso, 47 - 01100 - Viterbo, Italy
Tel.: +390761357727; Fax: +390761357707
e-mail: giulia.rotundo@uniroma1.it

S. Greco et al. (Eds.): Preferences and Decisions, STUDFUZZ 257, pp. 53–67.
springerlink.com

In this paper a theoretical microeconomic structural model is constructed and developed. We rely on time series of assets traded in a financial market and we address the issue of giving mathematical proof of the exact relation between model parameters evidencing the presence of long memory.

The literature on structural models for long-memory is not wide. Some references are Willinger et al. (1998), Box-Steffensmaier and Smith (1996), Byers et al. (1997), Tschernig (1995), Mandelbrot et al. (1997). The keypoint of the quoted references is to assume distributional hypotheses on parameters of models in order to detect the presence of long memory in time series.

We adopt the approach of the structural model of Kirman and Teyssiere (2002) is based on the assumption that the market is populated by interacting agents. The interaction among agents leads to an imitative behavior, that can affect the structure of the asset price dynamics. Several authors focus their research on describing the presence of an imitative behavior in financial markets (see, for instance, Avery and Zemsky (1998), Chiarella et al. (2003), Bischi et al. (2006)).

The traditional viewpoint on the agent-based models in economics and finance relies on the existence of representative rational agents. Two different behaviors of agents follow from the property of rationality: firstly, a rational agent analyzes the choices of the other actors and tends to maximize utility and profit or minimize the risk. Secondly, rationality consists in having rational expectations, i.e. the forecast on the future realizations of the variables are assumed to be identical to the mathematical expectations of the previous values conditioned on the available information set. Thus, rationality assumption implies agents' knowledge of the market's dynamics and equilibrium, and ability to solve the related equilibrium equations.

Simon (1957) argues that it seems to be unrealistic assuming the complete knowledge about the economic environment, because it is too restrictive. Moreover, if the equilibrium model's equations are nonlinear or involve a large number of parameters, it can be hard to find a solution.

An heterogeneous agent systems is more realistic, since it allows the description of agents' heterogeneous behaviors evidenced in the financial markets (see Kirman (2006) for a summary of some stylized facts supporting the agents' heterogeneity assumption). Moreover, heterogeneity implies that the perfect knowledge of agent beliefs is unrealistic, and then bounded rationality takes place (see Hommes, 2001).

Brock and Hommes (1997, 1998) propose an important contribution on this field. The authors introduce the learning strategies theory to discuss agents' heterogeneity in economic and financial models. More precisely, they assume that different types of agents have different beliefs about future variables's realizations and the forecast rules are commonly observable by all the agents.

Brock and Hommes (1998) consider an asset in a financial market populated by two typical investor types: fundamentalists and chartists. An agent is fundamentalist if he/she believes that the price of the aforementioned asset is determined by its fundamental value. In contrast, chartists perform a technical analysis of the market and do not take into account the fundamentals.

More recently, important contributions on this field can be found in Chiarella and He (2002), Föllmer et al. (2005), Alfarano et al. (2008), Chiarella et al. (2006). For an excellent survey of heterogeneous agents models, see Hommes (2006).

In this paper, heterogeneity is assumed to be involved within each single agent, that wears simultaneously two hats: the forecast of the assets' prices are driven by technical analysis of the market (chartist approach) but also by the fundamentals' value (fundamentalist point of view).

In our model each agent performs price forecasts following a short term approach, but the collective behavior can exhibit long memory property. In this context, we extend some existing results (see Zaffaroni 2004, 2007a, 2007b) about the arise of the long memory property due to the aggregation of micro units, by enlarging the class of probability densities of agents' parameters. The contribution of cross-correlation parameters among the agents to the long memory of the aggregate is shown. Furthermore, it is also evidenced that the presence of long memory in the asset price time series implies that the log returns have long memory as well.

The rest of this paper is organized as follows: section 2 introduces the model; section 3 provide the proof of long memory property of the prices. Section 4 provides the analysis of the returns, and section 5 is devoted to the conclusions. The Appendix contains some well-known definitions and results, for an easier reference.

2 The Model

The basic features of the market model, that we are going to set up, are the existence of two groups of agents, with heterogeneity inside each group.

Let us consider a market with N agents that can make an investment either in a risk free or in a risky asset. Furthermore, the risky asset has a stochastic interest rate $\rho_t \sim N(\rho, \sigma_t^2)$ and the risk free bond has a constant interest rate r. We suppose that $\rho > r$ for the model to be consistent.

Let $P_{i,t}$ be the estimate of the price of the risky asset done by the agent i at time t. The change of the price at time $t+1$ forecasted by the i-th agent, conditioned to his information at time t, I_t, is given by $\Delta P_{i,t+1}|I_{i,t}$.

Let us assume that the market is not efficient, i.e. we can write the following relationship:

$$\mathbf{E}(P_{t+1}|I_t) = \Delta P_{t+1}|I_t + P_t \tag{1}$$

where $\mathbf{E}$ is the expected value operator, as usual.

In this model, we suppose that the behavior of the investors is due to an analysis of the market data (by a typical chartist approach) and to the exploration of the behavior of market's fundamentals (by a fundamentalist approach). Moreover, the forecasts are influenced by an error term, common to all the agents:

$$(\Delta P_{i,t+1}|I_{i,t}) = (\Delta P^c_{i,t+1}|I_{i,t}) + (\Delta P^f_{i,t+1}|I_{i,t}) + u_t, \tag{2}$$

where $(\Delta P^c_{i,t+1}|I_{i,t})$ is the contribute of the chartist approach, $(\Delta P^f_{i,t+1}|I_{i,t})$ is associated to the fundamentalist point of view and u_t is a stochastic term representing an error in forecasts.

As a first step we assume that all the agents have the same weight in the market and that the price P_t of the asset in the market at time t is given by the mean of the asset price of each agent at the same time. So we can write

$$P_t = \frac{1}{N}\sum_{i=1}^{N} P_{i,t}. \tag{3}$$

The chartists catch information from the time series of market prices. The forecast of the change of prices performed by the agent i is assumed to be given by the following linear combination:

$$\Delta P^c_{i,t+1}|I_{i,t} = \alpha_i^{(1)}(P_{i,t} - P_{i,t-1}) + \alpha_i^{(2)}(P_t - P_{t-1}), \tag{4}$$

with $\alpha_i^{(1)}, \alpha_i^{(2)} \in \mathbf{R}, \forall i$. Formula (4) captures the idea of a stochastic relationship providing the estimate changes of prices by relying on a linear combination of the two previous price's forecasts, each of them adjusted to the actual market prices got at the relative time.

The fundamentalist approach takes in account the analysis made by the investors on the fundamental values of the market.

The fundamental variables $\bar{P}_{i,t}$ can be described by the following random walk:

$$\bar{P}_{i,t} = \bar{P}_{i,t-1} + \varepsilon_t, \quad \varepsilon_t \sim N(0, \sigma_\varepsilon^2). \tag{5}$$

The fundamental prices observed by the agent i at time t, $\tilde{P}_{i,t}$, are assumed to be biased by a stochastic error:

$$\tilde{P}_{i,t} = \bar{P}_{i,t} + \bar{\alpha}_{i,t}$$

with $\bar{\alpha}_{i,t} = \beta_i P_t$, where β_i, $i = 1, \dots, N$, are parameters drawn by sampling from the cartesian product $(1-\xi, 1+\xi)^N$, $\xi > 0$, equipped with the relative product probability measure. The definition of $\bar{\alpha}_{i,t}$ takes into account the fact that the error in estimating depends on the adjustment performed by each agent of the market price. More precisely, the observation of the fundamental prices is affected by the subjective opinion of the agents on the influence on the fundamental of the market price. If $\beta_i > 1$, then agent i guesses that market price is responsible of an overestimate of the fundamental prices. Otherwise, the converse consideration applies.

Moreover, the forecasts of the fundamentalist agents is based on the fundamental prices and his/her forecast on market prices at the previous data. So we can write

$$\Delta P^f_{i,t+1}|I_{i,t} = \nu(\tilde{P}_{i,t} - P_t), \tag{6}$$

with $\nu \in \mathbf{R}$. Thus

$$\Delta P^f_{i,t+1}|I_{i,t} = \nu \bar{P}_{i,t} + \nu(\beta_i - 1)P_t. \tag{7}$$

Let us define $d_{i,t}$ to be the demand of the risky asset of the agent i at the date t. Thus the wealth invested in the risky asset is given by $P_{t+1}d_{i,t}$ and, taking into account the stochastic interest rate ρ_{t+1}, we have that the wealth grows as $(1+\rho_{t+1})P_{t+1}d_{i,t}$. The remaining part of the wealth, $(W_{i,t} - P_t d_{i,t})$ is invested in risk free bonds and thus gives $(W_{i,t} - P_t d_{i,t})(1+r)$ (Cerqueti and Rotundo, 2003).

The wealth of the agent i at time $t+1$ is given by $W_{i,t+1}$, and it can be written as

$$W_{i,t+1} = (1+\rho_{t+1})P_{i,t+1}d_{i,t} + (W_{i,t} - P_{i,t}d_{i,t})(1+r).$$

The expression of $W_{i,t+1}$ can be rewritten as

$$W_{i,t+1} = (1+\rho_{t+1})\Delta P_{i,t+1}d_{i,t} + W_{i,t}(1+r) - (r-\rho_{t+1})P_{i,t}d_{i,t} \tag{8}$$

Each agent i at time t optimizes the mean-variance utility function

$$U(W_{i,t+1}) = \mathbf{E}(W_{i,t+1}) - \mu V(W_{i,t+1}),$$

where $\mathbf{E}$ and V are the usual mean and variance operators and thus:

$$\mathbf{E}(W_{i,t+1}|I_{i,t}) = (1+\rho)(\Delta P_{i,t+1}|I_{i,t})d_{i,t} + W_{i,t}(1+r) - (r-\rho)P_{i,t}d_{i,t}$$

and

$$V(W_{i,t+1}|I_{i,t}) = V[(1+\rho_{t+1})(P_{i,t+1}|I_{i,t})](d_{i,t})^2.$$

Each agent i maximizes his expected utility with respect to his demand $d_{i,t}$, conditioned to his information at the date t. For each agent i the first order condition is

$$(1+\rho)(\Delta P_{i,t+1}|I_{i,t}) - (r-\rho)P_{i,t} - 2\mu V[(1+\rho_{t+1})(P_{i,t+1}|I_{i,t})]d_{i,t} = 0,$$

By the first order conditions we obtain

$$d_{i,t} = b_{i,t}P_{i,t} + g_{i,t}(\Delta P_{i,t+1}|I_{i,t})$$

with

$$b_{i,t} = \frac{\rho - r}{2\mu V((P_{i,t+1}|I_{i,t})(1+\rho_{t+1}))}; \quad g_{i,t} = \frac{\rho+1}{2\mu V((P_{i,t+1}|I_{i,t})(1+\rho_{t+1}))}.$$

Let $X_{i,t}$ be the supply function at time t for the agent i. Then

$$X_{i,t} = b_{i,t}P_{i,t} + g_{i,t}(\Delta P_{i,t+1}|I_{i,t}). \tag{9}$$

Let us denote

$$\gamma_{i,t} = \frac{X_{i,t}}{b_{i,t}}, \quad c = \frac{1+\rho}{r-\rho} = \frac{g_{i,t}}{b_{i,t}}, \quad \lambda_i := \frac{-c\alpha_i^{(2)}}{1+c\alpha_i^{(1)}}. \tag{10}$$

By (2), (4), (7) and (9) we get:

$$P_{i,t} = \frac{1}{1+c} \cdot \frac{1-\lambda_i}{1-\lambda_i L} \left(\gamma_{i,t} - cv\bar{P}_{i,t} \right) - \frac{c}{1+c} \cdot \frac{1-\lambda_i}{1-\lambda_i L} u_t -$$

$$- \frac{c}{1+c} \cdot \frac{1-\lambda_i}{1-\lambda_i L} \left[v(\beta_i - 1) - \alpha_i \right] P_t - \frac{\lambda_i}{1-\lambda_i L} P_{t-1}, \tag{11}$$

where L is the backward time operator.

Condition (3) and equation (11) allow to write the market price as

$$P_t = \frac{1}{N} \sum_{i=1}^{N} \left\{ \frac{1}{1+c} \cdot \frac{1-\lambda_i}{1-\lambda_i L} \left(\gamma_{i,t} - cv\bar{P}_{i,t} \right) - \frac{c}{1+c} \cdot \frac{1-\lambda_i}{1-\lambda_i L} u_t - \right.$$

$$\left. - \frac{c}{1+c} \cdot \frac{1-\lambda_i}{1-\lambda_i L} \left[v(\beta_i - 1) - \alpha_i \right] P_t - \frac{\lambda_i}{1-\lambda_i L} P_{t-1} \right\}. \tag{12}$$

3 Long Term Memory of Prices

This section shows the long term memory property of market price time series. Equation (12) evidences the contribution of each agent to the market price formation.

Each agent is fully characterized by her/his parameters, and it is not allowed to change them. Parameters are independent with respect to the time and they are not random variables, but they are fixed at the start up of the model in the overall framework of independent drawings.

The heterogeneity of the agents is obtained by sampling α_i, $i = 1, \ldots, N$ from the cartesian product $\mathbf{R}^N$ with the relative product probability measure. No hypotheses are assumed on such a probability up to this point.

In order to proceed and to examine the long term memory property of the aggregate time series, the following assumption is needed:

Assumption (A)

$$\alpha_i = v(\beta_i - 1) < -\frac{1}{c}. \tag{13}$$

This Assumption thus introduces a correlation in the way in which actual prices P_t play a role in the fundamentalists' and chartists' forecasts, and meets the chartists' viewpoint that market prices reflect the fundamental values. Moreover, a relationship between the parameters of the model describing the preferences and the strategies of the investors, α_i and v, and the interest rates of the risky asset and risk free bond (combined in the parameter c) is evidenced.

By a pure mathematical point of view, since $\rho > r$ (and, consequently, $c < -1$), the variation range of α_i is, in formula (13), respected.

We assume that Assumption (A) holds hereafter.

By (12) and (13), market's price can be disaggregated and written as

$$P_t = \frac{1}{N} \cdot \frac{1}{1+c} \sum_{i=1}^{N} \frac{1-\lambda_i}{1-\lambda_i L} \gamma_{i,t} - \frac{1}{N} \cdot \frac{c}{1+c} \sum_{i=1}^{N} \frac{1-\lambda_i}{1-\lambda_i L} u_t -$$

$$- \frac{1}{N} \cdot \frac{cv}{1+c} \sum_{i=1}^{N} \frac{1-\lambda_i}{1-\lambda_i L} \bar{P}_{i,t} - \frac{1}{N} \sum_{i=1}^{N} \frac{\lambda_i}{1-\lambda_i L} P_{t-1} =: A_t^1 + A_t^2 + A_t^3 + A_t^4, \quad (14)$$

and $\lambda_i \in (0,1)$, for each $i = 1,\dots,N$.

Equation (14) fixes the role of the parameters of the model in the composition of the price.

The theoretical analysis of the long term memory of the time series (14) is carried on through two steps:

- long memory is detected for each component of P_t;
- the terms are aggregated.

3.1 The Idiosyncratic Component

A_t^1 is the idiosyncratic component of the market, and it gives the impact of the supply over market's prices, filtered through agents' forecasts parameters.

The degree of long term memory can be fixed through a direct analysis of the rate of decay of the correlation function. In the next result a sufficient condition for the long term memory property of A_t^1 is shown.

Theorem 1. *Let us assume that there exists $a, b \in (0,+\infty)$ such that $\lambda_i \in [0,1]$ and λ_i are sampled by a $B(a,b)$ distribution.*

Fixed $i = 1,\dots,N$, let $\gamma_{i,t}$ be a stationary stochastic process such that

$$\mathbf{E}[\gamma_{i,t}] = 0, \qquad \forall i \in \{1,\dots,N\},\ t \in \mathbf{N}; \quad (15)$$

$$\mathbf{E}[\gamma_{i,u}\gamma_{j,v}] = \delta_{i,j}\delta_{u,v}\sigma_\gamma^2, \qquad \forall i,j \in \{1,\dots,N\},\ u,v \in \mathbf{N}.^1 \quad (16)$$

Then, as $N \to +\infty$, the long term memory property for A_t^1 holds, with Hurst's exponent H_1, in the following cases:

- *$b > 1$ implies $H_1 = 1/2$;*
- *$b \in (0,1)$ and the following equation holds:*

$$\sum_{h=-\infty}^{+\infty} \mathbf{E}[A_t^1 A_{t-h}^1] = 0, \quad (17)$$

imply $H_1 = (1-b)/2$. In this case it results $H_1 < 1/2$, and the process is mean reverting.

[1] $\delta_{i,j}$ is the usual Kronecker symbol, e.g. $\delta_{i,j} = 1$ for $i = j$; $\delta_{i,j} = 0$ for $i \neq j$.

Proof. First of all, we need to show that

$$\mathbf{E}\left[A_t^1 A_{t-h}^1\right] \sim h^{-1-b}, \qquad \text{as } N \to +\infty. \tag{18}$$

Let us examine $A_t^1 A_{t-h}^1$.

$$A_t^1 A_{t-h}^1 = \frac{1}{N^2(1+c)^2} \sum_{i=1}^{N} \frac{1-\lambda_i}{1-\lambda_i L} \gamma_{i,t} \sum_{j=1}^{N} \frac{1-\lambda_j}{1-\lambda_j L} \gamma_{j,t-h} =$$

$$= \frac{1}{N^2(1+c)^2} \sum_{i=1}^{N} (1-\lambda_i) \Big[\sum_{l=0}^{\infty} (\lambda_i L)^l \Big] \gamma_{i,t} \cdot \sum_{j=1}^{N} (1-\lambda_j) \Big[\sum_{m=0}^{\infty} (\lambda_j L)^m \Big] \gamma_{j,t-h}.$$

The terms of the series are positive, and so it is possible to exchange the order of the sums:

$$A_t^1 A_{t-h}^1 = \frac{1}{(1+c)^2} \sum_{m=0}^{\infty} \sum_{l=0}^{\infty} \frac{1}{N^2} \sum_{i=1}^{N} \sum_{j=1}^{N} (1-\lambda_i)\lambda_i^l (1-\lambda_j)\lambda_j^m \gamma_{i,t-m} \gamma_{j,t-h-l}. \tag{19}$$

In the limit as $N \to +\infty$ and setting $x := \lambda_i$, $y := \lambda_j$, (19) becomes:

$$A_t^1 A_{t-h}^1 = \frac{1}{(1+c)^2} \sum_{m=0}^{\infty} \sum_{l=0}^{\infty} \int_0^1 \int_0^1 (1-x)x^l (1-y)y^m \gamma_{x,t-m} \gamma_{y,t-h-l} \mathrm{d}F(x,y), \tag{20}$$

where F is the joint distribution over x and y.

Taking the mean w.r.t. the time and by using the hypothesis (16), we get

$$\mathbf{E}\left[A_t^1 A_{t-h}^1\right] = \frac{1}{(1+c)^2} \sum_{m=0}^{\infty} \sum_{l=0}^{\infty} \int_0^1 \int_0^1 (1-x)x^l (1-y)y^m \delta_{x,y} \delta_{m,l+h} \sigma_\gamma^2 \mathrm{d}F(x,y) = \tag{21}$$

$$= \frac{1}{\beta(a,b)} \cdot \frac{\sigma_\gamma^2}{(1+c)^2} \sum_{l=0}^{\infty} \int_0^1 (1-x)^{1+b} x^{2l+h+a-1} \mathrm{d}x. \tag{22}$$

By using the distributional hypothesis on λ_i, for each i, we get

$$\mathbf{E}\left[A_t^1 A_{t-h}^1\right] = \frac{1}{\beta(a,b)} \cdot \frac{\sigma_\gamma^2}{(1+c)^2} \sum_{l=0}^{\infty} \frac{\Gamma(h+a+2l)\Gamma(b+2)}{\Gamma(h+a+b+2l+2)} \sim$$

$$\sim \frac{1}{\beta(a,b)} \cdot \frac{\sigma_\gamma^2}{(1+c)^2} h^{-1-b}. \tag{23}$$

Now, the rate of decay of the autocorrelation function related to A^1 is given by (23). By using the results in Rangarajan and Ding (2000) on such rate of decay and the Hurst's exponent of the time series, we obtain the thesis.

3.2 The Common Component

A_t^2 describes the common component of the market. In fact, A_t^2 represents the portion of the forecast driven by an external process independent by the single investor.

Theorem 2. *Let us assume that u_t is a stationary stochastic process, with*

$$\mathbf{E}[u_t] = 0;$$

$$\mathbf{E}[u_s u_t] = \delta_{s,t}\sigma_u^2. \tag{24}$$

Moreover, let us assume that there exists $a, b \in (0, +\infty)$ such that the parameters λ_i are drawn by a $B(a,b)$ distribution.

Then, as $N \to +\infty$, the long term memory property for A_t^2 holds, with Hurst's exponent H_2, with the following distinguishing:

- *$b > 1$ implies $H_2 = 1/2$;*
- *$b \in (0,1)$ and the following equation holds:*

$$\sum_{h=-\infty}^{+\infty} \mathbf{E}[A_t^2 A_{t-h}^2] = 0, \tag{25}$$

imply $H_2 = (1-b)/2$. In this case it results $H_2 < 1/2$, and the process is mean reverting.

Proof. The proof is similar to the one of Theorem 1.

3.3 The Component Associated to the Perception of the Fundamentals' Value

A_t^3 is a term typically linked to the perception of the fundamentals' value by the agents.

By the definition of $\bar{P}$ given in (5), we can rewrite A_t^3 as

$$A_t^3 = \frac{1}{N}\sum_{i=1}^{N} \frac{-c}{1+c}\frac{1-\lambda_i}{1-\lambda_i L}\left[\sum_{j=0}^{t-1} \varepsilon_{t-j} + \bar{P}_{i,0}\right], \tag{26}$$

where $\varepsilon \sim N(0, \sigma_\varepsilon^2)$ and $\{\bar{P}_{i,0}\}_{i=1,\dots,N}$ is a set of normal random variable i.i.d. with mean 0 and variance $\sigma_{\bar{P}}$, for each $i = 1, \dots, N$.

The stability of the gaussian distribution implies that

$$\sum_{j=0}^{t-1} \varepsilon_{t-j} + \bar{P}_{i,0} =: \Gamma_t \sim N(0, \sigma_\Gamma^2). \tag{27}$$

In particular, Γ_t is a stationary stochastic process.

By (26) and (27), we can write

$$A_t^3 = \frac{1}{N}\sum_{i=1}^{N}\frac{-c}{1+c}\frac{1-\lambda_i}{1-\lambda_i L}\Gamma_t, \tag{28}$$

The long memory property is formalized in the following result.

Theorem 3. *Suppose that λ_i are parameters drawn by a $B(a,b)$ distribution, for each $i = 1,\ldots,N$, and $a,b > 0$.*

Then, as $N \to +\infty$, the long term memory property for A_t^3 holds, with Hurst's exponent H_3, with the following distinguishing:

- *$b > 1$ implies $H_3 = 1/2$;*
- *$b \in (0,1)$ and the following equation holds:*

$$\sum_{h=-\infty}^{+\infty} \mathbf{E}[A_t^3 A_{t-h}^3] = 0, \tag{29}$$

imply $H_3 = (1-b)/2$. In this case it results $H_3 < 1/2$, and the process is mean reverting.

Proof. The proof is similar to the one provided for Theorem 1.

3.4 The Component Associated to the Empirical Analysis of the Previous Data of the Market's Price

A_t^4, finally, takes in account that the behavior of the investors at time t in strongly influenced by the situation of the market's price observed at time $t-1$. The analysis of the previous results is subjectively calibrated, and this fact explains the presence in this term of a coefficient dependent on i.

In order to treat this case, we need to point out that P_t is a stationary process, since it can be viewed recursively as a sum of stationary processes. Therefore, the following result holds:

Theorem 4. *Suppose that λ_i are parameters drawn by a $B(a,b)$ distribution, for each $i = 1,\ldots,N$, and $a,b > 0$.*

Then, as $N \to +\infty$, the long term memory property for A_t^4 holds, with Hurst's exponent H_4, with the following distinguishing:

- *$b > 1$ implies $H_4 = 1/2$;*
- *$b \in (0,1)$ and the following equation holds:*

$$\sum_{h=-\infty}^{+\infty} \mathbf{E}[A_t^4 A_{t-h}^4] = 0, \tag{30}$$

imply $H_4 = (1-b)/2$. In this case it results $H_4 < 1/2$, and the process is mean reverting.

Proof. The proof is similar to the one provided for Theorem 1.

3.5 Aggregation of the Components

In this part of the work we want just summarize the results obtained for the disaggregate components of the market's forecasts done by the investors.

Theorem 5. *Suppose that λ_i are sampled by a $B(a,b)$ distribution, for each i, with $b \in \mathbf{R}$.*

Then, for $N \to +\infty$, we have that P_t has long memory with Hurst's exponent H given by

$$H = \max\left\{H_1, H_2, H_3, H_4\right\}, \tag{31}$$

Proof. It is well-known that, if X is a fractionally integrated process or order $d \in [-1/2, 1/2]$, then X exhibits the long term memory property, with Hurst's exponent $H = d + 1/2$. Therefore, using Proposition 1, by Theorems 1, 2, 3 and 4, we obtain the thesis.

Remark 1. Theorem 5 provides the long term memory measure of P_t. The range of the Hurst's exponent includes as particular case $H = 1/2$, that correspond to brownian motion. Thus the model can describe periods in which the efficient market hypothesis is fulfilled as well as periods that exhibit antipersistent behavior. Moreover, the long term memory property can not be due to the occurrence of shocks in the market. This finding is in agreement with the impulsive nature of market shocks, not able to drive long-run equilibria in the aggregates.

4 Analysis of Returns

This section aims at mapping the long memory exponent of price time series generated by the model into long memory of log-returns. In order to achieve this goal, we analyze the effect of log-transformation of a long-memory process. Dittman and Granger (2002) provide theoretical results on the long memory degree of nonlinear transformation of $I(d)$ processes only if the transformation can be written a finite sum of Hermite polynomials. Therefore they cannot be used for examining log-returns, which the logarithms is involved in.

The same authors provide further results through numerical analysis. Let $\{X_t\}_t$ be $I(d)$, $Y_t = g(X_t)$ with $g(\cdot)$ a transcendental transformation. Numerical estimates of the degree of long memory of Y_t, d', suggest the following behaviour:

1. $-\frac{1}{2} < d < 0$ antipersistence is destroyed by non-odd transformations, hence $d' = 0$;
2. $d = 0$ uncorrelated processes remain uncorrelated under any transformation: $d' = 0$;
3. $0 < d < \frac{1}{2}$ stationary long memory processes. The size of the long memory of stationary long memory processes ($0 < d < \frac{1}{2}$) diminishes under any transformation ($d' \leq d$). The higher is the Hermite rank of the transforming function,

the bigger is the decrease, even if none of the functions examined can be written as a finite sum of Hermite polynomials. If the transforming function has Hermite rank J and it can be written as a finite sum of Hermite polynomials, then $d' = max\{0, (d-0.5)J + 0.5\}$. Therefore, if $J = 1$, then $d' = d$;

4. $d \geq \frac{1}{2}$ nonstationary processes. The size of the long memory diminishes under any transformation. $d' \leq d$

Extensive simulations reported in Chen et al., (2005) on the effects of nonlinear filters on the estimate of long term memory provide further confirmation the results reported above. In particular, they show that in case of logarithm transformation, the degree of long memory is not changing in the interval $(-0.1, 0.8)$. Discrepancy from $(0, 1/2)$ could rise from precision and biases of the numerical estimate. Small changes in the degree of long memory were expected, due to the violation of the hypothesis of the transforming function being a finite sum of Hermite polynomials, but they aren't got from the analysis of Chen et al., (2005).

Remark 2. From the usual results on differencing, we remark that if $log(P_t)$ is $I(d)$ then the log-returns time series $r_t = log(P_t) - log(P_{t-1})$ is $d' = d - 1$.

We can state the following

Theorem 6. *If the price history is $I(d)$, then returns are $I(d')$, where*

1. *if $-1/2 < d \leq 0$, then $d' = -1$*
2. *if $0 < d < 1/2$, then $d' = d - 1$*
3. *$(d > 1/2)$ the degree of long memory diminishes, but no analytical expressions are available.*

Corollary 1. *Uncorrelated returns($d' = 0$) are obtained if $d = 1$.*

Corollary 2. *Long memory in returns ($d' > 0$) is obtained if $d > 1$.*

5 Conclusions and Further Developments

In this paper a theoretical microeconomic model for time series of assets traded in a financial market is constructed. The market is assumed to be populated by heterogeneous agents. We provide mathematical results concerning the presence of long memory in prices and log-returns.

Our work extends Zaffaroni (2004, 2007a, 2007b), discussing the long term memory property in an agents' aggregation framework by enlarging the class of probability densities of agents' parameters.

Moreover, we study the shift of the memory property from the asset price time series to the log-returns. In particular, it is also evidenced that the presence of long memory in the asset price time series implies that the log returns have long memory as well.

The model allows also for the correlation between the agents and its approach can be useful for modeling also other kind of interaction between the agents.

References

1. Hamburger, C.: Quasimonotonicity, regularity and duality for nonlinear systems of partial differential equations. Ann. Mat. Pura. Appl. 169, 321–354 (1995)
2. Alfarano, S., Lux, T., Wagner, F.: Time variation of higher moments in a financial market with heterogeneous agents: An analytical approach. Journal of Economic Dynamics & Control 32, 101–136 (2008)
3. Avery, C., Zemsky, P.: Multidimensional Uncertainty and Herd Behavior in Financial Markets. American Economic Review 88(4), 724–748 (1988)
4. Bischi, G.I., Gallegati, M., Gardini, L., Leombruni, R., Palestrini, A.: Herd Behavior and Nonfundamental Asset Price Fluctuations in Financial Markets. Macroeconomic Dynamics 10, 502–528 (2006)
5. Box-Steffenmaier, J.M., Smith, R.M.: The Dynamics of Aggregate Partisanship. American Political Science Review 90, 567–580 (1996)
6. Brock, W.A., Hommes, C.H.: Rational route to randomness. Econometrica 65, 1059–1095 (1997)
7. Brock, W.A., Hommes, C.H.: Heterogeneous beliefs and routes to chaos in a simple asset pricing model. Journal of Economic Dynamics & Control 22, 1235–1274 (1998)
8. Byers, D., Davidson, J., Peel, D.: Modelling Political Popularity: An Analysis of Long Range Dependence in Opinion Poll Series. Journal of the Royal Statistical Society, Series A, 471–490 (1997)
9. Cerqueti, R., Rotundo, G.: Microeconomic modeling of financial time series with long term memory. In: Proceedings of IEEE International Conference on Computational Intelligence for Financial Engineering (CIFEr 2003), Hong Kong, pp. 191–198 (2003)
10. Chen, Z., Hu, K., Carpena, P., Bernaola-Galvan, P., Stanley, H.E., Ivanov, P.C.: Effect of nonlinear filters on detrended fluctuation analysis. Physical Review E 71, 101–104 (2005)
11. Chiarella, C., Dieci, R., Gardini, L.: Asset price and wealth dynamics in a financial market with heterogeneous agents. Journal of Economic Dynamics & Control 30, 1755–1786 (2006)
12. Chiarella, C., Gallegati, M., Leombruni, R., Palestrini, A.: Asset Price Dynamics among Heterogeneous Interacting Agents. Computational Economics 22(2), 213–223 (2002)
13. Chiarella, C., He, X.: Heterogeneous beliefs, risk and learning in a simple asset pricing model. Computational Economics 19, 95–132 (2002)
14. Dittman, I., Granger, C.W.J.: Properties of nonlinear transformations of fractionally integrated processes. Journal of Econometrics 110, 113–133 (2002)
15. Föllmer, H., Horst, U., Kirman, A.: Equilibria in financial markets with heterogeneous agents: a probabilistic perspective. Journal of Mathematical Economics 41, 123–155 (2005)
16. Granger, C.W.J.: Long memory relationships and the aggregation of dynamic models. Journal of Econometrics 14, 227–238 (1980)
17. Hommes, C.H.: Financial Markets as nonlinear adaptive evolutionary systems. Quantitative Finance 1, 149–167 (2001)
18. Hommes, C.H.: Heterogeneous agent models in economics and finance. In: Tesfatsion, L., Judd, K.L. (eds.) Handbook of Computational Economics, vol. 2, pp. 1109–1186. Elsevier, North Holland (2006)
19. Hurst, H.E.: Long term storage of reservoirs. Transactions of the American Society of Civil Engineers 116, 770–799 (1951)
20. Hurst, H.E.: A suggested statistical model of some time series that occur in nature. Nature 180, 494 (1957)

21. Jouini, E.: Price functionals with bidask spreads: an axiomatic approach. Journal of Mathematical Economics 34, 547–558 (2000)
22. Kirman, A.P.: Heterogeneity in economics. Journal of Economic Interaction and Coordination 1, 89–117 (2006)
23. Kirman, A.P., Teyssiére, G.: Microeconomic models for long-memory in the volatility of financial time series. Studies in Nonlinear Dynamics and Econometrics 5, 281–302 (2002)
24. MacLeod, A.I., Hipel, K.W.: Preservation of the rescaled adjusted range, I: A reassessment of the Hurst phenomenon. Water Resources Research 14, 491–508 (1978)
25. Madhavan, A., Smidt, S.: An analysis of changes in specialist inventories and quotations. The Journal of Finance 48, 1595–1628 (1993)
26. Mandelbrot, B.B.: Statistical methodology for non periodic cycles: From the covariance to R/S analysis. Annals of Economic and Social Measurement, 259–290 (1972)
27. Mandelbrot, B.B., Wallis, J.: Noah, Joseph and operational hydrology. Water Resources Research 4, 909–918 (1968)
28. Mandelbrot, B.B., Fischer, A., Calvet, L.: A Multifractal Model of Asset Returns. Cowles Foundation Discussion Paper 1164 (1997)
29. Roger, P.: Properties of bid and ask reservation prices in the rank-dependent expected utility model. Journal of Mathematical Economics 34, 269–285 (2000)
30. Rangarajan, G., Ding, M.: Integrated approach to the assessment of long range correlation in time series data. Physical Review E 61(5), 4991–5001 (2000)
31. Simon, H.: Models of man. John Wiley & Sons, New York (1957)
32. Tscherning, R.: Long Memory in Foreign Exchange Rates Revisited. Journal of International Financial Markets, Institutions & Money 5, 53–78 (1995)
33. Willinger, W., Paxson, V., Taqqu, M.S.: Self Similarity and Heavy Tails: Structural Modeling of Network Traffic. In: A Practical Guide To Heavy Tails: Statistical Techniques and Applications. Birkhäuser, Boston (1998)
34. Zaffaroni, P.: Contemporaneous aggregation of linear dynamic models in large economies. Journal of Econometrics 120, 75–102 (2004)
35. Zaffaroni, P.: Memory and aggregation for models of changing volatility. Journal of Econometrics 136, 237–249 (2007a)
36. Zaffaroni, P.: Contemporaneous aggregation of GARCH processes. Journal of Time Series Analysis 28, 521–544 (2007b)

A Appendix

A.1 Beta Distribution

We recall in this subsection the beta distribution.

Definition 1. If Z is an ordinary beta-distributed random variable which can take values between 0 and 1, the probability density function of Z is

$$p(z) = \frac{1}{\beta(a,b)} z^{a-1}(1-z)^{b-1},\ 0 < z < 1, \tag{32}$$

where a e b are positive parameters and

$$\beta(a,b) = \int_0^1 z^{a-1}(1-z)^{b-1}\mathrm{d}z.$$

We refer to this distribution as $B(a,b)$.

A.2 Sum of Integrated Processes

We recall a result due to Granger, (1980):

Proposition 1. *If* $\{X_t\}_t$ *and* $\{Y_t\}_t$ *are independent integrated processes of order, respectively,* d_X *and* d_Y*, then the sum* $Z_t := X_t + Y_t$ *is an integrated process of order* d_Z*, where*

$$d_Z = \max\left\{d_X, d_Y\right\}.$$

From Comparative Degrees of Belief to Conditional Measures

Giulianella Coletti and Barbara Vantaggi

Abstract. Aim of this paper is to give a contribute to the discussion about the "best" definition of conditional model for plausibilty functions and its subclass of the possibility functions.

We propose to use the framework of the theory of measurements: by studying the comparative structure underling different conditional models. This approach gives an estimate of the "goodness" and "effectiveness" of the model, by pointing out the rules necessarily accepted by the user. Moreover, the results related to the characterization of comparative degree of belief by means conditional uncertainty measures can be used in decision theory. They are in fact necessary when we need a model for a decision maker interested in choosing by taking into account, at the same moment, different scenarios.

1 Introduction

Alternative models to expected utility theory have been proposed in decision theory under risk and uncertainty, since expected utility theory performs poorly in some situations ([35]). These more flexible theories are based on non-additive measures [7, 26, 40, 46]). The attention has been focused mainly on classes of probabilities whose lower (or upper) envelope is a measure with "nice" properties such as convex capacity, belief function, possibility, probability sophistication and so on (see for instance [6, 7, 23, 31, 36, 39, 44]). In fact, in these cases, the minimum (maximum) of expected utility can be also expressed as specific integrals with respect to the lower (upper) probability.

Giulianella Coletti
Dept. Matematica e Informatica, Univ. of Perugia, via Vanvitelli 1, Perugia, Italy
e-mail: coletti@dipmat.unipg.it

Barbara Vantaggi
Dept. Me.Mo.Mat., Univ. La Sapienza of Rome, via Scarpa 16, Rome, Italy
e-mail: vantaggi@dmmm.uniroma1.it

S. Greco et al. (Eds.): Preferences and Decisions, STUDFUZZ 257, pp. 69–84.
springerlink.com

The systematic study of qualitative frameworks is also faced on decision theory where Multiple Priors [31] and Choquet Expect Utility (CEU) [40] models are nowadays widely adopted. In particular, starting from [31, 44, 46], many characterizations of preferences representable by the minimum (or maximum) of a class of expected utilities over a set of probabilities have been presented. In particular, a natural counterparts to the expected utility criterion is the pair of possibilistic optimistic and pessimistic criteria, originally introduced by Yager [47] and Whalen [42]. These criteria were axiomatized in the setting of Von-Neuman and Morgenstern theory, based on the comparison of possibilistic lotteries by Dubois, Prade Godo, etc [25] and in the Savagean setting of acts under uncertainty by Dubois Prade and Sabbadin [29].

These works propose a foundation to qualitative decision making in a static world. But the important issue of a qualitative decision theory when new input information can be received was left open. Actually, to perform decision models, which can be updated, conditional measures need to be considered.

In the relevant literature, following the Kolmogorovian probabilistic model, a conditional measure is usually defined starting from an unconditional one. But this is a very restrictive view of conditioning, trivially corresponding to just a modification of the "world". It is instead essential to regard conditioning events as "variables" or, in other words, as uncertain events which can be either true or false. This point of view gives the opportunity to the decision maker or the field expert to take into account at the same time all the possible scenarios (represented by the conditioning events of interest).

Then, a conditional model needs to deal with ordinal relations $\preceq$ defined on an arbitrary set of conditional events.

This topic has been faced in [14] by using as reference model conditional probability and generalized decomposable measures. Moreover, in [13, 16] relations representable by a conditional possibility and conditional necessity have been characterized. This class of models is interesting in decision theory also for the fact that in the unconditional case conditions on acts (in the style of Savage) assuring a possibilistic representation have been given (in [29] for the unconditional case and in [24] for the conditional case) and in a optimistic attitude of decision maker has been carried out.

In this paper we recall the above quoted characterization of ordinal relations, that is been used for the model proposed in [24] with the aim to bridge the gap between qualitative conditional possibility and the axiomatization of possibilistic preference functionals, thus paving the way toward possibilistic decision under uncertainty in a dynamic epistemic environment.

Furthermore, since we are interested on belief function setting we provide a characterization of ordinal relations defined on an arbitrary finite set of conditional events representable by a conditional plausibility (belief). These characterizations are valid also for partial relations, i.e. not complete and defined on a set without a structure.

The provided characteristic axioms recall that one introduced in [14] for comparative conditional probabilities and they have a natural interpretation in terms of betting.

2 Conditional Uncertainty Measures

We recall explicitly the definitions of conditional possibility as introduced in [3] and conditional plausibility (see for instance [9]). The relevant dual conditional functions are defined by duality as follow: given a conditional possibility [plausibility] $\varphi(\cdot|\cdot)$, a conditional necessity [belief function] $\psi(\cdot|\cdot)$ is, for every event $E|H \in \mathbf{C}$:

$$\psi(E|H) = 1 - \varphi(E^c|H)$$

2.1 *Plausibility*

Definition 1. A function Pl defined on $\mathbf{A} \times \mathbf{H}$ is a conditional plausibility if satisfies the following conditions

a1) $Pl(E|H) = Pl(EH|H)$;
a2) $Pl(\cdot|H)$ is a plausibility function for every $H \in \mathbf{H}$;
a3) For every $E \in \mathscr{A}$ and $H, K \in \mathbf{H}$

$$Pl(E|K) = Pl(E|H) \cdot Pl(H|K).$$

We note that the conditional belief function $Bel_D(\cdot|\cdot)$ obtained by duality as mentioned before, i.e. $Bel_D(E|H) = 1 - Pl(E^c|H)$, is the natural generalization of that given by Dempster in [21] deeply studied in [22].

In the literature there are many other definitions of conditional belief (and plausibility) (see for instance [43, 33, 41, 2, 12].

In particular, we denote by $Bel_P(\cdot,\cdot)$ a conditional belief obtained through the product rule,

$$Bel_P(E \wedge H|K) = Bel_P(E|H \wedge K)Bel_P(H|K)$$

for any $E, H \in \mathbf{A}$ and $H \wedge K, K \in \mathbf{H}$, and in the case $\Omega \in \mathbf{H}$, then for any conditioning event B such that $Bel(B) > 0$ one has $Bel_P(A|B) = \frac{Bel(A \wedge B)}{Bel(B)}$.

On the other hand, a conditional belief obtained through Bayes rule is denoted by $Bel_B(\cdot,\cdot)$ and is obtained when $\Omega \in \mathbf{H}$ for any pair of events $A, B \in \mathbf{A}$ such that $Bel(A \wedge B) + Pl(A^c \wedge B) > 0$ as

$$Bel(A|B) = \frac{Bel(A \wedge B)}{Bel(A \wedge B) + Pl(A^c \wedge B)}.$$

Finally we recall the following characterization result, directly proved in [9], which is in fact a particular case of a general result given in [10, 11]:

Theorem 1. *Let* $\mathbf{E} = \mathbf{A} \times \mathbf{H}$ *be a finite set of conditional events* $E|H$ *such that* $\mathbf{A}$ *is an algebra and* $\mathbf{H}$ *an additive set with* $\mathbf{H} \subset \mathbf{A}$ *and* $\emptyset \notin \mathbf{H}$. *For a function* Pl *on* $\mathbf{E}$ *the following statements are equivalent:*

(a) Pl *is a conditional plausibility on* $\mathbf{E}$*;*
(b) There exists (at least) a class $\{Pl_\alpha\}$ *of plausibility functions such that, called* H_0^α *the greatest set of* $\mathbf{H}$ *for which* $Pl_{(\alpha-1)}(H_1^\alpha) = 0$, *results* $Pl_\alpha(H_1^\alpha) = 1$ *and* $H_1^\alpha \subset H_1^\beta$ *for all* $\beta < \alpha$. *Moreover, for every* $E_i|F_i$, *there exists an* α *such that,* $Pl_\beta(F_i) = 0$ *for all* $\beta < \alpha$, *and* $Pl_\alpha(F_i) > 0$ *and*

$$Pl(E_i|F_i) = \frac{Pl_\alpha(E_i^c \wedge F_i)}{Pl_\alpha(F_i)}. \tag{1}$$

Moreover any conditional belief Bel_D on $\mathbf{E}$, obtained through the product rule, is related to a dual conditional plausibility, and so for any conditional event $E_i|F_i$, there exists an α such that, $Pl_\beta(F_i) = 0$ for all $\beta < \alpha$, and $Pl_\alpha(F_i) > 0$ and

$$Bel_D(E_i|F_i) = 1 - \frac{Pl_\alpha(E_i^c \wedge F_i)}{Pl_\alpha(F_i)}. \tag{2}$$

The class of (unconditional) plausibilities Pl_α in condition *(b)* of Theorem 1 necessarily contains more than one element whenever in $\mathbf{H}$ there are events with zero plausibility. We can say that Pl_1 gives a refinement of those events judged with zero plausibility under Pl_0. This shows that conditional belief functions, as well as conditional plausibility, are more general than belief functions (or, respectively, plausibility).

In [17] the construction of the class $\{Pl_\alpha\}$ characterizing (in the sense of the above result) a conditional belief is explicitly shown in an example.

2.2 Possibility

As it is well known possibility measures are a particular interesting subclass of plausibility measures.

In possibility theory the notion of conditioning is a problem of long-standing interest, in fact various definitions of conditional possibility have been introduced (see, e.g., [18, 19, 23, 27, 28, 32, 48]) mainly by analogy with Kolmogorovian probabilistic case. In all proposal in fact a T-conditional possibility $\Pi(A|B)$ is "essentially" defined as a solution of the equation

$$\Pi(A \wedge B) = T(\Pi(B), x), \tag{3}$$

where T is any t-norm, (the most common t-norm is $T = \min$).

Nevertheless, two problems arise from this definition:

(a) equation (3) can have no solution for some pairs $\{\Pi(A \wedge B), \Pi(B)\}$;

(b) equation (3) can admit more than one solution for some pairs $\{\Pi(A \wedge B), \Pi(B)\}$.

An arbitrary solution needs not to be a normalized possibility: it happens, for example, choosing Zadeh's conditioning rule [48] where $\Pi(A|B) = \Pi(A \wedge B)$ for any $A \in \mathbf{B}$. Therefore some additional condition must be required in order to avoid such problem.

A proposal [26] is to take the greatest solution (known as minimum specificity principle), i.e.

$$\Pi(A|B) = \begin{cases} \Pi(A \wedge B) & \text{if } \Pi(A \wedge B) < \Pi(B) \\ 1 & \text{if } \Pi(A \wedge B) = \Pi(B). \end{cases} \quad (4)$$

According to the above definition, if A and B are incompatible (i.e. $A \wedge B = \emptyset$) and $\Pi(B) = 0$, then $\Pi(A|B)$ is equal to 1 (according to (4)) instead of 0 as it would be natural (being $A|B = \emptyset|B$). Therefore, it is too strong to chose a unique value of $\Pi(A|B)$ in $[0,1]$ for any event A such that $\Pi(A \wedge B) = \Pi(B)$: however the choice of value 1 for (at least) one atom $C \subseteq B$ is necessary (see [1]) to get a normalized possibility. The above problem comes out for any t-norm, even strictly monotone ones. All these problems can be solved by introducing a direct definition of conditional possibility as in [3]:

Definition 2. Let $\mathbf{F} = \mathbf{B} \times \mathbf{H}$ be a set of conditional events such that $\mathbf{B}$ is a Boolean algebra and $\mathbf{H}$ an additive set (i.e. closed with respect to finite logical sums), with $\mathbf{H} \subseteq \mathbf{B} \setminus \{\emptyset\}$. Let T be a t-norm, function $\Pi : \mathbf{F} \to [0,1]$ is a T-conditional possibility if it satisfies the following properties:

1. $\Pi(E|H) = \Pi(E \wedge H|H)$, for every $E \in \mathbf{B}$ and $H \in \mathbf{H}$;
2. $\Pi(\cdot|H)$ is a possibility, for any $H \in \mathbf{H}$;
3. $\Pi(E \wedge F|H) = T\{\Pi(E|H), \Pi(F|E \wedge H)\}$, for any $H, E \wedge H \in \mathbf{H}$ and $E, F \in \mathbf{B}$.

Conditional necessity function $N(\cdot|\cdot)$ is obtained by duality, as mentioned before, i.e. $N(E|H) = 1 - \Pi(E^c|H)$,

Nevertheless the problem to choose the best t-norm to make conditioning remain open, in fact due to peculiarity of "max" operator, any T-norm is syntactically correct (since the distributivity is assured). In the following we try to give a contribution to this discussion from a different perspective by studying the comparative framework underling a conditional model.

2.3 Coherent Conditional Plausibilities

In the following we denote by $\mathbf{F} = \{E_1|F_1, E_2|F_2, \ldots, E_m|F_m\}$ an arbitrary finite set of conditional events, by $\mathbf{A}$ the algebra generated by $\{E_1, F_1, \ldots, E_m, F_m\}$ and by $\mathbf{H}$ the additive set generated by the set of the conditioning events $\{F_1, \ldots, F_m\}$.

Definition 3. A function $f(\cdot|\cdot)$ on an arbitrary finite set $\mathbf{F}$ is a coherent conditional plausibility (belief) if there exists $\mathbf{C} \supset \mathbf{F}$, with $\mathbf{C} = \mathbf{A} \times \mathbf{H}$ such that $f(\cdot|\cdot)$ can be extended from $\mathbf{F}$ to $\mathbf{C}$ as a conditional plausibility (belief).

The following theorem (proved in [9]) characterizes coherent conditional plausibility (belief) functions in terms of a class of plausibilities $\{Pl_1, ..., Pl_m\}$.

Theorem 2. *Let* $\mathbf{F} = \{E_1|F_1, E_2|F_2, \ldots, E_m|F_m\}$ *be an arbitrary finite set of conditional events and denote by* $\mathbf{A} = \{H_1, H_2, ..., H_n\}$ *the algebra spanned by* $\{E_1, \ldots, E_m, F_1, \ldots, F_m\}$, $\mathbf{H}$ *the additive set generated by* $\{F_1, \ldots, F_m\}$ *and* $H_0^0 = \vee_{j=1}^m F_j$. *For a real function* Pl (Bel_D) *on* $\mathbf{F}$ *the following statements are equivalent:*

(a) Pl (Bel_D) *is a coherent conditional plausibility (belief) assessment;*
(b) there exists (at least) a class $\mathscr{P} = \{Pl_\alpha\}$ *of plausibility functions such that* $Pl_\alpha(H_0^\alpha) = 1$ *and* $H_0^\alpha \subset H_0^\beta$ *for all* $\beta < \alpha$, *where* H_0^α *is the greatest element of* $\mathbf{H}$ *for which* $Pl_{(\alpha-1)}(H_0^\alpha) = 0$.

Moreover, for every $E_i|F_i$, *there exists an index* α *such that* $Pl_\beta(F_i) = 0$ *for all* $\alpha > \beta$, $Pl_\alpha(F_i) > 0$ *and*

$$Pl(E_i|F_i) = \frac{Pl_\alpha(E_i \wedge F_i)}{Pl_\alpha(F_i)}, \tag{5}$$

$$\left(Bel_D(E_i|F_i) = 1 - \frac{Pl_\alpha(E_i^c \wedge F_i)}{Pl_\alpha(F_i)}\right), \tag{6}$$

(c) all the following systems (S^α), *with* $\alpha = 0, 1, 2, ..., k \leq n$, *admit a solution* $\mathbf{X}^\alpha = (x_k^\alpha)$:

$$(S^\alpha) = \begin{cases} \sum\limits_{H_k \wedge F_i \neq \emptyset} x_k^\alpha \cdot Pl(E_i|F_i) = \sum\limits_{H_k \wedge E_i \wedge F_i \neq \emptyset} x_k^\alpha, \ \forall F_i \subseteq H_0^\alpha \\ \sum\limits_{H_k \in H_0^\alpha} x_k^\alpha = 1 \\ x_k^\alpha \geq 0, \qquad \forall H_k \subseteq H_0^\alpha \end{cases}$$

$$\left((S^\alpha) = \begin{cases} \sum\limits_{H_k \wedge F_i \neq \emptyset} x_k^\alpha \cdot [1 - Bel_D(E_i|F_i)] = \sum\limits_{H_k \wedge E_i^c \wedge F_i \neq \emptyset} x_k^\alpha, \ \forall F_i \subseteq H_0^\alpha \\ \sum\limits_{H_k \in H_0^\alpha} x_k^\alpha = 1 \\ x_k^\alpha \geq 0, \qquad \forall H_k \subseteq H_0^\alpha \end{cases} \right)$$

where H_0^α *is the greatest element of* $\mathbf{H}$ *such that* $\sum\limits_{H_i \wedge H_0^\alpha \neq \emptyset} x_i^{(\alpha-1)} = 0$.

Condition *(c)* stresses that this measure can be written in terms of a suitable class of basic assignments, instead of just one as in the classical case where all the conditioning events have positive plausibility.

Note that every class $\mathscr{P}$ (condition (b) of Theorem 2) is said to be agreeing with both the conditional belief Bel_D and its dual conditional plausibility Pl. Whenever there are events in $\mathbf{H}$ with zero plausibility the class of unconditional plausibilities consists on more than one element and we can say that Pl_1 gives a refinement of those events judged with zero plausibility under Pl_0.

Results in the same style of the above theorem, characterizing conditional possibility and necessity in terms of a class of unconditional possibilities, have been given in [3, 4, 15, 16, 30].

3 Ordinal Relations

Let **A** be an algebra of events: denote by $\preceq$ a binary relation on **A** and, as usual, $A \prec B$ denotes that $A \preceq B$ and $\neg(B \preceq A)$; while $A \sim B$ stands for $A \preceq B$ and $B \preceq A$. If we give the sentence "φ representing $\preceq$" the meaning of , "φ being strictly monotone with $\preceq$", then for any choice of a *capacity* function φ as numerical framework of reference, it is necessary that $\preceq$ satisfies the following conditions:

(1)$\preceq$ is a weak order on the algebra **A**;
(2)for any $A \in \mathbf{A}$, $\emptyset \preceq A$ and $\emptyset \prec \Omega$;
(3)for any $A, G \in \mathbf{A}$

$$A \subset G \Rightarrow A \preceq G.$$

When we specialize the capacity function (probability, belief, plausibility, and so on) representing $\preceq$, then we need to add to the above axioms a specific relevant condition, which essentially expresses a (more or less strong) sort of "comparative additivity". The first (and the best known) additivity axiom (de Finetti [20], Koopman [34]) is the following

(p) for $E, F, H \in \mathbf{A}$, with $E \wedge H = F \wedge H = \emptyset$, both the following implications hold:

$$E \preceq F \Rightarrow E \vee H \preceq F \vee H$$

$$E \prec F \Rightarrow E \vee H \prec F \vee H$$

In fact the above axiom (p) is a necessary condition for the representability of $\preceq$ with an *additive function* with values in a totally ordered set (also, for instance, the set of nonstandard real numbers). If we refer instead to more general measures of uncertainty, such as belief functions, plausibilities and so on, then it is easy to see that (p) can be violated.

Nevertheless, also in this case a weaker additivity axiom is necessary; see, for this aspect, the following conditions (b) and (pl) given in [45, 5, 14]: the first one characterizing relations representable by a belief function and the second one by plausibility. Further conditions (PO), (NEC) introduced in [23], characterize relations representable by a possibility and a necessity respectively:

(b) if $E, F, H \in \mathbf{A}$, with $E \subseteq F$ and $F \wedge H = \emptyset$, then

$$E \prec F \Rightarrow (E \vee H) \prec (F \vee H)$$

(pl) if $E, F, H \in \mathbf{A}$, with $E \subseteq F$ and $F \wedge H = \emptyset$, then

$$E \sim F \Rightarrow (E \vee H) \sim (F \vee H)$$

(PO) for every $A, G, H \in \mathbf{A}$

$$A \preceq G \Rightarrow (A \vee H) \preceq (G \vee H)$$

(NEC) for every $A, B, C \in \mathbf{A}$

$$A \preceq B \Longrightarrow (A \wedge C) \preceq (B \wedge C).$$

4 Comparative Conditional Plausibilities

The aim of this section is to characterize relations representable by a conditional plausibility. Before to introduce comparative conditional plausibility, we need to define an indicator function different from the standard one I,

$$\hat{I}(A) : \mathbf{A_F} \to \{0,1\}$$

associated to the event $A \in \mathbf{F}$, and $\hat{I}(A)$ is 0 on any event $E \in \mathbf{A_F}$ incompatible with A (i.e. $A \wedge E = \emptyset$) and 1 on any event $E \in \mathbf{A_F}$ such that $A \wedge E \neq \emptyset$.

Definition 4. Let $\mathbf{F}$ be an arbitrary set of conditional events. A comparative conditional plausibility is a relation on $\mathbf{F}$ satisfying the following condition:

(ccpl) for every $E_i|H_i \preceq F_i|K_i \in \mathbf{F}$ there exist $\alpha_i, \beta_i \in [0,1]$ with $\alpha_i \leq \beta_i$ with $\alpha_i < \beta_i$ for $E_i|H_i \prec F_i|K_i$, such that, for every $n \in \mathbb{N}$ and for every $E_i|H_i \preceq F_i|K_i$, $\lambda_i, \lambda_i', \geq 0, (i = 1, \ldots, n)$, one has:

$$\sup_{H^o} \left\{ \sum_i \left[\lambda_i'(\hat{I}_{F_i \wedge K_i} - \beta_i \hat{I}_{K_i}) + \lambda_i(\alpha_i \hat{I}_{H_i} - \hat{I}_{E_i \wedge H_i}) \right] \right\} \geq 0$$

where $H^o = (\bigvee_{i:\lambda_i'>0} K_i) \vee (\bigvee_{i:\lambda_i>0} H_i)$.

Condition (ccpl) is of the same kind of condition (ccp) introduced in [14] characterizing relations representable by a coherent conditional probability. The main difference consists in the fact that (ccp) is based on indicator of events, while (ccpl) is based on $\hat{I}$.

We note that if for any i we have $H_i = K_i = \Omega$, then taking $\lambda_i' = \lambda_i$ one has that

$$G = \left\{ \sum_i \left[\lambda_i(\hat{I}_{F_i} - \beta_i + \alpha_i - \hat{I}_{E_i}) \right] \right\} = \left\{ \sum_i \left[\lambda_i(\hat{I}_{F_i} - \hat{I}_{E_i} - \delta_i) \right] \right\} \geq 0$$

with $\delta_i > 0$ if the relation $\preceq$ is strict.

So (ccpl) reduces to the following condition (cpl)

$$\sup \left\{ \sum_i \left[\lambda_i(\hat{I}_{F_i} - \hat{I}_{E_i}) \right] \right\} \geq \sum_i \lambda_i \delta_i$$

which is equivalent, in a *finite* ambit, to the following condition (qpl), necessary and sufficient for the existence of a coherent plausibility representing $\preceq$, essentially introduced in [37]:

(qpl) for any $E_i, F_i \in \mathbf{F}$ there exists $\lambda_i \geq 0$ $(i = 1, ..., n)$ such that $E_i \preceq F_i$

$$\sup\left\{\sum_i \left[\lambda_i(\hat{I}_{F_i} - \hat{I}_{E_i})\right]\right\} \leq 0 \text{ implies } E_i \sim F_i$$

Nevertheless, as discussed, for probability theory, in [14], in an infinite ambit condition (ccpl) is stronger than condition (qpl). In fact, by following the same line suggested in [8], it is possible to prove that it is necessary and sufficient for the existence of a coherent plausibility representing $\preceq$ in an arbitrary (possibly infinite) set of events.

We finally recall that when the set of events is a finite Boolean algebra, it is possible to characterize the comparative degree of belief $\preceq$ representable by a plausibility measure, by means purely comparative axioms. In fact conditions axioms (1), (2), (3)and (pl) are necessary and sufficient, as proved in [45] and [5].

Note that (ccpl) can be interpreted in terms of coherent bets (analogously to the interpretation of (ccp) in [14]), moreover it does not require any logical structure for the class of conditional events and it applies also to not complete relations.

Remark 1. We start by giving an interpretation of (qpl) in term of bets (in the unconditional case): for every pair $E_i \preceq F_i$, one bets on F_i and versus E_i and obtains the following gain. Let C be an atom and k the number of events (in the relevant algebra) containing C. When an atom C occurs such that $C \subset E_i^c \wedge F_i$, the gain is equal to $k\lambda_i$; if $C \subset E_i \wedge F_i$ occurs, then the gain is equal to 0; if $C \subset E_i \wedge F_i^c$, the gain is equal to $-k\lambda_i$. What (qpl) requires is that, for any bet involving a finite set of pairs $E_i \preceq F_i$, the global gain is positive at least for an atom.

We give now an interpretation of condition (ccpl). For each pair of conditional events $E_i|H_i \preceq F_i|K_i$ we consider a bet on $F_i|K_i$ versus $E_i|H_i$. If the events H_i and K_i do not occur the bet is called off. When H_i occurs and K_i does not occur the gain is $k\lambda_i(\alpha_i - 1)$ if the atom C occurring is in $E_i \wedge H_i$ and $k\lambda_i\alpha_i$ if $C \subseteq E_i^c \wedge H_i$, where k is the number of events (in the relevant algebra) contained on H^o and containing C. Similarly, if K_i (and not H_i) occurs, if the atom C occurs, we obtain $k\lambda_i'(1 - \beta_i)$ or $-k\lambda_i\beta_i$,. Finally, if both H_i and K_i occur, then then the gain is the sum of the gains corresponding in the latter two situations, when an atom $C \subseteq H_i \wedge K_i$ occurs. The coherence condition (ccpl) requires that given n pairs of conditional events such that $E_i|H_i \preceq F_i|K_i$, there exists for each $F_i|K_i$ a value β_i (and for each $E_i|H_i$ a value α_i), with $\alpha_i \leq \beta_i$ such that the global gain is certainly not negative for every choice of positive numbers λ_i.

Note that it is essential to require in axiom (ccpl) that the sup is computed on the union of conditioning events H_i (or K_i) such that λ_i (or λ_i'), associated to $E_i|H_i$ (or $F_i|K_i$) is strictly greater than 0.

In fact, for example, considering the statement $\emptyset|H \prec \emptyset|K$ with H and K logical independent event, we consider $0 \leq \alpha < \beta \leq 1$ and for every non-negative λ', λ the gain is

$$\sup_{H \vee K} G = \sup \lambda'(-\beta \hat{I}_K) + \lambda(\alpha \hat{I}_H) \geq 0.$$

Since $H \wedge K^c$ is a possible event, G correspondingly assumes the value $\lambda\alpha$, which is greater than or equal to zero for any choice of λ, λ'.

Condition (ccpl) characterizes the relations representable by a conditional plausibility, as shown in the following result:

Theorem 3. *Let $\preceq$ be a binary relation on an arbitrary finite set of conditional events $\mathbf{F}_o = \{E_i|H_i, F_i|K_i\}_{i\in I}$. For a binary relation $\preceq$, the following statements are equivalent:*

- *$\preceq$ is a comparative conditional plausibility;*
- *there exists a coherent conditional plausibility $Pl(\cdot|\cdot)$ on $\mathbf{F}_o$ representing $\preceq$.*

Proof. Since Pl represents $\preceq$, for any $E_i|H_i \preceq F_i|K_i$ then $Pl(E_i|H_i) = \alpha_i \leq \beta_i = Pl(F_i|K_i)$, moreover if $E_i|H_i \prec F_i|K_i$ then $\alpha_i < \beta_i$, so from Theorem 1 it follows that there exists a sequence of compatible linear systems $\mathscr{S}_{\mathbf{F}_o}, \mathscr{S}_{\mathbf{F}_1}, \ldots, \mathscr{S}_{\mathbf{F}_k}$. Actually, these systems admit a semi-positive solution (i.e. $x_r^\alpha \geq 0$ and $\sum_r x_r^\alpha > 0$).

From a classic alternative theorem (see, e.g., Fenchel (1951)) the system

$$\mathscr{S}'_{\mathbf{F}_j} = \begin{cases} (\hat{I}_{F_j \wedge K_j} - \alpha_j \hat{I}_{K_j}) \times W \geq 0 & F_j|K_j \in \mathbf{F}_j \\ (-\hat{I}_{E_i \wedge H_i} + \alpha_i \hat{I}_{H_i}) \times W \geq 0 & E_i|H_i \in \mathbf{F}_j \end{cases}$$

has a semi-positive solution if and only if the following inequality

$$\sum_j \lambda'_j(\hat{I}_{F_j \wedge K_j} - \beta_j \hat{I}_{K_j})] + \lambda_i(-\hat{I}_{E_i \wedge H_i} + \alpha_i \hat{I}_{H_i}) < 0$$

does not admit a non-negative solution, which is equivalent to (ccpl).

Vice versa, assuming (ccpl), for any $\lambda_i > 0$ and $\lambda'_j > 0$, with $i \in I$ and $j \in J$,

$$\sup_{H^0} G = \sum_j \lambda'_j(\hat{I}_{F_j \wedge K_j} - \beta_j \hat{I}_{K_j}) + \sum_i \lambda_i(-\hat{I}_{E_i \wedge H_i} + \alpha_i \hat{I}_{H_i}) \geq 0,$$

with $H^0 = (\vee_{j\in J} K_j) \vee (\vee_{i\in I} H_i)$, implies that $\sup_{H^0} G \geq 0$ for any $\lambda_i \geq 0$ and $\lambda'_j \geq 0$. This condition is equivalent to the fact that

$$\sum_j \lambda'_j(\hat{I}_{F_j \wedge K_j} - \beta_j \hat{I}_{K_j})] + \sum_i \lambda_i(-\hat{I}_{E_i \wedge H_i} + \alpha_i \hat{I}_{H_i}) < 0 \tag{7}$$

does not admit a solution for any $\lambda'_j \geq 0, \lambda_i \geq 0$. From the aforementioned classic alternative theorem, the system of inequalities admits no non-negative solution iff the dual system $\mathscr{S}_{\mathbf{F}_o}$ has a semi-positive solution.

Then, there is a semi-positive solution W_o for $\mathscr{S}_{\mathbf{F}_o}$, then, let

$$\mathbf{F}_1 = \{E_i|H_i \in \mathbf{F}_o : \hat{I}_{H_i} \times W_o = 0\} \cup \{F_j|K_j \in \mathbf{F}_o : \hat{I}_{K_j} \times W_o = 0\},$$

if $\mathbf{F}_1$ is not empty, then there is a semi-positive solution W_1 for $\mathscr{S}_{\mathbf{F}_1}$ and so on till $\mathbf{F}_{k+1}$ is empty.

For any H_i (K_j) there is a unique W_α (W_β) such that $\hat{I}_{H_i} \times W_\alpha > 0$ ($\hat{I}_{K_j} \times W_\beta > 0$) and by putting $Pl(E_i|H_i) = \frac{\hat{I}_{E_i \wedge H_i} \times W_\alpha}{\hat{I}_{H_i} \times W_\alpha}$ $\left(Pl(F_j|K_j) = \frac{\hat{I}_{F_j \wedge K_j} \times W_\beta}{\hat{I}_{K_j} \times W_\beta} \right)$, it follows that the function Pl represents $\preceq$, and from Theorem 1, Pl on $\mathbf{F}_o$ is a coherent conditional plausibility.

Remark 2. We would remark that contrary to the unconditional case it it seems not possible to obtain purely comparative condition when the set of conditional events is a product of an algebra for an additive set contained on the algebra.

5 Comparative Conditional Beliefs

We introduce an axiom for comparative conditional beliefs (similar to (ccpl)), that characterizes relations representable by a conditional plausibility.

We need to define an indicator function different from I and $\hat{I}(A)$,

$$I^*(A) : \mathbf{A}_\mathbf{F} \rightarrow \{0, 1\}$$

associated to the event $A \in \mathbf{A}_\mathbf{F}$ and $I^*(A)$ is 1 on any event $E \in \mathbf{A}_\mathbf{F}$ included in A (i.e. $A \subseteq E$) and 0 otherwise.

Definition 5. Let $\mathbf{F}$ be an arbitrary finite set of conditional events. A comparative conditional belief is a relation on $\mathbf{F}$ satisfying the following condition:

(ccbel) for every $E_i|H_i \preceq F_i|K_i \in \mathbf{F}$ there exist $\alpha_i, \beta_i \in [0, 1]$ with $\alpha_i \leq \beta_i$ with $\alpha_i < \beta_i$ for $E_i|H_i \prec F_i|K_i$, such that, for every $E_i|H_i \preceq F_i|K_i$, $\lambda_i, \lambda_i', \geq 0, (i = 1, ..., n)$, one has:

$$\sup_{H^o} \left\{ \sum_i \left[\lambda_i'(I^*_{F_i \wedge K_i} - \beta_i)\hat{I}_{K_i} + \lambda_i(\alpha_i - I^*_{E_i \wedge H_i})\hat{I}_{H_i} \right] \right\} \geq 0$$

where $H^o = (\bigvee_{i:\lambda_i'>0} K_i) \vee (\bigvee_{i:\lambda_i>0} H_i)$.

Note that (ccbel) is based on I^* instead of I as (ccp) or $\hat{I}$ as (ccpl). Also (ccbel) can be interpreted in terms of coherent bets: essentially we can adopt the same interpretation made for (ccpl), by changing only the definition of the numbers k and h.

The next result is similar to Theorem 3 and shows that condition (ccbel) characterizes the relations representable by a conditional belief:

Theorem 4. *Let $\preceq$ be a binary relation on an arbitrary finite set of conditional events $\mathbf{F} = \{E_i|H_i, F_i|K_i\}_{i \in I}$. For a binary relation $\preceq$, the following statements are equivalent:*

- *$\preceq$ is a comparative conditional belief;*
- *there exists a coherent conditional belief $Bel_D(\cdot|\cdot)$ on $\mathbf{F}$ representing $\preceq$.*

Proof. Since Bel_D represents $\preceq$, for any $E_i|H_i \preceq F_i|K_i$ then $Bel_D(E_i|H_i) = \alpha_i \leq \beta_i = Bel_D(F_i|K_i)$, moreover if $E_i|H_i \prec F_i|K_i$ then $\alpha_i < \beta_i$, so $Pl(E_i^c|H_i) = 1-\alpha_i \geq 1-\beta_i = Pl(F_i^c|K_i)$.

From Theorem 2 it follows that there exists a sequence of compatible linear systems $\mathscr{S}_{\mathbf{F}_o}, \mathscr{S}_{\mathbf{F}_1}, ..., \mathscr{S}_{\mathbf{F}_k}$.

Actually, these systems admit a semi-positive solution (i.e. $x_r^\alpha \geq 0$ and $\sum_r x_r^\alpha > 0$).

From a classic alternative theorem (see, e.g., Fenchel (1951)) the system

$$\mathscr{S}'_{\mathbf{F}_j} = \begin{cases} (\hat{I}_{E_i^c} - 1 + \alpha_i)\hat{I}_{H_i}) \times W \geq 0 & E_i|H_i \in \mathbf{F}_j \\ (-I^*_{F_j^c} + 1 - \beta_j)\hat{I}_{K_j}) \times W \geq 0 & F_j|K_j \in \mathbf{F}_j \end{cases}$$

and so

$$\mathscr{S}'_{\mathbf{F}_j} = \begin{cases} (\alpha_i - I^*_{E_i})\hat{I}_{H_i}) \times W \geq 0 & E_i|H_i \in \mathbf{F}_j \\ (I^*_{F_j} - \beta_j)\hat{I}_{K_j}) \times W \geq 0 & F_j|K_j \in \mathbf{F}_j \end{cases}$$

has a semi-positive solution if and only if

$$\sum_j \lambda'_j(I^*_{F_j \wedge K_j} - \beta_j)\hat{I}_{K_j}] + \lambda_i(-I^*_{E_i \wedge H_i} + \alpha_i)\hat{I}_{H_i} < 0$$

does not admit a non-negative solution , which is equivalent to (ccbel).

Vice versa, assuming (ccbel), for any $\lambda_i > 0$ and $\lambda'_j > 0$, with $i \in I$ and $j \in J$,

$$\sup_{H^0} G = \sum_j \lambda'_j(I^*_{F_j \wedge K_j} - \beta_j)\hat{I}_{K_j} + \sum_i \lambda_i(-I^*_{E_i \wedge H_i} + \alpha_i)\hat{I}_{H_i} \geq 0,$$

with $H^0 = (\vee_{j \in J} K_j) \vee (\vee_{i \in I} H_i)$, implies that $\sup_{H^0} G \geq 0$ for any $\lambda_i \geq 0$ and $\lambda'_j \geq 0$. This condition is equivalent to the fact that

$$\sum_j \lambda'_j(I^*_{F_j \wedge K_j} - \beta_j)\hat{I}_{K_j} + \sum_i \lambda_i(-I^*_{E_i \wedge H_i} + \alpha_i)\hat{I}_{H_i} < 0 \tag{8}$$

does not admit a solution for any $\lambda'_j \geq 0, \lambda_i \geq 0$. From the aforementioned classic alternative theorem, the system of inequalities admits no non-negative solution iff the dual system $\mathscr{S}_{\mathbf{F}_o}$ has a semi-positive solution.

Considering the family $\mathbf{F}_o$, one gets that there is a semi-positive solution W_o for $\mathscr{S}_{\mathbf{F}_o}$, then, let

$$\mathbf{F}_1 = \{E_i|H_i \in \mathbf{F}_o : \hat{I}_{H_i} \times W_o = 0\} \cup \{F_j|K_j \in \mathbf{F}_o : \hat{I}_{K_j} \times W_o = 0\},$$

if $\mathbf{F}_1$ is not empty, then there is a semi-positive solution W_1 for $\mathscr{S}_{\mathbf{F}_1}$ and so on till $\mathbf{F}_{k+1}$ is empty.

For any H_i (K_j) there is a unique W_α (W_β) such that $\hat{I}_{H_i} \times W_\alpha > 0$ ($\hat{I}_{K_j} \times W_\beta > 0$) and by putting

$$Bel_D(E_i|H_i) = 1 - Pl(E_i^c|H_i) = 1 - \frac{Pl_\alpha(E_i^c \wedge H_i)}{Pl_\alpha(H_i)} = \frac{I^*_{E_i \wedge H_i} \times W_\alpha}{\hat{I}_{H_i} \times W_\alpha}$$

$\left(Bel_D(F_j|K_j) = \frac{I^*_{F_j \wedge K_j} \times W_\beta}{\hat{I}_{K_j} \times W_\beta} \right)$, it follows that the function Bel_D represents $\preceq$, and from Theorem 2, Bel_D on $\mathbf{F}_o$ is a coherent conditional belief.

6 Comparative Conditional Possibility

We consider now a particular class of plausibility: possibility measures. In [16] we provide a characterization of ordinal relations on $\mathbf{E} = \mathbf{A} \times \mathbf{H}$ representable by a conditional possibilities (T = min). In the sequel we recall the main results.

Definition 6. Let $\mathbf{E} = \mathbf{A} \times \mathbf{H}$. A binary relation $\preceq$ on $\mathbf{E}$ is called comparative conditional possibility iff the following conditions hold:

1. $\preceq$ is a total preorder;
2. for any $H, K \in \mathbf{H}$, $\emptyset|H \sim \emptyset|K \prec H|H \sim K|K$;
3. for any $A, B \in \mathbf{A}$ and $H, B \wedge H \in \mathbf{H}$,

$$A \wedge B|H \preceq A|B \wedge H$$

   and moreover if either $A \wedge B|H \prec B|H$ or $B|H \sim H|H$, then $A \wedge B|H \sim A|B \wedge H$;
4. for any $H \in \mathbf{H}$ and any $A, B, C \in \mathbf{A}$

$$A|H \preceq B|H \Rightarrow (A \vee C)|H \preceq (B \vee C)|H.$$

Condition (3) requires that when new information "B" is assumed the degree of belief of an event A (or better of $A \wedge B$) non-decreases. Moreover, if the new information is "almost sure", it means $B \sim \Omega$, then the degree of belief of an event A remains equal to its updated degree of belief.

Condition (4) is essentially that proposed by Dubois in [23], just reread on the hypothesis H. Moreover, condition (4) is equivalent (see [16]), under transitivity, to

$$A|H \preceq B|K \text{ and } C|H \preceq D|K \ \Rightarrow \ (A \vee C)|H \preceq (B \vee D)|K.$$

Theorem 5. *Let* $\mathbf{E} = \mathbf{A} \times \mathbf{H}$. *For a binary relation* $\preceq$ *on* $\mathbf{E}$ *the following statements are equivalent:*

i. $\preceq$ *is a comparative conditional possibility;*
ii. there exists a conditional possibility Π *on* $\mathbf{E}$ *representing* $\preceq$.

Obviously, among the comparative conditional possibilities there are also the ordinal relations representable by conditional possibilities satisfying minimum specifity principle, more precisely those satisfying a reinforcement of condition *3* of Definition 6, that is

(sc) for every $A, B \in \mathbf{A}$ and $H, B \wedge H \in \mathbf{H}$,

$$A \wedge B|H \preceq A|B \wedge H$$

and moreover if $A \wedge B \wedge H \neq \emptyset$ and $A \wedge B|H \sim B|H$, then $A|B \wedge H \sim H|H$.

We note that the axioms characterizing comparative conditional possibilities are purely qualitative. Nevertheless the above results are strictly dependent on the choice of T=min. In fact it is not possible to obtain similar results when T is a strict t-norm such as the product.

For the product in fact axiom (3), characterizing a comparative conditional possibility, is not necessary since if $A \wedge B|H \prec B|H$, the statement $A \wedge B|H \sim A|B \wedge H$ cannot hold. Actually such axiom should be modified as follows for any

$A, B \in \mathbf{A}$ and $H, B \wedge H \in \mathbf{H}$,

$$A \wedge B|H \preceq A|B \wedge H$$

and moreover if $B|H \sim H|H$, then $A \wedge B|H \sim A|B \wedge H$;

but it is necessary, but not sufficient for the representability of $\preceq$ with a T-conditional possibility (with T= $\cdot$).

7 Conclusion

We deal with the representability problem of ordinal relations by some well-known conditional measures such as possibility, and plausibility. This analysis gives a new perspective to conditioning operation: in fact, it puts in evidence that in possibility setting the conditioning obtained through the t-norm of minimum gives rise to ordinal relations satisfying conditions in a purely qualitative form, while by considering the t-norm of the product gives rise to conditions more difficult to be explained. Obviously, such problem comes out also in belief function setting.

Furthermore, we provide characterizations in terms of necessary and sufficient conditions for the representability of a binary relation by means of a conditional plausibility (belief) and we give an interpretation through betting scheme.

These characterizations give us an estimate of the "goodness" and "effectiveness" of the conditioning operations in these settings, and they allow to study this problem from a perspective different to that studied in [17] based on "local representability". Actually, we recall that for possibility and plausibility it comes out that a relation is locally representable if and only if it is representable by a strict positive unconditional measure. The above characterizations give rise to a fundamental difference among these measures and probability: binary relations representable by a strict positive probability are also locally representable by a conditional probability, while the converse is not true.

References

1. Amor, N.B., Mellouli, K., Benferhat, S., Dubois, D., Prade, H.: A theoretical framework for possibilistic independence in a weakly ordered setting. International Journal of Uncertainty, Fuzziness and Knowledge-Based Systems 10(2), 117–155 (2002)
2. Ben Yaghlane, B., Smets, P., Mellouli, K.: About conditional belief function independence. In: Benferhat, S., Besnard, P. (eds.) ECSQARU 2001. LNCS (LNAI), vol. 2143, pp. 340–349. Springer, Heidelberg (2001)

3. Bouchon-Meunier, B., Coletti, G., Marsala, C.: Independence and Possibilistic Conditioning. Annals of Mathematics and Artificial Intelligence 35, 107–124 (2002)
4. Bouchon-Meunier, B., Coletti, G., Marsala, C.: Conditional possibility and necessity. In: Bouchon-Meunier, G.-R., Magdalena, Y. (eds.) Technologies for Constructing Intelligent Systems, pp. 59–71. Springer, Heidelberg (2002) (Selected papers from IPMU 2000)
5. Capotorti, A., Coletti, G., Vantaggi, B.: Preferences representable by a lower expectation: some characterizations. Theory and Decision 64, 119–146 (2008)
6. Chateauneuf, A.: On the use of capacities in modeling uncertainty aversion and risk aversion. Journal of Mathematical Economics 20, 343–369 (1991)
7. Chateauneuf, A.: Decomposable capacities, distorted probabilities and concave capacities. Mathematical Social Sciences 31, 19–37 (1996)
8. Coletti, G.: Coherent qualitative probability. Journal of Mathematical Psychology 34, 297–310 (1990)
9. Coletti, G., Mastroleo, M.: Conditional belief functions: a comparison among different definitions. In: Proc. of WUPES 2006, Mikulov (Czech Rep.), pp. 24–35 (2006)
10. Coletti, G., Scozzafava, R.: From conditional events to conditional measures: a new axiomatic approach. Annals of Mathematics and Artificial Intelligence 32, 373–392 (2001)
11. Coletti, G., Scozzafava, R.: Probabilistic logic in a coherent setting. In: Trends in Logic, vol. 15. Kluwer, Dordrecht (2002)
12. Coletti, G., Scozzafava, R.: Toward a general theory of conditional beliefs. International Journal of Intelligent Systems 21(3), 229–259 (2006)
13. Coletti, G., Vantaggi, B.: Comparative Conditional Possibilities. In: Godo, L. (ed.) ECSQARU 2005. LNCS (LNAI), vol. 3571, pp. 872–883. Springer, Heidelberg (2005)
14. Coletti, G., Vantaggi, B.: Representability of ordinal relations on a set of conditional events. Theory and Decision 60, 137–174 (2006)
15. Coletti, G., Vantaggi, B.: Possibility theory: conditional independence. Fuzzy Sets and Systems 157, 1491–1513 (2006)
16. Coletti, G., Vantaggi, B.: Comparative models ruled by possibility and necessity: a conditional world. International Journal of Approximate Reasoning 45, 341–363 (2007)
17. Coletti, G., Vantaggi, B.: A view on conditional measures through local representability of binary relations. International Journal of Approximate Reasoning 47, 268–283 (2008)
18. De Baets, B., Tsiporkova, E., Mesiar, R.: Conditioning in possibility theory with strict order norms. Fuzzy Sets and Systems 106, 221–229 (1999)
19. de Cooman, G.: Possibility theory II: Conditional Possibility. International Journal of General Systems 25, 325–351 (1997)
20. de Finetti, B.: Sul significato soggettivo della probabilità. Fundamenta Matematicae 17, 293–329 (1931)
21. Dempster, A.P.: A generalizatin of Bayesian Inference. The Royal Stat. Soc. B 50, 205–247 (1968)
22. Denoeux, T., Smets, P.: Classification using Belief Functions: the Relationship between the Case-based and Model-based Approaches. IEEE Transactions on Systems, Man and Cybernetics B (2006) (to appear)
23. Dubois, D.: Belief structure, possibility theory and decomposable confidence measures on finite sets. Comput. Artificial Intelligence 5, 403–416 (1986)
24. Dubois, D., Fargier, H., Vantaggi, B.: An axiomatization of conditional possibilistic preference functionals. In: Mellouli, K. (ed.) ECSQARU 2007. LNCS (LNAI), vol. 4724, pp. 803–815. Springer, Heidelberg (2007)
25. Dubois, D., Godo, L., Prade, H., Zapico, A.: On the Possibilistic Decision Model: From Decision under Uncertainty to Case-based Decision. Int. J. Uncertainty, Fuzziness and Knowledge-Based Systems 7, 631–670 (1999)

26. Dubois, D., Prade, H.: Possibility Theory. Plenum Press, New York (1988)
27. Dubois, D., Prade, H.: The logical view of conditioning and its application to possibility and evidence theories. International Journal of Approximate Reasoning 4, 23–46 (1990)
28. Dubois, D., Prade, H.: Bayesian conditioning in possibility theory. Fuzzy Sets and Systems 92, 223–240 (1997)
29. Dubois, D., Prade, H., Sabbadin, R.: Decision-theoretic foundations of qualitative possibility theory. European Journal of Operational Research 128, 459–478 (2001)
30. Ferracuti, L., Vantaggi, B.: Independence and conditional possibilities for strictly monotone triangular norms. International Journal of Intelligent Systems 21(3), 299–323 (2006)
31. Gilboa, I., Schmeidler, D.: Maxmin expected utility with non-unique prior. Journal of Mathematical Economics 18, 141–153 (1989)
32. Hisdal, E.: Conditional possibilities independence and noninteraction. Fuzzy Sets and Systems 1, 283–297 (1978)
33. Jaffray, J.Y.: Bayesian updatingand Belief functions. IEEE Trans. on Systems, man and Cybernetic 5, 1144–1152 (1992)
34. Koopman, B.O.: The axioms and algebra of intuitive probability. Ann. Math. 41, 269–292 (1940)
35. Kraft, C.H., Pratt, J.W., Seidenberg, A.: Intuitive probability on finite sets. Annals of Mathematical Statistics 30, 408–419 (1959)
36. Machina, M.J., Schmeidler, D.: Bayes without Bernoulli: simple conditions for probabilistically sophisticated choice. Journal of Economic Theory 67(1), 106–128 (1995)
37. Regoli, G.: Rational comparisons and numerical representation. In: Rios, S. (ed.) Decision Theory and Analysis: Trends and Challenges, pp. 113–126. Kluwer, Boston (1994)
38. Savage, L.J.: The Foundations of Statistics. Wiley, New York (1954)
39. Schmeidler, D.: Integral representation without additivity. Proceedings of the American Mathematical Society 97, 255–261 (1986)
40. Schmeidler, D.: Subjective probability and expected utility without additivity. Econometrica 57, 571–587 (1989)
41. Smets, P.: Belief function and transferable belief model. Documentation of SIPTA (2000), `http://www.sipta.org/documentation/belief/belief.ps`
42. Whalen, T.: Decision Making under Uncertainty with Various Assumptions about Available Information. IEEE Trans. on Systems, Man and Cybernetics 14, 888–900 (1984)
43. Walley, P.: Belief function representations of statistical evidence. Ann. of Stat. 4, 1439–1465 (1987)
44. Wakker, P.: Continuous Subjective Expected Utility With Non-Additive Probabilities. Journal of Mathematical Economics 18, 1–27 (1989)
45. Wong, S.K.M., Yao, Y.Y., Bollmann, P., Burger, H.C.: Axiomatization of qualitative belief structure. IEEE Trans. Systems Man Cybernetics 21, 726–734 (1991)
46. Yaari, M.E.: The dual theory of choice under risk. Econometrica 55, 95–115 (1987)
47. Yager, R.R.: Possibilistic Decision Making. IEEE Trans. on Systems, Man and Cybernetics 9, 388–392 (1979)
48. Zadeh, L.A.: Fuzzy sets as a basis for a theory of possibility. Fuzzy Sets and Systems 1, 3–28 (1978)

Delay and Interval Effects with Subadditive Discounting Functions

Salvador Cruz Rambaud and María José Muñoz Torrecillas

Abstract. Delay effect appears as an anomaly of the traditional discounted utility model according to which a decrease of the discount rate is performed as waiting time increases. But, in this description, it is not clear if the benchmark (that is to say, the reference instant in the assessment process) or the discounted amount availability is fixed or variable. In this way, other authors use the term common difference effect (and immediacy effect, when the first outcome is available immediately) and this expression at least does implies a variable discounted amount availability. Read introduces another different effect, the interval effect: longer intervals lead to smaller values of the discount rate r. Taking into account the parameter δ (geometric mean of the discount factor), the interval effect implies larger values of δ. In this paper we try to clarify the concepts of delay and interval effect and we deduce some relationships between these concepts and certain subadditive discounting functions.

Keywords: Delay effect, interval effect, subadditivity.

1 Introduction

Several authors have included, in their studies of the discounted utility model, some failures when fitting the empirical data to this model that they have labelled as "anomalies". Among them we can find the *delay effect* consisting of the decrease of the discount rate as waiting time increases, that is, the discount rates tend to be higher in short intervals than in longer ones. Nevertheless, in this description, it is not clear what "waiting time" means, that is to say, it is not clear if the reference instant in the assessment process or the discounted amount availability is fixed or variable. As a consequence, in many studies, the delay is confounded with the interval or it is not properly distinguished and therefore the delay effect is not properly defined.

Salvador Cruz Rambaud · María José Muñoz Torrecillas
Departamento de Dirección y Gestión de Empresas
La Cañada de San Urbano, s/n (04120) Almería, Spain
University of Almería, Spain
e-mail: scruz@ual.es, mjmtorre@ual.es

S. Greco et al. (Eds.): Preferences and Decisions, STUDFUZZ 257, pp. 85–110.
springerlink.com

In financial studies the distinction between *delay* and *interval* in a discounting function was established, for the first time, by Cruz and Ventre (1998). In this paper, $F(t,a)$ is said to be the equivalent amount, at instant *t*, of a monetary unit at instant $t+a$. On the other hand, in psychological studies the distinction between delay and interval in a discounting function was introduced by Read (2001, 2003). In his studies, $F(d,t)$ gives the value of an outcome at the end of the interval, $d+t$, as a fraction at the beginning of the interval, *d*. So, *d* denotes the *delay* to the earlier outcome and *t* denotes the *interval* separating the outcomes. A comparison between the elements of the two researches can be shown in Table 1:

Table 1 Equivalence of terminologies from Cruz and Ventre's, and Read's papers

ELEMENTS OF A DISCOUNTING FUNCTION	
Cruz and Ventre (1998)	**Scholten and Read (2006)**
Amount	Reward/outcome
Initial instant	Earlier delay/beginning of the interval
Final instant	Last delay/end of the interval
Interval	Interval

An alternative consideration to *delay effect* is subadditive discounting whereby the discount in a long time interval is bigger when the delay is subdivided. Subadditive discounting means that the discount is higher when the interval is divided into subintervals. Subadditive discounting implies smaller values of the discounting function for more subdivided intervals. For example, the discounting function for one year will be greater than the product of the corresponding discounting function values for each month.

In this introductory section, we are going to analyze the literature on the delay effect and the interval effect, in order to define both effects and, in the case of experimental works, to check if the empirical works use the terms *delay* and *interval* in the same way. In this analysis we are going to answer the same following questions, in every revised paper:

1. How the delay effect is defined?
2. Is it named delay effect?
3. Is it an experimental work? If yes, how is presented the choice?
4. Is the interval effect also defined? If yes, how is defined the interval effect?

Prelec and Loewenstein (1991) define an anomaly of DU (Discounted Utility) model in intertemporal choice named *common difference effect*. "DU implies that a person's preference between two single-outcome temporal prospects should depend on the absolute time interval between delivery of the objects". However, these authors claim both intuition and experimental evidence (Thaler 1981, Benzion *et al.*, 1989) indicating that the impact of a constant time difference between two outcomes becomes less significant as both outcomes are made more remote (that is more delayed). They formalize this effect in the following way:

$$(x,d_1)\sim(y,d_2) \text{ implies } (x,d_1+\varepsilon)\prec(y,d_2+\varepsilon) \text{ for } y>x,$$

$$\varepsilon>0,$$

being x and y two equivalent outcomes available at instants d_1 and d_2, respectively. Thus, for example, a person will be indifferent between having 20 euros today and having 25 euros in one month, but will prefer 25 euros in 11 months to 20 euros in 10 months.

They propose the property of *decreasing absolute sensitivity for time*: "increasing the absolute magnitude of all values of an attribute by a common additive constant decreases the weight of the attribute". In the case of time: adding a constant to both delays diminishes the importance of time, thus shifting preference in favor of the prospect with the larger money outcome. As we can observe this property is inconsistent with stationarity. A special case of stationarity violation occurs when one of the outcomes is available immediately and it is accomplished that:

$$(x,d_1)\sim(y,d_2) \text{ implies } (x,d_1+\varepsilon)\prec(y,d_2+\varepsilon), \text{ for } d_1=0 \text{ and } y>x,\ \varepsilon>0.$$

This is called *immediacy effect* and means that decision-makers give special importance to the immediate results. Prelec and Lowenstein state that the immediacy effect can be formally included inside the common difference effect. But they also point out that many researchers, however, think that these phenomena are qualitatively different and justify a separate treatment.

The interval effect is not defined in Prelec and Lowenstein's work.

Thaler (1981) tests the following hypothesis, although he doesn't name it as *delay effect*: "Specifically, the hypothesis to be tested is that the discount rate implicit in choices will vary inversely with the length of time to be waited."

The experimental work of Thaler consisted of a set of questionnaires (four different forms were used: three for gains and one for losses) to be answered by a group of students at the University of Oregon. For the case of the gains[1], they were told that they had won some money in a lottery and they could take the money *now* or wait until late. They were asked how much they would require to make waiting just as attractive as getting the money now. In all cases subjects were instructed to assume that there was no risk of not getting the reward if they waited. The waiting time varied from 1 month to 10 years (specifically the time delays used were 1, 3 and 6 months, and 1, 3, 5 and 10 years). And the magnitudes of the outcomes (hypothetical) were 15, 75, 250, 1,000, 1,200 and 3,000 dollars.

As a result, he found that the implicit discount rates dropped sharply as the length of time increased. He gives an intuitive explanation to this result: "the

[1] As we are not going to analyze here the sign effect, we will only focus in the case of gains.

difference between today and tomorrow seem greater than the difference between a year from now and a year plus one day". Responses imply that the subjects have a discount function which is nonexponential.

The interval effect is not defined in Thaler's work.

Benzion, Rapoport and Yagil (1989) defined the delay effect in the following way, although they don't name it as *delay effect*: "The mean discount rates decrease monotonically in t, that is to say decline as the time necessary to wait increases".

In their experimental work, the subjects of the study were 204 undergraduate and graduate students of economics and finance of the Universtity of Haifa and the Technion-Israel Institute of Technology who participate on a voluntary basis. These students had to make intertemporal choices between hypothetical rewards of 40, 200, 1,000 and 5,000 dollars. The decision could been formalized in the following way: $(y,0)$ vs. (x, t), being t 0.5, 1, 2 and 4 years.

The results showed that the mean discount rates decreased monotonically in t (t is defined as the length of time to be waited). For example, the discount rates for deferring a 200-dollar amount were 0.428; 0.255; 0.230 and 0.195 for delays of 6 months, 1, 2, and 4 years, respectively. After their experiment, they concluded that the discount rates inferred from the riskless choices supported the previous findings reported by Thaler (1981): the discount rates declined as the time necessary to wait increased.

The interval effect is not defined in Benzion and Rapoport's work.

Chapman (1996) defines the delay effect stating that decision makers have very long discount rates over short delays but much smaller discount rates over longer delays. She refers to it as *delay effect* and she also states that delay effect implies that people will reverse their preferences over time. Chapman illustrates this effect with the following example: "suppose it is 1996 and you have a choice between \$200 in 2004 (8 years from now) and \$100 in 2002 (6 years from now) and that you prefer the first option. Six years later in 2002 you might prefer \$100 right away (in 2002) to \$200 2 years from now (in 2004). Note that both questions ask whether one wishes to delay payment for 2 years to double their money, but the second decision takes place 6 years after the first decision".

In her experimental work, the participants were 40 undergraduates at the University of Illinois who participated for class credit and had to answer a questionnaire that consisted of two parts: discounting questions and exchange rate questions. The discounting section contained 32 questions involving a choice between an outcome *now* and an outcome later. Participants were asked to specify the magnitude of the delayed outcome that would make the two options equally attractive. The outcomes were in two domains: money and health. For money questions, participants were told to imagine they had won a lottery and had a choice between two monetary prizes, for example: 500 dollars now or x dollars one year from now.

The conclusions were that health and money decisions revealed larger discount rates for short delays.

The interval effect is not defined in Chapman's work.

In Soman *et al.* (2005) theoretical work delay and interval effect are defined (together with other effects or anomalies in intertemporal choice, such as magnitude or sign effect[2]). "The delay effect suggests that the discount rate is smaller for larger delays (Thaler, 1981). And the interval effect suggests that the discount rate depends on the time interval between the two outcomes used to impute the discount rate-the greater the temporal interval, the smaller the discount rate (Read, 2001)".

Scholten and Read (2006) define delay effect: "discount rates tend to be higher the closer the outcomes are to the present" and point out that this delay effect violates the stationary axiom of normative theory. They name it *delay effect* and, more specifically, the *effect of the delay to interval onset*. They also define the interval effect: "discount rates tend to be higher the closer the outcomes are to one another". This interval effect violates the transitivity axiom of normative theory. They called it also the *effect of interval length*.

Nevertheless in a previous work of Read delay and interval effect are defined in the following way: the geometric mean of the discount factor (δ) will be larger (and the geometric mean of the discount rate, r, will be smaller) the longer the delay (that is the delay effect). This delay effect is generally attributed to some form of hyperbolic discounting, although it could equally be due to the interval effect. Interval effect: the difference between the delays to two outcomes is the interval between them. Longer interval leads to smaller values of r or larger values of δ (Read, 2004). Following Read (2004) if the first outcome is received immediately (that is $d_1 = 0$), then the delay and the interval are confounded, and the two effects can be confused.

Scholten and Read made two experiments to study the delay effect, among others (they studied also the subadditivity and superadditivity). One of the hypothesis tested in their experiments was that δ would be lower for early intervals than for later intervals of the same length (the delay effect). The participants were 53 students from the London School of Economics who were paid 5 pounds. They had to choose between hypothetical amounts of money available at different times. The intervals used were 1, 3 and 17 weeks and the delays 1, 2, 3, 4, 15, 16, 17 and 18 weeks. The hypothesis of delay effect was confirmed by the experiment.

We consider that, in order to test the delay effect, it is necessary to have the two rewards or outcomes delayed, that is to say, that d must be a value different from zero. This way, we will be testing what Prelec and Loewenstein named *common difference effect* (and most authors named *delay effect*), different from the immediacy effect, that appears when the delay to the first outcome is zero which coincides with interval effect. This delay to the first outcome has also been named front-end-delay (FED). A critical design feature in the empirical literature on hyperbolic discounting is the use of a time delay to the early payment option in order to control for any confounding effects from fixed premia due to transactions costs (Harrison and Lau, 2005). That is, to introduce the front-end-delay (FED). If individuals are more impatient about immediate delays than about future delays of the same length, they will demand a premium in order to accept a delay of any length.

---

[2] For a review of these other anomalies see Cruz and Muñoz (2004).

If we add a FED, both payments will be delayed and so the premium applies to both choices becoming irrelevant to a choice between them (Coller and Williams, 1999; Harrison and Lau, 2005).

In the following table we resume the revision of the most relevant literature on delay and interval effect that we have previously explained. We have added a column that shows if in the referred work the FED is considered, that is to say, if the delay is different to zero and so the two rewards or payment are delayed. When the choice is presented as \$X dollars now or \$Y dollars in a future instant of time there is no FED. This is the case in which delay coincides with interval and we cannot distinguish among delay and interval effects. Prelec and Loewenstein consider this distinction and define the common difference effect, when there is a FED, and the immediacy effect, when the first outcome is available immediately (that is, when $d_1 = 0$). In Sholten and Read experimental work is considered a FED and so the delay and the interval effects are properly distinguished. In the other experimental works, it is not possible to make this distinction because the first outcome included in the choice is available "now".

Table 2 Delay effect in the literature

Authors	**Delay effect is defined?**	**Name it as delay effect?**	**Is a FED presented in the experiments?**	**Define also interval effect?**
Benzion *et al.* (1989)	Yes	No	No	No
Chapman (1996)	Yes	Yes	No	No
Prelec and Loewenstein (1991)	Yes	No. *Common difference effect*	No experimental work	No
Scholten and Read (2006)	Yes	Yes	Yes	Yes
Soman *et al.* (2005)	Yes	Yes	No experimental work	Yes
Thaler (1981)	Yes	No	No	No

Before going into Section 2, we are going to make a remark regarding delay and interval notation. In intertemporal choice a first classification of time could be between the time as a period and the time as a point. Thus, for instance, an amount can be discounted for a week, a year, etc. In this case, we are considering the time as a period. But we can consider the discounted amount at January 1, 2009 of a reward for a month. Of course, in this case, we are considering the time as a point (date) and as a period (month). In what follows,

- If time is considered as a period, we will refer to it as an *interval*. In this case, it will be denoted by t. Here time has the algebraic structure of a convex cone, because time can be summed up and enlarged by multiplying by a convenient positive real number.
- If time is considered as a point (date), we will refer to it as a *delay*. In this case, it will be denoted by d. Here the two classes of time give rise to the structure of an affine space where points are the dates and where vectors are intervals.

In the following sections, we will define the framework in which delay is equal to interval and time is considered only as an interval (Section 2) and the framework in which time is considered as a delay and as an interval (Section 3). In these frameworks, we will define some properties of the discounting function or the discounting factor, according to different cases that arise when we consider different possible combinations of delay and interval. Finally, in Section 4 we will present some conclusions.

2 Discounting by Intervals

Consider an intertemporal framework in which time is considered only as an interval. In this case, a *discounting function* is a decreasing real-valued function

$$\begin{aligned} F : \Re^{+} \cup \{0\} &\to]0,1] \\ t &\mapsto F(t) \end{aligned}$$

such that $F(0)=1$. $F(t)$ represents the subjective value at time 0 of a \$1 reward which would be available after t periods of time. Observe that, in this case, time as delay does not exist because expression $F(t)$ does not depend on the instant at which it is applied. In effect, if $F(d,t)$ represents the subjective value at time d of a \$1 reward which would be available t periods of time after instant d, we can write

$$F(d,t) = F(t),$$

for every $t \in \Re^+ \cup \{0\}$. A discounting function verifying such condition will be called a *stationary* or *regular discounting function*. A paradigm of stationary discounting function is exponential discounting which follows the equation

$$F(t) = \exp(-kt).$$

On the other hand, a key concept for studying the intertemporal choice is the *instantaneous discounting rate*, defined as

$$\delta(t) := -\frac{\mathrm{d} \ln F(t)}{\mathrm{d}t}.$$

This magnitude represents the infinitesimal relative decrease of the discounting function with respect to time. In effect,

$$\delta(t) = \lim_{h \to 0} \frac{F(t) - F(t+h)}{h \cdot F(t)}.$$

Another important characteristic of the exponential discounting is *consistency*, that is to say, the relation of preference between two rewards does not change if their respective intervals are enlarged the same value. It is well-known that, in this case, the instantaneous discounting rate is constant. Therefore, a first case of *inconsistency* can be described when $\delta(t)$ is decreasing. A similar concept is *declining impatience* which means that, for a given interval, the more delayed is the interval, the higher is the discounting function (or the discount factor). Nevertheless, we are going to define below different types of declining impatience, also called *increasing patience*. But before, remember that we have to distinguish two very often confused concepts used in the literature: delay and interval. Consider $1 available at time $d + t$. It is well known that a discounting function $F(d, t)$ calculates an equivalent amount at time *d*, being time 0 the benchmark. The parameter *t* will be named the *interval* and $d + t$ the *delay*. Observe that the interval coincides with the delay when $d = 0$. That is to say, interval and delay effects coincide in this case.

Nevertheless, as we will see later, the so-defined declining impatience is a strong condition to describe the inconsistency of a discounting function, whereby we are going to define other weaker or stronger types of declining impatience. The first type of declining impatience we are going to consider is the following one:

2.1 "Weak" Declining Impatience or Declining Impatience of Type I

To explain the concept of the "weak" declining impatience, we are going to consider that $d = 0$ (so we have not to distinguish between interval and delay) and that the interval is variable ($t_1 < t_2$):

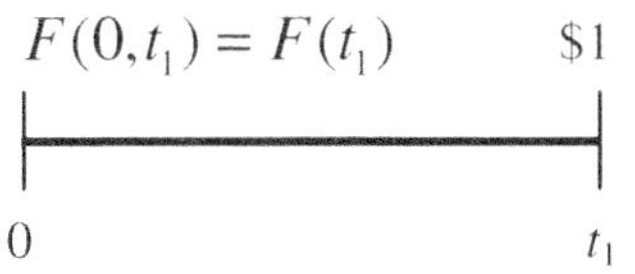

In general, it is verified that $F(t_1) > F(t_2)$, but in this kind of declining impatience it is verified that:

$$[F(t_1)]^{\frac{1}{t_1}} < [F(t_2)]^{\frac{1}{t_2}},$$

that is, the geometric mean of the discounting function in the interval $[0,t_1]$ is lower than the geometric mean of the discounting function in the interval $[0,t_2]$. In particular, when t_1 is the center of the interval $[0,t_2]$:

$$[F(t_1)]^{\frac{1}{t_1}} < [F(t_2)]^{\frac{1}{2t_1}},$$

which implies that[3]:

$$F(t_1) < \sqrt{F(t_2)}.$$

Observe that this definition is based on variable delays and so variable length intervals, where the lower point (0) is fixed. In this case, it makes sense talking about the average discounting function.

Theorem 1. "Weak" declining impatience or declining impatience of type I is verified if and only if the tangent at every point to the logarithmic factor crosses the straight line $x=0$ at a point with positive second component, that is to say, the instantaneous rate of discount is lower than the logarithmic density.

Proof. First, let us suppose that the condition is necessary. In effect, if "weak" declining impatience is verified, then:

$$[F(t_1)]^{\frac{1}{t_1}} < [F(t_2)]^{\frac{1}{t_2}}.$$

[3] This result has been presented in Read's work (2001) using the discount factor and was enunciated in the following way: $\sqrt{f_{0\to T}} > f_{0\to T/2}$.

Writing this discounting function by means of its instantaneous discount rate:

$$e^{-\frac{\int_0^{t_1}\delta(x)\mathrm{d}x}{t_1}} < e^{-\frac{\int_0^{t_2}\delta(x)\mathrm{d}x}{t_2}}.$$

Taking napierian logarithms in both members, it would remain:

$$\frac{-\int_0^{t_1}\delta(x)\mathrm{d}x}{t_1} < \frac{-\int_0^{t_2}\delta(x)\mathrm{d}x}{t_2}.$$

As $\int_0^{t}\delta(x)\mathrm{d}x = \varphi(t)$ is the logarithmic factor of $F(t)$, we will write:

$$\frac{-\varphi(t_1)}{t_1} < \frac{-\varphi(t_2)}{t_2},$$

from where:

$$\frac{\varphi(t_1)}{t_1} > \frac{\varphi(t_2)}{t_2}$$

or

$$\theta(t_1) < \theta(t_2),$$

being $\theta(t)$ the logarithmic density of the discounting function in the interval $[0,t]$. So, the function $\theta(x)$ is decreasing and its derivative, negative:

$$\theta'(x) = \frac{\delta(x)x - \varphi(x)}{x^2} < 0.$$

Therefore:

$$\delta(x)x - \varphi(x) < 0,$$

from which:

$$\delta(x) < \frac{\varphi(x)}{x},$$

or what is the same:

$$\delta(x) < \theta(x).$$

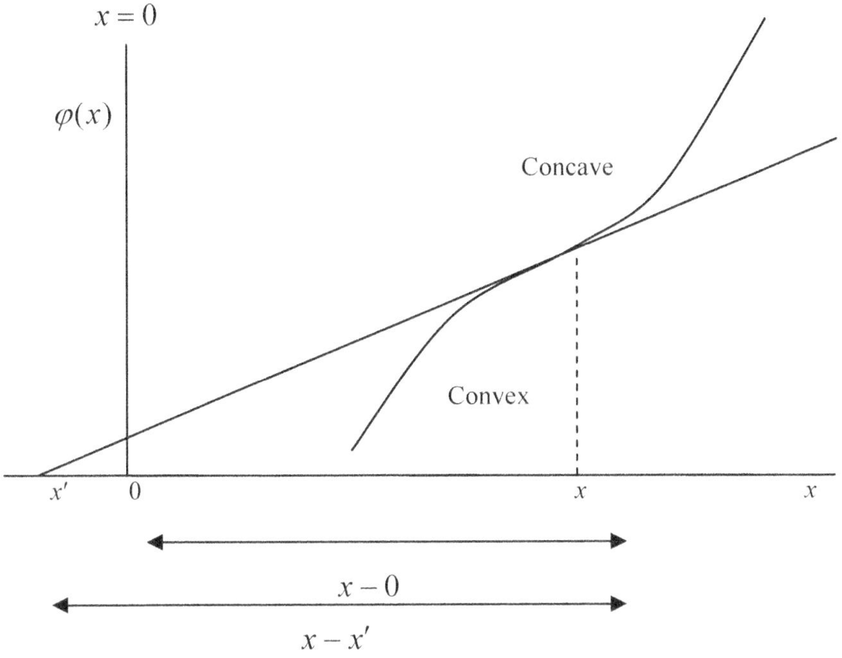

Fig. 1 Logarithmic factor in "weak" declining impatience

Observe that:

$$\delta(x) = \frac{d}{dx}\varphi(x) = \frac{\varphi(x)}{x - x'} < \frac{\varphi(x)}{x}.$$

Then the tangent line at x to the logarithmic factor crosses the straight line $x = 0$ at a point with positive second component.

Reciprocally, it can be shown that this condition is sufficient by making the reasoning in the opposite way. □

Remarks:

1. The logarithmic factor is always increasing, but it can be either concave or convex, so the instantaneous rate can be increasing or decreasing.
2. Hyperbolic discounting verifies "weak" declining impatience. In effect, starting from the hyperbolic discounting function defined by Ainslie (1992), Mazur (1987) and Rachlin (1989):

$$F(t) = \frac{1}{1 + k \cdot t}.$$

Let us consider, for example $k = 0.1$ and that intervals have lengths 3 and 5, respectively. By calculating the geometric average or mean discounting function per unit of time, we would have:

$$\left(\frac{1}{1+0.1\cdot 3}\right)^{\frac{1}{3}} = 0.916260327\,,$$

$$\left(\frac{1}{1+0.1\cdot 5}\right)^{\frac{1}{5}} = 0.922107911\,.$$

Then, we can see that "weak" declining impatience is verified.

3 Discounting by Delays and Intervals

Consider an intertemporal framework in which time is considered as a delay and as an interval. In this case, a *discounting function* is a real-valued function of two variables

$$F : \Re \times \Re^{+} \cup \{0\} \rightarrow]0,1]$$

$$(d,t) \mapsto F(d,t)$$

decreasing with respect to t and such that $F(d,0) = 1$. Such a discounting function will be called a *dynamic discounting function*. Now the *instantaneous discounting rate* is defined as

$$\delta(t) := -\frac{\mathrm{d}\ln F(d,t)}{\mathrm{d}t}\,.$$

This magnitude represents again the infinitesimal relative decrease of the discounting function with respect to time. In effect,

$$\delta(d,t) = \lim_{h\to 0}\frac{F(d,t) - F(d,t+h)}{h\cdot F(d,t)}\,.$$

It is also well-known that, in this case, a necessary and sufficient condition for consistency is that the instantaneous discounting rate only depends on t (not on d). In what follows, we will try to establish a parallelism between this case and the former one. Therefore, a first case of inconsistency can be described when $\delta(d,t)$ depends on d and thus we will assume that $\delta(d,t)$ is decreasing with respect to d, that is to say, keeping constant the value of t.

Remark. In the discounting function $F(t)$ the date of reference (benchmark) is 0, whereby $F(t)$ will the called a *spot discounting function*, while in the expression $F(d,t)$ the date of reference is any future date *d*, whereby $F(d,t)$ will the called a *forward discounting function*.

3.1 "Intermediate" Declining Impatience or Declining Impatience of Type II

Now we will represent a temporal line where the appraisal instant 0 is fixed, the delay is variable ($d_1 < d_2$) and the interval length is constant (*t*). Then, we can observe that, now, the interval does not match with the delay and that the second delay is greater than the first one. In this case, we will use the factor which is defined by transitivity as:

$$f(d,t) = \frac{F(d+t)}{F(d)}$$

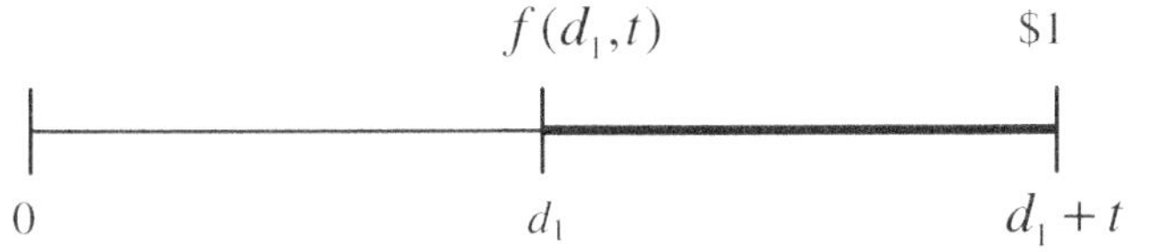

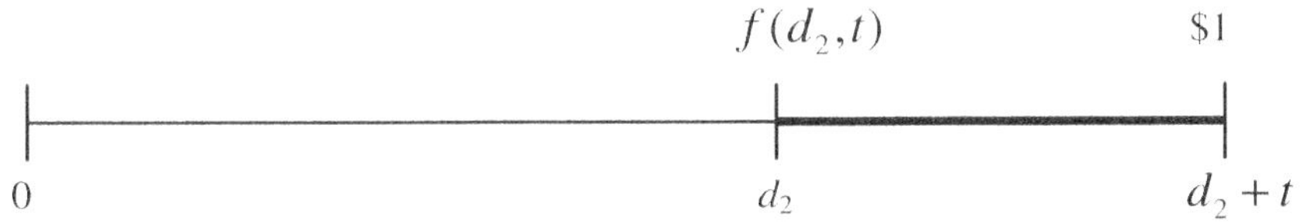

Now intermediate declining impatience or declining impatience of type II means:

$$f(d_1,t) < f(d_2,t).$$

Observe that this definition is based on different delays but equal length intervals whereby the lower endpoints differ in the same amount as the upper endpoints (delays). In this case, it is indifferent talking about the mean discount factors, because raising both members, respectively, to $\frac{1}{d_1+t-d_1}$ and to $\frac{1}{d_2+t-d_2}$ does not affect to the inequality, since both expressions are equivalent.

Theorem 2. "Intermediate" declining impatience or declining impatience of type II is verified if and only if the instantaneous discount rate is decreasing.

Proof. First, let us see that the condition is necessary. In effect, assume that:

$$f(d_1,t) < f(d_2,t),$$

that is to say, the discount factor is increasing with respect to *d*. Writing this factor according to the instantaneous rate:

$$e^{-\int_{d_1}^{d_1+t}\delta(x)\mathrm{d}x} < e^{-\int_{d_2}^{d_2+t}\delta(x)\mathrm{d}x},$$

from which, taking napierian logarithms in both members and changing the signs:

$$\int_{d_1}^{d_1+t}\delta(x)\mathrm{d}x > \int_{d_2}^{d_2+t}\delta(x)\mathrm{d}x.$$

We can break each one of the previous integrals into two, according to the subintervals in which we can divide each of the integration intervals of both members. Denoting $d_2 - d_1 := h$, the former scheme remains as follows:

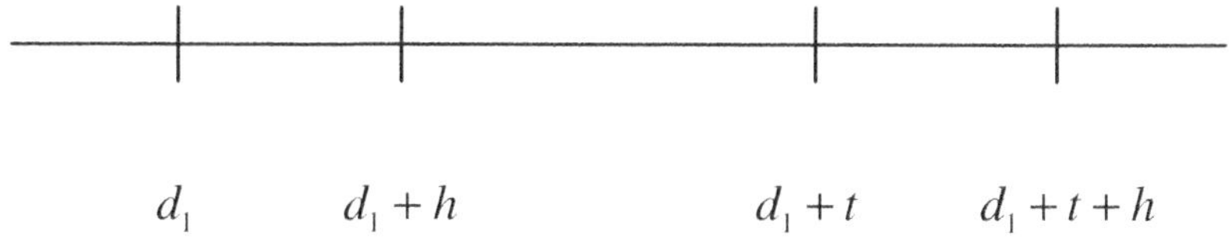

$$\int_{d_1}^{d_1+h}\delta(x)\mathrm{d}x + \int_{d_1+h}^{d_1+t}\delta(x)\mathrm{d}x > \int_{d_1+h}^{d_1+t}\delta(x)\mathrm{d}x + \int_{d_1+t}^{d_1+t+h}\delta(x)\mathrm{d}x$$

and simplifying and dividing by *h*, it would remain:

$$\frac{\int_{d_1}^{d_1+h}\delta(x)\mathrm{d}x}{h} > \frac{\int_{d_1+t}^{d_1+t+h}\delta(x)\mathrm{d}x}{h}.$$

Letting $h \to 0$:

$$\delta(d_1) > \delta(d_1+t),$$

Then $\delta(x)$ is declining. Let us see now that the condition is sufficient:

$$f(d_1,t) = e^{-\int_{d_1}^{d_1+t}\delta(x)\mathrm{d}x} < e^{-\int_{d_1}^{d_1+t}\delta(x+h)\mathrm{d}x} = e^{-\int_{d_1+h}^{d_1+t+h}\delta(x)\mathrm{d}x} = f(d_2,t).$$ □

Observe that hyperbolic discounting also verifies "intermediate" declining impatience. Taking the hyperbolic discounting function defined by Mazur (1984), that is the discount factor associated to the discounting function defined by Ainslie:

$$f(d,t) = \frac{1+kd}{1+k(d+t)}$$

and considering the intervals [3,5] and [7,9], we have:

$$\frac{1+0.1\cdot 3}{1+0.1\cdot 5} = 0.8\hat{6},$$

$$\frac{1+0.1\cdot 7}{1+0.1\cdot 9} = 0.894736842.$$

Theorem 3. "Intermediate" declining impatience implies "weak" declining impatience.

Proof. It is evident. □

Let us see now an example of discounting function verifying the "weak" declining impatience, but not the "intermediate". In effect, let us suppose that the logarithmic density has the following expression:

$$\varphi(x) = \frac{x^3+2x}{x^2+1},$$

where, as it can be observed, the appraisal instant is 0. It is verified that:

$$\varphi'(x) = \frac{x^4+x^2+2}{(x^2+1)^2} > 0,$$

which implies that $\varphi(x)$ is decreasing. Moreover, $\varphi(0)=0$.

On the other hand, $\delta(x) = \varphi'(x) = \frac{x^4+x^2+2}{(x^2+1)^2}$ is increasing in the interval $(0,\sqrt{3})$ and decreasing in $(\sqrt{3},\infty)$, as can be seen when making its derivative:

$$\delta'(x) = \frac{2x^3-6x}{(x^2+1)^3}.$$

Then:

$$F(t) = e^{-\frac{t^3+2t}{t^2+1}}$$

verifies the declining impatience of type I, but not the type II.

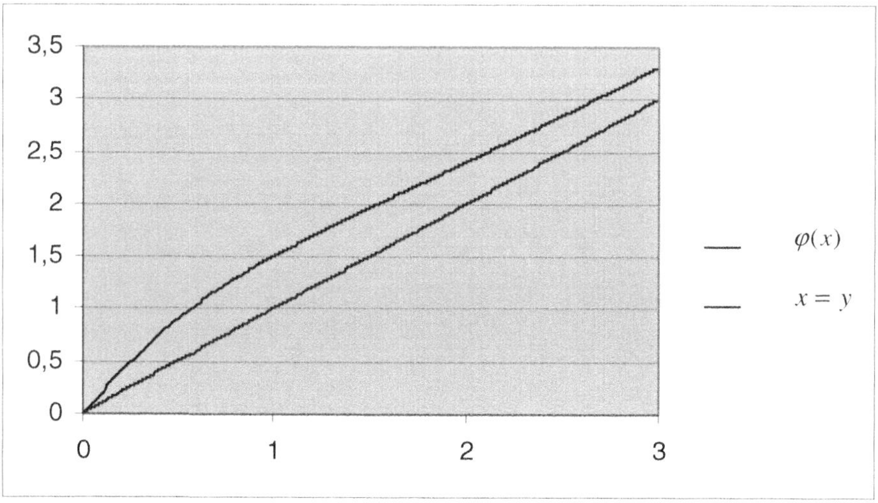

Fig. 2 Graphic representation of $\varphi(x) = \dfrac{x^3 + 2x}{x^2 + 1}$

Theorem 4. A discounting function $F(t)$ verifies the "intermediate" declining impatience if and only if

$$[f(d_1,t_1)]^{\frac{1}{t_1}} < [f(d_2,t_2)]^{\frac{1}{t_2}},$$

for all $d_1 < d_2$, and t_1 and t_2 such that $d_2 < d_1 + t_1$:

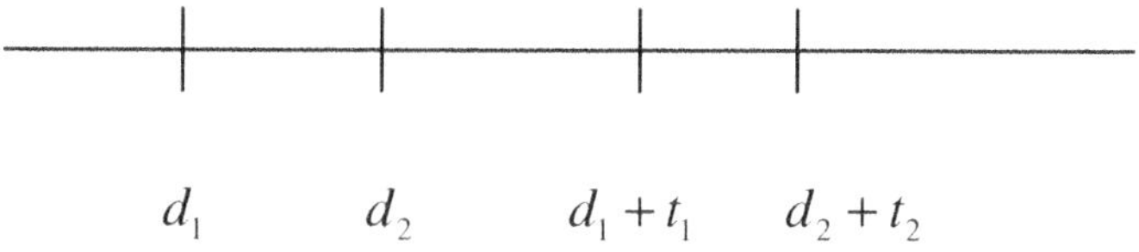

Proof. Firstly, starting from the previous inequality, $[f(d_1,t_1)]^{\frac{1}{t_1}} < [f(d_2,t_2)]^{\frac{1}{t_2}}$, it is obvious that the discounting function $F(t)$ verifies the "intermediate" declining impatience, taking $t_1 = t_2$. Secondly, if it is verified the "intermediate" declining impatience, then, by means of theorem 2, $\delta(t)$ is decreasing. Then, we have to show that:

$$\frac{\int_{d_1}^{d_1+t_1} \delta(x)\mathrm{d}x}{t_1} > \frac{\int_{d_2}^{d_2+t_2} \delta(x)\mathrm{d}x}{t_2}.$$

Taking d_1, d_2 and t_1, and denoting $t_2 = t_1 + x$, $x \in [d_1 - d_2, +\infty)$, let us study the behaviour of the auxiliary function:

$$g(x) = \frac{\int_{d_2}^{d_2+t_1+x} \delta(x)\mathrm{d}x}{t_1 + x} - \frac{\int_{d_1}^{d_1+t_1} \delta(x)\mathrm{d}x}{t_1}.$$

Making its derivative:

$$g'(x) = \frac{\delta(d_2 + t_1 + x)(t_1 + x) - \int_{d_2}^{d_2+t_1+x} \delta(x)\mathrm{d}x}{(t_1 + x)^2} =$$

(by applying the theorem of the mean value for integral calculus and simplifying)

$$= \frac{\delta(d_2 + t_1 + x) - \delta(\xi)}{t_1 + x} < 0,$$

since $\xi \in (d_2, d_2 + t_1 + x)$ and δ is decreasing. Therefore, $g(x)$ is decreasing. Let us see now the value of $g(d_1 - d_2)$:

$$g(d_2) = \frac{\int_{d_2}^{d_1+t_1} \delta(x)\mathrm{d}x}{d_1 + t_1 - d_2} - \frac{\int_{d_1}^{d_1+t_1} \delta(x)\mathrm{d}x}{t_1} = \frac{t_1 \int_{d_2}^{d_1+t_1} \delta(x)\mathrm{d}x - (d_1 + t_1 - d_2)\int_{d_1}^{d_1+t_1} \delta(x)\mathrm{d}x}{(d_1 + t_1 - d_2)t_1} =$$

$$= \frac{(d_1 + t_1 - d_2)\int_{d_2}^{d_1+t_1} \delta(x)\mathrm{d}x + (d_2 - d_1)\int_{d_2}^{d_1+t_1} \delta(x)\mathrm{d}x - (d_1 + t_1 - d_2)\int_{d_1}^{d_2} \delta(x)\mathrm{d}x - (d_1 + t_1 - d_2)\int_{d_2}^{d_1+t_1} \delta(x)\mathrm{d}x}{(d_1 + t_1 - d_2)t_1} =$$

(simplifying)

$$= \frac{(d_2 - d_1)\int_{d_2}^{d_1+t_1} \delta(x)\mathrm{d}x - (d_1 + t_1 - d_2)\int_{d_1}^{d_2} \delta(x)\mathrm{d}x}{(d_1 + t_1 - d_2)t_1} =$$

(applying the theorem of the mean value for integral calculus)

$$= \frac{(d_2 - d_1)(d_1 + t_1 - d_2)\delta(\xi) - (d_1 + t_1 - d_2)(d_2 - d_1)\delta(\eta)}{(d_1 + t_1 - d_2)t_1} =$$

(and simplifying, again)

$$= \frac{(d_2 - d_1)[\delta(\xi) - \delta(\eta)]}{t_1} < 0,$$

since $\eta < \xi$, $\eta \in (d_1, d_2)$ and $\xi \in (d_2, d_1 + t_1)$ and δ is decreasing. So $g(x)$ is always negative and so it is shown that the first addend is lower than the second one. □

In what follows, we will describe a procedure to generate logarithmic densities of discounting functions verifying the "weak" declining impatience. As the tangent at every point x to the logarithmic density y has to cross the $x = 0$ axis:

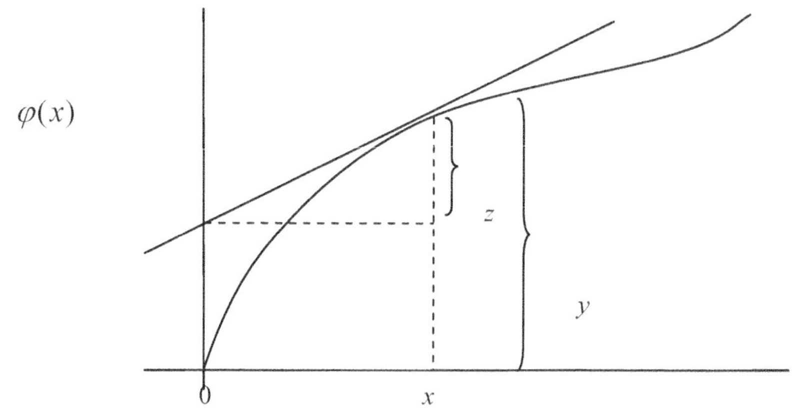

Fig. 3 Logarithmic density

$$z < y,$$

$$xy' < y,$$

$$xy' - y < 0$$

$$xy' - y = \phi(x), \text{ with } \phi(x) < 0.$$

Dividing both members of previous equation by x:

$$y' - \frac{1}{x}y = \frac{\phi(x)}{x},$$

where we will denote $-\frac{1}{x}$ by $f(x)$ and $\frac{\phi(x)}{x}$ by $g(x)$.

First of all, we will calculate the general solution of the homogeneous differential equation:

$$y' - \frac{1}{x}y = 0,$$

that is a differential equation of separable variables:

$$y' = \frac{1}{x}y,$$

$$\frac{y'}{y} = \frac{1}{x},$$

$$\ln y = \ln x + c,$$

from which we can find the value of y:

$$y = cx.$$

On the other hand, a particular solution to the complete differential equation is given by:

$$y_p(x) = c(x)e^{-\int f(x)dx},$$

where:

$$c(x) = \int g(x) e^{\int f(x)dx} dx =$$

(replacing $f(x)$ and $g(x)$ by their respective values)

$$= \int \frac{\phi(x)}{x} e^{-\ln x} dx = \int \frac{\phi(x)}{x} \frac{1}{x} dx = \int \frac{\phi(x)}{x^2} dx .$$

Making $\frac{1}{x} = z$ and $-\frac{1}{x^2} dx = dz$, one has:

$$c(z) = -\int \phi\left(\frac{1}{z}\right) dz = \psi(z) ,$$

Then $c(x) = \psi\left(\frac{1}{x}\right)$ and the general solution of the complete differential equation will be:

$$y = cx + \psi\left(\frac{1}{x}\right) x .$$

For example, if $\phi(x) = -xe^{-x}$, then:

$$c(x) = \int \frac{-xe^{-x}}{x} dx = e^{-x} .$$

So,

$$y_p(x) = e^{-x} x ,$$

which implies that:

$$y = cx + e^{-x} x = x\left(c + \frac{1}{e^x}\right) .$$

In effect, observe that the solution found in the previous differential equation can be expressed as:

$$y = cx + \frac{x}{e^x} .$$

And the graphic representation of $y = \frac{x}{e^x}$ is:

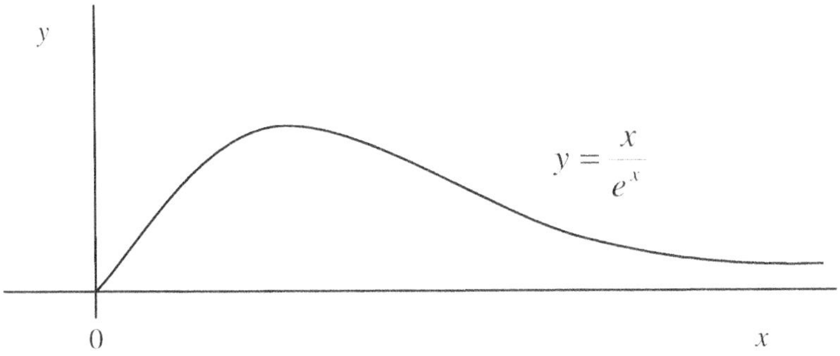

Fig. 4

therefore, if we add the linear function $y = cx$, it will be verified the differential inequality for some given value of c:

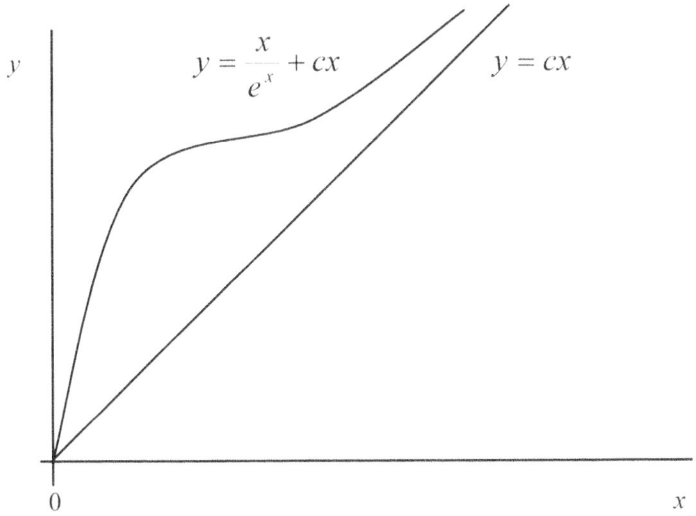

Fig. 5

Let us see for $c = 1$:

$$y = x + xe^{-x} = x(1 + e^{-x}).$$

It is verified that the image of 0 is $y(0)=0$ and y is increasing, because:

$$y' = (1+e^{-x}) + x(-e^{-x}) = e^{-x}(1-x)+1>0 .$$

In effect, the graphic representation of the function $y_1 = \dfrac{1-x}{e^x}$ is:

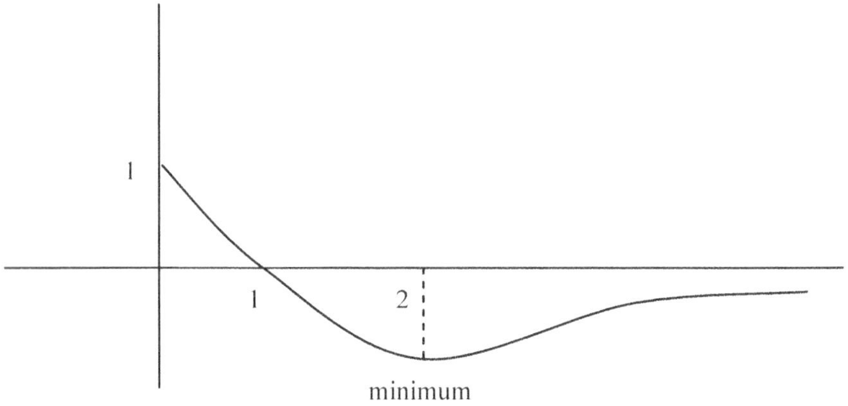

Fig. 6 Function y_1

since:

$$y_1' = y'' = \frac{-e^x-(1-x)e^x}{(e^x)^2} = \frac{x-2}{e^x},$$

so it has a minimum at $x=2$, whose second component is $\dfrac{1-2}{e^2}=\dfrac{-1}{e^2}$. As $\dfrac{1}{e^2}<1$, y' is positive, so the logarithmic density is increasing and, therefore, the instantaneous rate is positive. But, second derivative of y reveals that the rate can be either increasing or decreasing.

3.2 "Strong" Declining Impatience or Declining Impatience of Type III

Let us represent two intervals with the same length, but different appraisal instants. In the first case, the interval will be valued at moment 0 and the second, at moment d, both coinciding with the lower endpoint of each interval.

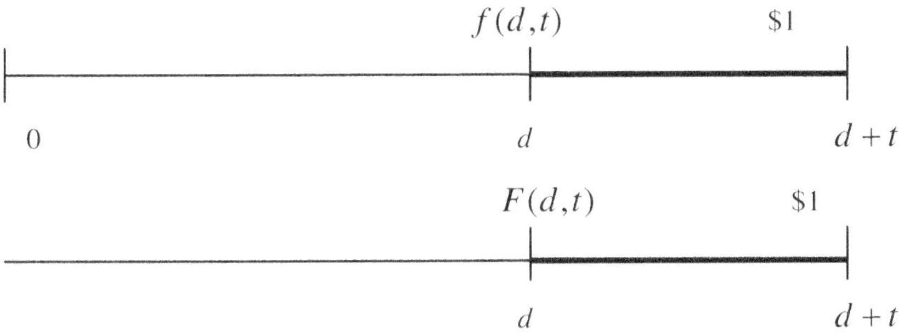

In this case, the "strong" declining impatience means that:

$$F(d,t) < f(d,t),$$

that it to say,

$$F(d,t) < \frac{F(d+t)}{F(d)},$$

from where we can deduce:

$$F(d) \cdot F(d,t) < F(d+t),$$

that is the concept of subadditivity. Observe that, in this case, the discount rate decreases if the interval $[d, d+t]$ is valued from 0. Taking mean discounting functions:

$$F(d)^{\frac{1}{d+t}} \cdot F(d,t)^{\frac{1}{d+t}} < F(d+t)^{\frac{1}{d+t}}$$

$$\left[F(d)^{\frac{1}{d}}\right]^{\frac{d}{d+t}} \cdot \left[F(d,t)^{\frac{1}{t}}\right]^{\frac{t}{d+t}} < F(d+t)^{\frac{1}{d+t}}.$$

Denoting the average discounting function in the interval $[t_1, t_2]$ as $\delta_{t_1 \to t_2}$, depending on the appraisal interval, we obtain:

$$\delta_{0 \to d+t} > \left(\delta_{0 \to d}\right)^{\frac{d}{d+t}} \cdot \left(\delta_{d \to d+t}\right)^{\frac{t}{d+t}},$$

that is to say, the average discounting function in the interval $[0, d+t]$ is higher than the product of the weighted geometric mean of the discounting functions in the intervals $[0, d]$ and $[d, d+t]$, using $\frac{d}{d+t}$ and $\frac{t}{d+t}$ as weights. Note that the sum of these weights is 1:

$$\frac{d}{d+t} + \frac{t}{d+t} = 1\,.$$

In the particular case in which d is the center of the interval $[0, d+t]$:

$$d = \frac{d+t}{2},$$

it is verified that:

$$t = \frac{d+t}{2}\,.$$

Then:

$$\delta_{0 \to d+t} > \left(\delta_{0 \to d}\right)^{\frac{1}{2}} \cdot \left(\delta_{d \to d+t}\right)^{\frac{1}{2}},$$

that is Read's (2003) definition of subadditivity.

3.3 *"Very Strong" Declining Impatience or Declining Impatience of Type IV*

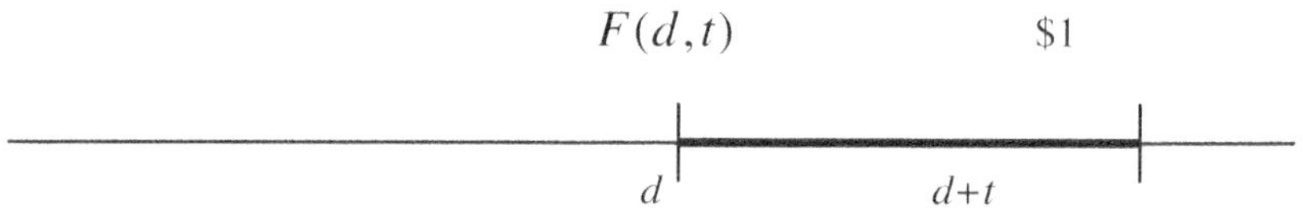

$$F(t) < F(d,t)\,.$$

This is what has been defined as a *contractive discounting function*. Also, in the farthest interval, instantaneous discount rate is lower. On the other hand, a discounting function is said to be *expansive* if $F(t) \geq F(d,t)$.

Theorem 5. An expansive discounting function verifying the declining impatience of type II , also verifies the declining impatience of type III.

Proof. It is well-known that, for an interval *t*, it is verified that:

$$f(0,t) = F(t).$$

By declining impatience of type II and the expansiveness of $F(t)$,

$$f(d,t) > f(0,t) = F(t) \geq F(d,t).$$

Therefore, it is verified the declining impatience of type III. □

Actually, we can not strictly speak about neither a longer interval nor farther postponed delay, so it could be analyzed under stationarity assumption, finding the following conclusion:

Corollary. In the case of a stationary discounting function, the concept of declining impatience of type II also implies declining impatience of type III.

Proof. In effect, if the declining impatience of type II is verified, taking into account that $F(t) = F(d,t)$, following the proof of Theorem 5, the declining impatience of type III would be verified. □

4 Conclusions

Several authors have included in their papers and in their experiments the concept of *delay effect*. Nevertheless in some studies it is not clearly established the difference between *delay* and *interval* and, as a consequence, the *delay effect* has been identified with the *interval effect*. This occurs, specially, when the delay is equal to the interval, and so *delay effect* and *interval effect* coincide. Taking into account this difference, we have introduced two frameworks. In the first one (Section 2: Discounting by intervals) we have only considered time as an interval. In this case, a discounting function is a decreasing-real valued function. In the second case (Section 3: Discounting by delays and intervals) we have considered time as a delay and as an interval and so the discounting function is a real-valued function of two variables. In these frameworks, we have established four types of what we have called "declining impatience" (DI): *weak*, *intermediate*, *strong* and *very strong* (the first type corresponds to the framework in which delay and interval coincide, as $d = 0$). And we have shown different properties of the discounting function/factor, depending on different possibilities of relationship between interval and delay (theorems 1 to 5). We could point out, from these relationships, the following results relating declining impatience to hyperbolic and subbaditive discounting functions. The weak and the intermediate DI are verified by the hyperbolic discounting. The strong DI leads to the concept of subadditivity and the very strong DI leads to a contractive discounting function.

References

Ainslie, G.: Picoeconomics: The strategic interaction of successive motivational states within the person. Cambridge University Press, Cambridge (1992)

Benzion, U., Rapoport, A., Yagil, J.: Discount rates inferred from decisions: An experimental study. Management Science 35(3), 270–284 (1989)

Chapman, G.: Temporal discounting and utility for health and money. Journal of Experimental Psychology: Learning, Memory and Cognition 22(3), 771–791 (1996)

Coller, M., Williams, M.: Eliciting individual discount rates. Experimental Economics 2, 107–127 (1999)

Cruz, S., Muñoz, M.J.: An analysis of the anomalies in traditional discount-ing models. International Journal of Psychology and Psychological Therapy 4(1), 105–128 (2004)

Cruz, S., Ventre, A.: Decomposable financial laws and profitability. International Journal of Uncertainty, Fuzziness and Knowledge-based Systems 6(4), 329–344 (1998)

Harrison, G., Lau, M.: Is the Evidence for Hyperbolic Discounting in Humans Just An Experimental Artefact? Behavioral and Brain Sciences 28(5), 657 (2005)

Mazur, J.: Tests of an equivalence rule for fixed and variable reinforcer delays. Journal of Experimental Psychology: Animal Behavior Processes 10, 426–436 (1984)

Mazur, J.: An adjustement procedure for studying delayed reinforcement. In: Commons, M., Mazur, J., Nevins, J., Rachlin, H. (eds.) Quantitative analyses of behavior: The effect of delay and of intervening events on reinforcement value, ch. 2, Erlbaum, Hillsdale (1987)

Prelec, D., Loewenstein, G.: Decision making over time and under uncertainty: A common approach. Management Science 37, 770–786 (1991)

Rachlin, H.: Judgement, decision and choice: A cognitive/behavioral synthesis. Freeman, New York (1989)

Read, D.: Is time-discounting hyperbolic or subadditive? Journal of Risk and Uncertainty 23(1), 5–32 (2001)

Read, D.: Subadditive intertemporal choice. In: Loewenstein, G., Read, D., Baumeister, R. (eds.) Time and decision: economic and psychological perspectives on intertemporal choice. Russell Sage Foundation, New York (2003)

Read, D.: Intertemporal choice. In: Koehler, D., Harvey, N. (eds.) Blackwell Handbook of Judgment and Decision Making. Blackwell, Oxford (2004)

Sholten, M., Read, D.: Discounting by intervals: A generalized model of in-tertemporal choice. Management Science 52(9), 1424–1436 (2006)

Soman, D., Ainslie, G., Frederick, S., Li, X., Lynch, J., Moreau, P., Mitchell, A., Read, D., Sawyer, A., Trope, Y., Wertenbroch, K., Zauberman, G.: The psychology of intertemporal discounting: Why are distant events valued differently from proximal ones? Marketing Letters 16(3/4), 347–360 (2005)

Thaler, R.: Some empirical evidence on dynamic inconsistency. Economic Letters 8, 201–207 (1981)

Pairwise Comparison Matrices: Some Issue on Consistency and a New Consistency Index

Bice Cavallo, Livia D'Apuzzo, and Gabriella Marcarelli

Abstract. In multicriteria decision making, the pairwise comparisons are an useful starting point for determining a ranking on a set $X = \{x_1, x_2, ..., x_n\}$ of alternatives or criteria; the pairwise comparison between x_i and x_j is quantified in a number a_{ij} expressing how much x_i is preferred to x_j and the quantitative preference relation is represented by means of the matrix $A = (a_{ij})$. In literature the number a_{ij} can assume different meanings (for instance a ratio or a difference) and so several kind of pairwise comparison matrices are proposed. A condition of consistency for the matrix $A = (a_{ij})$ is also considered; this condition, if satisfied, allows to determine a weighted ranking that perfectly represents the expressed preferences. The shape of the consistency condition depends on the meaning of the number a_{ij}. In order to unify the different approaches and remove some drawbacks, related for example to the fuzzy additive consistency, in a previous paper we have considered pairwise comparison matrices over an abelian linearly ordered group; in this context we have provided, for a pairwise comparison matrix, a general definition of consistency and a measure of closeness to consistency. With reference to the new general unifying context, in this paper we provide some issue on a consistent matrix and a new measure of consistency that is easier to compute; moreover we provide an algorithm to check the consistency of a pairwise comparison matrix and an algorithm to build consistent matrices.

Bice Cavallo
University of Naples Federico II, Via Toledo 402, Naples, Italy
e-mail: bice.cavallo@unina.it

Livia D'Apuzzo
University of Naples Federico II, Via Toledo 402, Naples, Italy
e-mail: liviadap@unina.it

Gabriella Marcarelli
University of Sannio, Via delle Puglie 82, Benevento, Italy
e-mail: marcarel@unisannio.it

S. Greco et al. (Eds.): Preferences and Decisions, STUDFUZZ 257, pp. 111–122.
springerlink.com

1 Introduction

Let $X = \{x_1, x_2, ..., x_n\}$ be a set of alternatives or criteria. An useful tool to determine a weighted ranking on X is a *pairwise comparison matrix* (PCM)

$$A = \begin{pmatrix} a_{11} & a_{12} & \dots & a_{1n} \\ a_{21} & a_{22} & \dots & a_{2n} \\ \dots & \dots & \dots & \dots \\ a_{n1} & a_{n2} & \dots & a_{nn} \end{pmatrix} \tag{1}$$

which entry a_{ij} expresses how much the alternative x_i is preferred to alternative x_j. A condition of *reciprocity* is assumed for the matrix $A = (a_{ij})$ in such way that the preference of x_i over x_j expressed by a_{ij} can be exactly read by means of the element a_{ji}. Under a suitable condition of *consistency* for $A = (a_{ij})$, stronger than the reciprocity, X is totally ordered and there exists a *consistent vector* $\underline{w}$, that perfectly represents the preferences over X; then $\underline{w}$ provides the proper weights for the elements of X.

The shape of the reciprocity and consistency conditions depends on the different meaning given to the number a_{ij}, as the following well known cases show.

Multiplicative case [12, 13]: $a_{ij} \in]0, +\infty[$ is a preference ratio and the conditions of *multiplicative reciprocity* and *consistency* are given respectively by

$$a_{ji} = \frac{1}{a_{ij}} \quad \forall\, i, j = 1, \dots, n,$$
$$a_{ik} = a_{ij} a_{jk} \forall\, i, j, k = 1, \dots, n.$$

A consistent vector is a positive vector $\underline{w} = (w_1, w_2, ..., w_n)$ verifying the condition $\frac{w_i}{w_j} = a_{ij}$.

Additive case [1, 11]: $a_{ij} \in]-\infty, +\infty[$ is a preference difference and *additive reciprocity* and *consistency* are expressed as follows

$$a_{ji} = -a_{ij} \quad \forall\, i, j = 1, \dots, n,$$
$$a_{ik} = a_{ij} + a_{jk} \forall\, i, j, k = 1, \dots, n.$$

A consistent vector is a vector $\underline{w} = (w_1, w_2, ..., w_n)$ verifying the condition $w_i - w_j = a_{ij}$.

Fuzzy case [1, 10]: $a_{ij} \in [0, 1]$ measures the distance from the indifference that is expressed by 0.5; in this case the following conditions of *fuzzy reciprocity* and *fuzzy additive consistency* are considered

$$a_{ji} = 1 - a_{ij} \quad \forall\, i, j = 1, \dots, n,$$
$$a_{ik} = a_{ij} + a_{jk} - 0.5 \forall\, i, j, k = 1, \dots, n.$$

A consistent vector is a vector $\underline{w} = (w_1, w_2, ..., w_n)$ verifying the condition $w_i - w_j = a_{ij} - 0.5$.

The multiplicative *PCMs* play a basic role in the Analytic Hierarchy Process, a procedure developed by T.L. Saaty at the end of the 70s [12, 13, 14, 15]; properties of multiplicative PCMs are analyzed in [2, 3, 4, 5, 8] in order to determine the actual ranking of the alternatives and find ordinal and cardinal evaluation vectors. The pairwise comparisons are an useful starting point for quantifying judgments expressed in verbal terms; the multiplicative scale suggested by Saaty translates the verbal comparisons into preference ratios a_{ij} belonging to the set $S^* = \{1, 2, 3, 4, 5, 6, 7, 8, 9, \frac{1}{2}, \frac{1}{3}, \frac{1}{4}, \frac{1}{5}, \frac{1}{6}, \frac{1}{7}, \frac{1}{8}, \frac{1}{9}\}$. But the assumption of the Saaty scale restricts the Decision Maker's possibility to be consistent: indeed if the Decision Maker (DM) expresses the following preference ratios $a_{ij} = 5$ and $a_{jk} = 3$ then he will not be consistent because $a_{ij}a_{jk} = 15 > 9$. Analogously for the fuzzy case, the assumption that $a_{ij} \in [0, 1]$, restricts the possibility to respect the fuzzy additive consistency: indeed, if the DM claims $a_{ij} = 0.9$ and $a_{jk} = 0.8$, then he will not be consistent because $a_{ij} + a_{jk} - 0.5 = 1.7 - 0.5 > 1$.

In order to unify the several approaches to PCMs and remove the mentioned drawbacks, in [6], we introduce PCMs over an abelian linearly ordered group (*alo-group*) $\mathcal{G} = (G, \odot, \leq)$. The reciprocity and consistency conditions are expressed in terms of the group operation $\odot$ and this allow us to remove the drawbacks linked to the usual definitions of consistency; a notion of distance $d_{\mathcal{G}}$, linked to $\mathcal{G}$, is also introduced. The assumption of *divisibility* for $\mathcal{G}$ allows us to introduce the mean $m_{\odot}(a_1, ..., a_n)$ of n elements and associate a *mean vector* $\underline{w}_{m_{\odot}}$ to a PCM. Then, we define the consistency index $I_{\mathcal{G}}(A)$ as mean of the distances $d_{\mathcal{G}}(a_{ik}, a_{ij} \odot a_{jk})$, with $i < j < k$.

In this paper, we add the fourfold original contribution:

1. we analyze properties of consistent PCMs;
2. we provide an algorithm to check whenever or not a matrix is consistent;
3. we provide a new consistency index for a PCM, obtained from an optimization of the index $I_{\mathcal{G}}(A)$ provided in [6];
4. we provide an algorithm to build a consistent matrix by means of $n-1$ comparisons.

2 Preliminaries

Let us recall some notions and results related to an alo-group (see [6] for details). These results will be useful in Section 3 to define a PCM over an alo-group and to define in this context a new consistency index.

Let G be a non empty set provided with a total weak order $\leq$ and a binary operation $\odot : G \times G \to G$. $\mathcal{G} = (G, \odot, \leq)$ is called *alo-group*, if and only if $(G, \odot)$ is an abelian group and the the following implication holds:

$$a < b \Rightarrow a \odot c < b \odot c,$$

where $<$ is the strict simple order associated to $\leq$.

If $\mathscr{G} = (G, \odot, \leq)$ is an alo-group, then we assume that: e denotes the *identity* of $\mathscr{G}$, $x^{(-1)}$ the *symmetric* of $x \in G$ with respect to $\odot$, $\div$ the *inverse operation* of $\odot$ defined by "$a \div b = a \odot b^{(-1)}$".

For a positive integer n, the *(n)-power* $x^{(n)}$ of $x \in G$ is defined as follows

$$x^{(1)} = x$$
$$x^{(n)} = \bigodot_{i=1}^{n} x_i, \; x_i = x \;\forall i = 1, \ldots, n, \; n \geq 2$$

If $b^{(n)} = a$, then we say that b is the *(n)-root* of a and write $b = a^{(1/n)}$. $\mathscr{G}$ is *divisible* if and only if for each positive integer n and each $a \in G$ there exists the (n)-root of a.

Proposition 2.1. *[6] A non trivial alo-group $\mathscr{G} = (G, \odot, \leq)$ has neither the greatest element nor the least element.*

The interval $[0, 1]$ and the Saaty set S*, embodied with the usual order $\leq$ on R, have minimum and maximum; so, by Proposition 2.1, there is no operation structuring $[0, 1]$ or S* as an alo-group.

Proposition 2.2. *[6] Let $\mathscr{G} = (G, \odot, \leq)$ be an alo-group and $||a \div b|| = (a \div b) \vee (b \div a)$. Then, the operation*

$$d_{\mathscr{G}} : (a, b) \in G^2 \to ||a \div b|| \in G \tag{2}$$

verifies the conditions:

1. $d_{\mathscr{G}}(a, b) \geq e$;
2. $d_{\mathscr{G}}(a, b) = e \Leftrightarrow a = b$;
3. $d_{\mathscr{G}}(a, b) = d_{\mathscr{G}}(b, a)$;
4. $d_{\mathscr{G}}(a, b) \leq d_{\mathscr{G}}(a, c) \odot d_{\mathscr{G}}(b, c)$.

Definition 2.1. *The operation $d_{\mathscr{G}}$ in (2) is a $\mathscr{G}$-*metric *or $\mathscr{G}$-*distance.

Definition 2.2. *Let $\mathscr{G} = (G, \odot, \leq)$ be a divisible alo-group. Then, the $\odot$-* mean *$m_{\odot}(a_1, a_2, \ldots, a_n)$ of the elements $a_1, a_2, \ldots, a_n$ of G is defined by*

$$m_{\odot}(a_1, a_2, \ldots, a_n) = \begin{cases} a_1 & n = 1, \\ (\bigodot_{i=1}^{n} a_i)^{(1/n)} & n \geq 2. \end{cases}$$

Definition 2.3. *An* isomorphism *between two alo-groups $\mathscr{G} = (G, \odot, \leq)$ and $\mathscr{G}' = (G', \circ, \leq)$ is a bijection $h : G \to G'$ that is both a lattice isomorphism and a group isomorphism, that is:*

$$x < y \Leftrightarrow h(x) < h(y)$$
$$h(x \odot y) = h(x) \circ h(y).$$

Proposition 2.3. *[6] Let $h : G \to G'$ be an isomorphism between the alo-groups $\mathcal{G} = (G, \odot, \leq)$ and $\mathcal{G}' = (G', \circ, \leq)$. Then,*

$$d_{\mathcal{G}'}(a', b') = h(d_{\mathcal{G}}(h^{-1}(a'), h^{-1}(b'))).$$

Moreover, $\mathcal{G}$ is divisible if and only if $\mathcal{G}'$ is divisible and, under the assumption of divisibility:

$$m_{\circ}(y_1, y_2, ..., y_n) = h\big(m_{\odot}(h^{-1}(y_1), h^{-1}(y_2), ..., h^{-1}(y_n))\big).$$

2.1 *Real Alo-Groups*

An alo-group $\mathcal{G} = (G, \odot, \leq)$ is a *real* alo-group if and only if G is a subset of the real line R and $\leq$ is the total order on G inherited from the usual order on R. Let $+$ and $\cdot$ be the usual addition and multiplication on R and $\otimes :]0,1[^2 \to]0,1[$ the operation defined by $x \otimes y = \frac{xy}{xy + (1-x)(1-y)}$. Then examples of real divisible alo-groups are the following:

Multiplicative alo-group
$\mathbf{]0,+\infty[} = (]0,+\infty[, \cdot, \leq)$; then $e = 1$, $x^{(-1)} = 1/x$, $x^{(n)} = x^n$ and $x \div y = \frac{x}{y}$. So

$$d_{\mathbf{]0,+\infty[}}(a,b) = \frac{a}{b} \vee \frac{b}{a}$$

and $m_{\cdot}(a_1, ..., a_n)$ is the geometric mean: $\left(\prod_{i=1}^{n} a_i\right)^{\frac{1}{n}}$.
Additive alo-group
$\mathcal{R} = (R, +, \leq)$; then $e = 0$, $x^{(-1)} = -x$, $x^{(n)} = nx$, $x \div y = x - y$. So

$$d_{\mathcal{R}}(a,b) = |a - b| = (a - b) \vee (b - a)$$

and $m_{+}(a_1, ..., a_n)$ is the arithmetic mean: $\frac{\sum_i a_i}{n}$.
Fuzzy alo-group
$\mathbf{]0,1[} = (]0,1[, \otimes, \leq)$; then $e = 0.5$, $x^{(-1)} = 1 - x$, $x \div y = \frac{x(1-y)}{x(1-y)+(1-x)y}$ and

$$d_{\mathbf{]0,1[}}(a,b) = \frac{a(1-b)}{a(1-b)+(1-a)b} \vee \frac{b(1-a)}{b(1-a)+(1-b)a}.$$

Remark 2.1. *The group operation $\otimes$ considered for a fuzzy alo-group is a restriction to $]0,1[$ of a well-known representable uninorm over the closed interval $[0,1]$(see [9]).*

The bijection

$$h : x \in]0,+\infty[\to \log x \in R$$

is an isomorphism between $\mathbf{]0,+\infty[}$ and $\mathcal{R}$ and

$$v : t \in]0,+\infty[\to \frac{t}{t+1} \in]0,1[$$

is an isomorphism between **]0,+∞[** and **]0,1[**. So, by Proposition 2.3, the mean $m_{\otimes}(a_1,\ldots,a_n)$ related to the fuzzy alo-group can be computed as follows:

$$m_{\otimes}(a_1,\ldots,a_n) = v((\prod_{i=1}^{n} v^{-1}(a_i))^{\frac{1}{n}}).$$

3 Pairwise Comparison Matrices over a Divisible Alo-Group

In this section, $\mathscr{G} = (G,\odot,\leq)$ is a divisible alo-group and $A = (a_{ij})$ in (1) is a PCM over $\mathscr{G}$, that is $a_{ij} \in G$, $\forall i,j \in \{1,\ldots,n\}$. The symbols $\underline{a}_1,\underline{a}_2,\ldots,\underline{a}_n$ indicate the rows of A and $\underline{a}^1,\underline{a}^2,\ldots,\underline{a}^n$ indicate the columns of A. The *mean vector* associated to A is the vector

$$\underline{w}_{m_{\odot}}(A) = (m_{\odot}(\underline{a}_1), m_{\odot}(\underline{a}_1),\cdots, m_{\odot}(\underline{a}_n)). \tag{3}$$

We say that A is *reciprocal* PCM, with respect to $\odot$, if and only if [6]:

$$a_{ji} = a_{ij}^{(-1)} \quad \forall\, i,j = 1,\ldots,n, \quad (reciprocity). \tag{4}$$

If $A = (a_{ij})$ is reciprocal, then $a_{ii} = e$ for each $i = 1,2,\ldots,n$ and $a_{ij} \odot a_{ji} = e$ for $i,j \in \{1,2,\ldots,n\}$.

3.1 Consistent PCMs

In [6], we provide the following definition:

Definition 3.1. $A = (a_{ij})$ *is a* consistent *matrix with respect to* $\odot$, *if and only if:*

$$a_{ik} = a_{ij} \odot a_{jk} \quad \forall i,j,k. \tag{5}$$

Moreover, $\underline{w} = (w_1,\ldots,w_n)$, *with* $w_i \in G$, *is a* consistent *vector for* $A = (a_{ij})$ *if and only if*

$$w_i \div w_j = a_{ij} \quad \forall i,j = 1,2,\ldots,n. \tag{6}$$

Proposition 3.1. *A consistent PCM is also a reciprocal matrix.*

Proof. Let $A = (a_{ij})$ be a consistent matrix. By (5) and the properties of the group operation $\odot$, for each choice of i, we get: $a_{ii} = a_{ii} \odot e = a_{ii} \odot (a_{ii} \odot a_{ii}^{(-1)}) = (a_{ii} \odot a_{ii}) \odot a_{ii}^{(-1)} = a_{ii} \odot a_{ii}^{(-1)} = e$; and, for every choice of i and j: $a_{ij} \odot a_{ji} = a_{ii} = e$ so that $a_{ji} = a_{ij}^{(-1)}$.

Proposition 3.2. *A reciprocal PCM* $A = (a_{ij})$ *is consistent if and only if there exists a consistent vector* $\underline{w} = (w_1, w_2,\ldots,w_n)$.

Proof. The condition of consistency together whit the condition of reciprocity claims that

$$a_{ik} \div a_{jk} = a_{ij} \quad \forall i,j,k. \tag{7}$$

Then, for every choice of $k \in \{1,2,...,n\}$ the column $\underline{a}^k = (a_{1k}, a_{2k}, \ldots, a_{nk})$ verifies (6). Viceversa if $\underline{w} = (w_1, w_2, ..., w_n)$ verifies (6), then

$$a_{ik} = w_i \div w_k = w_i \odot w_k^{(-1)} = w_i \odot w_j^{(-1)} \odot w_j \odot w_k^{(-1)} = a_{ij} \odot a_{jk};$$

so A is a consistent matrix.

Proposition 3.3. *[6] Let $A = (a_{ij})$ be consistent. Then the mean vector $\underline{w}_{m_\odot}$ in (3) is a consistent vector.*

3.2 How to Check the Consistency of a Reciprocal Matrix

From now on, we assume that $A = (a_{ij})$ is a reciprocal PCM over an alo-group. We provide an algorithm of computational complexity order equal to $O(n^2)$ to check whenever or not t he matrix is consistent.

Proposition 3.4. *[6] $A = (a_{ij})$ is a* consistent *matrix with respect to $\odot$, if and only if:*

$$a_{ik} = a_{ij} \odot a_{jk} \quad \forall\, i,j,k : i < j < k.$$

Proposition 3.5. *[6] $A = (a_{ij})$ is a* consistent *matrix with respect to $\odot$, if and only if:*

$$d_{\mathscr{G}}(a_{ik}, a_{ij} \odot a_{jk}) = e \quad \forall\, i,j,k : i < j < k.$$

By previous propositions, we obtain new characterizations of a consistent PCM. By Proposition 3.4 and associativity of $\odot$, we have the following proposition:

Proposition 3.6. *A is consistent if and only if, $\forall i,k : i < k-1$, the following equality holds:*

$$a_{ik} = a_{i\,i+1} \odot a_{i+1\,i+2} \odot \ldots \odot a_{k-1\,k}.$$

By Proposition 3.6 and associativity of $\odot$, we have the following proposition:

Proposition 3.7. *A is consistent if and only if, $\forall i,k : i < k-1$, the following equality holds:*

$$a_{ik} = a_{i\,i+1} \odot a_{i+1\,k}.$$

Thus, in order to check whenever or not a matrix is consistent, we provide Algorithm 1, for which computational complexity order is equal to $O(n^2)$.

3.3 Improving the Consistency Index

Let T be the set $\{(a_{ij}, a_{jk}, a_{ik}), i < j < k\}$ and n_T the cardinality of T. By Proposition 3.5 $A = (a_{ij})$ is inconsistent if and only if $d_{\mathscr{G}}(a_{ik}, a_{ij} \odot a_{jk}) > e$ for some

Algorithm 1. Checking consistency

```
i = 1;
ConsistentMatrix=true;
while i ≤ n − 2 and ConsistentMatrix do
   k = i + 2;
   while k ≤ n and ConsistentMatrix do
      if a_ik ≠ a_i i+1 ⊙ a_i+1 k then
         ConsistentMatrix=false;
      end if
      k = k + 1;
   end while
   i = i + 1;
end while
```

triple $(a_{ij}, a_{jk}, a_{ik}) \in T$. Thus, in [6] we have provided the following definition of consistency index:

Definition 3.2. *The* consistency index *of A is given by:*

$$I_{\mathscr{G}}(A) = \begin{cases} d_{\mathscr{G}}(a_{13}, a_{12} \odot a_{23}) & n = 3 \\ \left(\bigodot_T d_{\mathscr{G}}(a_{ik}, a_{ij} \odot a_{jk}) \right)^{\left(\frac{1}{n_T}\right)} & n > 3, \end{cases}$$

with $n_T = \frac{n(n-2)(n-1)}{6}$.

Remark 3.1. *$I_{\mathscr{G}}(A) \geq e$ and A is consistent if and only if $I_{\mathscr{G}}(A) = e$.*

As corollary of Proposition 2.3 we get the following

Proposition 3.8. *Let $\mathscr{G}' = (G', \circ, \leq)$ be a divisible alo-group isomorphic to $\mathscr{G}$ and $A' = (h(a_{ij}))$ the transformed of $A = (a_{ij})$ by means of the isomorphism $h : G \to G'$. Then $I_{\mathscr{G}}(A) = h^{-1}(I_{\mathscr{G}'}(A'))$.*

At the light of Proposition 3.7, it is reasonable to define a new consistency index, considering only the $\odot -$ *mean* of the distances $d_{\mathscr{G}}(a_{ik}, a_{i\,i+1} \odot a_{i+1\,k})$, with $i < k-1$. Let T^* be the set $\{(a_{i\,i+1}, a_{i+1\,k}, a_{ik}), i < k-1\}$ and n_{T^*} the cardinality of T^*, then we provide the following definition:

Definition 3.3. *The* consistency index *of A is given by:*

$$I^*_{\mathscr{G}}(A) = \begin{cases} d_{\mathscr{G}}(a_{13}, a_{12} \odot a_{23}) & n = 3 \\ \left(\bigodot_{T^*} d_{\mathscr{G}}(a_{ik}, a_{i\,i+1} \odot a_{i+1\,k}) \right)^{\left(\frac{1}{n_{T^*}}\right)} & n > 3, \end{cases}$$

with $n_{T^*} = \frac{(n-2)(n-1)}{2}$.

Assertions analogous to ones in Remark 3.1 and Proposition 3.8 follow.

Remark 3.2. *$I^*_{\mathscr{G}}(A) \geq e$ and A is consistent if and only if $I^*_{\mathscr{G}}(A) = e$.*

Proposition 3.9. *Let $\mathscr{G}' = (G', \circ, \leq)$ be a divisible alo-group isomorphic to $\mathscr{G}$ and $A' = (h(a_{ij}))$ the transformed of $A = (a_{ij})$ by means of the isomorphism $h : G \to G'$. Then $I^*_{\mathscr{G}}(A) = h^{-1}(I^*_{\mathscr{G}'}(A'))$.*

3.4 Building a Consistent Matrix

Given $X = \{x_1, x_2, \dots, x_n\}$ the set of alternatives, Proposition 3.7 allow us to build a consistent PCM starting from a fixed alternative x_i and the $n-1$ comparisons between x_i and x_j, for $j \neq i$. These comparisons are expressed by one of the following sequences:

1. $a_{i1}, \dots a_{i\,i-1}, a_{i\,i+1}, \dots a_{in}$,
2. $a_{1i}, \dots a_{i-1\,i}, a_{i+1\,i}, \dots a_{ni}$.

For instance, we provide Algorithm 2 to build a consistent PCM starting from the sequence $\{a_{12}, a_{13}, \dots, a_{1n}\}$; we use the equality $a_{i+1\,k} = a_{i+1\,i} \odot a_{i\,k}$, $\forall i,k$ such that $i < k-1$, obtained from Proposition 3.7.

Algorithm 2. Building a consistent matrix

for $k = 2, \dots, n$ **do**
  $a_{k1} = a_{1k}^{(-1)}$
end for
for $i = 1, \dots, n$ **do**
  $a_{ii} = e$
end for
for $i = 1, \dots n-2$ **do**
  for $k = i+2 \dots n$ **do**
    $a_{i+1\,k} = a_{i+1\,i} \odot a_{i\,k}$
    $a_{k\,i+1} = a_{i+1\,k}^{(-1)}$
  end for
end for

By Proposition 3.7, we can also build a consistent PCM starting from one of the following sequences:

3. $a_{12}, a_{23}, \dots a_{n-1\,n}$,
4. $a_{21}, a_{32}, \dots a_{n\,n-1}$.

For the fuzzy case, the authors in reference [7] build a consistent matrix by means of sequence 3.

In the following, in Example 1 and in Example 2, we show how to build a consistent matrix by means of Algorithm 2; the examples are related to the multiplicative case and the fuzzy case respectively.

Example 1. *Let $\{x_1, x_2, x_3, x_4, x_5\}$ be a set of alternatives. We suppose that the DM prefers x_1 to each other alternative and expresses the following preference ratios: $a_{12} = 2$, $a_{13} = 4$, $a_{14} = 5$ and $a_{15} = 6$. By means of Algorithm 2, we obtain:*

$$a_{21} = \frac{1}{2}, \quad a_{31} = \frac{1}{4}, \quad a_{41} = \frac{1}{5}, \quad a_{51} = \frac{1}{6},$$

$$a_{11} = a_{22} = a_{33} = a_{44} = a_{55} = 1,$$

$$a_{23} = a_{21}a_{13} = 2, \quad a_{32} = \frac{1}{2},$$

$$a_{24} = a_{21}a_{14} = \frac{5}{2}, \quad a_{42} = \frac{2}{5},$$

$$a_{25} = a_{21}a_{15} = 3, \quad a_{52} = \frac{1}{3},$$

$$a_{34} = a_{32}a_{24} = \frac{5}{4}, \quad a_{43} = \frac{4}{5},$$

$$a_{35} = a_{32}a_{25} = \frac{3}{2}, \quad a_{53} = \frac{2}{3},$$

$$a_{45} = a_{43}a_{35} = \frac{6}{5}, \quad a_{54} = \frac{5}{6},$$

and therefore:

$$A = \begin{pmatrix} 1 & 2 & 4 & 5 & 6 \\ \frac{1}{2} & 1 & 2 & \frac{5}{2} & 3 \\ \frac{1}{4} & \frac{1}{2} & 1 & \frac{5}{4} & \frac{3}{2} \\ \frac{1}{5} & \frac{2}{5} & \frac{4}{5} & 1 & \frac{6}{5} \\ \frac{1}{6} & \frac{1}{3} & \frac{2}{3} & \frac{5}{6} & 1 \end{pmatrix}$$

Example 2. *Let $\{x_1, x_2, x_3, x_4, x_5\}$ be a set of alternatives. We suppose that the following values express how much the DM prefers x_1 to each other alternative: $a_{12} = 0.6$, $a_{13} = 0.7$, $a_{14} = 0.8$ and $a_{15} = 0.9$. By means of Algorithm 2, we obtain:*

$$a_{21} = 0.4, \quad a_{31} = 0.3, \quad a_{41} = 0.2, \quad a_{51} = 0.1,$$

$$a_{11} = a_{22} = a_{33} = a_{44} = a_{55} = 0.5,$$

$$a_{23} = \frac{a_{21}a_{13}}{a_{21}a_{13} + (1 - a_{21})(1 - a_{13})} = 0.609, \quad a_{32} = 0.391,$$

$$a_{24} = \frac{a_{21}a_{14}}{a_{21}a_{14} + (1 - a_{21})(1 - a_{14})} = 0.727, \quad a_{42} = 0.273,$$

$$a_{25} = \frac{a_{21}a_{15}}{a_{21}a_{15} + (1 - a_{21})(1 - a_{15})} = 0.857, \quad a_{52} = 0.143,$$

$$a_{34} = \frac{a_{32}a_{24}}{a_{32}a_{24} + (1 - a_{32})(1 - a_{24})} = 0.632, \quad a_{43} = 0.368,$$

$$a_{35} = \frac{a_{32}a_{25}}{a_{32}a_{25} + (1 - a_{32})(1 - a_{25})} = 0.794, \quad a_{53} = 0.206,$$

$$a_{45} = \frac{a_{43}a_{35}}{a_{43}a_{35} + (1-a_{43})(1-a_{35})} = 0.692, \quad a_{54} = 0.308,$$

and therefore:

$$A = \begin{pmatrix} 0.5 & 0.6 & 0.7 & 0.8 & 0.9 \\ 0.4 & 0.5 & 0.609 & 0.727 & 0.857 \\ 0.3 & 0.391 & 0.5 & 0.632 & 0.794 \\ 0.2 & 0.273 & 0.368 & 0.5 & 0.692 \\ 0.1 & 0.143 & 0.206 & 0.308 & 0.5 \end{pmatrix}$$

4 Conclusion and Future Work

We have provided an algorithm to check whenever or not a reciprocal PCM over an alo-group is consistent, together with an algorithm to build a consistent matrix by means of $n-1$ comparisons. Moreover, starting from a consistency index $I_{\mathscr{G}}(A)$ defined in [6], and naturally linked to a notion of distance, we have provided a new consistency index, easier to compute, that is an optimization of $I_{\mathscr{G}}(A)$.

Our future work will be directed to find the link between the new consistent index and the mean vector $\underline{w}_{m_{\cdot}} = (w_1, w_2, \cdots, w_n)$ associated to the matrix in order to state the reliability of this vector as a weighting priority vector.

Moreover, following the results in [3, 4, 5, 8], for the multiplicative case, our aim is to investigate, in the general case, the following problems related to a PCM:

- to individuate the conditions on a PCM inducing a qualitative ranking (*actual ranking*) on the set X;
- to individuate the conditions ensuring the existence of vectors representing the actual ranking at different levels.

References

[1] Barzilai, J.: Consistency measures for pairwise comparison matrices. J. MultiCrit. Decis. Anal. 7, 123–132 (1998)

[2] Basile, L., D'Apuzzo, L.: Ranking and weak consistency in the a.h.p. context. Rivista di matematica per le scienze economiche e sociali 20(1), 99–110 (1997)

[3] Basile, L., D'Apuzzo, L.: Weak consistency and quasi-linear means imply the actual ranking. International Journal of Uncertainty, Fuzziness and Knowledge-Based Systems 10(3), 227–239 (2002)

[4] Basile, L., D'Apuzzo, L.: Transitive matrices, strict preference and intensity operators. Mathematical Methods in Economics and Finance 1, 21–36 (2006)

[5] Basile, L., D'Apuzzo, L.: Transitive matrices, strict preference and ordinal evaluation operators. Soft Computing 10(10), 933–940 (2006)

[6] Cavallo, B., D'Apuzzo, L.: A general unified framework for pairwise comparison matrices in multicriterial methods. International Journal of Intelligent Systems 24(4), 377–398 (2009)

[7] Chiclana, F., Herera-Viedma, E., Alonso, S., Herera, F.: Cardinal Consistency of Reciprocal Preference Relations: A Characterization of Multiplicative Transitivity. IEEE Transaction on fuzzy sistems 17(1), 14–23 (2009)

[8] D'Apuzzo, L., Marcarelli, G., Squillante, M.: Generalized consistency and intensity vectors for comparison matrices. International Journal of Intelligent Systems 22(12), 1287–1300 (2007)
[9] Fodor, J., Yager, R., Rybalov, A.: Structure of uninorms. International Journal of Uncertainty, Fuzziness and Knowledge-Based Systems 5(4), 411–427 (1997)
[10] Herrera-Viedma, E., Herrera, F., Chiclana, F., Luque, M.: Some issue on consistency of fuzzy preferences relations. European journal of operational research 154, 98–109 (2004)
[11] Lootsma, F.A.: Multi-Criteria Decision Analysis via Ratio and Difference Judgement. Applied Optimization, vol. 29. Kluwer Academic Publishers, Dordrecht (1999)
[12] Saaty, T.L.: A scaling method for priorities in hierarchical structures. J. Math. Psychology 15, 234–281 (1977)
[13] Saaty, T.L.: The Analytic Hierarchy Process. McGraw-Hill, New York (1980)
[14] Saaty, T.L.: Axiomatic foundation of the analytic hierarchy process. Management Science 32(7), 841–855 (1986)
[15] Saaty, T.L.: Decision Making for Leaders. University of Pittsburgh (1988)

On a Decision Model for a Life Insurance Company Rating

Fabio Baione, Paolo De Angelis, and Riccardo Ottaviani

Abstract. A rating system is a decision support tool for analysts, regulators and stakeholders in order to evaluate firm capital requirements under risky conditions. The aim of this paper is to define an actuarial model to measure the Economic Capital of a life insurance company; the model is developed under Solvency II context, based on option pricing theory.

In order to asses a life insurance company Economic Capital it is necessary to involve coherent risk measures already used in the assessment of banking Solvency Capital Requirements, according to Basel II standards. The complexity of embedded options in life insurance contracts requires to find out operational solutions consistent with Fair Value principle, as defined in the International Accounting Standards (IAS).

The paper is structured as follows: Section 1 describes the development of the Insurance Solvency Capital Requirement standards; Section 2 introduces the theoretical framework of Economic Capital related to risk measures; Section 3 formalizes the actuarial model for the assessment of a life insurance company rating; Section 4 offers some results due to an application of the actuarial model to a portfolio of surrendable participating policies with minimum return guaranteed and option to annuitise.

Keywords: Fair Value, Option Pricing, Rating Models, Solvency, Insurance Risk Management.

JEL CLASSIFICATION: D81 · G13 · G22 · G24.

Fabio Baione · Paolo De Angelis
Ma. D.E.F.A. Department
University of Rome, "La Sapienza"
Via del Castro Laurenziano, 9 - 00161 Rome
e-mail: fabio.baione@uniroma1.it, paolo.deangelis@uniroma1.it

Riccardo Ottaviani
S.E.A.D. Department
University of Rome, "La Sapienza"
Via Regina Elena, 295 - 00161 Rome
e-mail: riccardo.ottaviani@uniroma1.it

S. Greco et al. (Eds.): Preferences and Decisions, STUDFUZZ 257, pp. 123–138.
springerlink.com

1 Capital Requirements for Regulatory and Solvency Purposes: A Survey

With reference to insurance and reinsurance contracts, the issue of the International Financial Reporting Standard 4 (IFRS 4) by the International Accounting Standards Board (IASB) in March 2004 established the end of Phase I of the IAS project.

Due to the absence of a liquid market where trading insurance contracts, a Fair Valuation of an insurance contract is an hard task, so the IAS project has been developed in two phases. In general, the Fair Value of financial assets/liabilities is equal to its market value when a sufficiently liquid market exists, otherwise it is necessary to estimate the market value by using a theoretical model consistent with market transactions. During Phase I, the fair valuation of Italian insurance contracts has been applied exclusively to assets backing technical provisions, thus postponing to phase II the achievement of a theoretical market-consistent model for the liability component.

At the same time, in order to introduce new systems for measuring and monitoring solvency requirements for insurance firms, the European Commission has introduced the Solvency II project; in particular, the project is structured on 3 pillars, as follows:

- Pillar I (Capital Requirements): defines the criteria for setting minimum capital requirements for regulatory and solvency purposes in insurance risks management;
- Pillar II (Risk Management): increases the internal audit function by means of Risk Management function to implement risks controlling and monitoring;
- Pillar III (Transparency and market discipline): increases market discipline through the adoption of common standards for disclosure and best practices in the market.

The European Commission has requested to CEIOPS (Committee of European Insurance and Occupational Pensions Supervisors) to define methodological principles to compute minimum capital requirements for European insurance undertakings. For these purposes, CEIOPS has promoted several quantitative impact studies (QIS) designed to assess the practicability, implications and possible impacts of the different methodologies for the entity capital allocation.

Solvency II and IAS projects consider Fair Valuation as the starting point for insurance liabilities assessment. Reserving Risk Capital (i.e. risk capital measure of technical provisions) is Fair Valuated by means of a Best Estimate Liability (BEL) and a Risk Margin Value (RMV) [1].

Over recent years IASB and CEIOPS have continued side by side to produce respectively technical documentations -for the Fair Valuation of technical provisions- and quantitative studies - four at the date of this paper - to monitor the technicality level achieved by insurance companies in setting solvency capital requirements. By means of the quantitative impact studies CEIOPS has set itself the goal of identifying standards for the assessment and measurement of Minimum Capital Requirements (MCR), for regulatory purposes, and Solvency Capital Requirements (SCR) for solvency purposes, assessing the applicability and possible effects on balance sheet.

In QIS I the target security level to be taken in the evaluation of technical provisions has been defined and compared with predetermined confidence levels. It has also been provided the definition of best estimate and early models for the evaluation of the Risk Margin Value and Economic Capital, in reference to a certain security level via Quantile Approach.

In QIS II a deterministic parametric model has been proposed to evaluate the SCR and MCR. In particular, the computation of SCR is obtained by means of an additive procedure involving the amount of capital absorbed by each risk source and the relative correlation.

With QIS3 an extension of QIS2 model has been exposed. The study is focused on further information about the practicability and suitability of the calculations and it looks for quantitative information about the possible impact on the balance sheets, and the amount of capital that might be needed, if the approach and the calibration set out in the QIS3 specification were to be adopted as the Solvency II standard.

QIS4 has provided all stakeholders with detailed information on the quantitative impact on insurers and reinsurers' solvency capital requirements, comparing

---

[1] RMV reflects the market price of the main sources of uncertainty in insurance risk management. The international debate on a fair valuation methodology seems to converge on two approaches:

- Quantile approach: technical provisions is defined at a predefined percentile level (typically 75°-90° safety-level) and risk margin is obtained as difference with BEL;
- Cost of Capital approach: risk margin is defined as a percentage of Solvency Capital Requirement.

These approaches are used in the standard formula proposed by CEIOPS in QIS framework; it is worth noting that CoC approach is similar to the Swiss Solvency Test (SST) approach. In SST context, solvency capital requirement, as defined by Swiss Regulator (FOPI), is based on the expected shortfall of change of risk bearing capital over a 1 year time horizon on a 99% confidence level; therefore risk margin value is set as 6% of CoC. The threshold 6% is to be interpreted as the spread between the expected rate of return, requested by the stakeholders, and a risk-free rate. The selection of 6% level is based on the probability of default, estimated from market data and associated with different credit ratings classes. The expected shortfall on a 99% level of confidence, corresponds approximately to a 99.6% to 99.8% Value at Risk which implies a BBB rating company so that 6% over risk-free was chosen [Swiss Federal Office of Private Insurance (2004)].

results under Solvency II with Solvency I. In particular, the European Commission has asked CEIOPS to work on the development of simplifications in computing technical provisions and SCR, coherent with the proportionality principle.

The "deterministic" hypothesis at the base of the risk capital model proposed by CEIOPS have been reported in the guidelines attached to each QIS. Regarding to technical provisions such hypothesis have to be estimated using a market-consistent model and a best estimate approach plus a RMV, respectively for diversifiable and non-diversifiable risks.

The exercise of the quantitative impact studies inspired the development of internal models by undertakings in respect of minimal attributes in terms of methodology and information; in particular, the internal models must meet minimum information standards on the probability distribution, or statistical information (mean, variance, percentiles, moments of higher order, etc.) taking into account correlations of each uncertainty source of risk.

2 Economic Capital and Risk Measures

The management of an insurance company is a complex equilibrium system between the policyholders' expectations, measured in terms of benefits, and stakeholders return on investment. The risky nature of insurance business needs to cover insurance liabilities with capital in excess of assets assumed to back liabilities, measured as the mean of the probability distribution of the random variable company's profits; unexpected deviation of benefits, compared with balance sheet estimates, have to find adequate coverage using free assets (share capital and reserves).

From the shareholders point of view the investment risk, measured in terms of potential loss of invested capital, is assessed in terms of expected return on investment. Risk, capital and return are closely related because higher insurance risk involves higher solvency capital requirement and, therefore, higher shareholders' expectation on rate of return. A change in the Solvency Capital Requirement level by the Regulator is reflected by a side in a larger level of security for policyholders, but the other in a decreasing of shareholders' rate of return and therefore less attraction in insurance market investment.

An actuarial model for the assessment of the rating of a life insurance company requires a specific definition of variables involved in its implementation.

It is suitable to define Economic Capital for an insurance company. For this purpose it is possible refer to the definition introduced by the Society of Actuaries (SOA - Specialty Guide on Economic Capital, March 2004) *"Economic Capital is defined as the excess of the market value of the assets over the liabilities of required to ensure that obligations can be satisfied at a given level of risk tolerance, over a specified time horizon"* where is highlighted the close relationship between the market valuation and the Economic Capital, already mentioned in the previous paragraph. Basic values necessary to determine Economic Capital are:

Best Estimate Liability (BEL)

BEL corresponds to the amount of assets that the insurance company should hold to pay its commitments in a scenario with "best estimate" assumptions. The BEL can be calculated with a prospective approach as the present value of the insurance cash flows discounted using the expected asset return (at market value).

Scenario Needed Assets (SNA)

It is the amount of assets required to back liabilities in a specific scenario.

Needed Assets (NA)

It is the SNA required to back an adverse scenario corresponding to a specific security level (Security Factor).

Security Factor

It is the probability level corresponding to the insurance company "solvability" on a fixed time horizon.

Long-term Solvency (Tier 1)

It is the amount of assets required over the long term (20-30 years) in addition to BEL to back policyholder benefits at a fixed Security Factor. It can be expressed as:

Tier1 = NA-BEL.

Intermediate Solvency (Tier 2)

It is defined as the cost to borrow capital necessary to back the eventual insolvency in interim periods (observed on the projection period) to a predetermined Security Factor.

Economic Capital (EC)

It can be expressed as:

Economic Capital = Tier1 + Tier 2

Economic Excess Assets

It represents free shareholders' assets in excess of Economic Capital.

Figures below show a representation of the basic values defined respectively in a Quantile (Figure 1) and in a Balance Sheet approach (Figure 2).

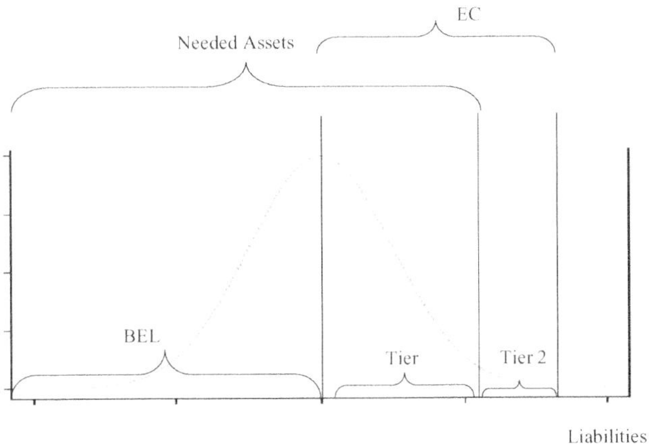

Fig. 1 Quantile Approach

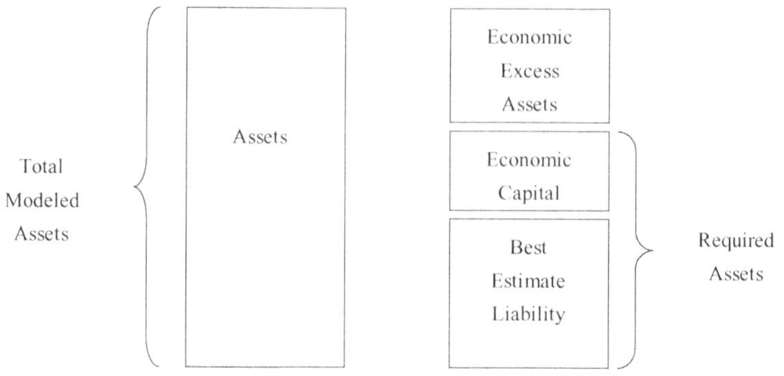

Fig. 2 Balance Sheet

The variables defined above are consistent with the Economic Capital framework, respectively, within the Quantile Approach and the Cost of Capital Approach; according to these approaches a mathematical formula of the EC is derived by means of risk measure arguments.

From a theoretical point of view the Economic Capital is:

$$K(X) = \rho_\alpha(X) - FV(X) \tag{3.1}$$

where, $K(X)$ is the Economic Capital (SCR in QIS), $\rho_\alpha(X)$ is a risk measure[2] sets at a predefined security level (Security Factor) and $FV(X)$ the Fair Value (BEL + RMV) of insurance liabilities.

In financial literature several risk measures, coherent and not, have been defined; based on banking experiences we can classify these measures into the following:

<u>*Value at Risk (VaR):*</u> defined as the maximum (or minimum) potential loss in a given time horizon with an alpha% (or 1-alpha%) of best cases (or worst):

$$VaR_\alpha[X] = \sup\{x \in \Re | \Pr[X \le x] \le \alpha\} = \inf\{x \in \Re | \Pr[X > x] \le 1 - \alpha\} \tag{3.2}$$

The VaR measure takes into account the default probability but not the magnitude; it is not a coherent risk measure because it doesn't respect the subadditive condition,

---

[2] A risk measure is a function that assigns an amount of capital to the extreme values of the probability density function of the r.v. "loss". A risk measure is coherent in the sense of Artzner et al. (1998), if it respects the following conditions:

- Monotonicity: if $X \ge Y$ then $\rho(X) \ge \rho(Y)$
- Positive Homogeneity: if $\lambda \ge 0$ then $\rho(\lambda X) = \lambda\rho(X)$
- Translation invariance $\rho(X + a) = \rho(X) + a$
- Subadditivity $\rho(X+Y) \le \rho(X) + \rho(Y)$.

so it may increase in case of risk diversification. It is comparable to Risk of Ruin (Probability of Ruin) of Risk Theory.

Standard Deviation: sets the measure as the expected value of the loss plus a pre-determined multiple of the standard deviation:

$$SD[X] = E[X] + \delta \cdot StDev[X] \tag{3.3}$$

Tail Conditional Expectation (TCE) or Tail VaR (TVAR): defined as the expected loss in a given time horizon in (1-alpha)% of worst case:

$$TCE_\alpha[X] = E[X|X > VaR_\alpha[X]] \tag{3.4}$$

The TCE is always greater than the corresponding VaR:

$$TCE_\alpha[X] = VaR_\alpha[X] + E[X - VaR_\alpha[X]|X > VaR_\alpha[X]] \tag{3.5}$$

3 An Actuarial Model for a Life Insurance Company Rating

An actuarial model to assess the rating of an insurance company needs a structured evaluation process split up several stages, as follows: firstly, to define a mathematical model for a technical provisions fair valuation of contracts, according to IAS 39 principle; secondly, to individuate typical insurance random processes describing the dynamics of risk sources, in order to determine probability distribution of technical provisions value, necessary for EC assessment; thirdly, to choice an operational approach to set the rating level in accordance with methodologies used for.

A definition of Fair Value is proposed by *International Accounting Standards Committee (IASC): "The amount for which an asset could be exchanged or liability settled, between knowledgeable, willing parties in an arm's length transaction"*; therefore, under markets efficiency conditions, a Fair Value of an insurance policy is equal to its equilibrium-price. In absence of an efficient market, a Fair Value could be estimated through a consistent theoretical bid/ask model, joined with similar assets and liabilities.

The selected valuation technique makes maximum use of market inputs and relies as little as possible on entity-specific inputs. It incorporates all factors that market participants would consider in setting a price and it is consistent with accepted economic methodologies for pricing financial instruments. The Fair Value has to be based on:

- observable current market transactions in similar instruments;
- valuation techniques whose variables include primarily observable market data and calibrated periodically on observable current market transactions in similar instrument or to other observable current market data;
- valuation techniques that are commonly used by market participants to price similar instruments and have demonstrated to provide realistic estimates of prices obtained in market transactions.

3.1

Hereafter it is briefly reported an actuarial model for the Fair Value of the technical provision of a life insurance contract proposed by Baione-De Angelis-Fortunati (2006). The operational approach, based on Option Pricing Theory, uses a Monte-Carlo procedures to reproduce uncertainty implied by demographic and financial risks.

In order to present the actuarial model, it is supposed to operate in a traditional efficient market. It is assumed, in fact, that financial and insurance markets are perfectly competitive, frictionless and free of arbitrage opportunities. Moreover, all agents are supposed to be rational and non-satiated, and to share the same information.

The technical provision Fair Value at time $t \in [0, s]$ for a general life insurance contract can be expressed as:

$$V(r_t, \mu_t, S_t, t) = \hat{E}\left[\left(\sum_{\tau \in (t,s)} \varphi(t,\tau) CFL_\tau - \sum_{\tau \in (t,s)} \varphi(t,\tau) CFA_\tau\right) \middle| F_t^{r,\mu,S}\right] \tag{3.1.1}$$

where $\hat{\mathrm{E}}$ denotes the usual expectation under the risk-neutral probability measure, $\varphi(t,\tau)$ is the stochastic discount factor dependent on the spot-rate dynamic between t and τ, $\mathbf{CFL}_\tau$ and $\mathbf{CFA}_\tau$ respectively the annual random cash flows of the insurance company and policyholder, jointly dependent on the spot rate and the force of mortality dynamics; F_t^r, F_t^μ and $F_t^{r,\mu}$ are σ-algebras associated with the above defined filtrations.

In the original work, the model is proposed for a surrendable participating endowment policy with option to annuitise; the contribution of each embedded option is presented with the relative sampling distribution.

A surrendable participating endowement policy with option to annuitise can be shared in four components as follows:

Basic Contract

The basic contract is a standard endowment policy with benefit C_0, net constant annual premium[3] P, technical rate i, maturity n, written on a x aged male. The Fair Value is:

$$FV_M = \hat{\mathrm{E}}\left\{\left(C\left[\sum_{\tau=1}^{n-t} \varphi(t,t+\tau)\, q_{x+t}(\tau-1,\tau) + \varphi(t,n-t)\, p_{x+t}(n-t)\right] - P\sum_{\tau=1}^{m-t} \varphi(t,t+\tau-1)\, p_{x+t}(\tau-1)\right) \middle| F_t^{r,\mu}\right\} \tag{3.1.2}$$

where

[3] The annual premium is computed using first order technical basis.

$\varphi(t,t+\tau)=e^{-\int_0^\tau r_{t+u}du}$ is the stochastic discount factor between t and $t+\tau$;

$q_{x+t}(\tau-1,\tau)=e^{-\int_0^{\tau-1}\mu_{x+t+u:u}du}\left(1-e^{-\int_{\tau-1}^{\tau}\mu_{x+t+u:u}du}\right)$ is the death stochastic probability at $x+t$ age, over $t+\tau$ period;

$p_{x+t}(\tau)=e^{-\int_0^\tau \mu_{x+t+u:u}du}$ is the survival stochastic probability at $x+t$ age, over τ periods.

Participation option

From a financial point of view Italian participating contracts are similar to indexed notes: the minimum guarantee could be expressed as an European Option. In particular, since the minimum guarantee is attached at the end of each year, the relative embedded option is a cliquet type [De Felice Moriconi (2001)]. Defined ρ_t as the participating policy-owners rate of return on benefits and premiums at time t, the Fair Value FV_{MR} of a non-surrandable participating contract is:

$$FV_{MR}=\hat{E}\left[\left(\sum_{\tau=1}^{n-t}C_{t+\tau-1}\,\varphi(t,t+\tau)\,q_{x+t}(\tau-1,\tau)+C_{t}\,\psi(t,n-t)\,p_{x+t}(n-t)\right)-P_t\sum_{\tau=1}^{m-t}\psi(t,t+\tau-1)\,p_{x+t}(\tau-1)\,\middle|\,F_t^{r,\mu,S}\right] \tag{3.1.3}$$

where $\psi(t,t+\tau)=\left[\prod_{k=1}^{\tau}(1+\rho_{t+k})e^{-\int_0^\tau r_{t+u}du}\right]$ is the stochastic value of an indexed zero coupon bond with maturity $t+\tau$ and notional amount equal to one.

Annuity option

Some traditional Italian policies enable the policyholder to convert cash benefit at maturity into a guaranteed annuity payable throughout the remaining lifetime, with annuity conversion rate fixed at issue. The guaranteed annuity option pay-out at maturity is expressed as [Ballotta and Haberman (2003)]:

$$\hat{E}\left(\max(C_n;GC_nRVA_{x+n})\middle|F_n^{r,\mu}\right)=C_n+GC_n\hat{E}\left(\max(RVA_{x+n}-K;0)\middle|F_n^{r,\mu}\right) \tag{3.1.4}$$

where C_n is the benefit at maturity, G is the annuity conversion rate, GC_n is the guaranteed annuity, RVA_{x+n} is the present value random variable of the life annuity of 1 euro per year, paid at time n and age $x+n$; $\max(RVA_{x+n}-K;0)$ is the annuity option pay-out at maturity with strike price $K=1/G$.

At time $t \in [0,n]$ the Fair Value of the Option to Annuitise (OTA) is:

$$OTA = \hat{E}\left\{p_{x+t}(n-t)\varphi(t,n-t)\left[GC_n \max\left(RVA_{x+n} - K;0\right)\right]\middle| F_t^{r,\mu,S}\right\} \tag{3.1.5}$$

Surrender Option

If an insurance contract provides to the policyholder a contractual right to claim the benefit before the maturity, this right is called surrender option. A surrender option is an American-style option[4] that enables the policyholder to surrender the policy and receive the so called surrender value.

The Fair Value of a contract, including a surrender option, can be evaluated by means of a backward recursive procedure running from time *s-1* to time *0*; at any time *t=1,2,..,s-1*, the surrender option is exercised if and only if the surrender value is higher then the continuation value of the contract.

In order to compute the surrender option we use the Least Squares Monte Carlo Approach following Andreatta and Corradin (2003); the Fair Value of the surrender option is evaluated by means of the difference between the value of a surrendable contract and a non-surrendable one.

3.2

To implement the model described above it is necessary to define the stochastic processes of demographic and financial sources of risk.

The spot rate $\{r_t; t = 1,2,....\}$ is defined by means of a mean reverting square root diffusion equation as in Cox, Ingersoll and Ross (1985):

$$dr_t = k(\theta - r_t)dt + \sigma_r \sqrt{r_t}\, dZ_t^r \tag{3.2.1}$$

where k is the mean reversion coefficient, θ is the long term rate, σ_r is the volatility parameter and $\{Z_t^r\}$ is a standard Brownian motion.

The force of mortality $\{\mu_{x+t}; t = 1,2,....\}$ is described by means of a Mean-Reverting Brownian Gompertz (MRBG) model

$$\mu_{x+t:t} = \mu_{x:0} e^{g_x t + \sigma_\mu Y_t}, \text{ con } g_x, \sigma_\mu, \mu_{x:0} > 0 \tag{3.2.2}$$

where $g_{x,s}$ resumes on time s the deterministic correction due to age x and the effect of longevity risk; $\{Y_t\}$ is a stochastic process to model random variations in the forecast trends; σ_x represents the standard deviation of the process $\{\mu_{x+t:t}; t = 1,2,....\}$; in particular the stochastic process $\{Y_t\}$ is described by means of a mean reverting diffusion process:

[4] According to the recent actuarial literature on this argument a contract with a surrender option is called American, a contract without a surrender option is called European [Bacinello (2003), Andreatta and Corradin (2003)].

$$dY_t = -bY_t dt + dW_t^Y, \qquad Y_0 = 0, \quad b \geq 0 \tag{3.2.3}$$

where b is the mean reversion coefficient and $\{W_t\}$ is a standard Brownian motion.

Fair pricing of an insurance participating policy depends also by reference portfolio's dynamic; in particular we assume to work in a Black-Scholes economy where the reference portfolio is compounded mainly by a bond index and a minority by a stock index. The two components are described by the following equation:

$$dS_t^{(i)} = r_t S_t^{(i)} dt + \sigma^{(i)} S_t^{(i)} dZ_t^{(i)} \qquad i = \begin{cases} 1: \text{stock index} \\ 2: \text{bond index} \end{cases} \tag{3.2.4}$$

where $S_t^{(i)}, \sigma^{(i)}$ e $\{Z_t^{(i)}\}$ are, for each reference portfolio's component, market price, volatility parameter and a Wiener process. At last, the three sources of financial uncertainty are correlated:

$$dZ_t^{(k)} dZ_t^{(j)} = \rho_{k,j} dt \qquad k, j = 1,2,r \tag{3.2.5}$$

hence, reference portfolio could be expressed as a combination of the random variables introduced above

$$S_t = (1-\alpha) S_t^{(1)} + \alpha S_t^{(2)} \tag{3.2.6}$$

The development of a stochastic model via Monte Carlo procedures allows to define the sampling distribution of the value of technical provisions and embedded options previously defined. Under the independence condition of insurance contracts in a life insurance portfolio, it is possible to extend the analysis to the entire portfolio. In such a way, it can be determined the sampling distribution of the portfolio value and the related risk measures, as defined in the previous section.

3.3

Once defined a fair valuation model of a life insurance portfolio, the Economic Capital provides a straight image of the rating level associated to the insurance company. In fact the Economic Capital is equivalent to the amount needed to prevent the ruin of the company in a specified percentage of possible outcome. The Risk of Ruin (RoR) is the probability of the company's inability to pay back the policyholder benefits. This probability is theoretically comparable with the risk of default of a corporate bond issued by the insurance company. Policyholders are therefore exposed to the same default risk as well as investors in a corporate bond with a specified rating. The default probability of these securities traded on regulated financial markets is estimated on observed market data. The difference between one and the Risk of Ruin represents the security level of the firm solvency ability (Security Factor); it is therefore possible to establish a mathematical relationship between the Security Factor (SF) and the default probability, directly observable on the market data, as follows:

$$SF = \left(1 - p\_deafult_t^{rating}\right)^D \tag{3.3.1}$$

Where $p\_deafult_t^{rating}$ is the annual default probability over t years and D is a time measure of exposure, coherent with the run-off of liabilities.
Formula (3.3.1) could be used alternatively, to set:

- the rating level, by management or rating agency, and define not arbitrarily the Security Factor to base the Economic Capital assessment;
- the Security Factor, by management or Regulator, to derive the default probability, comparing with the same value observed on the market, and to associate the relative rating class.

4 Some Results

The rating model described in Section 3 is implemented with reference to a portfolio of participating endowment policies; for ease of computation, the portfolio is characterized by a single model point, whose characteristics are described in Table 1. The Fair Value of the whole contract and relative embedded options it is obtained via Monte Carlo simulation, implementing a code by means of Matlab and VBA tools. The numerical results are obtained by means of 100,000 replications for each random process involved in the managed portfolio.

Table 1 reports some contract features used for numerical analysis.

Table 1 Model Point Features

Sex	Male
Age	45
Duration	15
Technical rate	1.00%
Mortality Table	SIM 92
Sum Assured at inception	100 €
Sum Assured at valuation date	100 €
Participation coefficient	85.0%
Guaranteed rate for annuity	4.85%
Annually compounded surrender discount rate	3.00%
Reference portfolio participation coefficient:	
Stock index	10%
Bond Index	90%

Table 2 shows parameters estimated on market data and used in mortality and financial risk diffusion processes.

Table 2 Set of Estimated Parameters

r_t		μ_{x+t}		S_t	
r_0	0.034023	b	0.5	S_0	0.11
k	0.248499	σ	0.07	σ_S	0.017
θ	0.048360	g	0.19	$\rho_{r,S}$	-0.1
σ_r	0.049026				

With reference to CIR model, risk-adjusted parameters have been calibrated on market values of euro swap interest rates observed on 30/06/2006 and estimated by means of Brown and Dybvig (1986) framework. Parameters for MRGB model have been calibrated on the force of mortality derived from ISTAT life tables and the projected one called "RG48". At last, reference fund parameters have been estimated on daily market value of Emu-Bond Index (3-5 years) and MSCI World Index observed between 2003 and 2006.

Table 3 reports, for few relevant years a comparison of technical provisions computed respectively following Italian local rules and the Fair Value approach.

Table 3 Mathematical Reserve and Fair Value of the Model Point

Year	Mathematical Reserve	Basic Contract	Participation Option	Option to Annuities	Surrender Option	Whole Contract
0	0.00	-18.76	18.47	-	0.35	0.06
5	34.75	12.15	22.81	-	0.48	35.44
10	79.31	50.61	29.03	-	0.72	80.36
15	137.24	100.00	37.24	-	0.00	137.24

Figure 3 exhibits the sampling distribution of the FV of the whole contract at time $t = 5$, compared with a Normal density distribution (solid line).

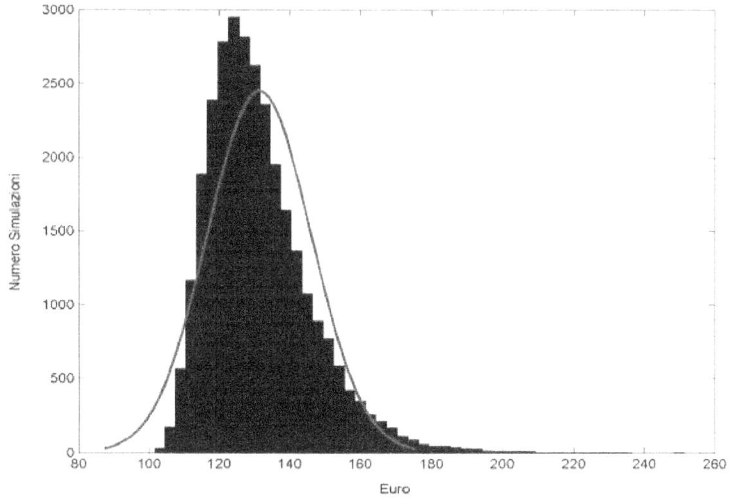

Fig. 3 Sampling distribution of FV whole contract

The risk measure is estimated with reference to market default probabilities on different rating classes referred to several time exposures in years, published on 30/06/2006 by Standard & Poor's CreditPro®.

Table 4 Default Probabilities

Rating	Y1	Y5	Y10	Y15
AAA	0.000	0.096	0.443	0.583
AA	0.010	0.293	0.815	1.276
A	0.041	0.586	1.831	2.847
BBB	0.274	2.831	5.824	8.320
BB	1.117	10.653	18.294	21.576
B	5.383	24.161	32.377	37.181
CCC/C	27.021	47.560	53.047	55.896

From (3.3.1) it is possible to derive the EC by means of VaR and TCE measures. Security Factor, for each rating class, needed to set the hypothesis on the annual default probability and the risk exposure period. A simple way is to set:

- an uniform distribution of default probability over t years;
- a risk exposure period equal to the insurance contract maturity[5].

[5] The risk exposure period, whatever the risk source considered, could be improved if calculated as the Macauly Duration or as stochastic Duration of the insurance contracts liabilities.

For instance, at time $t = 5$, the Security Factor of AAA rating is obtained as:

$$SF = \left(1 - \frac{0.443}{10 \cdot 100}\right)^{10} \cong 0.9956 \text{ and } \quad RoR = 1 - SF = 1 - 0.9956 = 0.0044$$

Associated the Security Factor with each rating class, it is easy to compute the relative EC by means of VAR and TCE measures, as shown in Table 5.

Table 5 Economic Capital

Rating	Security Factor	VaR	EC - VaR	TCE	EC - TCE
AAA	99.56%	69.63	34.18	76.19	40.75
AA	99.19%	63.82	28.38	70.87	35.43
A	98.18%	57.59	22.15	65.26	29.82
BBB	94.33%	49.27	13.82	56.96	21.52
BB	83.14%	41.87	6.43	49.56	14.12
B	71.96%	38.41	2.97	45.52	10.08
CCC/C	57.98%	36.07	0.63	42.72	7.28

Results exposed above may be compared with the Minimum Solvency Capital Requirement provided by the current rules under Solvency I criteria; in particular, for the model point analysed the Minimum Solvency Capital Requirement accounts[6] 1.39€.

It is worth noting that the comparison between the MSCR under Solvency I and values exposed in Table 5 places the portfolio solvency level between B and CCC/C rating class, using VaR measure or lower with TCE; hence, the current Minimum Solvency Margin seems to be an inadequate solvency level in reference to Solvency II criteria.

This situation is also well known to the Italian Regulator that constantly requires life insurance companies a solvency ratio, obtained as the ratio between free assets and Minimum Solvency Capital Requirement, not lower than 150%-200%.

5 Conclusions

The development of internal models for assessing the Economic Capital of a life insurance company and the relative rating requires a high quality of methodological

[6] Under Solvency I criteria the MSCR is defined as 4% of mathematical provisions plus 0.3% of sum at risk.

and technical structure. Actually only the most important insurance groups have suitable expertise to address the issues in the IAS and Solvency II projects.

So it is necessary to invest in technology and knowledge in this area to improve the efficiency of the private insurance system; the selection of efficient internal models shall enable a more efficient capital allocation which by a side could improve insurance benefits and the other get better shareholders' return on assets.

Acknowledgments

This work has been developed within the National Research project (PRIN-2007) entitled "Risk Evaluation Models for the Solvency of Insurance Companies and Pension Funds" issued by the Italian Ministry of University and Research (MIUR) and coordinated by Prof. P. De Angelis.

Particular thanks to Dr. Andrea Fortunati who helped to develop the actuarial model and the numerical application proposed.

References

Andreatta, G., Corradin, S.: Valuing the Surrender Options Embedded in a Portfolio of Italian Life Guaranteed Participating Policies: a Least Squares Monte Carlo Approach. In: Real option theory meets practice. 8th Annual International Conference, Montrèal Canada, June 17-19 (2004)

Artzner, P., Delbaen, F., Eber, J.M., Heath, D.: Coherent Measures of Risk. Math. Finance. 3, 203–228 (1999)

Baione, F., De Angelis, P., Fortunati, A.: Actuarial Models for a Fair Evaluation of Life Insurance Policies. Meeting MTISD 2004. University of Sannio. In: Amenta, P., D'Ambra, L., Squillante, M., Ventre-Metodi, A. (eds.) Modelli e Tecnologie dell'informazione a supporto delle decisioni, pp. 31–55. Franco Angeli (2004) ISBN 88-464-7440-6

Baione, F., De Angelis, P., Fortunati, A.: On a Fair Value Model for Partecipating Life Insurance Policies. Investment Management and Financial Innovations 2, 105–115 (2006)

Ballotta, L., Haberman, S.: Guaranteed annuity conversion options and their valuation (2003), http://www.casact.org

Brown, J.S., Dybvig, P.H.: The empirical implications of the Cox, Ingersoll, Ross theory of the term structure of interest rates. Journal of Finance 4, 1–32 (1986)

CEIOPS: EU Solvency II project – the first Quantitative Impact Study (2005), http://www.ceiops.org

CEIOPS: Quantitative Impact Study 2 (2006), http://www.ceiops.org

CEIOPS: Quantitative Impact Study 3 (2007), http://www.ceiops.org

CEIOPS: Quantitative Impact Study 4 (2008), http://www.ceiops.org

Cox, J.C., Ingersoll, J.E., Ross, S.A.: A theory of the term structure of interest rates. Econometrica, 53–408 (1985)

De Angelis, P.: Gli Standards Internazionali per la valutazione del Fair Value delle compagnie di Assicurazione: implicazioni attuariali. In: Seminar of Italian Institute of Actuary, Rome (2001)

De Felice, M., Moriconi, F.: Finanza dell'assicurazione sulla vita. Research Group on: Modelli per la finanza matematica. Working paper 40 (2001)

IASB: IFRS 4: Insurance Contracts (2004), http://www.iasb.org

Swiss Federal Office of Private Insurance: White Paper of the Swiss Solvency Test (2004)

Qualitative Bipolar Decision Rules: Toward More Expressive Settings

Didier Dubois and Hélène Fargier

Abstract. An approach to multicriteria decision-making previously developed by the authors is reviewed. The idea is to choose between alternatives based on an analysis of the pros and the cons, i.e. positive or negative arguments having various strengths. Arguments correspond to criteria or affects of various levels of importance and ranging on a very crude value scale containing only three elements: good, neutral or bad. The basic decision rule in this setting is based on two ideas: focusing on the most important affects, and when comparing the merits of two alternatives considering that an argument against one alternative can be counted as an argument in favour of the other. It relies on a bipolar extension of comparative possibility ordering. Lexicographic refinements of this crude decision rule turn out to be cognitively plausible, and to generalise a well-known choice heuristics. It can also be encoded in Cumulative Prospect Theory. The paper lays bare several lines of future research, especially an alternative to the bicapacity approach to bipolar decision-making, that subsumes both Cumulative Prospect Theory and our qualitative bipolar choice rule. Moreover, an extension of the latter to non-Boolean arguments is outlined.

1 Introduction

It is known from many experiments in cognitive psychology that humans often evaluate alternatives or objects for the purpose of decision-making by considering positive and negative aspects separately. Under this bipolar view, comparing two decisions comes down to comparing pairs of sets of arguments or features, namely, the set of pros and cons pertaining to one decision versus the set of pros and cons

Didier Dubois
IRIT, CNRS & Université de Toulouse, France
e-mail: dubois@irit.fr

Hélène Fargier
CNRS & Université de Toulouse, France
e-mail: fargier@irit.fr

S. Greco et al. (Eds.): Preferences and Decisions, STUDFUZZ 257, pp. 139–158.
springerlink.com

pertaining to the other. Such kind of information involving negative and positive features is called *bipolar*. Psychologists have shown [35, 8, 38] that the simultaneous presence of positive and negative arguments prevents decisions from being simple to make, except when their strengths have different orders of magnitude.

In the family of numerical decision models, classical utility theory does not exploit bipolarity. Utility functions are defined up to an increasing affine transformation (i.e., they rely on an interval scale), and the separation between positive and negative evaluations has no special meaning. Cumulative Prospect Theory (CPT, for short) proposed by [39] is an attempt to explicitly account for positive and negative evaluations in the numerical setting. It computes the so-called net predisposition for a decision, viewed as the difference between two numbers, the first one measuring the importance of the group of positive features, the second one the importance of the group of negative features. Such group importance evaluations are modelled by non-additive set functions called capacities. More general numerical models, namely bi-capacities[28] and bipolar capacities [30] encompass situations where positive and negative criteria are not independent from each other.

However, Gigerenzer and Todd [25] have argued that human decisions are often made on the basis of an ordinal ranking of the strength of criteria rather than on numerical evaluations, hence the qualitative nature of the decision process. People choose according to the most salient arguments in favour of a decision or against the others. They seldom resort to explicit numerical computations of figures of merit. This idea is also exploited in Artificial Intelligence in qualitative decision theory [11]. See [14] for a recent survey of qualitative decision rules under uncertainty. So-called conditional preference networks (CP-nets) [7] allow for an easier representation of preference relations on multidimensional sets of alternatives, using local conditional preference statements interpreted ceteris paribus.

Most qualitative approaches use preference relations that express statements like "decision *a* is better than decision *b* for an agent." However, people also need to express that some decision is good or bad for them, a notion that simple preference relations cannot express. Using a simple preference relation, the best available choice may fail to be really suitable for the decision-maker. In other circumstances, even the worst ranked option remains somewhat acceptable. To discriminate between these two situations, one absolute landmark or reference point expressing neutrality or indifference, and explicitly separating the positive and the negative judgments, must appear in the model. Even ordinal decision methods need to inject some form of bipolarity. Note that multicriteria decision methods based on the merging of outranking relations use concordance and discordance tests between criteria, where the notion of veto prevents the choice of alternatives that rate too low with respect to some criteria. It can be viewed as an attempt to capture the idea of bipolar preference [36].

In this chapter, we review a bipolar and qualitative decision-making model based on the symmetric comparison of positive and negative arguments, first proposed in [13]. The proposed ordinal and bipolar decision rules have been recently axiomatized in terms of properties they satisfy[5] and their empirical validity tested[6]. Unsurprizingly, these choice rules are strongly related to possibility theory [32, 12, 15], and rely on a bipolar extension of possibility measures tailored to the comparison of sets with elements having positive or negative polarity.

The paper is structured as follows. Section 2 recalls the basic typology of bipolar information. Section 3 present the formal framework for a qualitative approach to bipolar decision-making. Section 4 recalls the main qualitative bipolar decision rules and their properties in the case of Boolean bipolar criteria. Section 5 discusses the links between these decision rules and related works relying on a numerical approach. Section 6 suggests a joint extension of the CPT and the qualitative bipolar framework that differs from the bicapacity setting of Grabisch and Labreuche. Finally an attempt to generalize the qualitative setting beyond all-or-nothing criteria is outlined.

2 Bipolarity in Information Management

Dubois and Prade [16, 17] provide a general discussion on the bipolar representation of information, showing that bipolarity can be at work in reasoning, learning and decision processes. This section outlines various types of bipolarity.

2.1 Value Scales

The representation of bipolarity depends on the proper interpretation of value scales. A bipolar scale $(L, >)$ is a totally ordered set with a prescribed interior element e called *neutral*, separating the positive evaluations $\lambda > e$ from the negative ones $\lambda < e$. Mathematically, if the scale is equipped with a binary operation $\star$ (an aggregation operator), e is an idempotent element for $\star$, possibly acting as an identity. When the scale is bounded, the bottom (fully negative) is denoted 0_L and the top (fully positive) 1_L.

Examples:

- The most obvious quantitative bipolar scale is the (completed) real line equipped with the standard addition, where 0 is the neutral level. Isomorphic to it is the unit interval equipped with an associative uninorm like $\frac{xy}{xy+(1-x)(1-y)}$. Then the neutral point is 0.5, 0 plays the same role as $-\infty$ and 1 as $+\infty$ in the real line. Also the interval $[-1,1]$ is often used as a bipolar scale;
- The simplest qualitative bipolar scale contains three elements: $\{-,0,+\}$.

In bipolar scales, the negative side of the scale is the mirror image of the positive one. An object is evaluated on such a bipolar scale as being either positive or negative or neutral. It cannot be positive and negative at the same time. This is called a *univariate bipolar* framework.

Another type of bipolar framework uses two distinct totally ordered scales L^+ and L^- for separately evaluating the positive and the negative information. This is the *bivariate unipolar* framework. Here each scale is unipolar in the sense that the neutral level is at one end of the scale. In a *positive* scale the bottom element is neutral. In a *negative* scale the top element is neutral. A bipolar scale can be viewed as the union of a positive and a negative scale $L^+ \cup L^-$ extending the ordering relations on each scale so $\forall \lambda^+ \in L^+, \lambda^- \in L^-, \lambda^+ > \lambda^-$. The symmetrisation of finite unipolar scales is incompatible with associative operations [26] : only infinite bipolar scales seem to support such operations!

2.2 The Three Forms of Bipolarity

Three forms of bipolarity can be found at work in the literature, we call types I, II, III for simplicity [16, 17]. The two last types use two unipolar bivariate scales.

- **Type I : Symmetric univariate bipolarity.** It relies on the use of bipolar scales. The two truth-values *true* and *false* of classical logic offer a basic view of bipolarity. However, the neutral value only appears in three-valued logics. Probability theory exhibits a type I bipolarity as the probability of an event is clearly living on a bipolar scale $[0,1]$ whose top means *totally sure* and bottom *impossible* (not to be confused with *true* and *false*). The neutral value is 0.5 and refers to the total uncertainty about whether an event or its contrary occurs (not to be confused with *half-true*). In decision theory, Cumulative Prospect Theory [39] uses the real line as a bipolar scale. It is numerical, additive, and bipolar. It measures the importance of positive affects and negative affects *separately*, by two monotonic set functions σ^+, σ^- and finally computes a net predisposition $N = \sigma^+ - \sigma^-$.
- **Type II : Symmetric bivariate bipolarity.** It works with a unipolar scale L where $0_L = \mathbf{0}$ is neutral. Here, an item is judged according to two independent evaluations:
  - one in favour of the item, say $\alpha^+ \in L$
  - one in disfavour of the item, say $\alpha^- \in L$.

  Positive and negative strengths are computed similarly on the basis of the same data and can be conflicting. If L is equipped with an order-reversing map ν, a constraint $\alpha^+ \leq \nu(\alpha^-)$ may limit the allowed conflict between positive and negative information: it prevents the pair of ratings (α^+, α^-) from expressing the idea of being fully in favour of an option and fully against it. An example of such an evaluation framework with limited conflict stems an imprecise rating on a univariate bipolar scale Λ in the form of an interval $[\alpha_*, \alpha^*]$. It is clear that lower bounds of ratings on Λ live on a positive unipolar scale ($L^+ = \Lambda$, interpreting $0_L \in \Lambda$ as neutral in L^+) and upper bounds of ratings on Λ live on a negative unipolar scale ($L^- = \Lambda$, interpreting $1_L \in L$ as neutral in L^-). Then let $\alpha^+ = \alpha_*$ and $\alpha^- = \nu(\alpha *)$, where ν is the order-reversing map in Λ. Well-known examples of such a bipolarity can be found in formal frameworks for argumentation where reasons for asserting a proposition and reasons for refuting it are collected [1]. Argumentation frameworks can be used to explain decisions made according to several criteria [2]. This also typical of uncertainty theories [18] leaving room for incomplete information where each event A is evaluated by an interval $[C(A), Pl(A)]$, $C(A)$ and $Pl(A)$ reflecting their certainty and plausibility

respectively. Then, $\alpha^+ = C(A)$, and $\alpha^- = C(A^c) = 1 - Pl(A)$. A good example of certainty/plausibility pairs displaying this kind of bipolarity are belief and plausibility functions of Shafer [37].
- **Type III : Asymmetric bipolarity.** In this form of bipolarity, the negative part of the information is not of the same nature as the positive part, while in type II bipolarity only the polarities are opposite. When merging information in the Type III setting, negative and positive pieces of information will not be aggregated using the same principles. The positive side is not a mirror image of the negative side either. Nevertheless, positive and negative information cannot be completely unrelated. They must obey minimal consistency requirements. This third kind of bipolarity is extensively discussed in [19] within the framework of possibility theory. In uncertainty modeling or knowledge representation heterogeneous bipolarity corresponds to the pair (knowledge, data). Knowledge is negative information in the sense that it expresses constraints on how the world behaves, by ruling out impossible or unlikely relations: laws of physics, common sense background knowledge (claims like "birds fly"). On the contrary, data represent positive information because it represents examples, actual observations on the world. A not yet observed event is not judged impossible; observing it is a positive token of support. Accumulating negative information leads to ruling out more possible states of the world (the more constraints, the less possible worlds). Accumulating positive information enlarges the set of possibilities as being guaranteed by empirical observation. In decision-making, heterogeneous bipolarity concerns the opposition between constraints (possibly flexible ones) that state which solutions to a problem are unfeasible, and goals or criteria, that state which solutions are preferred [3].

The qualitative approach described in this paper is based on type II bipolarity, and is close to argumentation frameworks.

3 A Qualitative Framework for Choice Based on Pros and Cons

A formal elementary framework for bipolar multicriteria decision analysis requires

- a finite set $\mathscr{D}$ of potential decisions $a, b, c, \ldots$;
- a set X of criteria, viewed as mappings x with domain $\mathscr{D}$ ranging on a bipolar scale V;
- and a totally ordered scale L expressing the relative importance of criteria or groups of criteria.

In the simplest qualitative setting, criteria are valued on a bipolar scale $V = \{-, 0, +\}$, whose elements reflect negativity, neutrality and positivity respectively: each value $x(a)$ expresses that x stands as an argument in favour of $a \in \mathscr{D}$ or in disfavour of a or yet is irrelevant to a (when $x(a) = 0$) in the spirit of cooperative bigames [4] or formal argumentation theories [1]. For each alternative a, let $A = \{x, x(a) \neq 0\}$ be the set of relevant (non-indifferent) arguments in favour of decision a. Then

- $A^- = \{x, x(a) = -\}$ is the set of arguments against decision a
- $A^+ = \{x, x(a) = +\}$ is the set of arguments in favour of a.

It comes down to enumerating the pros and the cons of a. So, comparing decisions a and b comes down to comparing the pairs of disjoint sets (A^-, A^+) and (B^-, B^+).

Even if in our setting, arguments or criteria are Boolean, they can be more or less important. Levels of importance of individual criteria are expressed by a function $\pi : X \mapsto L$., where the scale L is unipolar positive. It has top 1_L (full importance) and bottom $0_L = e$ (no importance). Within a qualitative approach, L is typically finite. $\pi(x) = 0_L$ means that the decision maker is indifferent to criterion x ; 1_L is the highest level of attraction or repulsion (according to whether it applies to a positive or negative argument). Assignment π is supposed to be non trivial, i.e., at least one x has a positive level of importance.

In a nutshell, each argument $x(a)$ is of the all-or-nothing bipolar kind. It has

- *a polarity*: the presence of $x(a)$ is either good or bad (its absence is always neutral).
- *a degree of importance* $\pi(x) \in L$ that does not depend on the alternative.

Since we are looking for *qualitative* decision rules, the approach relies on two modelling assumptions:

- *The use of finite qualitative importance scales*: the qualitativeness of L means that there is a big step between one level of merit and the next lower one. Arguments are ranked in terms of the *order of magnitude* of their figures of importance by means of the mapping π.
- *Focus effect* : the order of magnitude of the importance of a group A of arguments is the one of the most important argument, in the group. This assumption perfectly suits the intuition of a qualitative scale as it means that several weaker arguments are always negligible compared with a single stronger one.

The second hypothesis made here implies that the importance of a group of criteria only depends on the importance of the individual ones. It means that there is no significant synergy between criteria, which are in some sense considered as redundant with respect to one another. The focus effect thus drastically reduces the type of interaction allowed between criteria, but it facilitates computations when evaluating decisions. Technically, it enforces a possibility measure [15] for measuring the importance of a set A of arguments relevant to a decision:

$$OM(A) = \max_{x \in A} \pi(x). \tag{1}$$

In fact, we only use comparative possibility relations [32, 12], interpreted in terms of *order of magnitude* of importance, hence the notation OM, here. When comparing an alternative a with an alternative b, it all comes down to comparing pairs of evaluations $(OM(A^-), OM(A^+))$ and $(OM(B^-), OM(B^+))$. Note that one may have $A^- \cap B^+ \neq \emptyset$ and $A^+ \cap B^- \neq \emptyset$. However, in many examples a qualitative criterion comes down to a Boolean feature whose absence is considered neutral and whose presence is either positive or negative per se. When this is the case, X is made of two subsets: X^+ is the set of arguments positive when present (taking their value in the set $\{0,+\}$), X^- is the set of arguments negative when present (taking their value in the set $\{-,0\}$). Elements of X can be called *affects*, a positive affect relevant for an option being an argument in favour of this option, and likewise for negative relevant affects acting as arguments against options. Then $A^- \cap B^+ = \emptyset$ and $A^+ \cap B^- = \emptyset$. This simplifying assumption can be made without loss of generality and will not affect the validity of our results in the ordinal setting. Indeed, any criterion x whose range is the full domain $\{-,0,+\}$ can be duplicated, leading to a positive affect x_+ in X^+ and a negative affect x_- in X^-. This transformation moreover enlarges the framework so as to allow an affect to be positive and negative simultaneously (e.g., "eating chocolate" can have both a positive and a negative aspect).

The proposed framework is clearly of type II in the bipolarity typology.

Example

Luc has to choose a holiday destination and considers two options for which he has listed the pros and cons. Option a is in a very attractive region (a strong pro); but it is very expensive, and the plane company has a terrible reputation (two strong cons). Option b is in a non-democratic country, and Luc considers it a strong con. On the other hand, Option b includes a tennis court, a disco, and a swimming pool. These are three pros, but not very decisive: they do matter, but not as much as the other arguments.

Note that Luc can only provide a rough evaluation of how strong a pro or a con is. He can only say that the attractiveness of the region, the price, the reputation of the company, and the fact of being in a non-democratic country are four arguments of comparable importance; and that swimming pool, tennis and disco are three positive arguments of comparable, but much lesser importance. Formally, let:

- $X^+ = \{tennis^+, swimming^+, disco^+\}$ be the subset of pros ;
- $X^- = \{price^{--}, company^{--}, non-demo^{--}\}$ be the subset of cons.

By convention in the above lists, doubling the sign symbol appearing in superscript indicates higher importance ($L = \{0_L, +, ++\}$). Available decisions are described by :

Option $a : A^+ = \{region^{++}\}; A^- = \{company^{--}, price^{--}\}$.
Option $b : B^+ = \{tennis^+, swimming^+, disco^+\}; B^- = \{non-demo^{--}\}$.

The kind of bipolarity accounted for here differs from the one considered in [3] where negative preferences refer to prioritized constraints while positive preferences refer to goals or desires. Constraints expressed as logical formulas have a prominent role and first select the most tolerated decisions; positive preferences (goals and desire) then act to discriminate among this set of tolerated decisions. Hence a positive evaluation, even if high, can never outperform a negative evaluation even if very weak. In this approach, negative features prevail over positive ones. The latter matter only when no constraint is violated. In the approach proposed here, positive and negative arguments play symmetric roles.

4 Some Qualitative Bipolar Outranking Relations

The aim of decision rules is to build an outranking relation $\succeq$ describing preference between between alternatives in $\mathscr{D}$, based on comparing sets of relevant affects A and B. It comes down to building a crisp preference relation over 2^X. $A \succeq B$ meaning that decision (with relevant affects forming set) A is at least as good as decision (with relevant affects forming set) B. Any outranking relation $\succeq$ includes:

- a symmetric part : $A \sim B \Leftrightarrow A \succeq B$ and $B \succeq A$;
- an asymmetric part: $A \succ B \Leftrightarrow A \succeq B$ and $not(B \succeq A)$;
- an incomparability relation : $A \asymp B \Leftrightarrow not(A \succeq B)$ and $not(B \succeq A)$.

Relations used here are not necessarily complete nor transitive: They are supposed to be quasi-transitive (their strict part $\succ$ is transitive).

However, as the set $X = X^+ \cup X^-$ of affects is partitioned into positive and negative ones, the relation on the powerset 2^X is said to be bipolar, in the sense that it will satisfy basic properties specific to the bipolar case:

Definition 1. A relation on a power set 2^X, where $X = X^+ \cup X^-$, is a *monotonic bipolar set relation* iff it is reflexive, quasi-transitive and satisfies the properties

1. *Non-Triviality*: $X^+ \succ X^-$.
2. *Positive monotony* : $\forall C, C' \subseteq X^+, \forall A, B : A \succeq B \Rightarrow C \cup A \succeq B \setminus C'$
3. *Negative monotony* : $\forall C, C' \subseteq X^-, \forall A, B : A \succeq B \Rightarrow C \setminus A \succeq B \cup C'$
4. *Weak Unanimity* : $\forall A, B, A^+ \succeq B^+$ and $A^- \succeq B^- \Rightarrow A \succeq B$.

Non-triviality says that an alternative that has all pros is preferred to one having all cons. Positive monotony is usual monotony in the positive case, while monotony is reversed for sets of negative affects. The weak unanimity says if an an alternative is not worse than another one from the point of view of its pros and its cons then the former should be weakly preferred to the latter. There is a strong unanimity version of this property, enforcing strict preference if, moreover, the preference either on the positive or the negative side is strict.

4.1 Pareto-Dominance

The first idea that comes to our mind for comparing alternatives a and b using pairs of evaluations $(OM(A^-), OM(A^+))$ and $(OM(B^-), OM(B^+))$ is to use Pareto-Dominance:

$$A \succeq^P B \iff OM(A^+) \geq OM(B^+) \text{ and } OM(A^-) \leq OM(B^-) \tag{2}$$

$\succeq^P$ collapses to Wald's pessimistic ordering if $X = X^-$ (choosing based on the worst feature [40]), and to its optimistic max-based counterpart if $X = X^+$. On Luc's example, there is a strong argument for Option a, but only weak arguments for Option b : $OM(A^+) > OM(B^+)$. In parallel, there are strong arguments both against A and against B: $OM(A^-) = OM(B^-)$. Luc will choose Option a.

This decision rule is in fact not very satisfactory. The bipolar outranking relation $\succeq^P$ concludes to incomparability in some cases when a preference would sound more natural. When A has both pros and cons, it is incomparable with the empty set of affects even if the importance of the cons in A is negligible in front of the importance of its pros. Another drawback of this rule can observed when the two decisions have the same evaluation on one of the two polarities. Namely, if $OM(A^-) = OM(B^-) > OM(A^+) > 0 = OM(B^+)$, then $A >^P B$, and this despite the fact $OM(A^+)$ may be very weak. In other terms, this rule does not completely obey the principle of focalisation on the most important arguments, especially when A^- and B^- do not contain the same number of affects. In Luc 's example, Option a is preferred despite its two major drawbacks.

4.2 *The Bipolar Possibility Relation*

The problem with the bipolar Pareto-dominance is that it does not account for the fact that the two evaluations *share a common importance scale*. The next decision rule for comparing A and B focuses on arguments of maximal strength in $A \cup B$, i.e., those at level $\max_{y \in A \cup B} \pi(y) = OM(A \cup B)$. The principle at work in this rule is simple: any argument against A (resp. against B) is an argument pro B (resp., pro A) and conversely. The most supported decision is then preferred.

Definition 2 (Bipolar Possibility Relation).
$A \succeq^{B\Pi} B \iff OM(A^+ \cup B^-) \geq OM(B^+ \cup A^-)$.

This rule decides that A is at least as good as B as son as there are maximally important arguments in favour of A or attacking B; $A \succ^{B\Pi} B$ if and only if at the highest level of importance, there is no argument against A and none for B. Obviously, $\succeq^{B\Pi}$ collapses to Wald's pessimistic ordering if $X = X^-$ and to its optimistic counterpart when $X = X^+$. In some sense, this definition is the most straightforward generalisation of possibility relations [32, 12] to the bipolar case.

In Luc's example, Option $a = \{region^{++}, company^{--}, price^{--}\}$ and Option $b = \{non-demo^{--}, tennis^+, swimming^+, disco^+\}$ are considered as very bad and thus indifferent since one trusts a bad airplane company and the other takes place in a non-democratic country.

The bipolar possibility relation satisfies the following properties

1. It is complete and quasi-transitive.
2. The restriction of $\succeq^{B\Pi}$ to single affects is a weak order (defining their relative importance).
3. *Ground Monotony*: $\forall A, B, x, x'$ such that $A \cap \{x, x'\} = \emptyset$ and $\{x'\} \succeq^{B\Pi} \{x\}$:
   $A \cup \{x\} \succ B \Rightarrow A \cup \{x'\} \succ B$; $A \cup \{x\} \sim B \Rightarrow A \cup \{x'\} \succeq B$;
   $B \succ A \cup \{x'\} \Rightarrow B \succ A \cup \{x\}$; $B \sim A \cup \{x'\} \Rightarrow B \succeq A \cup \{x\}$.
4. *Positive Cancellation*: $\forall x, z \in X^+, y \in X^-$, $\{x, y\} \sim \emptyset$ and $\{z, y\} \sim \emptyset \Rightarrow \{x\} \sim \{z\}$.
5. *Negative Cancellation* : $\forall x, z \in X^-, y \in X^+$, $\{x, y\} \sim \emptyset$ and $\{z, y\} \sim \emptyset \Rightarrow \{x\} \sim \{z\}$.

6. *Strict negligibility*: $\forall A,B,C,D : A \succ B$ and $C \succ D \Rightarrow A \cup C \succ B \cup D$.
7. *Idempotent Negligibility*[1]: $\forall A,B,C,D : A \succeq B$ and $C \succeq D \Rightarrow A \cup C \succeq B \cup D$.

Note that relation $\succeq^{B\Pi}$ is generally not transitive. Properties 2 and 3 are self-explanatory. Properties 4 and 5 express a form of anonymity. It is required when a positive argument blocks a negative argument of the same strength: this blocking effect should not depend on the arguments themselves, but on their position in the importance scale only. The two last properties are direct consequences of working with importance levels that are orders of magnitude. $A \succ B$ means that A is much better than $B0$, so much so as there is no way of overthrowing A by sets of weaker arguments (property 6). The last property expresses that several arguments of the same strength are worth just one.

The above properties turn out to be characteristic of the bipolar possibility rule [5]. They imply the existence of the importance scale, and the importance assignment to elementary affects as a possibility distribution π.

4.3 Bipolar Lexicographic Outranking Relations

The last property (Idempotent Negligibility) of the bipolar possibility rule is by far the most debatable feature of $\succeq^{B\Pi}$. It causes a drowning effect, usual in standard possibility theory. For instance, when B is strictly included in a set of positive affects A, then A is not strictly preferred to B when affects in $A \setminus B$ are of equal or lesser importance than those in B.

A tempting way of refining $\succeq^{B\Pi}$, noticing that the bipolar possibility relation basically relies on computing the maximum of π over subsets, is to use a leximax relation instead [10]. Then the number of arguments of equal strength on each side is taken into account. Among the two basic axioms of qualitative modeling, it comes down to giving up Idempotent Negligibility, while retaining Strict Negligibility. Choosing can then be based on counting arguments of the same strength, but we still do not allow an important argument to be superseded by several less important ones, however large their number be. The arguments in A and B are scanned top down, until a level is reached such that the numbers of positive and negative arguments pertaining to the two alternatives are different; then, the option with the least number of negative arguments and the greatest number of positive ones is preferred.

There are two such decision rules respectively called "Bivariate Levelwise Tallying" and (univariate) "Levelwise Tallying" [6], according to whether positive and negative arguments are treated separately or not [13].

For any importance level $\lambda \in L$, let $A_\lambda = \{x \in A, \pi(x) = \lambda\}$ be the λ-section of A, the set of affects of strength λ in A. Let $A^+_\lambda = A_\lambda \cap X^+$ (resp., $A^-_\lambda = A_\lambda \cap X^-$) be its positive (resp., negative) λ-section. Let $\delta(A,B)$ be the maximal level of importance where either the positive or the negative λ-sections of A and B differ, namely:

$$\delta(A,B) = \max\{\lambda : |A^+_\lambda| \neq |B^+_\lambda| \text{ or } |A^-_\lambda| \neq |B^+_-|\}.$$

$\delta(A,B)$ is called the *decisive level* pertaining to (A,B).

[1] It is called "Closeness" in [5].

Definition 3 (Bivariate Levelwise Tallying).
$A \succeq^{BL} B \iff |A^+_{\delta(A,B)}| \geq |B^+_{\delta(A,B)}|$ and $|A^-_{\delta(A,B)}| \leq |B^-_{\delta(A,B)}|$

It is easy to show that $\succeq^{BL}$ is reflexive, transitive, but not complete. Indeed, $\succeq^{BL}$ concludes to an incomparability if and only if there is a conflict between the positive view and the negative view at the decisive level. From a descriptive point of view, this range of incomparability is a good point in favour of $\succeq^{BL}$. On Luc's example for instance, the difficulty of the dilemma is clearly pointed point by this decision rule: at the highest level, Option a involves 3 strong arguments ($price^{--}$, $company^{--}$ and $attractive-region^{++}$) while Option b only involves one ($non-demo^{--}$). $\delta(A,B) = ++$ and Option a has more strong negative arguments than Option b, while Option b has one strong positive argument while Option a has none. An incomparability results, revealing the difficulty of the choice.

Now, if one can assume a compensation between positive and negative arguments at each importance level, one argument canceling another one on the other side, the following refinement of relation $\succeq^{BL}$ can be obtained:

Definition 4 (Univariate Levelwise Tallying).
$$A \succeq^{L} B \iff \exists \lambda \in L \setminus 0_L \text{ s. t. } \begin{cases} \forall \gamma > \lambda, & |A^+_\gamma| - |A^-_\gamma| = |B^+_\gamma| - |B^-_\gamma| \\ \text{and} & |A^+_\lambda| - |A^-_\lambda| > |B^+_\lambda| - |B^-_\lambda| \end{cases}$$
or $|A^+_\gamma| - |A^-_\gamma| = |B^+_\gamma| - |B^-_\gamma|, \forall \lambda \in L \setminus 0_L$ (the latter case is $A \sim^L B$).

Interestingly, $\succeq^{L}$ is closely related to the decision rule originally proposed two centuries ago by Benjamin Franklin [22]. On Luc's dilemma, the strong pro of Option a is now cancelled by one of its strong cons - they are discarded. A strong con on each side remains. Because of a tie at level $++$, the procedure then examines the second priority level: there are 3 weak pros for Option b (no cons) and no pro nor con w.r.t. Option a : Option b is thus elected.

The two decision rules proposed in this section obviously generate monotonic bipolar outranking relations. Each of them refines $\succeq^{B\Pi}$. The most decisive one is $\succeq^{L}$, which is moreover complete and transitive. This relation is the refinement of $\succeq^{B\Pi}$ that is a weak order and that satisfies the principle of preferential independence without introducing any bias on the importance order elementary affects (that is, preserving the restriction of $\succeq^{B\Pi}$ to single affects). See [5] for such an axiomatisation. It turns out that Levelwise Tallying is the most likely decision rule to be used by people as empirical studies suggest [6].

5 Bridging the Gap between Qualitative Choice Heuristics and Cumulative Prospect Theory

Qualitative choice heuristics were extensively studied and advocated by Gigerenzer and his colleagues [25]. In the "Take the best" approach [24], each criterion x is supposed to have a positive side (it generates a positive argument x^+) and a negative side (it generates a negative argument x^-): fulfilling the criterion is a pro, missing it is a con, all the worse as the criterion is important. The criteria are then supposed to be totally ranked and of very different importance: there does not exist x, y such that $\pi(x) = \pi(y)$. So, we can scan elementary criteria top down from the stronger to the weaker when comparing alternatives. So, doing, as soon as we find a criterion in favour of decision a and in disfavour of decision b, a is preferred to b (hence the name "Take the best"). The Levelwise Tallying choice rule applied to such linearly ranked criteria coincides with "Take the best". But the former is capable of accounting for more decision situations than the latter heuristic — e.g., when several criteria share the same (and highest) degree of importance. In this sense, Levelwise Tallying is a natural extension of the "Take the best" qualitative rule advocated by psychologists.

In contrast to the "Take the best" approach, Cumulative Prospect Theory [39] accounts for positive and negative arguments using quantitative evaluations of criteria importance. CPT assumes that the strength of reasons supporting a decision and the strength of reasons against it can be measured by means of two numerical capacities σ^+ and σ^- respectively mapping subsets of X^+ and X^- to the unipolar scale $[0, +\infty)$. The capacity σ^+ reflects the importance of the group of positive arguments, and σ^- the importance of the group of negative arguments. The higher $\sigma^+(A^+)$, the more convincing the set of positive arguments and conversely the higher $\sigma^-(A^-)$, the more deterring is the set of negative arguments.

This approach moreover admits that it is possible to combine these evaluations by subtracting them and building a so-called "net predisposition" score expressed on a bipolar numerical scale (the real line):

$$\forall A \subseteq X, NP(A) = \sigma^+(A^+) - \sigma^-(A^-)$$

where $A^+ = A \cap X^+$, $A^- = A \cap X^-$. Alternatives are then ranked according to this net predisposition:

$$A \succeq^{CPT} B \iff \sigma^+(A^+) - \sigma^-(A^-) \geq \sigma^+(B^+) - \sigma^-(B^-). \tag{3}$$

It turns out that the Levelwise Tallying heuristics $\succeq^L$ can be encoded in the CPT model. To this effect, we can the use classical encoding of the leximax (unipolar) procedure by comparing sums of suitably chosen weights forming a super-increasing sequence [33]. It is easy to show that actually both outranking relations $\succeq^L$ and $\succeq^{BL}$ can be encoded by means of numerical capacities:

Proposition 1. *There exist two capacities σ^+ and σ^- such that:*
$A \succeq^L B \Leftrightarrow \sigma^+(A^+) - \sigma^-(A^-) \geq \sigma^+(B^+) - \sigma^-(B^-)$

$$A \succeq^{BL} B \Leftrightarrow \text{ and } \begin{cases} \sigma^+(A^+) \geq \sigma^-(B^+) \\ \sigma^+(B^-) \geq \sigma^-(A^-) \end{cases}$$

For instance, denoting $\lambda_1 = 0_L < \lambda_2 < \cdots < \lambda_l = 1_L$ the l elements of L, we can use the capacity

$$\sigma^+(A) = \sum_{\lambda_i \in L} |A^+_{\lambda_i}| \cdot |X|^i.$$

We can similarly use a second sum of super-increasing numbers to represent the importance of sets of negative arguments:

$$\sigma^-(A) = \sum_{\lambda_i \in L} |A^-_{\lambda_i}| \cdot |X|^i$$

This proposition clearly shows that the $\succeq^L$ ranking of decisions is a particular case of the CPT decision rule (using big-stepped probabilities). In summary, $\succeq^L$ complies with the spirit of qualitative bipolar reasoning while being efficient. In the meantime, it has the advantages of numerical measures (transitivity and representability by a pair of functions). In other words, our framework bridges the gap between Take The Best and CPT, the two main antagonistic approaches to bipolar decision-making.

6 Extensions of the Qualitative Bipolar Setting

The CPT approach and its variants assume a kind of independence between X^+ and X^-. But this assumption does not always hold. So it is natural to extend it to more general ways of computing net predisposition indices.

However there are two ways of moving toward a more general approach, because there are two equivalent ways of defining the CPT relations, that suggest different generalisations:

$$A \succeq^{CPT} B \iff \text{any of} \begin{cases} \sigma^+(A^+) - \sigma^-(A^-) \geq \sigma^+(B^+) - \sigma^-(B^-), \\ \sigma^+(A^+) + \sigma^-(B^-) \geq \sigma^+(B^+) + \sigma^-(A^-). \end{cases} \quad (4)$$

Moreover the assumption of Boolean criteria, made in previous sections, can also be relaxed in the qualitative case.

6.1 Univariate and Bivariate Bicapacities

The first inequality in (4) is naturally extended by means of *bicapacities*. Bicapacities were introduced by Grabisch and Labreuche [28, 29] so as to handle non separable bipolar preferences: a bicapacity BC is defined on $\mathcal{Q}(X) := \{(A^+, A^-) \in 2^X, A^+ \cap A^- = \emptyset\}$ and increase (resp., decrease) with the addition of elements in A^+ (resp., A^-).

Definition 5. A bicapacity is a function $BC(\cdot,\cdot) : \mathcal{Q}(X) \to \mathbb{R}$ such that $BC(\emptyset,\emptyset) = 0$, and $BC(A^+, A^-) \geq BC(B^+, B^-)$ whenever $A^+ \supseteq B^+$ and $A^- \subseteq B^-$.

Note that the outranking relation $\succeq^{BC}$ on alternatives, induced by BC is complete and transitive, positive and negative monotonic. It does not necessary satisfy unanimity. To get it, one must request the property:

$$BC(A^+,\emptyset) \geq BC(B^+,\emptyset) \text{ and } BC(\emptyset,A^-) \geq BC(\emptyset,B^-) \Longrightarrow BC(A^+,A^-) \geq BC(B^+,B^-)$$

Nontriviality is ensured by the normalisation $BC(X^+, \emptyset) = 1, BC(\emptyset, X^-) = -1$. Likewise, Ground Monotony, Positive and Negative Cancellation properties are to be requested if needed for $\succeq^{BC}$.

Originally, bi-capacities stem from bi-cooperative games [4], where players are divided into two groups, the "pros" and the "cons": player x is sometimes in favour, sometimes against, but cannot be both simultaneously. In our modeling context, we typically restrict to $A^+ \subset X^+$ and $A^- \subset X^-$. The net predisposition of CPT is recovered letting $BC(A^+, A^-) = \sigma^+(A^+) - \sigma^-(A^-) = NP(A)$. If moreover the capacities σ^+ and σ^- are probability measures, then $\succeq^{BC}$ satisfies unanimity, i.e. is a monotonic bipolar set relation, that moreover satisfies Ground Monotony, Positive and Negative Cancellation properties.

Since the comparison of net predispositions and more generally bi-capacities systematically provides a complete and transitive outranking relation, it can fail to capture a large range of decision-making attitudes: contrasting affects make decision difficult, so why should the comparison of objects having bipolar evaluations systematically yield a complete and transitive relation ? It might imply some incompatibilities. That is why bicapacities were generalized by means of bipolar capacities [30]. The idea is to use two measures, a measure of positiveness (that increases with the addition of positive arguments and the deletion of negative arguments) and a measure of negativeness (that increases with the addition of negative arguments and the deletion of positive arguments), however without combining then. In other terms, a bipolar capacity σ is defined by a pair of bicapacites σ^+ and σ^-, namely by: $\sigma(A) = (\sigma^+(A^+, A^-), \sigma^-(A^-, A^+))$. Then A is preferred to B iff it is the case with respect to both measures – i.e. according to Pareto-dominance. This allows for the representation of conflicting evaluations and leads to a partial order. The bipolar qualitative Pareto-dominance rule of Section 4.1, as well as $\succeq^{BL}$, obviously belong to the family of decision rules based on bipolar capacities. See [27] for a comparative discussion of bicapacities and bipolar capacities, in the setting of conjoint measurement.

6.2 An Alternative Generalisation of CPT

The second inequality expressing the $\succeq^{CPT}$ relation in (4) is more directly related to the bipolar possibility relation. More precisely, $\succeq^{B\Pi}$ can be viewed as the natural qualitative counterpart of $\succeq^{CPT}$; indeed, the bipolar possibility decision rule comes down to changing $+$ into max in $\sigma^+(A^+) + \sigma^-(B^-) \geq \sigma^+(B^+) + \sigma^-(A^-)$. However, being only quasi-transitive, the relation $\succeq^{B\Pi}$ cannot be represented by means of a bicapacity.

So, there is another track for generalizing the CPT framework, turning possibility measures into standard capacities $\kappa : 2^X \to [0,1]$ defining:

$$A \succeq^{\kappa} B \iff \kappa(A^+ \cup B^-) \geq \kappa(B^+ \cup A^-) \quad (5)$$

adopting the view that an argument against alternative a is an argument in favour of b in the pairwise comparison of alternatives.

The following properties clearly hold for $\succeq^{\kappa}$: it is complete, and the restriction to single arguments is a weak order. However it is not clearly transitive, not even quasi-transitive in the general case. And while the non triviality, and both positive and negative monotony properties hold, the weak unanimity property, that would make $\succeq^{\kappa}$ a bipolar monotonic set relation, requires that κ satisfy an additional property on top of inclusion-monotonicity of capacities [9]:

Weak additivity: Let $A, B, C, D \subseteq X$ such that $A \cap C = \emptyset, B \cap D = \emptyset$; if $\kappa(A) \geq \kappa(B)$ and $\kappa(C) \geq \kappa(D)$ then $\kappa(A \cup C) \geq \kappa(B \cup D)$.

This property is, for capacities, equivalent to the following property involving only three subsets $A, B, C = D$ [12] :

If $\kappa(A) \geq \kappa(B)$ then $\kappa(A \cup C) \geq \kappa(B \cup C)$, provided that $(A \cup B) \cap C = \emptyset$.

This property implies that κ is a decomposable measure [9], that is, there exists an operation $\star$ acting on the image of 2^X by κ such that if $A \cap B = \emptyset$, $\kappa(A \cup B) = \kappa(A) \star \kappa(B)$. Due to compatibility with the underlying Boolean algebra of events, it is natural to consider that $\star$ is a co-norm. Choosing an Archimedean continuous co-norm, it is clear that $\succeq^{\kappa}$ can verify additional properties:

- *Transitivity*: $\kappa(A^+ \cup B^-) \geq \kappa(B^+ \cup A^-)$ and $\kappa(B^+ \cup C^-) \geq \kappa(C^+ \cup B^-)$ imply $\kappa(A^+ \cup C^-) \geq \kappa(C^+ \cup A^-)$. Indeed the preconditions imply

$$\kappa(A^+) \star \kappa(B^-) \star \kappa(B^+) \star \kappa(C^-) \geq \kappa(B^+) \star \kappa(A^-) \star \kappa(B^-) \star \kappa(C^+)$$

  which yields the results by simplification (if $\star$ is a strict t-norm or κ is properly normalized). This simplification cannot be made if $\star = \max$.
- *Ground monotony* holds under the same assumptions about $\star$.
- *Positive and negative cancellation* properties reduce to the trivial statement that $\kappa(\{x\}) = \kappa(\{y\}$ and $\kappa(\{z\}) = \kappa(\{y\}$ imply $\kappa(\{x\}) = \kappa(\{z\}$.

In fact, relation $A \succeq^{\kappa} B$ in (5) is a conjoint generalisation of $\succeq^{CPT}$ and $\succeq^{B\pi}$ that either comes down to one of them ($\succeq^{CPT}$ is obtained, if $\star$ is a nilpotent Archimedean t-norm and κ is properly normalized, or a strict co-norm, taking the logarithm of κ) or a combination of them (if $\star$ is an ordinal sum of the basic conorms $\alpha + \beta - \alpha\beta$, $\min(1, \alpha + \beta)$, max) up to a rescaling. So, the obtained approach is not very general.

One way of getting a more general model would be to consider again two capacities κ^+ on X^+ and κ^- on X^- and a standard binary aggregation operation $\oplus$ which is monotonically increasing in both places. Then the CPT model can be extended as follows:

$$A \succeq^{\kappa^+, \kappa^-, \oplus} B \iff \kappa^+(A^+) \oplus \kappa^-(B^-) \geq \kappa(B^+) \oplus \kappa(A^-) \qquad (6)$$

This set-relation is clearly bipolar monotonic (weak unanimity holds) and its transitivity is ensured if $\oplus$ is associative and strictly monotonic. The validity of ground monotony depends on the properties of the capacity. For instance, positive cancellation reads:

$$(\kappa^+(\{x\}) \oplus \kappa^-(\{y\}) = 0, \kappa^+(\{z\}) \oplus \kappa^-(\{y\}) = 0) \implies \kappa^+(\{x\}) = \kappa^+(\{z\}),$$

which holds if $\oplus$ is continuous and strictly monotonic. More insight is needed to figure out the potential of this approach.

6.3 Qualitative Non-boolean Criteria

Our basic qualitative setting is clearly simpler than usual MCDM frameworks where each $x \in X$ is a full-fledged criterion rated on a bipolar utility scale like $V = [-1, +1]$ containing a neutral value 0. Then comparing alternatives comes down to comparing vectors **a** and **b** in $V^{|X|}$ in the sense of a bicapacity BC by means of suitable adaptation of Choquet integral. In particular, the CPT model then computes the net predisposition in terms of a difference of Choquet integrals [31].

In the qualitative framework, we stick to a set X containing positive and negative affects, but we admit that each alternative a can be affected by x to a degree. Now we define degrees of satisfaction of criteria x in the same positive unipolar scale L as the one for importance. The convention is that if $x \in X^+$, the value $\mu_a(x) \in L$ evaluates how good a fares with respect to the positive affect x, and if $x \in X^-$, $\mu_a(x)$ evaluates how bad a fares with respect to the negative affect x. In this context, the set A of relevant affects for a becomes an L-fuzzy set. Using the notion of possibility of a fuzzy event [42], we can define the overall rating of an alternative a as

$$OM(A) = \sup_{x \in X} \min(\pi(x), \mu_a(x)) = \max(OM(A^+), OM(A^-)),$$

where A^+ is the restriction of μ_a to X^+, and A^- is the restriction of μ_a to X^-. The comparison of alternatives comes down to comparing pairs of ratings of the form $(OM(A^+), OM(A^-))$, where each element is a qualitative optimistic possibilistic utility functional [20, 14], namely $OM(A^+) = \sup_{x \in X^+} \min(\pi(x), \mu_a(x))$, $OM(A^-) = \sup_{x \in X^-} \min(\pi(x), \mu_a(x))$. The extension of the bipolar possibility decision rule to this framework simply reads

$$\mathbf{a} \succeq^{FB\Pi} \mathbf{b} \iff \max(OM(A^+), OM(B^-)) \geq \max(OM(B^+), OM(A^-)). \quad (7)$$

It is easy to see that OM is a monotonic bipolar *fuzzy*-set-function where monotony is with respect to fuzzy set inclusion on both the positive and negative parts of X. Relation $\succeq^{FB\Pi}$ satisfies bipolar pointwise Pareto-dominance defined by:

$$\mathbf{a} \geq \mathbf{b} \iff \forall x \in X^+, \mu_a(x) \geq \mu_b(x) \text{ and } \forall x \in X^-, \mu_a(x) \leq \mu_b(x),$$

namely, $\mathbf{a} \geq \mathbf{b}$ implies $\mathbf{a} \succeq^{FB\Pi} \mathbf{b}$. Unanimity, completeness and quasi-transitivity still hold. Strict and idempotent negligibility properties still hold.

Another bipolar setting for qualitative decision was proposed by Giang and Shenoy[23] who use, as a utility scale, a totally ordered set of possibility measures on a two-element set $\{0,1\}$ containing the values of the best and the worst rewards. Each such possibility distribution represents a qualitative lottery. Let $L_\Pi = \{(\alpha,\beta), \max(\alpha,\beta) = 1_L, \alpha,\beta \in L\}$. Coefficient α represents the degree of possibility of obtaining the worst reward, and coefficient β the degree of possibility of obtaining the best. This set can be viewed as a bipolar value scale ordered by the following complete preordering relation:

$$(\alpha,\beta) \geq_V (\gamma,\delta) \text{ if and only if } (\alpha \leq \gamma \text{ and } \beta \geq \delta)$$

The fact this relation is complete is due to the fact that pairs (α,β) and (γ,δ) such that neither $(\alpha,\beta) \geq_V (\gamma,\delta)$ nor $(\gamma,\delta) \geq_V (\alpha,\beta)$ hold cannot both lie in L_Π since it implies that either $\max(\alpha,\beta) < 1_L$ or $\max(\gamma,\delta) < 1_L$. The bottom of this bipolar scale is $(1_L, 0_L)$, its top is $(0_L, 1_L)$ and its neutral point $(1_L, 1_L)$ means "indifferent".

The canonical example of such a scale is the set of pairs $(\Pi(A^c), \Pi(A))$ of degrees of possibility for an event A and its complement A^c. An inequality such as $(\Pi(A^c), \Pi(A)) >_V (\Pi(B^c), \Pi(B))$ means that A is more likely (certain or plausible) than B (because it is equivalent to $\Pi(A) > \Pi(B)$ or $\Pi(B^c) > \Pi(A^c)$). In fact the induced likelihood ordering between events

$$A \succeq_{L\Pi} B \text{ if and only if } (\Pi(A^c), \Pi(A)) \geq_V (\Pi(B^c), \Pi(B))$$

is self-adjoint, that is, $A \succeq_{L\Pi} B$ is equivalent to $B^c \succeq_{L\Pi} A^c$.

Couching the Giang and Shenoy approach in our terminology, each alternative a is supposed to have a utility value for affect x in the form of (α_x, β_x) in L_Π. If $x \in X^+$, we have $\mu_a(x) = (\lambda, 1_L)$, and if $x \in X^-$, we have $\mu_a(x) = (1_L, \lambda)$. The proposed preference functional maps alternatives, viewed as tuples of pairs in L_Π indexed by elements of X, to L_Π itself. The importance of affects is also described by possibility weights $\pi(x)$. The so-called *binary possibilistic utility* of an alternative a is computed as the pair

$$W_{GS}(a) = (\max_{x \in X} \min(\pi(x), \alpha_x), \max_{x \in X} \min(\pi(x), \beta_x)),$$

which by construction lies in L_Π. This form results from simple and very natural axioms on possibilistic lotteries, which are counterparts to the Von Neumann and Morgenstern axioms [34]: a complete preorder of alternatives, increasingness in the wide sense according to the ordering in L_Π, substitutability of indifferent lotteries, and the assumption that the overall worth of alternatives is valued on L_Π [23]. Later on, Weng [41] proposed a Savage-style axiomatization of such binary possibilistic utility. It puts together the axiomatizations of the optimistic and the pessimistic possibilistic criteria by Dubois *et al.* [20], adding, to the axioms justifying Sugeno integral, two conditions: (i) the self-adjointness of the outranking relation on binary acts, and (ii) an axiom laying bare an act h that plays the role of a neutral point separating favourable from unfavourable alternatives.

Noticing that $\alpha_x = 1$ if $x \in X^-$ and $\beta_x = 1$ if $x \in X^+$, it is clear that the binary possibilistic utility is of the form:

$$W_{GS}(a) = (\max(OM(X^-), OM(A^+)), \max(OM(A^-), OM(X^+)),$$

and that this approach belongs to the type I bipolar setting, since L_Π is a bipolar scale. Compared to the relation $\succeq^{FB\Pi}$, W_{GS} has a major drawback: whenever $OM(X^-) = OM(X^+)$, that is, the most important positive and negative affects have the same importance, then $W_{GS}(a) = (1_L, 1_L), \forall a \in \mathscr{D}$ results, expressing indifference between all alternatives. Note that the approach proposed by Giang and Shenoy is tailored to decision under uncertainty, while we adapted it to decision with pros and cons. In the latter setting, the outranking relation $\succeq^{FB\Pi}$ looks more promising.

7 Conclusion

This chapter has proposed an overview of a qualitative argument-based approach to multi-aspect decision making, showing to what extent it bridges the gap between, on the one hand, purely heuristic descriptions of how people make decisions without measuring value or importance, and, on the other hand, refined numerical approaches that compare differences of evaluations of positive and negative rewards such as Cumulative Prospect Theory. The original point made by this paper is that the qualitative approach proposed here suggests another generalisation of Cumulative Prospect Theory that does not rely on the use of bicapacities. An extension of the bipolar possibility relation to fuzzy events has been used to account for arguments whose appropriateness to an alternative can be a matter of degree on top of grading their respective importance. This approach is shown to differ from the bipolar decision theory proposed by Giang and Shenoy. Besides, using lexicographic refinements of qualitative optimistic possibilistic utility functionals [21], it is clear that adaptations of the Levelwise Tallying decision rules to this refined situation can be envisaged. This is left for further research.

References

1. Amgoud, L., Cayrol, C., Lagasquie-Schiex, M.-C., Livet, P.: On bipolarity in argumentation frameworks. Int. J. of Intelligent Systems 23(10), 1062–1093 (2008)
2. Amgoud, L., Prade, H.: Using arguments for making and explaining decisions. Artificial Intelligence 173(3-4), 413–436 (2009)
3. Benferhat, S., Dubois, D., Kaci, S., Prade, H.: Bipolar possibility theory in preference modeling: Representation, fusion and optimal solutions. Information Fusion 7, 135–150 (2006)
4. Bilbao, J.M., Fernandez, J.R., Jiménez Losada, A., Lebrón, E.: Bicooperative games. In: Bilbao, J.M. (ed.) Cooperative games on combinatorial structures, pp. 23–26. Kluwer Academic Publishers, Dordrecht (2000)
5. Bonnefon, J.-F., Dubois, D., Fargier, H., Prade, H.: On the qualitative comparison of decisions having positive and negative features. J. Artificial Intelligence Research 32, 385–417 (2008)

6. Bonnefon, J.F., Dubois, D., Fargier, H., Leblois, S.: Qualitative heuristics for balancing the pros and cons. Theory and Decision 65, 71–85 (2008)
7. Boutilier, C., Brafman, R.I., Domshlak, C., Hoos, H.H., Poole, D.: CP-nets: A tool for representing and reasoning with conditional ceteris paribus preference statements. J. Artificial Intelligence Research (JAIR) 21, 135–191 (2004)
8. Cacioppo, J.T., Berntson, G.G.: Relationship between attitudes and evaluative space: A critical review, with emphasis on the separability of positive and negative substrates. Psychological Bulletin 115, 401–423 (1994)
9. Chateauneuf, A.: Decomposable capacities, distorted probabilities and concave capacities. Mathematical Social Sciences 31(1), 19–37 (1996)
10. Deschamps, R., Gevers, L.: Leximin and utilitarian rules: a joint characterization. J. of Economic Theory 17, 143–163 (1978)
11. Doyle, J., Thomason, R.: Background to qualitative decision theory. The AI Magazine 20(2), 55–68 (1999)
12. Dubois, D.: Belief structures, possibility theory and decomposable confidence measures on finite sets. Computers and Artificial Intelligence 5(5), 403–416 (1986)
13. Dubois, D., Fargier, H.: On the qualitative comparison of sets of positive and negative affects. In: Godo, L. (ed.) ECSQARU 2005. LNCS (LNAI), vol. 3571, pp. 305–316. Springer, Heidelberg (2005)
14. Dubois, D., Fargier, H., Prade, H., Sabbadin, R.: A survey of qualitative decision rules under uncertainty. In: Bouyssou, D., Dubois, D., Pirlot, M., Prade, H. (eds.) Decision-Making Process- Concepts and Methods, ch. 11, pp. 435–473. Wiley, Chichester (2009), http://www.wiley.com/
15. Dubois, D., Prade, H.: Possibility Theory. Plenum Press, New York (1988)
16. Dubois, D., Prade, H.: Bipolar representations in reasoning, knowledge extraction and decision processes. In: Greco, S., Hata, Y., Hirano, S., Inuiguchi, M., Miyamoto, S., Nguyen, H.S., Słowiński, R. (eds.) RSCTC 2006. LNCS (LNAI), vol. 4259, pp. 15–26. Springer, Heidelberg (2006)
17. Dubois, D., Prade, H.: An introduction to bipolar representations of information and preference. Int. J. Intelligent Systems. 23(8), 866–877 (2008)
18. Dubois, D., Prade, H.: Formal representations of uncertainty. In: Bouyssou, D., Dubois, D., Pirlot, M., Prade, H. (eds.) Decision-Making Process- Concepts and Methods, ch. 3, pp. 85–156. Wiley, Chichester (2009), http://www.wiley.com/
19. Dubois, D., Prade, H.: An overview of the asymmetric bipolar representation of positive and negative information in possibility theory. Fuzzy Sets and Systems 160, 1355–1366 (2009)
20. Dubois, D., Prade, H., Sabbadin, R.: Decision-theoretic foundation of qualitative possibility theory. European J. of Operational Research 128, 478–495 (2001)
21. Fargier, H., Sabbadin, R.: Qualitative decision under uncertainty: Back to expected utility. Artificial Intelligence 164, 245–280 (2005)
22. Franklin, B.: Letter to J. B. Priestley, 1772. In: Bigelow, J. (ed.) The Complete Works. Putnam, New York (1887)
23. Giang, P.H., Shenoy, P.P.: Two axiomatic approaches to decision-making using possibility theory. European J. of Operational Research 162, 450–467 (2005)
24. Gigerenzer, G., Goldstein, D.G.: Reasoning the fast and frugal way: Models of bounded rationality. Psychological Review 103, 650–669 (1996)
25. Gigerenzer, G., Todd, P.M., the ABC group: Simple heuristics that make us smart. Oxford University Press, Oxford (1999)
26. Grabisch, M.: The Moebius transform on symmetric ordered structures and its application to capacities on finite sets. Discrete Math 28(1-3), 17–34 (2004)

27. Grabisch, M., Greco, S., Pirlot, M.: Bipolar and bivariate models in multicriteria decision analysis: Descriptive and constructive approaches. Int. J. Intelligent Systems. 23(9), 930–969 (2008)
28. Grabisch, M., Labreuche, C.: Bi-capacities for decision making on bipolar scales. In: EUROFUSE 2002 Workshop on Information Systems, pp. 185–190 (2002)
29. Grabisch, M., Labreuche, C.: Bi-capacities — parts I and II. Fuzzy Sets and Systems 151(2), 211–260 (2005)
30. Greco, S., Matarazzo, B., Slowinski, R.: Bipolar Sugeno and Choquet integrals. In: EUROFUSE Workshop on Information Systems, Varenna, pp. 191–196 (2002)
31. Labreuche, C., Grabisch, M.: Generalized Choquet-like aggregation functions for handling bipolar scales. Eur. J. of Operational Research 172(3), 931–955 (2006)
32. Lewis, D.: Counterfactuals. Basil Blackwell, London (1973)
33. Moulin, H.: Axioms of Cooperative Decision Making. Wiley, New-York (1988)
34. Von Neumann, J., Morgenstern, O.: Theory of games and economic behavior. Princeton University Press, Princeton (1947)
35. Osgood, C.E., Suci, G.J., Tannenbaum, P.H.: The Measurement of Meaning. Univ. of Illinois Press, Chicago (1957)
36. Öztürk, M., Tsoukiàs, A.: Bipolar preference modeling and aggregation in decision support. Int. J. Intelligent Systems. 23(9), 970–984 (2008)
37. Shafer, G.: A Mathematical Theory of Evidence. Princeton University Press, Princeton (1976)
38. Slovic, P., Finucane, M., Peters, E., MacGregor, D.G.: Rational actors or rational fools? implications of the affect heuristic for behavioral economics. The J. of Socio-Economics 31, 329–342 (2002)
39. Tversky, A., Kahneman, D.: Advances in prospect theory: Cumulative representation of uncertainty. J. of Risk and Uncertainty 5, 297–323 (1992)
40. Wald, A.: Statistical Decision Functions. Wiley, Chichester (1950)
41. Weng, P.: An axiomatic approach in qualitative decision theory with binary possibilistic utility. In: Proc. of the 17th European Conference on Artificial Intelligence (ECAI 2006), Riva del Garda, Italy, pp. 467–471. IOS Press, Amsterdam (2006)
42. Zadeh, L.A.: Fuzzy sets as a basis for a theory of possibility. Fuzzy Sets and Systems 1, 1–28 (1978)

The Dynamics of Consensus in Group Decision Making: Investigating the Pairwise Interactions between Fuzzy Preferences

Mario Fedrizzi, Michele Fedrizzi, R.A. Marques Pereira, and Matteo Brunelli

Abstract. In this paper we present an overview of the soft consensus model in group decision making and we investigate the dynamical patterns generated by the fundamental pairwise preference interactions on which the model is based.

The dynamical mechanism of the soft consensus model is driven by the minimization of a cost function combining a collective measure of dissensus with an individual mechanism of opinion changing aversion. The dissensus measure plays a key role in the model and induces a network of pairwise interactions between the individual preferences.

The structure of fuzzy relations is present at both the individual and the collective levels of description of the soft consensus model: pairwise preference intensities between alternatives at the individual level, and pairwise interaction coefficients between decision makers at the collective level.

The collective measure of dissensus is based on non linear scaling functions of the linguistic quantifier type and expresses the degree to which most of the decision makers disagree with respect to their preferences regarding the most relevant alternatives. The graded notion of consensus underlying the dissensus measure is central to the dynamical unfolding of the model.

The original formulation of the soft consensus model in terms of standard numerical preferences has been recently extended in order to allow decision makers to express their preferences by means of triangular fuzzy numbers. An appropriate notion of distance between triangular fuzzy numbers has been chosen for the construction of the collective dissensus measure.

Mario Fedrizzi · Michele Fedrizzi · Ricardo Alberto Marques Pereira
University of Trento, Department of Computer and Management Science,
via Inama 5, 38122 Trento, Italy
e-mail: mario.fedrizzi@unitn.it, michele.fedrizzi@unitn.it,
ricalb.marper@unitn.it

Matteo Brunelli
IAMSR, Turku Centre for Computer Science, Abo Akademi University,
Joukahainengatan 3-5A, FIN-20520 Abo, Finland and University of Trento,
Department of Computer and Management Science, via Inama 5, 38122 Trento, Italy
e-mail: matteo.brunelli@abo.fi

S. Greco et al. (Eds.): Preferences and Decisions, STUDFUZZ 257, pp. 159–182.
springerlink.com

In the extended formulation of the soft consensus model the extra degrees of freedom associated with the triangular fuzzy preferences, combined with non linear nature of the pairwise preference interactions, generate various interesting and suggestive dynamical patterns. In the present paper we investigate these dynamical patterns which are illustrated by means of a number of computer simulations.

1 Introduction

In the study of aggregational models of group decision making the central notions of interaction and consensus have been the subject of a great deal of investigation. Fundamental contributions in this general area of research have been made by: Shapley (1953) on cooperative game theory [65]; French (1956) and Harary (1959) on social power theory [31] [44]; DeGroot (1974), Chatterjee and Seneta (1977), Berger (1981), Kelly (1981), and French (1981) on DeGroot's consensus formation model [16] [13] [7] [53] [32]; Sen (1982) on models of choice and welfare [64]; Wagner (1978, 1982) and Lehrer and Wagner (1981) on the rational choice model [73] [55] [74]; Anderson and Graesser (1976), Anderson (1981, 1991), and Graesser (1991) on the information integration model [4] [2] [3] [42]; Davis (1973, 1996) on the social decision scheme model [14] [15]; and Friedkin (1990, 1991, 1993, 1998, 1999, 2001), Friedkin and Johnsen (1990, 1997, 1999), and Marsden and Friedkin (1993, 1994) on social influence network theory [33] [38] [34] [35] [59] [60] [39] [36] [37] [40] [41].

In the classical literature stream indicated above the notion of consensus has conventionally been understood in terms of strict and unanimous agreement. However, since decision makers typically have different and conflicting opinions to a lesser or greater extent, the traditional strict meaning of consensus is often unrealistic. The human perception of consensus is typically 'softer', and people are generally willing to accept that consensus has been reached when most actors agree on the preferences associated to the most relevant alternatives.

In this different perspective, and in parallel with the traditional approach mostly formulated on a probabilistic basis, Ragade (1976) and Bezdek, Spillman, and Spillman (1977, 1978, 1979, 1980) proposed to conceptualize consensus within the fuzzy framework [63] [8] [9] [10] [66] [67] [68]. A few years later, combining the fuzzy notion of consensus with the expressive power of linguistic quantifiers, Kacprzyk and Fedrizzi (1986, 1988, 1989) and Kacprzyk, Fedrizzi, and Nurmi (1992, 1993, 1997) developed the so-called soft consensus measure in the context of fuzzy preference relations [47] [48] [49] [50] [19] [51] and considered various interesting implications of the model in the context of decision support, see Fedrizzi, Kacprzyk, and Zadrozny (1988) and Carlsson et al. (1992) [18] [12].

The soft consensus paradigm proposed by Kacprzyk and Fedrizzi was subsequently reformulated by the Trento research group [20] [21] [22] [23] [25] [24] [26] [27] [61] [28] [29] [30] . The linguistic quantifiers in the original soft consensus measure were substituted by smooth scaling functions with an analogous role and a dynamical model was obtained from the gradient descent optimization of a soft consensus cost function, combining a soft measure of collective dissensus with an individual mechanism of opinion changing aversion. The resulting soft consensus dynamics acts on the network of single preference structures by a combination of a collective process of diffusion and an individual mechanism of inertia.

Introduced as an extension of the crisp model of consensus dynamics described in [27], the fuzzy soft consensus model [29] substitutes the standard crisp preferences by fuzzy triangular preferences. The fuzzy extension of the soft consensus model is based on the use of a distance measure between triangular fuzzy numbers. In analogy with the standard crisp model, the fuzzy dynamics of preference change towards consensus derives from the gradient descent optimization of the new cost function of the fuzzy soft consensus model.

In the meantime a number of different fuzzy approaches have been proposed. The linguistic approach [79] is applicable when the information involved either at individual level or at group level present qualitative aspects that cannot be effectively represented by means of precise numerical values. Innovative approaches to the modelling of consensus in fuzzy environments were developed under linguistic assessments and the interested reader is referred, among others, to [45] [46] [5] [62] [11] [77]. The typical problem addressed is that in which decision makers have different levels of knowledge about the alternatives and use linguistic term sets with different cardinality to assess their preferences. This is the so-called group decision making problem in a multigranular fuzzy linguistic context.

Another different approach to the analysis of consensus under fuzziness, based on a distance from consensus, has been proposed in [69] using intuitionistic fuzzy preferences. In that paper, taking into account Atanasov's hesitation margin, the approach to consensus in [9] [10] and [68] has been extended to individual preferences represented by interval values. This approach has been further developed in [70] introducing a similarity measure to compare the distances between intuitionistic fuzzy relations. More recently, a new and more effective similarity measure has been introduced and applied to consensus analysis in the context of interval-valued intuitionistic fuzzy set theory [78].

The paper is organized as follows. In section 2 we briefly review the soft consensus model proposed in [27] and we show how to derive the soft consensus dynamics on the basis of a cost function W combining a soft measure of collective dissensus with an individual mechanism of opinion changing aversion. In section 3, assuming fuzzy triangular preferences as in [29], we describe the new distance measure and introduce the cost function W of the fuzzy soft consensus model. In section 4 we derive the dynamical laws of the fuzzy soft consensus model as applied to fuzzy triangular preferences. Section 5 contains the main contribution of the paper: we present and discuss a number of computer simulations in order to illustrate the complex and suggestive dynamical patterns generated by the dynamics of the fuzzy soft consensus model. Finally, in section 6 we present some concluding remarks and notes on future research.

2 The Soft Dissensus Measure and the Consensus Dynamics

In this section we present a brief review of the original soft consensus model introduced in [27]. Our point of departure is a set of individual fuzzy preference relations. If $A = \{a_1, \ldots, a_m\}$ is a set of decisional alternatives and $I = \{1, \ldots, n\}$ is a set of individuals, then the fuzzy preference relation R_i of individual i is given by its membership function $R_i : A \times A \to [0,1]$ such that

$$\begin{array}{lll} R_i(a_k,a_l) = & 1 & \text{if } a_k \text{ is definitely preferred over } a_l \\ R_i(a_k,a_l) \in & (0.5,1) & \text{if } a_k \text{ is preferred over } a_l \\ R_i(a_k,a_l) = & 0.5 & \text{if } a_k \text{ is considered indifferent to } a_l \\ R_i(a_k,a_l) \in & (0,0.5) & \text{if } a_l \text{ is preferred over } a_k \\ R_i(a_k,a_l) = & 0 & \text{if } a_l \text{ is definitely preferred over } a_k, \end{array}$$

where $i = 1, \ldots, n$ and $k,l = 1, \ldots, m$. Each individual fuzzy preference relation R_i can be represented by a matrix $[r^i_{kl}]$, $r^i_{kl} = R_i(a_k, a_l)$ which is commonly assumed to be reciprocal, that is $r^i_{kl} + r^i_{lk} = 1$. Clearly, this implies $r^i_{kk} = 0.5$ for all $i = 1, \ldots, n$ and $k = 1, \ldots, m$.

The general case $A = \{a_1, \ldots, a_m\}$ for the set of decisional alternatives is discussed in [27] and [29]. Here, for the sake of simplicity, we assume that the alternatives available are only two ($m = 2$), which means that each individual preference relation R_i has only one degree of freedom, denoted by $x_i = r^i_{12}$.

In the framework of the soft consensus model, assuming $m = 2$, the degree of dissensus between individuals i and j as to their preferences between the two alternatives is measured by

$$V(i,j) = f((x_i - x_j)^2)\,, \tag{1}$$

where f is a scaling function defined as

$$f(x) = -\frac{1}{\beta}\, ln(1 + e^{-\beta(x-\alpha)})\,. \tag{2}$$

In the scaling function formula above, $\alpha \in (0,1)$ is a threshold parameter and $\beta \in (0,\infty)$ is a free parameter. The latter controls the polarization of the sigmoid function $f' : [0,1] \to (0,1)$ given by

$$f'(x) = 1/(1 + e^{\beta(x-\alpha)})\,. \tag{3}$$

In the soft consensus model [27] each decision maker $i = 1, \ldots, n$ is represented by a pair of connected nodes, a primary node (dynamic) and a secondary node (static). The n primary nodes form a fully connected subnetwork and each of them encodes the individual opinion of a single decision maker. The n secondary nodes, on the other hand, encode the individual opinions originally declared by the decision makers, denoted $s_i \in [0,1]$, and each of them is connected only with the associated primary node.

The dynamical process of preference change corresponds to the gradient descent optimization of a cost function W, depending on both the present and the original network configurations. The value of W combines a measure V of the overall dissensus in the present network configuration with a measure U of the overall change from the original network configuration.

The various interactions involving node i are modulated by interaction coefficients whose role is to quantify the strength of the interaction. The consensual interaction between primary nodes i and j is modulated by the interaction coefficient $v_{ij} \in (0,1)$, whereas the inertial interaction between primary node i and the associated secondary node is modulated by the interaction coefficient $u_i \in (0,1)$. In the soft consensus model the values of these interaction coefficients are given by the derivative f' of the scaling function according to

$$v_{ij} = f'((x_i - x_j)^2), \quad v_i = \sum_{j(\neq i)=1}^{n} v_{ij}/(n-1), \quad u_i = f'((x_i - s_i)^2). \tag{4}$$

The average preference $\bar{x}_i$ is given by

$$\bar{x}_i = \sum_{j(\neq i)=1}^{n} v_{ij}x_j / \sum_{j(\neq i)=1}^{n} v_{ij} \tag{5}$$

and represents the average preference of the remaining decision makers as seen by decision maker $i = 1, \dots, n$.

The construction of the cost function W that drives the dynamics of the soft consensus model is as follows. The individual dissensus cost $V(i)$ is given by

$$V(i) = \sum_{j(\neq i)=1}^{n} V(i,j)/(n-1), \quad V(i,j) = f((x_i - x_j)^2) \tag{6}$$

and the individual opinion changing cost $U(i)$ is

$$U(i) = f((x_i - s_i)^2). \tag{7}$$

Summing over the various decision makers we obtain the collective dissensus cost V and inertial cost U,

$$V = \frac{1}{4}\sum_{i=1}^{n} V(i), \quad U = \frac{1}{2}\sum_{i=1}^{n} U(i) \tag{8}$$

with conventional multiplicative factors of $1/4$ and $1/2$. The full cost function W is then $W = (1-\lambda)V + \lambda U$ with $0 \le \lambda \le 1$.

The consensual network dynamics, which can be regarded as an unsupervised learning algorithm, acts on the individual opinion variables x_i through the iterative process

$$x_i \rightsquigarrow x_i' = x_i - \gamma \frac{\partial W}{\partial x_i}. \tag{9}$$

Analyzing the effect of the two dynamical components V and U separately we obtain

$$\frac{\partial V}{\partial x_i} = v_i(x_i - \bar{x}_i) \tag{10}$$

where the coefficients v_i were defined in (4) and the average preference $\bar{x}_i$ was defined in (5), and therefore

$$x_i' = (1 - \gamma v_i)x_i + \gamma v_i \bar{x}_i \,. \tag{11}$$

On the other hand, we obtain

$$\frac{\partial U}{\partial x_i} = u_i(x_i - s_i)\,, \tag{12}$$

where the coefficients u_i were defined in (4), and therefore

$$x_i' = (1 - \gamma u_i)x_i + \gamma u_i s_i \,. \tag{13}$$

The full dynamics associated with the cost function $W = (V + U)/2$ acts iteratively according to

$$x_i' = (1 - \gamma(v_i + u_i))x_i + +\gamma v_i \bar{x}_i + \gamma u_i s_i \,. \tag{14}$$

and the decision maker i is in dynamical equilibrium, in the sense that $x_i' = x_i$, if the following stability equation holds,

$$x_i = (v_i \bar{x}_i + u_i s_i)/(v_i + u_i) \tag{15}$$

that is, if the present opinion x_i coincides with an appropriate weighted average of the original opinion s_i and the average opinion value $\bar{x}_i$.

3 The Fuzzy Soft Dissensus Measure

Let us now assume that the decision makers preferences are expressed by means of fuzzy numbers, see for instance [17] [80], in particular by means of triangular fuzzy numbers. Then, in order to measure the differences between the decision makers preferences, we need to compute the distances between the fuzzy numbers representing those preferences. Let

$$\mathbf{x} = \{\varepsilon_L, x, \varepsilon_R\} \quad \mathbf{y} = \{\theta_L, y, \theta_R\} \tag{16}$$

be two triangular fuzzy numbers, where x is the central value of the fuzzy number $\mathbf{x}$ and ε_L, ε_R are its left and right spread, respectively. Analogously for the triangular fuzzy number $\mathbf{y}$.

Various definitions of distance between fuzzy numbers are considered in the literature [43] [52] [71] [72]. Moreover, the question has been often indirectly addressed in papers regarding the ranking of fuzzy numbers, see [75] [76] for a detailed review.

In our model we refer to a distance, indicated by $D^*(\mathbf{x},\mathbf{y})$, which belongs to a family of distances introduced in [43]. This distance is defined as follows.

For each $\alpha \in [0,1]$, the α–level sets of the two fuzzy numbers $\mathbf{x}$ and $\mathbf{y}$ are respectively

$$[x_L(\alpha), x_R(\alpha)] = [x - \varepsilon_L + \varepsilon_L\alpha \ , \ x + \varepsilon_R - \varepsilon_R\alpha] \tag{17}$$

$$[y_L(\alpha), y_R(\alpha)] = [y - \theta_L + \theta_L\alpha \ , \ y + \theta_R - \theta_R\alpha] \, . \tag{18}$$

The distance $D^*(\mathbf{x},\mathbf{y})$ between $\mathbf{x}$ and $\mathbf{y}$ is defined by means of the differences between the left boundaries of (17), (18) and the differences between the right boundaries of (17), (18). More precisely, the left integral I_L is defined as the integral, with respect to α, of the squared difference between the left boundaries of (17) and (18),

$$I_L = \int_0^1 (x_L(\alpha) - y_L(\alpha))^2 d\alpha \tag{19}$$

and the right integral I_R is defined as the integral, with respect to α, of the squared difference between the right boundaries of (17), (18),

$$I_R = \int_0^1 (x_R(\alpha) - y_R(\alpha))^2 d\alpha \, . \tag{20}$$

Finally, the distance $D^*(\mathbf{x},\mathbf{y})$ is defined as

$$D^*(\mathbf{x},\mathbf{y}) = \left(\frac{1}{2}(I_L + I_R)\right)^{1/2} \, . \tag{21}$$

The distance (21) is obtained by choosing $p = 2$ and $q = 1/2$ in the family of distances introduced in [43]. In order to avoid unnecessarily complex computations, we skip the square root and we use, in our model, the simpler expression

$$D(\mathbf{x},\mathbf{y}) = (D^*(\mathbf{x},\mathbf{y}))^2 \ = \ \frac{1}{2}(I_L + I_R) \, . \tag{22}$$

Note that expression (22), except for the numerical factor $1/2$, has been introduced, independently from [43], also in [57]. It has been then pointed out in [1] that (22) is not a distance, as it does not always satisfy the triangular inequality. Nevertheless, as long as optimization is involved, expression (22) can be equivalently used in place of the distance (21) [58]. In any case, for simplicity, in the following we shall use the term distance when referring to (22). Developing (19) and (20), we obtain

$$D(\mathbf{x},\mathbf{y}) = d^2 + \frac{1}{6}\delta_L^2 + \frac{1}{6}\delta_R^2 + \frac{d}{2}(\delta_R - \delta_L), \tag{23}$$

where $d = x - y$, $\delta_L = \varepsilon_L - \theta_L$ and $\delta_R = \varepsilon_R - \theta_R$.

As explained in the previous section, the preferences of the n decision makers are expressed by pairwise comparing the alternatives $a_1, a_2, ..., a_m$. Given a pair of alternatives, we assume that the preference of the first over the second alternative is represented, for decision maker i, by a triangular fuzzy number indicated by

$$\mathbf{r}^i = \{\varepsilon_L^i, r^i, \varepsilon_R^i\}, \tag{24}$$

where, as in (16), r^i is the central value of the fuzzy number $\mathbf{r}^i$, whereas ε_L^i and ε_R^i are its left and right spreads respectively. Analogously, let $\mathbf{r}^j$ be the triangular fuzzy number of type (24) representing the preference of the first alternative over the second given by decision maker j.

Following definition (22), the distance between the fuzzy preference of decision maker i and the one of decision maker j becomes

$$D(\mathbf{r}^i, \mathbf{r}^j) = d^2 + \frac{1}{6}\delta_L^2 + \frac{1}{6}\delta_R^2 + \frac{d}{2}(\delta_R - \delta_L), \tag{25}$$

where $d = r^i - r^j$, $\delta_L = \varepsilon_L^i - \varepsilon_L^j$ and $\delta_R = \varepsilon_R^i - \varepsilon_R^j$.

As assumed in the previous section, we consider a problem with $m = 2$ alternatives and we define the dissensus measure between two decision makers by applying the scaling function f to $D(\mathbf{r}^i, \mathbf{r}^j)$,

$$V(i,j) = f(D(\mathbf{r}^i, \mathbf{r}^j)). \tag{26}$$

The dissensus measure of decision maker i with respect to the rest of the group is given by the arithmetic mean of the various dissensus measures $V(i,j)$,

$$V(i) = \sum_{j(\neq i)=1}^{n} V(i,j)/(n-1). \tag{27}$$

Finally, the global dissensus measure of the group is defined by

$$V = \frac{1}{4}\sum_{i=1}^{n} V(i), \tag{28}$$

thus obtaining

$$V = \frac{1}{4}\sum_{i=1}^{n} \sum_{j(\neq i)=1}^{n} f(D(\mathbf{r}^i, \mathbf{r}^j))/(n-1). \tag{29}$$

Denoting by $\mathbf{s}^i = \{\theta_L^i, s^i, \theta_R^i\}$ the triangular fuzzy number describing the initial preference of decision maker i, the cost for changing the initial preference $\mathbf{s}^i$ into the actual preference $\mathbf{r}^i$ is given by

$$U(i) = f(D(\mathbf{r}^i, \mathbf{s}^i)). \tag{30}$$

The global opinion changing aversion component U of the group is given by

$$U = \frac{1}{2}\sum_{i=1}^{n} U(i). \tag{31}$$

As mentioned before, the global cost function W is defined as a convex combination of the components V and U,

$$W = (1-\lambda)V + \lambda U, \tag{32}$$

and the parameter $\lambda \in [0,1]$ represents the relative importance of the inertial component U with respect to the dissensus component V.

4 The Dynamics of the Fuzzy Soft Consensus Model

In [29] the original consensus dynamics described in section 2 was extended to the case in which preferences are expressed by means of triangular fuzzy numbers. In the consensus dynamics, the global cost function $W = W(\mathbf{r}^i) = W(\varepsilon_L^i, r^i, \varepsilon_R^i)$ is minimized through the gradient descent method. This implies that in every iteration the new preference $\mathbf{r}'$ is obtained from the previous preference $\mathbf{r}$ in the following way (we skip the index i for simplicity)

$$\mathbf{r} \to \mathbf{r}' = \mathbf{r} - \gamma \nabla W. \tag{33}$$

The consensus dynamics (33) will gradually update the three preference values $(\varepsilon_L, r, \varepsilon_R)$ according to

$$r \to r' = r - \gamma \frac{\partial W}{\partial r}, \quad \varepsilon_L \to \varepsilon_L{}' = \varepsilon_L - \gamma \frac{\partial W}{\partial \varepsilon_L}, \quad \varepsilon_R \to \varepsilon_R{}' = \varepsilon_R - \gamma \frac{\partial W}{\partial \varepsilon_R} \tag{34}$$

We can consider separately the effect of the two components V and U of W, since ∇W is a convex combination of ∇V and ∇U,

$$\nabla W = (1-\lambda)\nabla V + \lambda \nabla U. \tag{35}$$

Let us first consider the component V. Taking again into account the index i, we have

$$\frac{\partial V}{\partial r^i} = v_i \left((r^i - \bar{r}^i) + \frac{1}{4}\left(\varepsilon_R^i - \bar{\varepsilon}{}_R^i - \varepsilon_L^i + \bar{\varepsilon}{}_L^i\right) \right) \tag{36}$$

where

$$v_i = \sum_{j(\neq i)=1}^{n} v_{ij}/(n-1); \quad v_{ij} = f'(D(\mathbf{r}^i, \mathbf{r}^j)) \tag{37}$$

$$\bar{r}^i = \frac{\sum_{j(\neq i)=1}^{n} v_{ij} r^j}{\sum_{j(\neq i)=1}^{n} v_{ij}}, \quad \bar{\varepsilon}{}_L^i = \frac{\sum_{j(\neq i)=1}^{n} v_{ij} \varepsilon_L^j}{\sum_{j(\neq i)=1}^{n} v_{ij}}, \quad \bar{\varepsilon}{}_R^i = \frac{\sum_{j(\neq i)=1}^{n} v_{ij} \varepsilon_R^j}{\sum_{j(\neq i)=1}^{n} v_{ij}}, \tag{38}$$

Analogously, we compute

$$\frac{\partial V}{\partial \varepsilon_L^i} = v_i \left(\frac{1}{6}(\varepsilon_L^i - \bar{\varepsilon}{}_L^i) - \frac{1}{4}(r^i - \bar{r}^i)\right), \quad \frac{\partial V}{\partial \varepsilon_R^i} = v_i \left(\frac{1}{6}(\varepsilon_R^i - \bar{\varepsilon}{}_R^i) + \frac{1}{4}(r^i - \bar{r}^i)\right). \tag{39}$$

Let us now consider the inertial component U. We obtain

$$\frac{\partial U}{\partial r^i} = u_i((r^i - s^i) + \frac{1}{4}(\varepsilon_R^i - \theta_R^i - \varepsilon_L^i + \theta_L^i)) \tag{40}$$

where

$$u_i = f'(D(\mathbf{r}^i, \mathbf{s}^i)), \tag{41}$$

$$\frac{\partial U}{\partial \varepsilon_L^i} = u_i(\frac{1}{6}(\varepsilon_L^i - \theta_L^i) - \frac{1}{4}(r^i - s^i)) \tag{42}$$

and

$$\frac{\partial U}{\partial \varepsilon_R^i} = u_i(\frac{1}{6}(\varepsilon_R^i - \theta_R^i) + \frac{1}{4}(r^i - s^i)). \tag{43}$$

At this point we can summarize the effects of the two components obtaining

$$\frac{\partial W}{\partial r^i} = ((1-\lambda)v_i + \lambda u_i)\Delta r^i - (1-\lambda)v_i \Delta \bar{r}^i - \lambda u_i \Delta s^i \tag{44}$$

where

$$\Delta r^i = r^i + \frac{1}{4}(\varepsilon_R^i - \varepsilon_L^i), \quad \Delta \bar{r}^i = \bar{r}^i + \frac{1}{4}(\bar{\varepsilon}_R^i - \bar{\varepsilon}_L^i), \quad \Delta s^i = s^i + \frac{1}{4}(\theta_R^i - \theta_L^i). \tag{45}$$

The derivative of W with respect to the left spread becomes

$$\frac{\partial W}{\partial \varepsilon_L^i} = ((1-\lambda)v_i + \lambda u_i)\Delta \varepsilon_L^i - (1-\lambda)v_i \Delta \bar{\varepsilon}_L^i - \lambda u_i \Delta \theta_L^i \tag{46}$$

where

$$\Delta \varepsilon_L^i = \frac{1}{6}\varepsilon_L^i - \frac{1}{4}r^i, \quad \Delta \bar{\varepsilon}_L^i = \frac{1}{6}\bar{\varepsilon}_L^i - \frac{1}{4}\bar{r}^i, \quad \Delta \theta_L^i = \frac{1}{6}\theta_L^i - \frac{1}{4}s^i. \tag{47}$$

The derivative of W with respect to the right spread becomes

$$\frac{\partial W}{\partial \varepsilon_R^i} = ((1-\lambda)v_i + \lambda u_i)\Delta \varepsilon_R^i - (1-\lambda)v_i \Delta \bar{\varepsilon}_R^i - \lambda u_i \Delta \theta_R^i \tag{48}$$

where

$$\Delta \varepsilon_R^i = \frac{1}{6}\varepsilon_R^i + \frac{1}{4}r^i, \quad \Delta \bar{\varepsilon}_R^i = \frac{1}{6}\bar{\varepsilon}_R^i + \frac{1}{4}\bar{r}^i, \quad \Delta \theta_R^i = \frac{1}{6}\theta_R^i + \frac{1}{4}s^i. \tag{49}$$

Let us now present some numerical simulations in order to illustrate the dynamical behaviour of the fuzzy soft consensus model in some interesting cases.

5 Computer Simulations

In this section we present a number of computer simulations of the fuzzy soft consensus dynamics as applied to a single pair of preferences represented by triangular fuzzy numbers. Our goal is that of illustrating the various interesting dynamical patterns generated by the non linear nature of the pairwise interactions between preferences, given that these pairwise interactions are the fundamental elements of the soft consensus model.

The first four figures associated with each computer simulation (except the first) depict four successive configurations of the preference pair of triangular fuzzy numbers, corresponding to the following moments in time: the initial configuration $t = 0$, two intermediate configurations $t = 25$ and $t = 100$, and the final (quasi-asymptotic) configuration $t = 1000$. The two dots appearing in each of the four figures indicate the positions of the centers as they vary in time according to the original crisp version of the soft consensus model. The other three figures associated with each computer simulation show the time plot of the preference centers plus that of the left and right spreads.

In general we observe in the computer simulations two distinct dynamical phases, clearly illustrated by the graphical plots of the preference changes over time: a short phase with fast dynamics followed by a much longer phase with slow dynamics. Interestingly, the preference changes over time in each of these two phases are not always monotonic. Moreover, the computer simulations show that the dynamics of the fuzzy soft consensus model is generally faster than that of the original crisp model. The final (quasi-asymptotic) values of the preference centers in the fuzzy model show moderate but significant differences with respect to the corresponding final preference values in the original crisp model.

The distance $D(\mathbf{x}, \mathbf{y})$ between two fuzzy numbers $\mathbf{x}$ and $\mathbf{y}$ defined in (22) and involved in the construction of the cost functions V, U, W plays a key role in the fuzzy extension of the soft consensus model. In particular, the two distinct phases (fast and slow) observed in the consensus dynamics of the model can be understood in terms of the different magnitudes of the coefficients associated with the various terms in the decomposition formula (23). The fact that the coefficient associated with the distance between centers is three times larger than the coefficient associated with the distance between spreads (left and right together) produces initially a fast consensus dynamics of the centers, followed by a much slower adjustment dynamics of the spreads. Roughly speaking, the fast phase leads to an overlapping of the two fuzzy triangular numbers, whose shape is then adjusted by the slow dynamical phase.

In all computer simulations (except partially the first) the parameter choices are as follows: $\alpha = 0.3$, $\beta = 10$, $\lambda = 1/3$, and $\gamma = 0.01$.

- These figures illustrate the dynamics of the original crisp soft consensus model as applied to two crisp initial preferences 0.3 and 0.7, for three different choices of the parameter λ. This parameter controls the relative strength of the mechanism of opinion changing aversion with respect to the consensual aggregation mechanism. In the case $\lambda = 0$ the dynamics is purely consensual and thus, over time, the two preferences converge exactly to a common final value.

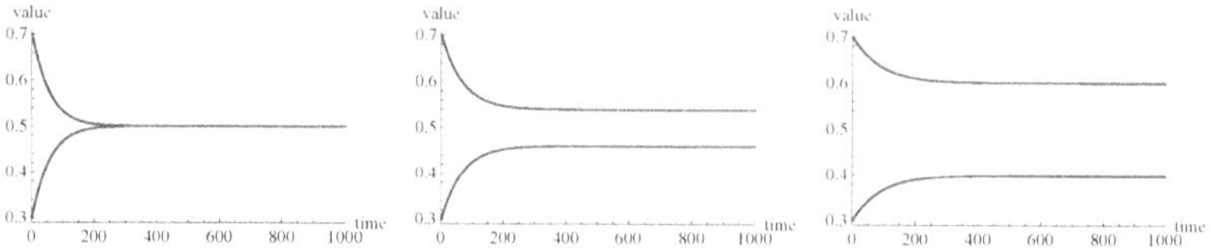

(a) $\lambda = 0$ (b) $\lambda = 1/3$ (c) $\lambda = 2/3$

Fig. 1 Crisp dynamics acting on two crisp preferences, for different values of λ

- These figures illustrate the dynamics of the fuzzy soft consensus model as applied to the same two crisp initial preferences 0.3 and 0.7 as before. Notice the two dynamical phases, initially fast and then slow, and the suggestive non monotonic behaviour of the internal spreads. Initially, in the fast phase, the centers approach rapidly and the internal spreads increase significantly whereas the external spreads remain essentially null, a sort of cooperative opening to the opposing preference. Then, in the slow phase, the centers keep on approaching very slowly while the internal spreads gradually decrease and the external spreads increase slightly, converging towards a nearly common final value. In the final configuration the spreads are once again very small (they were initially null) even though they reach much larger values during the transient "negotiation" process. This a suggestive reality effect of the non linear dynamics of the fuzzy soft consensus model.

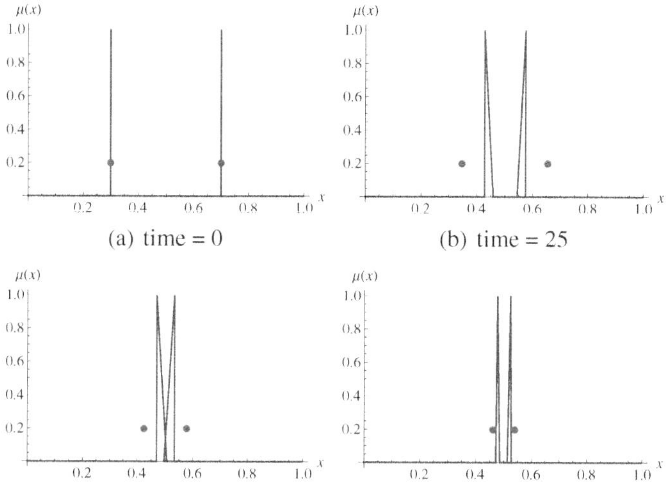

(a) time = 0 (b) time = 25
(c) time = 100 (d) time = 1000

Fig. 2 Fuzzy dynamics acting on two crisp preferences

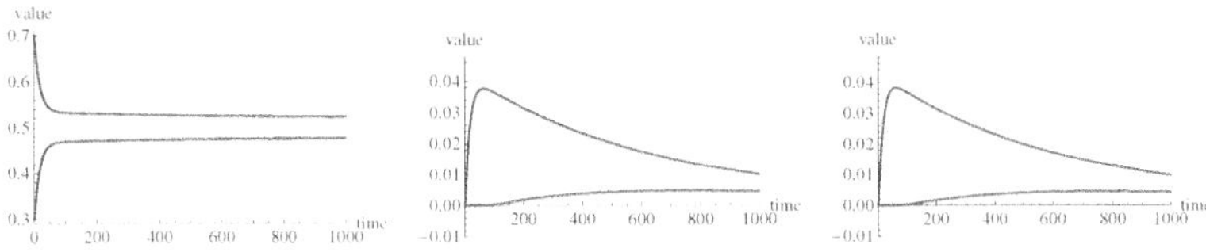

(a) Center plot (b) Left spread plot (c) Right spread plot

Fig. 3 Corresponding time plots of centers and spreads

- These figures illustrate the dynamics of the fuzzy soft consensus model as applied to two fuzzy initial preferences (isosceles triangles) centered at the usual values 0.3 and 0.7. Once again, notice the two dynamical phases, initially fast and then slow, and the suggestive non monotonic behaviour of the internal and external spreads. Initially, in the fast phase, the centers approach rapidly while the internal (resp. external) spreads increase (resp. decrease) significantly, again a sort of cooperative opening to the opposing preference. Then, in the slow phase, the centers keep on approaching very slowly while the internal (resp. external) spreads gradually decrease (resp. increase), converging towards a nearly common final value.

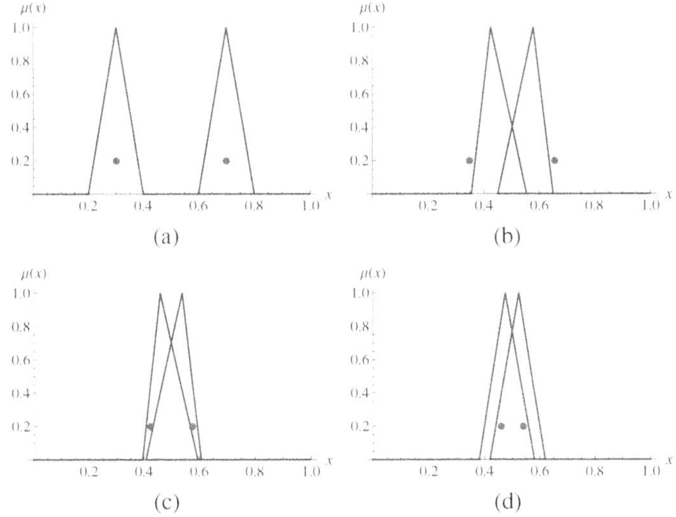

Fig. 4 Fuzzy dynamics acting on two isosceles triangles

- These figures illustrate the dynamics of the fuzzy soft consensus model as applied to two fuzzy initial preferences (right triangles facing each other) centered at the usual values 0.3 and 0.7. Once again, notice the two dynamical phases, initially fast and then slow, and the suggestive non monotonic behaviour of the

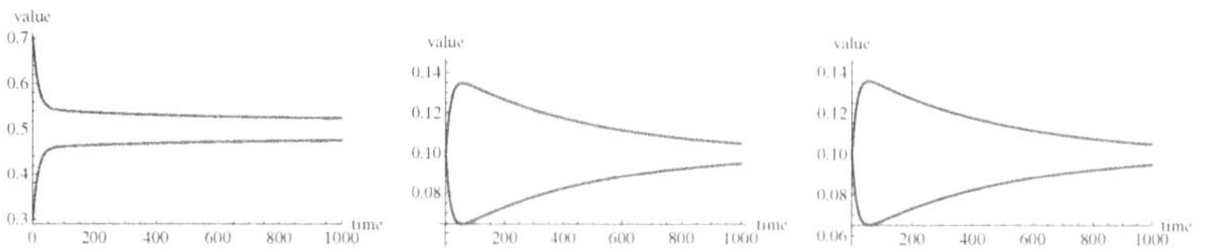

(a) Center plot (b) Left spread plot (c) Right spread plot

Fig. 5 Corresponding time plots of centers and spreads

internal spreads. Initially, in the fast phase, the centers approach rapidly and the internal spreads increase slightly whereas the external spreads remain essentially null, again a sort of cooperative opening to the opposing preference. Then, in the slow phase, the centers keep on approaching very slowly while the internal (resp. external) spreads gradually decrease (resp. increase), converging towards a nearly common final value.

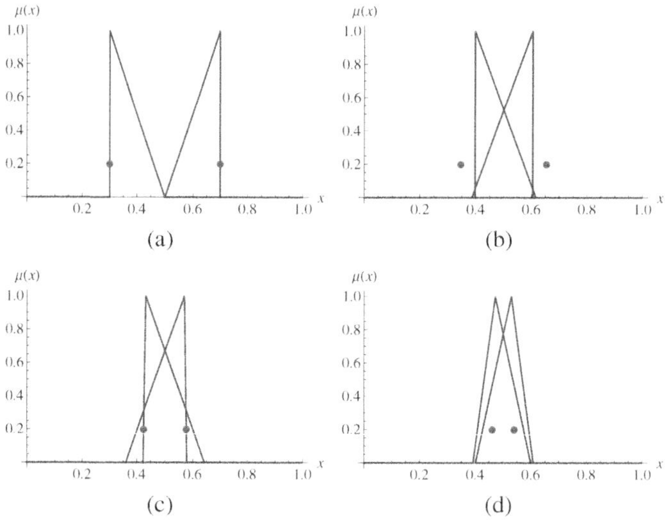

Fig. 6 Fuzzy dynamics acting on two right triangles facing each other

- These figures illustrate the dynamics of the fuzzy soft consensus model as applied to two fuzzy initial preferences (right triangles facing opposite to each other) centered at the usual values 0.3 and 0.7. Once again, notice the two dynamical phases, initially fast and then slow, and the suggestive non monotonic behaviour of the centers. Initially, in the fast phase, the centers approach rapidly (almost crossing) and the internal (resp. external) spreads increase (resp. decrease), again a sort of cooperative opening to the opposing preference. Then, in the slow phase, the centers adjust by moving away very slowly while the internal (resp. external)

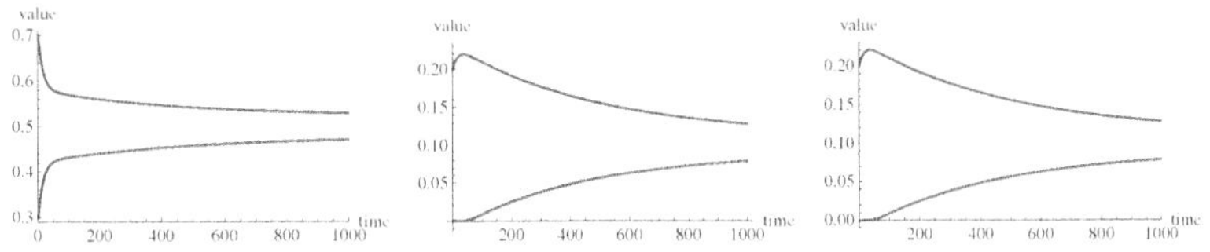

(a) Center plot (b) Left spread plot (c) Right spread plot

Fig. 7 Corresponding time plots of centers and spreads

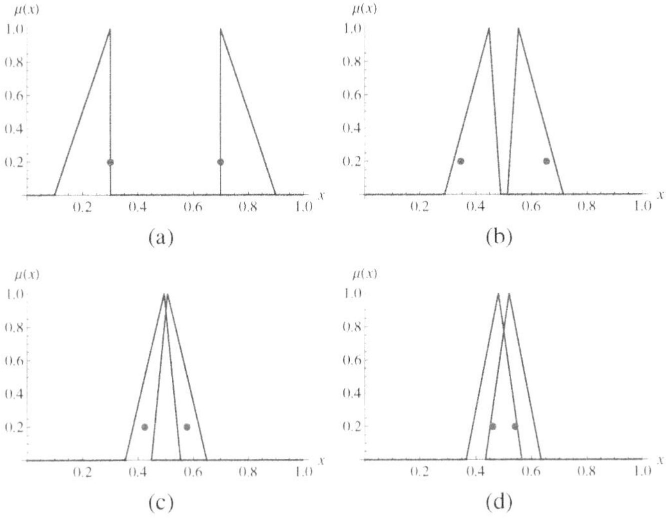

Fig. 8 Fuzzy dynamics acting on two right triangles facing opposite to each other

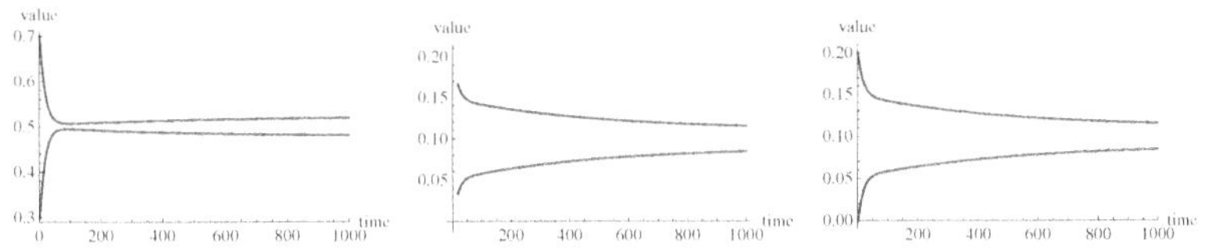

(a) Center plot (b) Left spread plot (c) Right spread plot

Fig. 9 Corresponding time plots of centers and spreads

spreads keep on gradually increasing (resp. decreasing), converging towards a nearly common final value.

- These figures illustrate the dynamics of the fuzzy soft consensus model as applied to two fuzzy initial preferences (different isosceles triangles) centered at the usual values 0.3 and 0.7. Once again, notice the two dynamical phases, initially fast and then slow, and the suggestive non monotonic behaviour of the internal and external spreads. Initially, in the fast phase, the centers approach rapidly and the internal (resp. external) spreads increase (resp. decrease) slightly on the right and significantly on the left. Then, in the slow phase, the centers keep on approaching very slowly while the internal (resp. external) spreads gradually decrease (resp. increase). In this case the dynamical pattern is more complex for the left spreads, with two crossings during the slow phase.

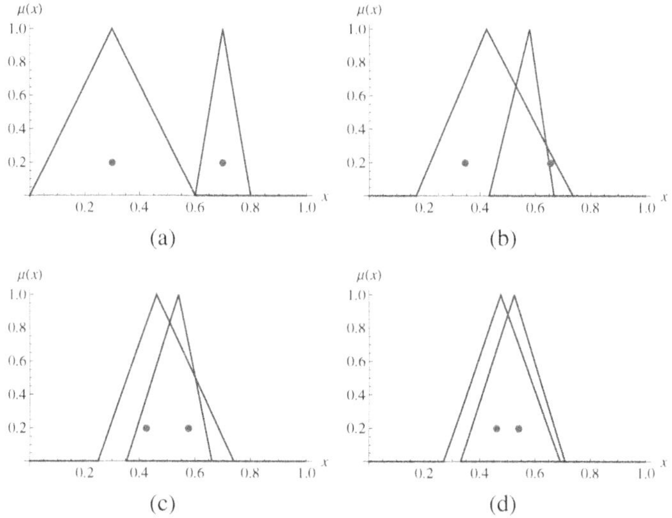

Fig. 10 Fuzzy dynamics acting on two different isosceles triangles

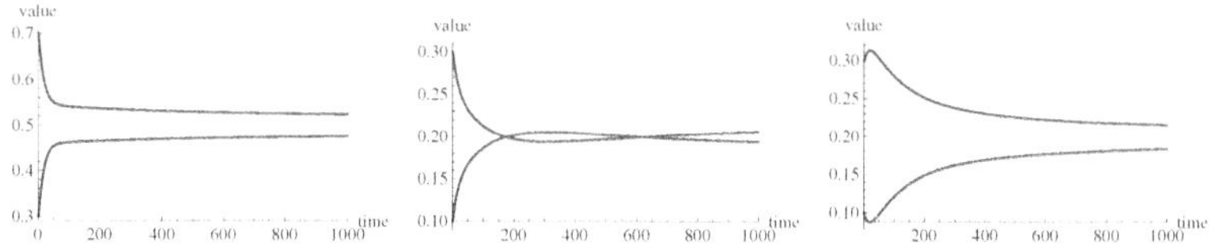

(a) Center plot (b) Left spread plot (c) Right spread plot

Fig. 11 Corresponding time plots of centers and spreads

- These figures illustrate the dynamics of the fuzzy soft consensus model as applied to one crisp and one fuzzy initial preferences (isosceles triangle the latter) centered at the usual values 0.3 and 0.7. Once again, notice the two dynamical phases, initially fast and then slow, and the suggestive non monotonic behaviour of the internal and external spreads. Initially, in the fast phase, the centers approach rapidly and the internal (resp. external) spreads increase (resp. stay null or decrease) slightly on the right and significantly on the left. Then, in the slow phase, the centers keep on approaching very slowly while the internal (resp. external) spreads gradually decrease (resp. increase). In this case the dynamical pattern is more complex for the left spreads, with one crossing between the two phases and another one during the slow phase.

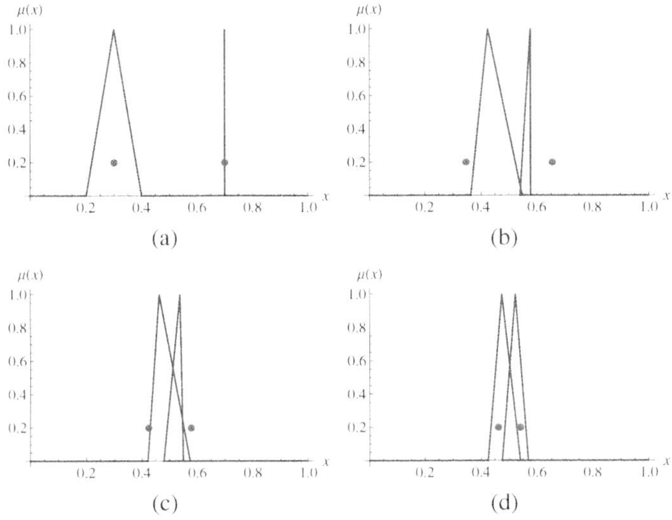

Fig. 12 Fuzzy dynamics acting on one crisp preference and one isosceles triangle

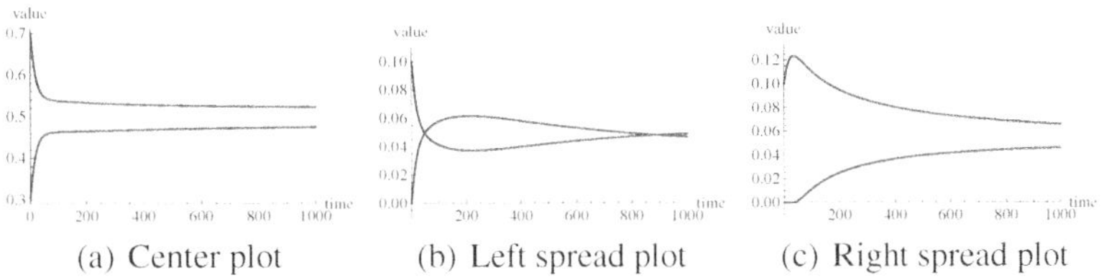

Fig. 13 Corresponding time plots of centers and spreads

- These figures illustrate the dynamics of the fuzzy soft consensus model as applied to one crisp and one fuzzy initial preferences (right triangle facing inward the latter) centered at the usual values 0.3 and 0.7. Once again, notice the two dynamical phases, initially fast and then slow, and the suggestive non monotonic behaviour of the internal and external spreads. Initially, in the fast phase, the centers approach rapidly and the internal (resp. external) spreads increase (resp. stay null) slightly on the right and significantly on the left. Then, in the slow phase, the centers keep on approaching very slowly while the internal (resp. external) spreads gradually decrease (resp. increase).

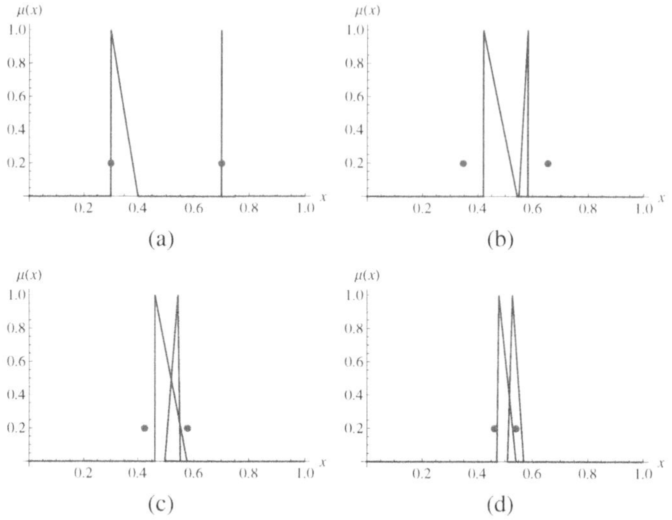

Fig. 14 Fuzzy dynamics acting on one crisp preference and one right triangle facing inward

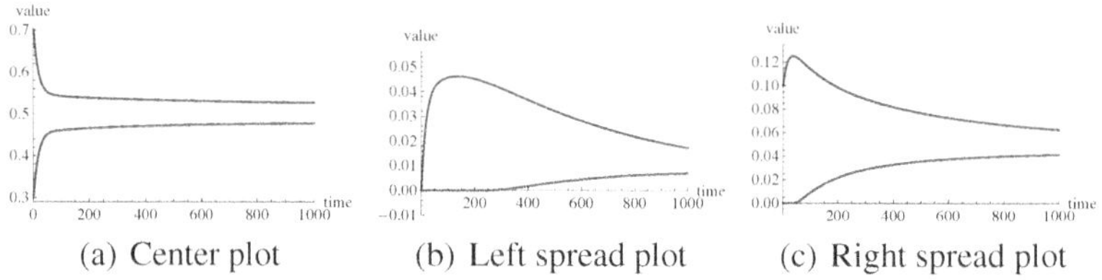

Fig. 15 Corresponding time plots of centers and spreads

- These figures illustrate the dynamics of the fuzzy soft consensus model as applied to one crisp and one fuzzy initial preferences (right triangle facing outward the latter) centered at the usual values 0.3 and 0.7. Once again, notice the two dynamical phases, initially fast and then slow, and the suggestive non monotonic behaviour of the internal and external spreads. Initially, in the fast phase, the centers approach rapidly and the internal (resp. external) spreads increase (resp. stay null or decrease) significantly on both sides. Then, in the slow phase, the centers keep on approaching very slowly while the internal (resp. external) spreads gradually decrease (resp. increase). In this case the dynamical pattern is more complex for the left spreads, with one crossing between the two phases and another one during the slow phase.

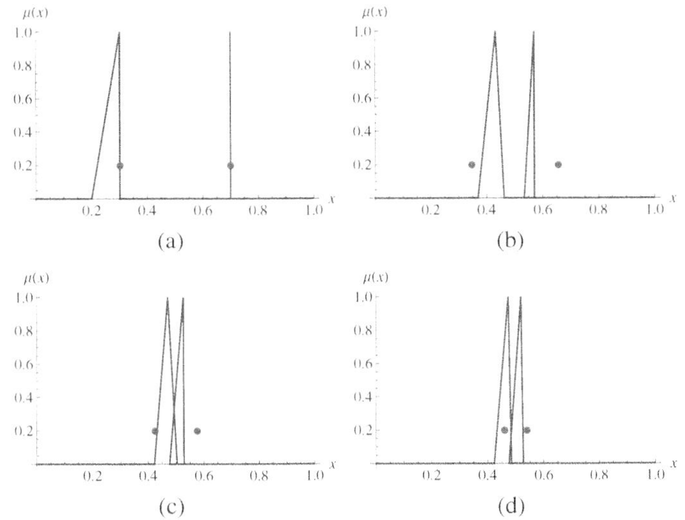

Fig. 16 Fuzzy dynamics acting on one crisp preference and one right triangle facing outward

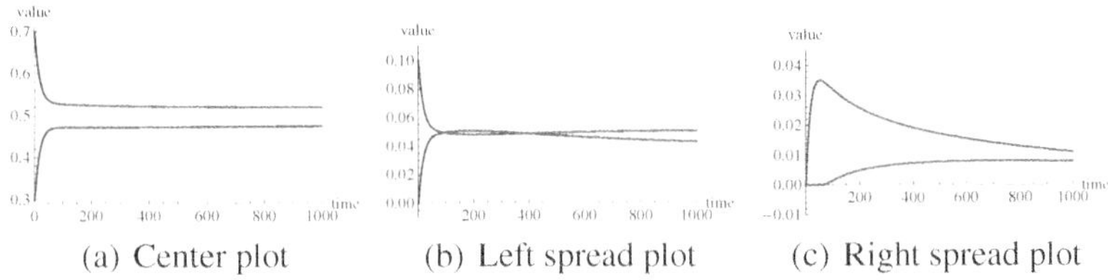

Fig. 17 Corresponding time plots of centers and spreads

- These figures illustrate the dynamics of the fuzzy soft consensus model as applied to the special case of two fuzzy initial preferences (right triangles facing opposite to each other) centered at 0.5 and 0.6. Once again, notice the two dynamical phases, initially fast and then slow, and the suggestive non monotonic behaviour of the centers. Initially, in the fast phase, the centers move rapidly towards each other, crossing and then moving away from each other. Then, in the slow phase, the centers adjust by slowly re-approaching, converging towards a nearly common final value. This is another interesting effect of the non linear dynamics of the fuzzy soft consensus model, due to the combined effect of the two mechanisms of consensus reaching and opinion changing aversion as they act on centers and spreads of the fuzzy triangular preferences.

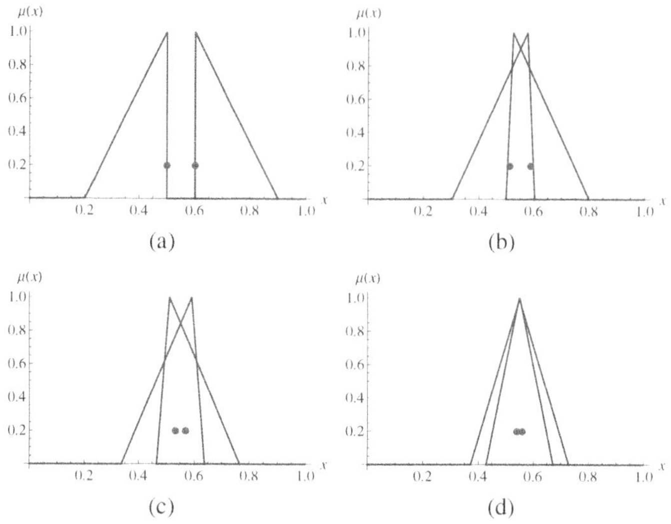

Fig. 18 Fuzzy dynamics acting on two right triangles facing opposite to each other

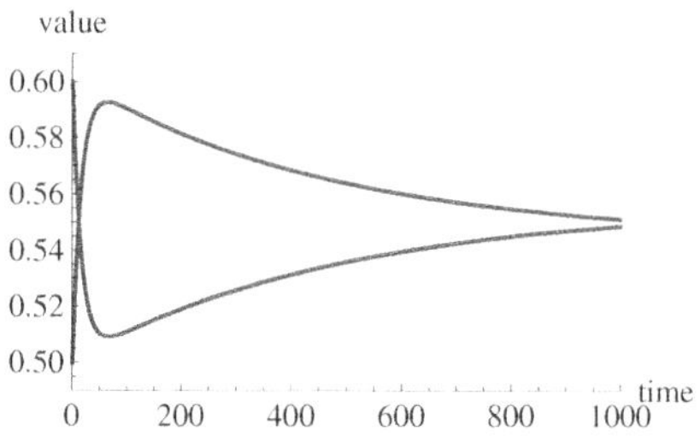

Fig. 19 Corresponding time plot of the centers

6 Concluding Remarks

We have illustrated by means of numerical simulations the dynamical behaviour of the fuzzy soft consensus model, in which the individual preferences are represented by triangular fuzzy numbers. A selection of these simulations is presented in section 5. The computer simulations provide clear evidence that the fuzzy soft consensus model exhibits interesting non standard opinion changing behaviour in relation to the original crisp version of the model. Future research should explore the particular features of the fuzzy soft consensus model and demonstrate the potential of the methodology as an effective support for the modelling of consensus reaching in multicriteria and multiagent decision making.

References

1. Abbasbandy, S., Asady, B.: Note on 'A new approach for defuzzification'. Fuzzy Sets and Systems 128, 131–132 (2002)
2. Anderson, N.H.: Foundations of Information Integration Theory. Academic Press, New York (1981)
3. Anderson, N.H.: Contributions to Information Integration Theory. Lawrence Erlbaum, Hillsdale (1991)
4. Anderson, N.H., Graesser, C.C.: An information integration analysis of attitude change in group discussion. Journal of Personality and Social Psychology 34, 210–222
5. Ben-Arieh, D., Chen, Z.: Linguistic labels aggregation and consensus measure for autocratic decision making using group recommendations. IEEE Transanctions on Systems, Man, and Cybernetics (A) 36, 558–568 (2006)
6. Ben-Arieh, D., Easton, T., Evans, B.: Minimum cost consensus with quadratic cost function. IEEE Transacitions on Systems, Man and Cybernetics (A) 39, 210–217 (2008)
7. Berger, R.L.: A necessary and sufficient condition for reaching a consensus using DeGroot's method. Journal of the American Statistical Association 76, 415–418 (1981)
8. Bezdek, J.C., Spillman, B., Spillman, R.: Fuzzy measures of preferences and consensus in group decision making. In: Fu, K.S. (ed.) Proc. IEEE Conference on Decision and Control, pp. 1303–1309 (1977)
9. Bezdek, J.C., Spillman, B., Spillman, R.: A fuzzy relation space for group decision theory. Fuzzy Sets and Systems 1, 255–268 (1978)
10. Bezdek, J.C., Spillman, B., Spillman, R.: Fuzzy relation spaces for group decision theory. Fuzzy Sets and Systems 2, 5–14 (1978)
11. Cabrerizo, F.J., Alonso, S., Herrera-Viedma, E.: A consensus model for group decision making problems with unbalanced fuzzy linguistic information. International Journal of Information Technology and Decision Making 8, 109–131 (2009)
12. Carlsson, C., Ehrenberg, D., Eklund, P., Fedrizzi, M., Gustafsson, P., Lindholm, P., Merkuryeva, G., Riissanen, T., Ventre, A.: Consensus in distributed soft environments. European Journal of Operational Research 61, 165–185 (1992)
13. Chatterjee, S., Seneta, E.: Towards consensus: some convergence theorems on repeated averaging. Journal of Applied Probability 14, 89–97 (1977)
14. Davis, J.H.: Group decision and social interaction: a theory of social decision schemes. Psychological Review 80, 97–125 (1973)
15. Davis, J.H.: Group decision making and quantitative judgments: a consensus model. In: Witte, E.H., Davios, J.H. (eds.) Understanding Group Behavior: Consensual Action by Small Groups. Lawrence Erlbaum, Mahwah (1996)
16. DeGroot, M.H.: Reaching a consensus. Journal of the American Statistical Association 69, 118–121 (1974)
17. Dubois, D., Prade, H.: Fuzzy Sets and Systems. Academic Press, San Diego (1980)
18. Fedrizzi, M., Kacprzyk, J., Zadrożny, S.: An interactive multi-user decision support system for consensus reaching processes using fuzzy logic with linguistic quantifiers. Decision Support Systems 4, 313–327 (1988)
19. Fedrizzi, M., Kacprzyk, J., Nurmi, H.: Consensus degrees under fuzzy majorities and fuzzy preferences using OWA (ordered weighted averaging) operators. Control and Cybernetics 22, 77–86 (1993)
20. Fedrizzi, M., Fedrizzi, M., Marques Pereira, R.A.: Consensual dynamics: an unsupervised learning model in group decision making. In: Proc. 3rd Intl. Conference on Fuzzy Logic, Neural Nets and Soft Computing, IIZUKA 1994, Iizuka, Japan (August 1994); published by Fuzzy Logic Systems Institute FLSI, Iizuka, Japan, 99–100 (1994)

21. Fedrizzi, M., Fedrizzi, M., Marques Pereira, R.A.: Consensual dynamics in group decision making. In: Paulo, S. (ed.) Proc. 6th Intl. Fuzzy Systems Association World Congress, IFSA 1995, vol. II, pp. 145–148. International Fuzzy Systems Association IFSA, Brazil (July 1995)
22. Fedrizzi, M., Fedrizzi, M., Marques Pereira, R.A., Zorat, A.: A dynamical model for reaching consensus in group decision making. In: Proc. 1995 ACM symposium on Applied computing, Nashville, Tennessee, USA, pp. 493–496 (1995)
23. Fedrizzi, M., Fedrizzi, M., Marques Pereira, R.A.: Consensus, dynamics, and group decision making. In: Proc. 20th Convegno dell' Associazione per la Matematica Applicata alle Scienze Economiche e Sociali AMASES 1996, pp. 243–254. University of Urbino, Urbino (September 1996)
24. Fedrizzi, M., Fedrizzi, M., Marques Pereira, R.A., Ribaga, C.: Consensual dynamics with local externalities. In: Proc. 1st International Workshop on Preferences and Decisions TRENTO 1997, Trento, June 5-7, pp. 62–69 (1997)
25. Fedrizzi, M., Fedrizzi, M., Marques Pereira, R.A.: An inertial equity principle in the dynamics of group decision making. In: Proc. VII International Fuzzy Systems Association World Congress IFSA 1997, Prague, June 25-29, pp. 95–98 (1997)
26. Fedrizzi, M., Marques Pereira, R.A., Molinari, F.: Optimal alternatives in consensus reaching processes. In: Proc. 7th Intl. Conference on Information Processing and Management of Uncertainty IPMU 1998, pp. 158–163. EDK, Paris (1998)
27. Fedrizzi, M., Fedrizzi, M., Marques Pereira, R.A.: Soft consensus and network dynamics in group decision making. Intl. Journal of Intelligent Systems 14, 63–77 (1999)
28. Fedrizzi, M., Fedrizzi, M., Marques Pereira, R.A.: On the issue of consistency in dynamical consensual aggregation. In: Bouchon-Meunier, B., Gutierrez-Rios, J., Magdalena, L., Yager, R.R. (eds.) Technologies for Contructing Intelligent Systems, vol. 1, pp. 129–137. Physica-Verlag (Springer), Heidelberg (2002)
29. Fedrizzi, M., Fedrizzi, M., Marques Pereira, R.A.: Consensus modelling in group decision making: a dynamical approach based on fuzzy preferences. New Mathematics and Natural Computation 3, 219–237 (2007)
30. Fedrizzi, M., Fedrizzi, M., Marques Pereira, R.A., Brunelli, M.: Consensual dynamics in group decision making with triangular fuzzy numbers. In: Proc. 41st Annual International Conference on System Sciences HICSS 2008, Waikoloa, Big Island, Hawaii, USA, pp. 70–78. IEEE Computer Society, Los Alamitos (2008)
31. French, J.R.P.: A formal theory of social power. The Psychological Review 63, 181–194 (1956)
32. French, S.: Consensus of opinion. European Journal of Operational Research 7, 332–340 (1981)
33. Friedkin, N.E.: Social networks in structural equation models. Social Psychology Quarterly 53, 316–328 (1990)
34. Friedkin, N.E.: Theoretical foundations for centrality measures. American Journal of Sociology 96, 1478–1504 (1991)
35. Friedkin, N.E.: Structural bases of interpersonal influence in groups: a longitudinal case study. American Sociological Review 58, 861–872 (1993)
36. Friedkin, N.E.: A Structural Theory of Social Influence. Cambridge University Press, Cambridge (1998)
37. Friedkin, N.E.: Choice shift and group polarization. American Sociological Review 64, 856–875 (1999)
38. Friedkin, N.E., Johnsen, E.C.: Social influence and opinions. Journal of Mathematical Sociology 15, 193–206 (1990)

39. Friedkin, N.E., Johnsen, E.C.: Social positions in influence networks. Social Networks 19, 209–222 (1997)
40. Friedkin, N.E., Johnsen, E.C.: Social influence networks and opinion change. Advances in Group Processes 16, 1–29 (1999)
41. Friedkin, N.E.: Norm formation in social influence networks. Social Networks 23, 167–189 (2001)
42. Graesser, C.C.: A social averaging theorem for group decision making. In: Anderson, N.H., Hillsdale, N.J. (eds.) Contributions to Information Integration Theory, vol. 2, pp. 1–40. Lawrence Erlbaum, Mahwah
43. Grzegorzewski, P.: Metrics and orders in space of fuzzy numbers. Fuzzy Sets and Systems 97, 83–94 (1998)
44. Harary, F.: A criterion for unanimity in French's theory of social power. In: Cartwright, D. (ed.) Studies in Social Power, pp. 168–182. Institute for Social Research, Ann Arbor
45. Herrera, F., Herrera-Viedma, E., Verdegay, J.L.: A model of consensus in group decision making under linguistic assessments. Fuzzy Sets and Systems 78, 73–87 (1996)
46. Herrera-Viedma, E., Martinez, L., Mata, F., Chiclana, F.: A consensus support system model for group decision making problems with multigranular linguistic preference relations. IEEE Transanctions on Fuzzy Systems 13, 644–658 (2005)
47. Kacprzyk, J., Fedrizzi, M.: 'Soft' consensus measures for monitoring real consensus reaching processes under fuzzy preferences. Control and Cybernetics 15, 309–323 (1986)
48. Kacprzyk, J., Fedrizzi, M.: A 'soft' measure of consensus in the setting of partial (fuzzy) preferences. European Journal of Operational Research 34, 316–325 (1988)
49. Kacprzyk, J., Fedrizzi, M.: A human-consistent degree of consensus based on fuzzy logic with linguistic quantifiers. Mathematical Social Sciences 18, 275–290 (1989)
50. Kacprzyk, J., Fedrizzi, M., Nurmi, H.: Group decision making and consensus under fuzzy preferences and fuzzy majority. Fuzzy Sets and Systems 49, 21–31 (1992)
51. Kacprzyk, J., Nurmi, H., Fedrizzi, M. (eds.): Consensus under Fuzziness. International Series in Intelligent Technologies. Kluwer Academic Publishers, Dordrecht (1997)
52. Kaufman, A., Gupta, M.M.: Introduction to fuzzy arithmetic, Van Nostrand Reinhold, New York (1991)
53. Kelly, F.P.: How a group reaches agreement: a stochastic model. Mathematical Social Sciences 2, 1–8 (1981)
54. Lee, H.–S.: Optimal consensus of fuzzy opinion under group decision making environment. Fuzzy Sets and Systems 132, 303–315 (2002)
55. Lehrer, K., Wagner, K.: Rational Consensus in Science and Society. Reidel, Dordrecht (1981)
56. Loewer, B. (guest ed.): Special Issue on Consensus, Synthese, vol. 62(1) (1985)
57. Ma, M., Kandel, A., Friedman, M.: A new approach for defuzzification. Fuzzy Sets and Systems 111, 351–356 (2000)
58. Ma, M., Kandel, A., Friedman, M.: Correction to A new approach for defuzzification. Fuzzy Sets and Systems 128, 133–134 (2002)
59. Mardsen, P.V., Friedkin, N.E.: Network studies of social influence. Sociological Methods Research 22, 127–151 (1993)
60. Mardsen, P.V., Friedkin, N.E.: Network studies of social influence. In: Wasserman, S., Galaskiewicz, J. (eds.) Advances in Social Network Analysis, vol. 22, pp. 3–25. Sage, Thousand Oaks (1994)
61. Marques Pereira, R.A., Bortot, S.: Consensual dynamics, stochastic matrices, Choquet measures, and Shapley aggregation. In: Proc. Linz Seminar 2001 on Valued Relations and Capacities in Decision Theory, LINZ 2001, Linz, Austria, February 2001. University of Linz (2001)

62. Mata, F., Martinez, L., Herrera-Viedma, E.: An adaptive consensus support model for group decision-making problems in a multigranular fuzzy linguistic context. IEEE Transactions on Fuzzy Systems 17, 279–290 (2009)
63. Ragade, R.K.: Fuzzy sets in communication systems and in consensus formation systems. Journal of Cybernetics 6, 21–38 (1976)
64. Sen, A.: Choice, Welfare, and Measurement. Basil Blackwell, Oxford (1982)
65. Shapley, L.S.: A value for n-person games. In: Kuhn, H.W., Tucker, A.W. (eds.) Contributions to the Theory of Games (II). Annals of Mathematical Studies, vol. 28, pp. 307–317. Princeton University Press, Princeton (1953)
66. Spillman, B., Bezdek, J.C., Spillman, R.: Coalition analysis with fuzzy sets. Kybernetes 8, 203–211 (1979)
67. Spillman, B., Spillman, R., Bezdek, J.C.: Development of an instrument for the dynamic measurement of consensus. Communication Monographs 46, 1–12 (1979)
68. Spillman, B., Spillman, R., Bezdek, J.C.: A fuzzy analysis of consensus in samll groups. In: Wang, P.P., Chang, S.K. (eds.) Fuzzy Sets Theory and Applications to Policy, Analysis, and Information Systems, pp. 291–308. Plenum Press, New York (1980)
69. Szmidt, E., Kacprzyk, J.: A consensus-reaching process under intuitionistic fuzzy preferences relations. International Journal of Intelligent Systems 18, 837–852 (2003)
70. Szmidt, E., Kacprzyk, J.: A similarity measure for intuitionistic fuzzy sets and its application in supporting medical diagnostics reasoning. In: Rutkowski, L., Siekmann, J.H., Tadeusiewicz, R., Zadeh, L.A. (eds.) ICAISC 2004. LNCS (LNAI), vol. 3070, pp. 388–393. Springer, Heidelberg (2004)
71. Tran, L., Duckstein, L.: Comparison of fuzzy numbers using a fuzzy distance measure. Fuzzy Sets and Systems 130, 331–341 (2002)
72. Voxman, W.: Some remarks on distances between fuzzy numbers. Fuzzy Sets and Systems 100, 353–365 (1997)
73. Wagner, C.: Consensus through respect: a model of rational group decision-making. Philosphical Studies 34, 335–349
74. Wagner, C.: Allocation, Lehrer models, and the consensus of probabilities. Theory and Decision 14, 207–220
75. Wang, X., Kerre, E.E.: Reasonable properties for the ordering of fuzzy quantities (I). Fuzzy Sets and Systems 118, 375–385 (2001)
76. Wang, X., Kerre, E.E.: Reasonable properties for the ordering of fuzzy quantities (II). Fuzzy Sets and Systems 118, 387–405 (2001)
77. Xu, Z.: Group decision making based on multiple types of linguistic preference relations. Information Sciences 178, 452–467 (2008)
78. Xu, Z., Yager, R.: Intuitionistic and interval-valued intuitonistic fuzzy preference relations and their measures of similarity for the evaluation of agreement within a group. Fuzzy Optimization and Decision Making 8, 123–139 (2009)
79. Zadeh, L.A.: The Concept of a Linguistic Variable and its Application to Approximate Reasoning I–II–III. Information Sciences 8, 199–249, 301-357 (1975), 9, 43–80 (1975)
80. Zimmermann, H.J.: Fuzzy Set Theory, 2nd edn. Kluwer Academic Publishers, Dordrecht (1991)

Fuzzy Preference Relations Based on Differences

János Fodor

Abstract. In this paper we introduce quaternary fuzzy relations in order to describe difference structures. Three models are developed and studied, based on three different interpretations of an implication. Functional forms of the quaternary relation are determined by solutions of functional equations of the same type.

1 Introduction

Preference modelling is a fundamental step of (multi-criteria) decision making, operations research, social choice and voting procedures, and has been studied extensively for several years. Typically, three binary relations (strict preference, indifference, and incomparability) are built up as a result of pairwise comparison of the alternatives. Then a single reflexive relation (the weak preference, or large preference) is defined as the union of the strict preference and indifference relations. All the three previous binary relations can be expressed in terms of the large preference in a unique way. Therefore, it is possible (and in fact, this is typical) to start from a reflexive binary relation , and build up strict preference, indifference and incomparability from it.

Some important classes of binary preferences have also been studied with respect to their representation by a real function (evaluation) of the alternatives [9]. As an illustration, consider a finite set of alternatives A, a binary relation P on A. Then there is a real-valued function f on A satisfying

$$aPb \iff f(a) > f(b) \tag{1}$$

if and only if P is asymmetric and negatively transitive. Such a P is called a *strict weak order*. If P is strict preference then a function f satisfying (1) is called a *utility*

János Fodor
Budapest Tech, Bécsi út 96/b, H-1034 Budapest, Hungary
e-mail: fodor@bmf.hu

S. Greco et al. (Eds.): Preferences and Decisions, STUDFUZZ 257, pp. 183–194.
springerlink.com

function [9]. In this situation we have an *ordinal scale*: transformations $\varphi : f(A) \to \mathbb{R}$ where $\varphi \circ f$ is also satisfies (1) are the strictly increasing functions.

The representation (1) arises in the measurement of temperature if P is interpreted as *"warmer then"*. According to the previous result, temperature is an ordinal scale — although it is well known that temperature is an interval scale. There is no contradiction: one can obtain this result by using judgments of comparative temperature *difference*.

To make this precise, one should introduce a quaternary relation D on a set A of objects whose temperatures are being compared. The relation $abDuv$ is interpreted as the difference between the temperature of a and the temperature of b is judged to be greater than that between the temperature of u and the temperature of v. We would like to find a real-valued function f on A such that for all $a,b,u,v \in A$ we have

$$abDuv \iff f(a) - f(b) > f(u) - f(v). \tag{2}$$

The main aim of the present paper is to study whether it is possible to extend, in a rational way, this approach to the use of a *quaternary fuzzy relation* on A. Note that the classical binary preference theory has successfully been extended in [4], and developed significantly further since that time (see the overview [2]).

The paper is organized as follows. In the next section we briefly summarize some results on difference measurement, especially on the representation (2). Some of these observations will guide us in the study of fuzzy extensions in Section 3. We will deal with the non-strict version W of D and investigate three models based on different forms of fuzzy implications. The functional form of an appropriate fuzzy difference operator will be given through solving some functional equation in each case. In Section 4 we study the strict quaternary relation D and the indifference E, based on W. We close the paper with concluding remarks.

2 Difference Measurement

In classical measurement theory the following situation has been studied in full details. The interested reader can find the notions and results in [9].

Let A be a set and D be a quaternary relation on A. In addition to the temperature interpretation, $abDuv$ makes sense also in preference: I like a over b more than I like u over v. We write $D(a,b,u,v)$ or, equivalently, $abDuv$. If the representation (2) holds then it is called *(algebraic) difference measurement.*

A representation theorem is known for (2). To this end we need to introduce some axioms. Before doing so, we define two quaternary relations E and W based on D as follows:

$$abEuv \iff [\text{not } abDuv \text{ and not } uvDab], \tag{3}$$

$$abWuv \iff [abDuv \text{ or } abEuv]. \tag{4}$$

Notice that in case of (2) we have

$$abEuv \iff f(a) - f(b) = f(u) - f(v), \tag{5}$$

$$abWuv \iff f(a) - f(b) \geq f(u) - f(v). \tag{6}$$

Definition 1. *Let $a_1, a_2, \ldots, a_i, \ldots$ be a sequence of elements from A. It is called a* standard sequence *if $a_{i+1}a_iEa_2a_1$ holds for all a_i, a_{i+1} in the sequence, and $a_2a_1Ea_1a_1$ does not hold.*

Definition 2. *A standard sequence is called* strictly bounded *if there exist $u, v \in A$ such that $uvDa_ia_1$ and a_ia_1Dvu for all a_i in the sequence.*

AXIOM D1. Suppose R is defined on $A \times A$ by

$$(a,b)R(u,v) \quad \iff \quad abDuv.$$

Then $(A \times A, R)$ is a strict weak order (i.e., it is asymmetric and negatively transitive).

AXIOM D2. For all $a, b, u, v \in A$, if $abDuv$ then $vuDba$.

AXIOM D3. For all $a, b, c, a', b', c' \in A$, if $abWa'b'$ and $bcWb'c'$ then $acWa'c'$.

AXIOM D4. For all $a, b, u, v \in A$, if $abWuv$ holds and $uvWxx$ holds, then there are $x, y \in A$ such that $axEuv$ and $ybEuv$.

AXIOM D5. Every strictly bounded sequence is finite.

Definition 3. *A relational system (A, D) satisfying axioms D1 through D5 is called an* algebraic difference structure.

The following result guarantees the existence of an appropriate real-valued function f, see [8].

Theorem 1 (Krantz et al. (1971)). *If (A, D) is an algebraic difference structure then there is a real-valued function f on A so that for all $a, b, u, v \in A$, representation (2) holds.*

In the next theorem, necessary and sufficient conditions are given when A is finite. We list the axioms first. For more details see [10].

AXIOM SD1. For all $a, b, u, v \in A$, $abWuv$ or $uvWab$.
AXIOM SD2. For all $a, b, u, v \in A$, $abDuv$ implies $vuDba$.
AXIOM SD3. If $n > 0$ and π and σ are permutations of $\{0, 1, \ldots, n-1\}$, and if $a_ib_iWa_{\pi(i)}b_{\sigma(i)}$ holds for all $0 < i < n$, then $a_{\pi(0)}b_{\sigma(0)}Wa_0b_0$ holds; this is true for all $a_0, a_1, \ldots, a_{n-1}, b_0, b_1, \ldots, b_{n-1} \in A$.

Theorem 2 (Scott (1964)). *Suppose A is a finite set, D is a quaternary relation on A, and E and W are defined by Eqs. (3) and (4). Then Axioms SD1–SD3 are necessary and sufficient for there to be a real-valued function f on A satisfying (2).*

As Köbberling writes in [7]: Preference differences play a key role in obtaining cardinal utility. They entail that an individual is not only able to decide which of some available commodity bundles is the most preferred, but also, the individual is able to compare the improvement between a first and a second commodity bundle to the improvement between a third and a fourth commodity bundle. Such comparable preference differences, accompanied by suitable axioms, are suf?cient to give cardinal utility.

Closing this section, we collect here some properties valid in the classical case. We would like to keep as many as possible for the fuzzy extension in the next section.

For all $a,b,u,v,x,y \in A$ we have

W1. $W(a,b,u,v) = W(a,u,b,v)$
W2. $W(a,b,u,v) = W(v,u,b,a)$
W3. $W(a,b,x,x) = W(a,b,y,y)$
W4. $W(a,b,u,v)$ implies $W(a,b,x,y)$ or $W(x,y,u,v)$.

3 Fuzzy Extensions

Our aim is to extend (6) to allow W to be a quaternary *fuzzy* relation. We fuzzify W and not D because of technical reasons on one hand. On the other hand, this way we can follow traditions in building preferences starting from a weak relation, and defining its strict part later on (see [4]).

Therefore, let I be any fuzzy implication, and Δ be any fuzzy difference operator. Then, define W by

$$W(a,b,u,v) = I(\Delta(f(u),f(v)),\Delta(f(a),f(b))), \tag{7}$$

for any $a,b,u,v \in A$.

At this formulation stage we require only the following general properties of I and Δ:

I1. I is a function from $[0,1]^2$ to $[0,1]$;
I2. I is nonincreasing in the first argument;
I3. I is nondecreasing in the second place;
I4. $I(0,0) = I(0,1) = I(1,1) = 1, I(1,0) = 0$ (that is, I an implication on $\{0,1\}$).

D1. Δ is a function from $\mathbb{R}^2$ to $[0,1]$;
D2. Δ is nondecreasing in the first argument;
D3. Δ is nondecreasing in the second place.

Notice that in the classical case (6) we have

$$I(u,v) = \begin{cases} 1 \text{ if } u \leq v, \\ 0 \text{ if } u > v. \end{cases}, \quad \text{and} \quad \Delta(x,y) = x - y,$$

where u,v and x,y can be any real number, not necessarily restricted to be in $[0,1]$.

In the sequel we study three models of the implication I. In any case, we restrict our investigations to implications defined from continuous Archimedean t-norms or t-conorms, or representable uninorms.

3.1 Model 1: The Use of R-Implications

Suppose now that $I = I_T$ is the R-implication defined from a continuous Archimedean t-norm T. An additive generator of T is denoted by t. Then I_T has the following functional form:

$$I_T(x,y) = t^{-1}(\max\{t(y) - t(x), 0\}). \tag{8}$$

Therefore, the quaternary fuzzy relation W is given by

$$W(a,b,u,v) = t^{-1}(\max\{t(\Delta(f(a),f(b))) - t(\Delta(f(u),f(v))), 0\}). \tag{9}$$

It is obvious now that property W1 is formulated as follows:

$$\Delta(f(u),f(v)) \le \Delta(f(a),f(b)) \quad \Longleftrightarrow \quad \Delta(f(b),f(v)) \le \Delta(f(a),f(u)),$$

and for $\Delta(f(u),f(v)) > \Delta(f(a),f(b))$ we have

$$t(\Delta(f(a),f(b))) - t(\Delta(f(u),f(v))) = t(\Delta(f(a),f(u))) - t(\Delta(f(b),f(v))).$$

In order to avoid complicated and heavy notation, we use simply the letters a,b,u,v, etc. to denote the function values $f(a),f(b),f(u),f(v)$. This can be done without any confusion.

Thus, the last equation can also be written as

$$t(\Delta(a,b)) + t(\Delta(b,v)) = t(\Delta(a,u)) + t(\Delta(u,v)). \tag{10}$$

For obtaining the general solution of this equation for Δ, we apply the following theorem.

Theorem 3. *The general solution of*

$$F(x,y) + F(y,z) = F(x,u) + F(u,z) \tag{11}$$

is

$$F(x,y) = h(y) - h(x) + C, \tag{12}$$

where h is any real function and C is any constant.

Proof. Notice first that (11) implies $F(x,x) = F(z,z)$ for all x,z. Indeed, let $y = x$, $u = z$ in (11). Then we get $F(x,x) + F(x,z) = F(x,z) + F(z,z)$, whence $F(x,x) = F(z,z)$ follows. Denote this common value by C.

Introducing a new function G by

$$G(x,y) := F(x,y) - C,$$

we can see that G satisfies

$$G(x,y) + G(y,z) = G(x,z). \tag{13}$$

Indeed, on the one hand we have

$$G(x,y) + G(y,z) = F(x,y) + F(y,z) - 2C.$$

On the other hand, $G(x,z) = F(x,z) - C$. Since $C = F(z,z)$, we get that (13) holds. The general solution of (13) is

$$G(x,y) = h(y) - h(x),$$

see Aczél [1], Theorem 1 in Section 5.1.2. Hence we get our statement. □

Since equation (10) is just that type in the theorem, we get the following result about Δ.

Theorem 4. *Assume that the implication I is represented by equation (8), where t is any additive generator of a continuous Archimedean t-norm. Define a quaternary fuzzy relation W by (9). Then W satisfies condition W1 (i.e., $W(a,b,u,v) = W(a,u,b,v)$) if and only if Δ is of the following form:*

$$\Delta(a,b) = t^{-1}(h(b) - h(a) + C), \tag{14}$$

where h is an appropriately chosen non-decreasing function and C is any positive constant.

In this case the quaternary relation W can be written as follows:

$$W(a,b,u,v) = t^{-1}(\max\{h(u) - h(v) - h(a) + h(b), 0\}). \tag{15}$$

Proof. Because (10) holds, we can apply Theorem 3 with $F(x,y) = t(\Delta(x,y))$. Thus, we must have

$$t(\Delta(a,b)) = h(b) - h(a) + C.$$

We have to guarantee that this equation is solvable. That is, the value $h(b) - h(a) + C$ must be in the range of the additive generator t. That range is either the set of non-negative real numbers, or an interval of $[0,\omega]$, with a finite ω.

In any case, let $\alpha < \beta$ be real numbers, and choose $C := \beta - \alpha$. Let h be a function from $\mathbb{R}$ to the bounded interval $[\alpha,\beta]$. This choice squezees the value of $h(b) - h(a) + C$ in the interval $[0,2C]$.

If the range of t is the set of non-negative real numbers then C can be any positive number. If the range of t is $[0,\omega]$, then any $C \le \omega/2$ is a good choice.

To ensure monotonicity of Δ (see conditions D2 and D3 above), the function h must be non-decreasing.

After these choices of h and C the form of Δ follows. Substituting this into equation (9) we obtain the rest of the proof. □

Notice the effect of the choice of C to the membership values (preference intensities). If the range of t is $[0,\omega]$, then $C < \omega/2$ implies we cannot reach zero degree of preference. This is always the case when $t(1) = +\infty$. The lowest membership degree can be arbitrarily close to zero, but it is always positive.

Fortunately, the quaternary relation W defined in the previous theorem satisfies all the four properties analogous to the classical case, as we prove it now.

Theorem 5. *The quaternary fuzzy relation W defined in (15) satisfies all the following properties:*

FW1. $W(a,b,u,v) = W(a,u,b,v)$
FW2. $W(a,b,u,v) = W(v,u,b,a)$
FW3. $W(a,b,x,x) = W(a,b,y,y)$
FW4. $W(a,b,u,v) \leq \max\{W(a,b,x,y), W(x,y,u,v)\}$.

Proof. FW1: this was our starting point to determine the functional form of W.

FW2 and FW3: Obvious from (15).

FW4: Suppose it is not true. Then there exist a,b,u,v,x,y such that $W(a,b,x,y) < W(a,b,u,v)$ and $W(x,y,u,v) < W(a,b,u,v)$. Taking into account the functional form of W, these inequalities imply that $W(a,b,x,y) < 1$, $W(x,y,u,v) < 1$, and thus

$$h(x) - h(y) - h(a) + h(b) > 0 \text{ and } h(u) - h(v) - h(x) + h(y) > 0.$$

These two inequalities together imply that

$$h(u) - h(v) > h(x) - h(y) > h(a) - h(b), \tag{16}$$

so we have

$$h(u) - h(v) - h(a) + h(b) > 0.$$

Thus, $W(a,b,u,v) = t^{-1}(h(u) - h(v) - h(a) + h(b))$. So $W(a,b,x,y) < W(a,b,u,v)$ and $W(x,y,u,v) < W(a,b,u,v)$ hold if and only if $h(u) - h(v) - h(a) + h(b) < h(x) - h(y) - h(a) + h(b)$ and $h(u) - h(v) - h(a) + h(b) < h(u) - h(v) - h(x) + h(y)$. These strict inequalities imply that

$$h(u) - h(v) < h(x) - h(y) < h(a) - h(b),$$

which contradicts to (16). This proves the theorem. □

Example. We would like to show an example. Consider the Łukasiewicz t-norm $T_{\mathrm{L}}(x,y) = \max\{x+y-1, 0\}$, which has an additive generator $t(x) = 1 - x$, so the inverse is $t^{-1}(x) = 1 - x$ ($x \in [0,1]$). The range of t is $[0,1]$, so let $\alpha = 0$, $\beta = 1$, $C = 1/2$, and

$$h(x) = \frac{e^x}{1+e^x} \quad (x \in \mathbb{R}).$$

Then, the quaternary fuzzy relation W has the following form:

$$W(a,b,u,v) = 1 - \max\left\{\frac{e^u}{1+e^u} - \frac{e^v}{1+e^v} - \frac{e^a}{1+e^a} + \frac{e^b}{1+e^b}, 0\right\}.$$

3.2 Model 2: The Use of S-Implications

Another broad class of fuzzy implications is based on a t-conorm S and a strong negation N:

$$I_{S,N}(x,y) = S(N(x),y) \quad (x,y \in [0,1]). \tag{17}$$

With the help of an S-implication, we can define the quaternary relation W as follows:

$$W(a,b,u,v) = S(N(\Delta(u,v)),\Delta(a,b)). \tag{18}$$

Suppose that S is a continuous Archimedean t-conorm with additive generator s (that is, we have $S(x,y) = s^{-1}(\min\{s(x)+s(y),s(1)\})$, and $N(x) = \varphi^{-1}(1-\varphi(x))$. Then (18) can be rewritten as

$$\begin{aligned} W(a,b,u,v) &= s^{-1}(\min\{s(N(\Delta(u,v))) + s(\Delta(a,b)), s(1)\}) \\ &= s^{-1}(\min\{s(\varphi^{-1}(1-\varphi(\Delta(u,v)))) + s(\Delta(a,b)), s(1)\}) \\ &= s^{-1}(\min\{g(1-\Gamma(u,v)) + g(\Gamma(a,b)), s(1)\}), \end{aligned} \tag{19}$$

where $g(x) = s(\varphi^{-1}(x))$ and $\Gamma(a,b) = \varphi(\Delta(a,b))$.

Now we formulate again property FW1 (see in Theorem 5) with the actual functional form of W:

$$\min\{g(1-\Gamma(u,v)) + g(\Gamma(a,b)), s(1)\} = \min\{g(1-\Gamma(b,v)) + g(\Gamma(a,u)), s(1)\}. \tag{20}$$

From this equality it follows that $g(1-\Gamma(u,v)) + g(\Gamma(a,b)) < s(1)$ if and only if $g(1-\Gamma(b,v)) + g(\Gamma(a,u)) < s(1)$. In this case (20) reduces to

$$g(1-\Gamma(u,v)) + g(\Gamma(a,b)) = g(1-\Gamma(b,v)) + g(\Gamma(a,u)). \tag{21}$$

Theorem 6. *Suppose s is an additive generator of a continuous Archimedean t-norm, and W is represented as in (18). Then property FW1 implies that*

$$\Delta(a,b) = s^{-1}(h(b) - h(a) + C), \tag{22}$$

where h is an appropriately chosen non-increasing function and C is a positive constant.

In this case

$$W(a,b,u,v) = s^{-1}(\min\{sNs^{-1}(h(v) - h(u) + C) + h(b) - h(a) + C, s(1)\}). \tag{23}$$

Proof. Notice that property FW3 implies $\Gamma(x,x) = \Gamma(y,y)$, and hence $\Delta(x,x) = \Delta(y,y)$. Denote this joint value by C.

Let $a = b$ and $u = v$ in (21). We obtain

$$g(1-C)+g(C)=g(1-\Gamma(a,u))+g(\Gamma(a,u)).$$

Using this equality, substitute $g(1-\Gamma(u,v))$ and $g(1-\Gamma(b,v))$ with the corresponding expressions to obtain

$$\begin{aligned}g(\Gamma(a,b))-[g(C)+g(1-C)-g(\Gamma(b,v))] &= g(\Gamma(a,u))-[g(C)+g(1-C)\\ &\quad -g(\Gamma(u,v))],\end{aligned}$$

whence we get equation (11) for F with $F(a,b)=g(\Gamma(a,b))$ now. Thus, from Theorem 3 we know that $g(\Gamma(a,b))=h(b)-h(a)+C$. By definition of g and Γ we get $g(\Gamma(a,b))=s(\Delta(a,b))$, so we need to solve

$$s(\Delta(a,b))=h(b)-h(a)+C$$

for Δ.

First of all, Δ is non-decreasing in the first variable and non-increasing in the second one if and only if h is a non-increasing function.

As in case of Theorem 4, we can guarantee the solvability of this equation by the appropriate choice of function h and constant C.

The value $h(b)-h(a)+C$ must be in the range of the additive generator s. That range is either the set of non-negative real numbers, or an interval of $[0,\omega]$, with a finite ω.

In any case, let $\alpha<\beta$ be real numbers, and choose $C:=\beta-\alpha$. Let h be a function from $\mathbb{R}$ to the bounded interval $[\alpha,\beta]$. This choice squezees the value of $h(b)-h(a)+C$ in the interval $[0,2C]$.

If the range of s is the set of non-negative real numbers then C can be any positive number. If the range of s is $[0,\omega]$, then any $C\leq\omega/2$ is a good choice.

Thus Δ indeed has the form (22).

Transforming back everything to the original functions s and Δ, finally we obtain the statement. □

As we stated, the form of W given in (23) is only necessary for having property FW1. In fact, we can have it only in very special cases, when the t-conorm S is nilpotent (i.e., when $s(1)<\infty$), and the strong negation N is generated also by s. Suppose $s(1)=1$. In this case we have

$$\begin{aligned}W(a,b,u,v) &= s^{-1}(\min\{sNs^{-1}(h(v)-h(u)+C)+h(b)-h(a)+C,s(1)\})\\ &= s^{-1}(\min\{1-[h(v)-h(u)+C]+h(b)-h(a)+C,1\})\\ &= s^{-1}(\min\{1-h(v)+h(u)+h(b)-h(a),1\}).\end{aligned}$$

Comparing this formula with the one coming from R-implications in equation (15), one can see that they are different, even in the case when S and T are duals (i.e., when $s(x)=t(1-x)$).

3.3 Model 3: The Implication Comes from a Representable Uninorm

In our third model we start from a representable uninorm (see [5]), and use its residual implication [3] (which is indeed an implication satisfying properties I1–I4) in the definition of W.

Let us recall that a *uninorm* is a function $U : [0,1] \times [0,1] \to [0,1]$ which is commutative, associative, nondecreasing, and has a neutral element $e \in [0,1]$ (i.e., $U(e,x) = x$ for all $x \in [0,1]$).

Representable uninorms can be obtained as follows. Consider $e \in]0,1[$ and a strictly increasing continuous $[0,1] \to \overline{\mathbb{R}}$ mapping g with $g(0) = -\infty$, $g(e) = 0$ and $g(1) = +\infty$. The binary operator U defined by

$$U(x,y) = g^{-1}(g(x) + g(y)), \quad \text{if } (x,y) \in [0,1]^2 \setminus \{(0,1),(1,0)\},$$

and either $U(0,1) = U(1,0) = 0$, or $U(0,1) = U(1,0) = 1$, is a uninorm with neutral element e (called *representable uninorm*). The function g is called an *additive generator* of U.

In case of a uninorm U, the residual operator I_U can be defined by

$$I_U(x,y) = \sup\{z \in [0,1] \mid U(x,z) \le y\}.$$

In some cases (for instance, when U is representable) I_U is an implication.

It is easily seen that in case of a representable uninorm U with additive generator function g the residual implication I_U is of the following form [3]:

$$I_U(x,y) = \begin{cases} g^{-1}(g(y) - g(x)) & \text{if } (x,y) \in [0,1]^2 \setminus \{(0,0),(1,1)\} \\ 1 & \text{, otherwise} \end{cases}. \tag{24}$$

Then, the quaternary fuzzy relation W can be introduced as follows:

$$W(a,b,u,v) = \begin{cases} 1 & \text{if } (\Delta(a,b),\Delta(u,v)) \in \{(0,0),(1,1)\}, \\ g^{-1}(g(\Delta(a,b)) - g(\Delta(u,v))) & \text{otherwise}. \end{cases} \tag{25}$$

Then, condition FW1 implies the functional equation

$$g(\Delta(a,b)) + g(\Delta(b,v)) = g(\Delta(a,u)) + g(\Delta(u,v)), \tag{26}$$

similarly to the previous two occurrences of the same type.

Theorem 7. *Assume that the implication I is represented by equation (24), where g is any additive generator of a representable uninorm. Define a quaternary fuzzy relation W by (25). Then W satisfies condition FW1 (i.e., $W(a,b,u,v) = W(a,u,b,v)$) if and only if Δ is of the following form:*

$$\Delta(a,b) = g^{-1}(h(b) - h(a) + C), \tag{27}$$

where h is any non-increasing function and C is any constant.

In this case the quaternary relation W can be written as follows:

$$W(a,b,u,v) = g^{-1}(h(u) - h(v) - h(a) + h(b)). \tag{28}$$

We emphasize that in the present case the equation

$$g(\Delta(a,b)) = h(b) - h(a) + C$$

has solution without any restriction to h or C, because the range of g is $\mathbb{R}$.

4 Strict Preference and Indifference

There exist axiomatic approaches to defining preference structures when we use binary fuzzy relations, see [4]. We try to apply those results in the present environment of quaternary fuzzy relations.

We start with a simple observation. According to the semantical meaning of $abWuv$ in the crisp case, it is obvious that W can be considered as a *binary relation* on $A \times A$. Thus, our task is easily completed.

First of all, recognize that W (as a binary relation on $A \times A$) is strongly complete: for any $a,b,u,v \in A$ we have either $abWuv$, or $uvWab$. Therefore, any axiomatization leads to the following unique formula for the strict preference D and indifference E (see [4]):

$$\begin{aligned} D(a,b,u,v) &= N(W(u,v,a,b)), \\ E(a,b,u,v) &= \min(W(a,b,u,v), W(u,v,a,b)), \end{aligned}$$

where N is a strong negation.

5 Conclusion

We have developed three approaches to quaternary fuzzy relations modelling difference measurement. Three simple formulas have been obtained which may be useful and attractive also in applications. We hope that the study can also be applied to fuzzy weak orders, where representations analogous to (1) could be proved only in two particular cases.

Acknowledgements. The author gratefully acknowledges the support obtained within the frames of Bilateral Romanian-Hungarian S&T Co-operation Project conducted by the Research and Technology Fund in Hungary, a support in part by OTKA K063405, and by COST Actions IC0602 and IC0702.

References

1. Aczél, J.: Lectures on Functional Equations and their Applications. Academic Press, New York (1966)
2. De Baets, B., Fodor, J.C.: Twenty years of fuzzy preference structures (1978–1997). Riv. Mat. Sci. Econom. Social. 20, 45–66 (1997)
3. De Baets, B., Fodor, J.C.: Residual operators of uninorms. Soft Computing 3, 89–100 (1999)
4. Fodor, J., Roubens, M.: Fuzzy Preference Modelling and Multicriteria Decision Support. Kluwer, Dordrecht (1994)
5. Fodor, J.C., Yager, R., Rybalov, A.: Structure of uninorms. Int. J. of Uncertainty, Fuzziness and Knowledge-Based Systems 5, 411–427 (1997)
6. Klement, E.P., Mesiar, R., Pap, E.: Triangular Norms. Kluwer, Dordrecht (2000)
7. Köbberling, V.: Strength of preference and cardinal utility. Economic Theory 27, 375–391 (2006)
8. Krantz, D.H., Luce, R.D., Suppes, P., Tversky, A.: Foundations of Measurement. Additive and polynomial representations, vol. I. Academic Press, New York (1971)
9. Roberts, F.S.: Measurement Theory. In: Encyclopedia of Mathematics and its Applications, vol. 7. Addison-Wesley, London (1979)
10. Scott, D.: Measurement models and linear inequalities. Journal of Mathematical Psychology 1, 233–247 (1964)

Indices of Collusion among Judges and an Anti-collusion Average

Cesarino Bertini, Gianfranco Gambarelli, and Angelo Uristani

Abstract. We propose two indices of collusion among Judges of objects or events in a context of subjective evaluation, and an average based on these indices. The aim is manifold: to serve as a reference point for appeals against the results of voting already undertaken, to improve the quality of scores summarized for awards by eliminating those that are less certain, and, indirectly, to provide an incentive for reliable evaluations. An algorithm for automatic computation is supplied. The possible uses of this technique in various fields of application are pointed out: from Economics to Finance, Insurance, Arts, artistic sports and so on.

Keywords: Average, Collusion, Index, Judges, Awards, Scores, Voting.

1 Introduction

In this paper we propose two indices of collusion among Judges of objects or events in a context of subjective evaluation, and an average based on these indices. The aim is manifold: to serve as a reference point for appeals against the results of voting already undertaken, to improve the quality of scores summarized

Cesarino Bertini
Assistant Professor, Department of Mathematics, Statistics, Computer Science and Applications
University of Bergamo, Faculty of Economics and Business Administration,
Via dei Caniana n.2, 24127 Bergamo, Italy
e-mail: cbertini@unibg.it

Gianfranco Gambarelli
Full Professor, Department of Mathematics, Statistics, Computer Science and Applications
University of Bergamo, Faculty of Economics and Business Administration,
Via dei Caniana n.2, 24127 Bergamo, Italy
e-mail: gambarex@unibg.it

Angelo Uristani
Doctorate on Computational Methods for Forecasting and Decisions in Economics and Finance, Department of Mathematics, Statistics, Computer Science and Applications
University of Bergamo, Faculty of Economics and Business Administration,
Via dei Caniana n.2, 24127 Bergamo, Italy
e-mail: angeluri@yahoo.it

S. Greco et al. (Eds.): Preferences and Decisions, STUDFUZZ 257, pp. 195–209.
springerlink.com

for awards by eliminating those that are less reliable and, indirectly, to provide an incentive for reliable evaluations. An algorithm for automatic computation is supplied.

This method can be applied to contexts in which those involved in judging are also involved in the outcome of their own evaluations: in Economics (e.g., project evaluation and problems of estimate), in Finance (e.g., company quotation), in the field of Insurance (e.g., providing customized insurance policies), in the Arts (e.g., judging singers and musicians whose record labels have links of some kind with the judges), in artistic sports (Rhythmic Gymnastics, Figure Skating, Diving) and so on.

2 Current Methods

A trait shared by a number of rules for data synthesis is that of giving little or no weight to data distribution tails, as the more central data are considered to be more reliable. Among the classic methods that drastically eliminate tail data, the trimmed means (and in particular the median) occupy a privileged position in cases where the need is felt to give those involved in subjective judgements an incentive to make correct evaluations so as to avoid their awards being dismissed. There is a downside to the aforementioned techniques when distribution is asymmetrical, since they may well eliminate reliable awards while taking unreliable ones into account. To avoid such disadvantages some weighted means can be used. The Coherent Majority Average in particular (Gambarelli 2008) is especially suited to cases where the set of possible scores is such that the distance between two consecutive numbers is constant. The correct application of this average relies on fair evaluations being supplied by a majority of judges. However, in cases where the adjudicating body is divided into sub-commissions, it may happen that a given sub-commission contains a number of colluding judges greater than the majority of this sub-commission. In such cases it is impossible to identify those involved within the context of a single evaluation, although conclusions may be drawn from the overall scores awarded during the entire process.

3 The New Idea

The new idea consists of constructing, once voting has taken place, a collusion index to be assigned to each subset of the set of judges. On the basis of this coalitional index, an individual collusion index is created for each judge (corresponding to the maximum coalitional index of the subsets to which each judge belongs). The set of judges is then subdivided into subsets with various levels of reliability (in terms of collusion). Finally, an average is calculated using only awards assigned by those judges considered to be more reliable in the context of this procedure.

Example

There are seven projects to be judged, proposed by six teams. The first team proposes two projects; all the other teams propose one each. There are four judges in all and they belong to the first four teams. Each adjudicating commission is made up of three judges, who must award each project a score from 1 to 10. Let us suppose that the scores awarded to the seven projects by the four judges are those shown in Table 1. This data gives rise to the impression that two judges (the third and fourth) are colluding among themselves, inasmuch as they award reciprocally high scores while awarding low scores to the others. This prompts the question as to whether a method exists to identify this supposed collusion objectively, and to take it into account to construct a fair average.

Table 1 The awards of the example

Teams	Projects	Judges			
		1	2	3	4
1	I	7	7	4	-
	II	7	7	-	4
2	I	-	7	4	4
3	I	-	8	10	9
4	I	6	-	9	9
5	I	7	7	4	-
6	I	7	6	-	5

4 The Data

Let O be a finite ordered set whose elements (to be called "objects") are shared in n (≥ 2) ordered classes (called "teams"). We call T the set of these classes and t_1, ..., t_n the elements of T.

Let J be an ordered set of n elements (the "judges") in biunivocal correspondence with T. We call j_1, ..., j_n the elements of J.

Let $\underline{J}$ be an ordered set of u elements (the "judges" lacking in corresponding objects) having empty intersection with J. We call $\underline{j}_1$, ..., $\underline{j}_u$ the elements of $\underline{J}$.

We define P the set of non-trivial parts of $J \cup \underline{J}$. For every $p \in P$ we call p' the complementary set of p with respect to P.

Let $\underline{O}$ be a finite ordered (even empty) set whose elements (to be called "objects lacking in corresponding judges") are shared in $\underline{n}$ (≥ 0) classes. We call $\underline{T}$ the (even empty) set of these classes and t_{n+1}, ..., $t_{n+\underline{n}}$ the ordered elements of $\underline{T}$.

Let Π be a finite set of rational positive numbers (the "points").

Remark. The use of positive scores is justified by the need to avoid null denominators. Obviously, suitable transformations may be applied in the case of voting systems that include zero.

We call a ("award") a function defined on $(J \cup \underline{J}) \times (T \cup \underline{T}) \times (O \cup \underline{O})$, taking values on $\Pi \cup \{\varnothing\}$, constructed as follows, on the basis of scores awarded by judges:

$$a(j,t,o)=\begin{cases}\varnothing & \text{if the } j\text{-th judge is unable to award any scores to the } o\text{-th element of the } t\text{-th class,}\\ \text{the element of } \Pi \text{ that the } j\text{-th judge awards to the } o\text{-th element of the } t\text{-th class} & \text{elsewhere}\end{cases}$$

We assume that each object receives at least one score and each judge gives at least one score.

Follow-up to our Example.

- $n = 4$ and $\underline{n} = 2$;
- O is made up of five projects: o_{1I} and o_{1II} both belonging to team t_1; o_{2I} belonging to t_2, and so on, up to o_{4I} belonging to t_4;
- $\underline{O}$ is made up of two projects: o_{5I} belonging to team t_5 and o_{6I} belonging to t_6;
- $J = (j_1, j_2, j_3, j_4)$;
- $\underline{J}$ is empty;
- $P = \{1, 2, 3, 4, 12, 13, 14, 23, 24, 34, 123, 124, 134, 234\}$;
- Π is the set of natural numbers from 1 to 10;
- the scores are: $a(1,1,1) = 7$, $a(1,1,2) = 7$, $a(1,2,1) = \varnothing$, $a(1,3,1) = \varnothing$, $a(1,4,1) = 6$ and so on, up to $a(4,6,1) = 5$.

5 The Means

To determine a coalitional collusion index, comparisons of the various arithmetic means of the scores supplied by each group of judges will be necessary:

- with objects whose classes correspond to such judges, and
- with other objects.

We start giving the following definitions, for every twin $x, y \in P$:

- $\underline{x}$ is the union between $\underline{O}$ and all objects belonging to all classes corresponding to x;
- $k(x, y)$ is the cardinality of the set of the numeric scores that all judges belonging to x assign to all objects belonging to all classes corresponding to y;
- $$m(x,y)=\begin{cases}\text{the arithmetic mean of the numeric scores that all judges belonging to } x \text{ assign to all objects belonging to all classes corresponding to } y & \text{if } k(x,y)>0\\ 0 & \text{if } k(x,y)=0\end{cases}$$

Convention. For simplicity, we will use the notation "*abc...*" to indicate the set $\{a, b, c, ...\}$, when it does not give rise to misunderstandings.

Follow-up to our Example.
For simplicity we give in Table 2 detailed illustrations only of cases involving p = 34 ($\rightarrow$ p' = 12).

Table 2 Detailed illustrations of cases involving p = 34

$m(p, p)$	$= m(34, 34)$	= (10+9+9+9)/4	= 37/4	= 9.25
$m(p', p)$	$= m(12, 34)$	= (6+8)/2	= 7	= 7.00
$m(p, \underline{p}')$	$= m(34, 1256)$	= (4+4+4+4+4+5)/6	= 25/6	$= 4.1\overline{6}$
$m(p', \underline{p}')$	$= m(12, 1256)$	= (7+7+7+7+7+7+7+7+6)/9	= 62/9	$= 6.\overline{8}$

6 The Coalitional Collusion Index

For every set of judges $p \in P$ we define:

$$\text{index of valuation of } p = \begin{cases} 1 & \text{if } m(p, p) = 0 \text{ or } m(p', p) = 0 \\ m(p, p)/m(p', p) & \text{elsewhere} \end{cases}$$

$$\text{index of valuation of } \underline{p}' = \begin{cases} 1 & \text{if } m(p, \underline{p}') = 0 \text{ or } m(p', \underline{p}') = 0 \\ m(p, \underline{p}')/m(p', \underline{p}') & \text{elsewhere} \end{cases}$$

The first index compares the average scores awarded by judges belonging to p to the objects belonging to their team, with the average scores that other judges have awarded to the same objects. The greater this index, the more the judges concerned are in agreement among themselves. In the examined case the index value is $m(34, 34) / m(12, 34) = 37/28$.

The second index compares the average scores awarded by judges belonging to p to the objects belonging to other teams, with the average scores that other judges have awarded to the same objects. The lower this index, the more the relevant judges are hostile to each other. In the examined case the index value is $m(34, 1256) / m(12, 1256) = 75/124$.

We call "coalitional collusion index of p" the ratio between the valuation index of p and the valuation index of $\underline{p}'$.

If all m are positive, then the coalitional collusion index of p is:

$$r(p) = \frac{m(p, p)}{m(p', p)} \cdot \frac{m(p', \underline{p}')}{m(p, \underline{p}')}$$

Elsewhere, the first and/or the second factor of the above product is 1.

This index encompasses a tendency towards solidarity among the judges of p, with hostility towards the other judges. Low values of $r(p)$ may be interpreted in terms of a lack in global collusion among the judges of p.

In our example the coalitional collusion index of p is $r(34) = (37/4)\cdot(62/9) / (7\cdot 25/6) = 1147/525$.

7 The Reliable Majority Assumption

It may happen in reality that a set of judges constituting a majority in a sub-commission have all colluded among themselves. But in our study we have to exclude the case that the collusion happens for sets of judges forming a majority in the commission as a whole: otherwise the solution to our problem no longer lies within the rules of computation. Thus what is given below is based on the following:

Assumption. There is no global collusion between all judges of every group constituting a majority in the commission as a whole.

We define P' as the collection of the sets of judges having a cardinality smaller than, or equal to, $n/2$.

Then we will consider that the only possible colluded coalitions are sets of P'.

8 The Individual Collusion Index

For every $j \in J \cup \underline{J}$ we call $c(j)$ ("individual collusion index of j") the maximum value of $r(p)$ for all $p \in P'$ to which judge j belongs.

Follow-up to our Example.

$n/2=2$,

$P' = \{(1), (2), (3), (4), (12), (13), (14), (23), (24), (34)\}$.

All coalitional collusion indices, with relevant components, are given in Table 3.

Regarding individual collusion indices, the maximum value of r is $1147/525 \approx 2.18$ (corresponding to 34), while the second r, in order of size, is $161/82 \approx 1.96$ (corresponding to 12). Therefore $c(1) = c(2) = 161/82$; $c(3) = c(4) = 1147/525$.

Table 3 The computation of the coalitional collusion indices

	Sets of judges									
	1	2	3	4	12	13	14	23	24	34
$m(p, p)$	7	7	10	9	7	7	33/5	29/4	20/3	37/4
$m(p', \underline{p}')$	41/6	88/13	90/14	46/7	23/3	19/3	46/7	13/2	46/7	62/9
$m(p', p)$	11/2	4	17/2	15/2	4	7	27/4	13/2	19/3	7
$m(p, \underline{p}')$	20/3	7	21/4	11/2	41/6	37/6	32/5	44/7	53/8	25/6
$r(p)$	287/220	22/13	1200/833	552/385	161/82	38/37	759/756	203/176	7360/7049	1147/525
$\approx$	1.30	1.69	1.44	1.43	1.96	1.03	1.003	1.15	1.04	2.18

9 The Classes of Reliability of Judges

A method has thus been established for assigning an unreliability index to each judge. This index may be used directly (for example, to weight the scores), or indirectly, to establish thresholds for less reliable judges. The latter use is especially suited to cases where an incentive for judges to give correct scores is deemed desirable, if a drastic exclusion of their evaluations is to be avoided. From here on, this will form the focus of our study. We shall begin by grouping judges into classes of equal individual collusion index.

We call C_1 ("judges at the first level of reliability") the subset of $J \cup \underline{J}$ made up of all j such that $c(j)$ is the minimum. For each integer $h>1$ we call C_h ("judges at the h-th level of reliability") the set made up of all $j \in (J \cup \underline{J}) \setminus (C_1 \cup \ldots \cup C_{h-1})$ such that $c(j)$ is the minimum.

10 The Anti-collusion Average

Now we need to find a criterion to determine which judges provide scores that are to be used (in this context in fact we exclude the scores of all other judges, to provide incentives for proper assessments). Of course we will use the scores of all judges of class C_1 i.e. at the first level of reliability. As for the other classes, we believe that a good compromise between representativeness and reliability is the set of best reliable classes, such that their total cardinality does not exceed half of the judges. Only in the case in which, using this selection, an object is lacking in evaluations, we consider, among the remaining classes, the one with the maximum level of reliability which is able to solve the problem.

We define:

- $S_k = \{ j \in C_1 \cup \ldots \cup C_k \}$
- $U_o = \{ j \in P \mid j \text{ assigned at least a score to } o \in O \cup \underline{O} \}$
- $K_o = \{ \min k \mid \text{at least a } j \in C_k \text{ assigned a score to } o \in O \cup \underline{O} \}$
- $R = \begin{cases} S_1 & \text{if } \mathrm{card}(S_1) \geq \dfrac{n}{2} \\ \{ S_k \mid \mathrm{card}(S_k) \leq \dfrac{n}{2} \text{ and } \mathrm{card}(S_{k+1}) > \dfrac{n}{2} \} & \text{otherwise} \end{cases}$
- $R_o = \begin{cases} \{ C_h \mid h = K_o \} & \text{if } R \cap U_o = \varnothing \\ R & \text{otherwise} \end{cases}$

The "anti-collusion average" (or, simply, "ACA") of each object o is the arithmetic mean of the scores that have been assigned to o by all judges belonging to R_o.

Follow-up to our Example.
$C_1 = \{1, 2\}$ ($c \approx 1.96$), $C_2 = \{3, 4\}$ ($c \approx 2.18$) and $C_h = \varnothing$ for all $h>2$. $R_o = \{1, 2\}$ for all objects and therefore only the related scores are used to calculate the ACA. Table 4 gives the result of this calculation, compared with the arithmetic mean and the median.

Table 4 Anti-Collusion Average and corresponding ranking

Project	R_o	Average			Ranking			
		ACA	Arithm. mean	Median	Order	ACA	Arithm. mean	Median
1_I	C_I	7	6	7	1	3	3	3, 4
1_{II}	C_I	7	6	7	2	1_I, 1_{II}, 2, 5	4	-
2	C_I	7	5	4	3	-	1_I, 1_{II}, 5, 6	1_I, 1_{II}, 5
3	C_I	8	9	9	4	-	-	-
4	C_I	6	8	9	5	-	-	-
5	C_I	7	6	7	6	6	-	6
6	C_I	6.5	6	6	7	4	2	2

Notice that the ranking corresponding to the ACA puts the fourth object in last place, while the other two averages had instead put it among the first two places.

11 An Algorithm

An algorithm for calculating the ACA is shown in the Appendix. The main characteristics are summarized below.

Input:
- the table of the awards;

Output:
- the ordered coalitional and individual collusion indices;
- the ACA;
- the ACA ranking with those of the arithmetic mean and median.

Procedure:
- Read and write the input data;
- Form all the coalitions having cardinality from 1 to l, where l is the largest integer smaller than or equal $n/2$;
- For each coalition, calculate the coalitional collusion index r;
- For each judge, calculate the individual collusion index c;
- On the basis of the individual collusion index, form the sets C_k of judges with the same c;
- Calculate the set R_o;
- For each object o, calculate the ACA;
- Write the output data.

Computational time increases exponentially with an increase in the number of judges n, as the number of sub-sets to be considered is equal to

$$\sum_{k=1}^{l}\frac{n!}{(n-k)!k!}$$

where l is the largest integer smaller than or equal to $n/2$. The calculation using a standard PC takes few instants for a dozen of judges.

12 Conclusion

In presenting our Example we said: "This data gives rise to the impression that two judges are colluding among themselves... This prompts the question as to whether a method exists to identify this supposed collusion objectively, and to take it into account to construct a fair average". As has been shown, the collusion indices here introduced and the ACA represent an answer to these questions.

Acknowledgments. This work is sponsored by MURST. The authors thank Valeria Ruggeri for help in computations; Silvio Giove, Benedetto Matarazzo, Paolo Pianca and Marina Piazza for useful suggestions.

Reference

Gambarelli, G.: The coherent majority average for juries evaluation processes. Journal of Sport Sciences 26, 1091–1095 (2008)

Appendix: Program Listing in Matlab

Variables (As example):

```
num_teams = 6;
num_judges = 4;
% the awards should be inserted as follows: first the projects belonging to E, then
% those that belong to E_bar
award = {[7 7 4 0;7 7 0 4]; [0 7 4 4]; [0 8 10 9]; [6 0 9 9]; [7 7 4 0]; [7 6 0 5]};
project_labels = {'1A' '1B' ' 2' ' 3' ' 4' ' 5' ' 6'};
E = [1 2 3 4];
E_bar = [5 6];
% G should be numbered with consecutive numbers after E_bar
G = [];
EG = [E G];
```

Function Main:

```
function main(num_teams, num_judges, award, project_labels, EG, G, E_bar)
% Display Input data
disp('- Input Data ----------------------------');
str = sprintf(' %4.0f  ', num_teams);
disp(strcat('Number of teams  : ', str));
```

```
str = [];
str = sprintf(' %4.0f    ', num_judges);
disp(strcat('Number of judges : ', str));
str = [];
disp(' ');
yy = 00;
for i = [1:1:size(award, 1)]
   matt = cat(1, award{i, :});
   for j = [1:1:size(matt, 1)]
      yy = yy + 1;
      str = sprintf('%4.0f \t', matt(j, :));
      disp(strcat('Award project _', project_labels{yy}, ': ', str));
      str = [];
   end
end
disp(' ');
str = sprintf(' %4.0f    ', EG);
disp(strcat('E  + G   : ', str));
str = [];
str = sprintf(' %4.0f    ', E_bar);
disp(strcat('E_bar : ', str));
str = [];
disp('----------------------------------------');
disp(' ');

% Evaluation of the possible juries
P = [];
P_label = [];
for nrich = [1:1:floor(num_judges/2)]
   P_temp = GenerateDistrib(nrich, num_judges);
   A = P_temp;
   for f = [1:1:num_judges]
      [xf, yf]=find(P_temp == f);
      A(xf, yf) = EG(f);
   end
   P = [P; GenerateDistrib(nrich, num_judges)];
   P_label = [P_label; A];
end

for i = [1:1:size(P, 1)]
   judge_p = find(P(i, :)>0);
   judge_p_first = setdiff([1:1:num_judges], judge_p);
   proj_p = find(P(i, [1:1:length(setdiff(EG, G))])>0);
   proj_p_first = setdiff(setdiff(EG, G), proj_p);
   proj_p_first_bar = union(E_bar, proj_p_first);
```

```
  if isempty(proj_p)
     self = 1;
  else
     if (ff_mean(judge_p, proj_p, award) == 0)|| ...
          ( ff_mean(judge_p_first, proj_p, award)==0)
        self = 1;
     else
        self = (ff_mean(judge_p, proj_p, award)/ ...
          ff_mean(judge_p_first, proj_p, award));
     end
  end
  if (ff_mean(judge_p, proj_p_first_bar, award) == 0)|| ...
          ( ff_mean(judge_p_first, proj_p_first_bar, award)==0)
     other = 1;
  else
     other = (ff_mean(judge_p, proj_p_first_bar, award)/ ...
          ff_mean(judge_p_first, proj_p_first_bar, award));
  end

  r(i) = self/other;
end


disp('- Coalitional Collusion Indices --------');
for jj = [1:1:length(r)]
  [xs] = find(P_label(jj, :)>0);
  ssss = P_label(jj, xs);
  str = sprintf('%d, ', ssss);
  stra = sprintf(': \t %s ', num2str(r(jj),'%2.3f'));
  disp(strcat('(',str(1:1:length(str)-2),')',stra));
end
disp('---------------------------------------');
disp(' ');

for nn = [1:1:num_judges]
  for i =[1:1:length(r)]
     aa(i) = ismember(nn, P(i, :));
     end
     c(nn) = max(r(aa));
  end

  o = 0;
  tol = 0.00001;
  while true
     o = o + 1;
     C{o} = find((min(c)<c+tol)&(min(c)>c-tol));
```

```
    C_Label{o} = EG(C{o});
    c(C{o}) = Inf;
    if sum(isinf(c)) == length(c)
        break
    end
end

% group formation for ACA evaluation
AC = [];
matr = cell2mat(award);
base_AC = C{1};
base_l = 1;
for l = [2:1:length(C)]
    if (length([base_AC C{l}]) <= (num_judges/2))
        base_AC = [base_AC C{l}];
        base_l = l;
    else
        break;
    end
end
for s = [1:1:size(matr, 1)]
    o = base_l;
    AC{s} = base_AC;
    while true
        if any(matr(s, AC{s})>0)
            break;
        end
        o = o + 1;
        AC{s} = union(AC{s}, C{o});
    end
end

mat = cat(1,award{:});
for s = [1:1:size(matr,1)]
    ACA(s) = sum(mat(s, AC{s})',1)./sum(mat(s, AC{s})'>0,1);
end
mean_for_comparison = sum(mat(:, :)')./sum(mat(:, :)'>0);
for u = [1:1:size(mat, 1)]
    median_for_comparison(u) = median(mat(u, mat(u, :)>0));
end

disp('- Classes of Reliability ----------------');
for i = [1:1:length(C)]
    str = sprintf('%2.0f° Class:    ', i);
    str2 = sprintf('%2.0f    ', C_Label{i});
    disp(strcat(str,'    ', str2));
```

```
    str = [];
    str2 = [];
  end
  disp('----------------------------------------');
  disp(' ');

  disp('- Comparison ----------------------------');
  str = sprintf('%s   \t', project_labels{:});
  disp(strcat('Projects : ', str));
  str = [];
  str = sprintf('%2.2f \t', ACA);
  str = strcat('ACA      : ', str);
  disp(str);
  str = [];
  str = sprintf('%2.2f \t', mean_for_comparison);
  str = strcat('Mean     : ', str);
  disp(str);
  str = [];
  str = sprintf('%2.2f \t', median_for_comparison);
  str = strcat('Median   : ', str);
  disp(str);

  %ranking
  o = 0;
  dummy_ACA = ACA;
  rank_no_ACA = [];
  while true
    o = o + 1;
    rank_no_ACA{o} = find(max(dummy_ACA) == dummy_ACA);
    dummy_ACA(rank_no_ACA{o}) = 0;
    if (sum(dummy_ACA) == 0)
      break
    end
  end

  o = 0;
  dummy_mean_comparison = mean_for_comparison;
  rank_no_mean = [];
  while true
    o = o + 1;
    rank_no_mean{o} = ...
        find(max(dummy_mean_comparison) == dummy_mean_comparison);
    dummy_mean_comparison(rank_no_mean{o}) = 0;
    if (sum(dummy_mean_comparison) == 0)
      break
    end
  end
```

```
o = 0;
dummy_median_comparison = median_for_comparison;
rank_no_median = [];
while true
   o = o + 1;
   rank_no_median{o} = ...
       find(max(dummy_median_comparison)== dummy_median_comparison);
   dummy_median_comparison(rank_no_median{o}) = 0;
   if (sum(dummy_median_comparison) == 0)
      break
   end
end

disp('----------------------------------------');
disp(' ');

disp('- Ranking (ACA) -------------------------');
for rr = [1:1:size(rank_no_ACA, 2)]
   disp(sprintf('  %s ', project_labels{rank_no_ACA{rr}}));
end

disp('- Ranking (Mean) ------------------------');

for rr = [1:1:size(rank_no_mean, 2)]
   disp(sprintf('  %s ', project_labels{rank_no_mean{rr}}));
end

disp('- Ranking (Median) ----------------------');
for rr = [1:1:size(rank_no_median, 2)]
   disp(sprintf('  %s ', project_labels{rank_no_median{rr}}));
end

end
```

Function *ff_mean*:

```
function [val_mean] = ff_mean(a, b, award)
   matr = cat(1, award{b});
   matri = matr(:, a);
   val_mean = mean(matri(find(matri(:)>0)));
end
```

Function *GenerateDistrib*:

```
function [Distr]=GenerateDistrib(nrich, n)
if (nrich == 1)
```

```
    Distr = eye(n, n) .* diag([1:1:n]);
    return
  end

  matrixComb = nchoosek([1:1:n], nrich);
  Distr = zeros(size(matrixComb,1), n);

  for cont = [1:1:n]
    [i, j] = find(matrixComb == cont);
    Distr(i, cont) = cont;
  end
  end
```

Scoring Rules and Consensus

José Luis García-Lapresta, Bonifacio Llamazares, and Teresa Peña

Abstract. In this paper we consider that voters rank order a set of alternatives and a scoring rule is used for obtaining a set of winning alternatives. The scoring rule we use is not previously fixed, but we analyze how to select one of them in such a way that the collective utility is maximized. In order to generate that collective utility, we ask voters for additional information: agents declare which alternatives are good and their degree of optimism. With that information and a satisfaction function, for each scoring rule we generate individual utility functions. The utility an alternative has for a voter should depend on whether this alternative is a winner for that scoring rule and on the position this alternative has in the individual ranking. Taking into account all these individual utilities, we aggregate them by means of an OWA operator and we generate a collective utility for each scoring rule. By maximizing the collective utility, we obtain the set of scoring rules that maximizes consensus among voters. Then, applying one of these scoring rules we obtain a collective weak order on the set of alternatives, thus a set of winning alternatives.

1 Introduction

Some group decision problems are designed for generating an order on the set of feasible alternatives or a set of winning alternatives from the orders that individuals provide on that set of alternatives. Within this approach, it is well-known that there does not exist perfect voting systems (see Arrow [1]). Thus, the problem is to devise group decision procedures satisfying some good properties but not all we may desire. In this contribution we focus on *scoring rules*, a class of voting systems where voters rank order the alternatives from best to worst and they associate a score to

José Luis García-Lapresta · Bonifacio Llamazares · Teresa Peña
PRESAD Research Group, Dept. of Applied Economics, University of Valladolid, Spain
e-mail: {lapresta,boni,maitepe}@eco.uva.es

S. Greco et al. (Eds.): Preferences and Decisions, STUDFUZZ 257, pp. 211–224.
springerlink.com

each alternative in a decreasing way. The alternatives are ordered by the collective scores obtained by adding up the individual scores[1].

Scoring rules only require voters to rank order all the alternatives, irrespectively whose of them are considered good or bad alternatives. On the other hand, *approval voting*, introduced by Brams and Fishburn [4], only requires voters to show the good alternatives. It is worth mentioning that two voters may declare the same ranking on the set of alternatives but they may have different opinions about which alternatives are good and bad. In this way, it is interesting the hybrid voting system *preference approval voting*, devised by Brams [3] and Brams and Sanver [5], that combines two informations provided for each voter: a ranking of all the alternatives and, additionally, the alternatives approved of[2]. We also consider this hybrid approach in our proposal for constructing individual utilities.

We are interested in finding a scoring rule that maximizes a collective utility function. So, the scoring rule is not fixed, but it depends on the individual preferences. In order to generate the collective utility function we try to maximize, we need to introduce individual utilities and an aggregation function. We assume that voters do not provide utility functions, but only three pieces of information for each individual:

1. A linear order on the set of alternatives.
2. The set of good alternatives.
3. A degree of optimism or riskiness.

By means of a satisfaction function and the above information, we introduce a utility function for each individual that assigns a utility value to each scoring rule. The satisfaction function depends on a mapping φ –the same for all individuals– that assigns a numerical value within the unit interval to each position in a decreasing manner. The satisfaction function of a voter is just φ for the positions associated with the good alternatives, being 0 for the other alternatives. Their values may be interpreted as measures of satisfaction for having an alternative in the set of winners of a scoring rule: $\varphi(j)$ is the satisfaction we consider a voter has whenever the j-th alternative is good for that voter and it is a winner[3]. If there are several winning alternatives, we take into account the degree of optimism or riskiness of that voter for aggregating the satisfactions associated with those winning alternatives. This aggregation phase will be conducted by means of OWA operators [12] with an attitudinal character for each voter according to the degree of optimism or riskiness.

Once the individual utility functions are constructed, we aggregate them through an OWA operator that generates a collective utility value for each scoring rule regarding the winning alternatives generated by that scoring rule, and their positions

[1] Scoring rules have been characterized by means of some interesting properties by Smith [11] and Young [13]. See Chebotarev and Shamis [7] for a referenced survey.

[2] Clearly, in preference approval voting a voter may approve of all alternatives, and at the other extreme of no alternatives. As pointed out by Brams [3], both extreme strategies are dominated from the game theory perspective.

[3] Notice again that we do not ask individuals about their utilities, but we interpret their utilities according to the information they provide and the general pattern given by φ.

in the individual rankings. Then, we find the scoring rules that maximize that collective utility function. Consequently, the scoring rule we apply maximizes consensus among the voters with respect to the specific opinions they declare.

The paper is organized as follows. Section 2 is devoted to introduce the notions we use. In Section 3, we include our proposal for maximizing consensus on what is the most appropriate scoring rule to use in the decision problem. Section 4 contains an illustrative example by using different OWA operators in the corresponding aggregation phases. Finally, Section 5 includes some concluding remarks.

2 Preliminaries

Consider a set of voters: $V = \{1,\ldots,m\}$, $m \geq 3$, showing their preferences on a set of alternatives $X = \{x_1,\ldots,x_n\}$, $n \geq 3$, by means of *linear orders* (reflexive, antisymmetric and transitive binary relations on X). The set of linear orders is denoted by $L(X)$. A *profile* is a vector $\boldsymbol{R} = (R_1,\ldots,R_m)$ of linear orders that contains the preferences of voters. The *position mapping* $o_i : X \longrightarrow \{1,\ldots,n\}$ assigns the position of each alternative in the linear order R_i (see the example of Table 1).

Table 1 Positions

R_i	$o_i(x_j)$
x_3	$o_i(x_1) = 2$
x_1	$o_i(x_2) = 4$
x_4	$o_i(x_3) = 1$
x_2	$o_i(x_4) = 3$

We denote by $x^i_{(j)}$ the j-th alternative of voter $i \in V$, i.e., $o_i\left(x^i_{(j)}\right) = j$.

2.1 *Scoring Rules*

A *scoring rule* is defined by a scoring vector $\boldsymbol{s} = (s_1,\ldots,s_n) \in \mathbb{R}^n$ satisfying $s_1 \geq \cdots \geq s_n$ and $s_1 > s_n$, where for each voter's ranking, s_1 points are assigned to the top-ranked alternative, s_2 points to the second-ranked alternative, and so on.

Consider the scoring rule associated with the scoring vector $\boldsymbol{s} = (s_1,\ldots,s_n)$ and the profile of linear orders $\boldsymbol{R} = (R_1,\ldots,R_m)$.

1. The *assignment of voter i* is defined by $r^i_{\boldsymbol{s}} : X \longrightarrow \{s_1,\ldots,s_n\}$, where

$$r^i_{\boldsymbol{s}}(x_j) = s_{o_i(x_j)} \text{ or, equivalently, } r^i_{\boldsymbol{s}}\left(x^i_{(j)}\right) = s_j.$$

2. The *collective assignment* is defined by $r_{\boldsymbol{s}} : X \longrightarrow \mathbb{R}$, where

$$r_{\mathbf{s}}(x_j) = \sum_{i=1}^{m} r_{\mathbf{s}}^i(x_j).$$

3. The *collective weak order* (complete and transitive binary relation) is defined by

$$x_j \succcurlyeq_{\mathbf{s}} x_k \Leftrightarrow r_{\mathbf{s}}(x_j) \geq r_{\mathbf{s}}(x_k).$$

The set of all the scoring vectors is denoted by

$$\mathscr{S} = \{\mathbf{s} = (s_1, \ldots, s_n) \in \mathbb{R}^n \mid s_1 \geq \cdots \geq s_n,\ s_1 > s_n\}.$$

2.2 Normalized Scoring Vectors

In order to normalize scoring vectors, we can consider the binary relation $\sim$ on $\mathscr{S}$ defined by $\mathbf{s} \sim \mathbf{s}'$ if there exist $a, b \in \mathbb{R}$ with $a > 0$ such that $s'_i = as_i + b$ for every $i \in \{1, \ldots, n\}$. It is easy to see that $\sim$ is an equivalence relation (reflexive, symmetric and transitive binary relation) on $\mathscr{S}$. Clearly, all the equivalent scoring vectors define the same scoring rule, because they produce the same social outcomes. For simplicity, in what follows we consider the following set of normalized scoring vectors

$$\mathscr{S}^0 = \{\mathbf{s} \in \mathscr{S} \mid s_1 = 1, s_n = 0\}.$$

It is important to note that for every $\mathbf{s} \in \mathscr{S}$ there exists $\mathbf{s}' \in \mathscr{S}^0$ such that $\mathbf{s} \sim \mathbf{s}'$:

$$s'_i = \frac{s_i - s_n}{s_1 - s_n} = \frac{1}{s_1 - s_n} s_i + \frac{-s_n}{s_1 - s_n}.$$

For $n = 3$ the set of normalized scoring vectors

$$\mathscr{S}^0 = \{(1, s, 0) \mid s \in [0, 1]\}$$

can be identified with the interval $[0, 1]$. Notice that $s = 0, \frac{1}{2}, 1$ correspond to plurality, the Borda rule and antiplurality, respectively (see Fig. 1).

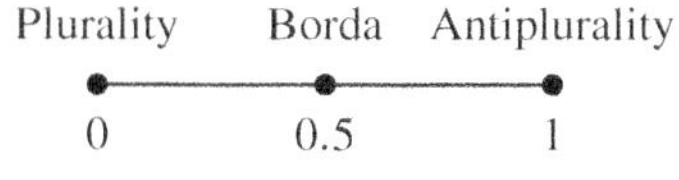

Fig. 1 The best-known scoring rules for $n = 3$

Analogously, for $n = 4$ the set of normalized scoring vectors

$$\mathscr{S}^0 = \{(1, s, t, 0) \mid 0 \leq t \leq s \leq 1\}$$

can be identified with the triangle $\{(s,t) \in [0,1]^2 \mid t \leq s\}$. Now, $(s,t) = (0,0)$, $(1,0)$, $(\frac{2}{3},\frac{1}{3})$, $(1,1)$ correspond to plurality, 2–approval voting, the Borda rule[4] and antiplurality, respectively (see Fig. 2).

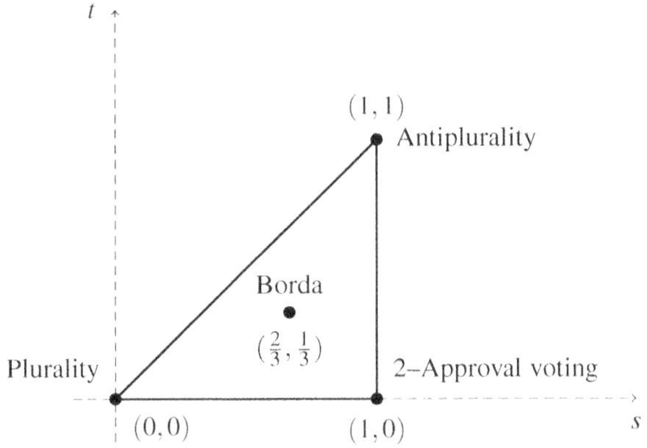

Fig. 2 The best-known scoring rules for $n = 4$

2.3 *Aggregation Functions and OWA Operators*

In our proposal we use aggregation functions (see Fodor and Roubens [8], Calvo *et al.* [6], Beliakov *et al.* [2] and Grabisch *et al.* [10]). An *aggregation function* is a continuous mapping $A : [0,1]^m \longrightarrow [0,1]$ that satisfies the following conditions:

1. *Monotonicity*: $A(x_1,\ldots,x_m) \leq A(y_1,\ldots,y_m)$ for all $(x_1,\ldots,x_m),(y_1,\ldots,y_m) \in [0,1]^m$ such that $x_i \leq y_i$ for every $i \in \{i,\ldots,m\}$.
2. *Unanimity* (or *idempotency*): $A(x,\ldots,x) = x$ for every $x \in [0,1]$.

It is easy to see that every aggregation function is *compensative*, i.e.,

$$\min\{x_1,\ldots,x_m\} \leq A(x_1,\ldots,x_m) \leq \max\{x_1,\ldots,x_m\},$$

for every $(x_1,\ldots,x_m) \in [0,1]^m$.

A class of anonymous aggregation functions that we will consider in our model are OWA operators (see Yager [12].

Given a weighting vector $\boldsymbol{w} = (w_1,\ldots,w_m) \in [0,1]^m$ such that $\sum_{i=1}^m w_i = 1$, the *OWA operator associated with* $\boldsymbol{w}$ is the mapping $A : [0,1]^m \longrightarrow [0,1]$ defined by

$$A(a_1,\ldots,a_m) = \sum_{i=1}^{m} w_i \cdot b_i$$

where b_i is the i-th greatest number of $\{a_1,\ldots,a_m\}$.

---

[4] Notice that $(\frac{2}{3},\frac{1}{3})$ is the baricenter of the triangle.

Some well-known aggregation functions are specific cases of OWA operators. For instance:

1. The *maximum*, given by the weighting vector $(1,0,\ldots,0)$.
2. The *minimum*, given by the weighting vector $(0,\ldots,0,1)$.
3. The *arithmetic mean*, given by the weighting vector $\left(\frac{1}{m},\ldots,\frac{1}{m}\right)$.
4. The *mid-range*, given by the weighting vector $(0.5,0,\ldots,0,0.5)$.
5. *Medians*, given by the following weighting vectors

   a. If m is odd
   $$w_i = \begin{cases} 1, \text{ if } i = \frac{m+1}{2}, \\ 0, \text{ otherwise.} \end{cases}$$

   b. If m is even
   $$w_i = \begin{cases} \theta, & \text{if } i = \frac{m}{2}, \\ 1-\theta, & \text{if } i = \frac{m}{2}+1, \\ 0, & \text{otherwise,} \end{cases}$$

   for some $\theta \in [0,1]$.

The *attitudinal character* (or *orness*) of an OWA operator $A : [0,1]^m \longrightarrow [0,1]$ associated with a weighting vector $\boldsymbol{w} = (w_1,\ldots,w_m)$ is defined by (see Yager [12]):

$$\alpha(A) = \frac{1}{m-1}\sum_{i=1}^{m}(m-i)w_i.$$

In some phases of our proposal we do not use a unique OWA operator, but a finite sequence of OWA operators, one for each dimension. In this way we will consider

$$A : \bigcup_{k=1}^{n}[0,1]^k \longrightarrow [0,1]$$

where the dimension will be determined by each specific context.

3 Maximizing Consensus

The first stage of the decision procedure consists of voters rank order the alternatives by means of linear orders, i.e., we have a profile $\boldsymbol{R} = (R_1,\ldots,R_m) \in L(X)^m$. At the same time, we fix a set of scoring vectors $\mathscr{S}^* \subseteq \mathscr{S}^0$. Then, every scoring rule associated with a scoring vector $\boldsymbol{s} \in \mathscr{S}^*$ generates a collective weak order $\succcurlyeq_{\boldsymbol{s}}$ on X for the considered profile $\boldsymbol{R}$. Since $\succcurlyeq_{\boldsymbol{s}}$ may be different depending on $\boldsymbol{s}$, our purpose is to select those scoring vectors $\boldsymbol{s}$ that maximize consensus among the voters.

Because of the notion of consensus is not unique, we need to introduce a proposal for measuring consensus between the profile $\boldsymbol{R}$ and the weak order $\succcurlyeq_{\boldsymbol{s}}$. In this way, we assume that voters are mainly interested in having their most preferred alternatives among the top ranked alternatives in $\succcurlyeq_{\boldsymbol{s}}$. So, given an arbitrary scoring rule with associated scoring vector $\boldsymbol{s} \in \mathscr{S}^*$, we will take into account the set of its maximums or winners

$$M_{\boldsymbol{s}}(\boldsymbol{R}) = \{x_j \in X \mid x_j \succcurlyeq_{\boldsymbol{s}} x_k \quad \forall k \in \{1,\dots,n\}\}.$$

Our proposal is based on two ingredientes:

1. A utility function $u_i : \mathscr{S}^* \longrightarrow [0,1]$ for each voter $i \in V$.
2. An OWA operator $A : [0,1]^m \longrightarrow [0,1]$.

We now define the *collective utility function* $u_A : \mathscr{S}^* \longrightarrow [0,1]$ by

$$u_A(\boldsymbol{s}) = A(u_1(\boldsymbol{s}),\dots,u_m(\boldsymbol{s})).$$

In order to reach consensus on the scoring rule to use in the decision problem, we propose to solve the problem

$$\max_{\boldsymbol{s} \in \mathscr{S}^*} u_A(\boldsymbol{s})$$

that is equivalent to find the set

$$\overline{\mathscr{S}}^* = \{\boldsymbol{s} \in \mathscr{S}^* \mid u_A(\boldsymbol{s}) \geq u_A(\boldsymbol{s}') \ \forall\, \boldsymbol{s}' \in \mathscr{S}^*\}.$$

Once calculated the set $\overline{\mathscr{S}}^*$, there are two possibilities:

- If $M_{\boldsymbol{s}}(\boldsymbol{R}) = M_{\boldsymbol{s}'}(\boldsymbol{R})$ for all $\boldsymbol{s},\boldsymbol{s}' \in \overline{\mathscr{S}}^*$, then we can choose any scoring vector $\boldsymbol{s} \in \overline{\mathscr{S}}^*$ and the winning alternatives are the elements of $M_{\boldsymbol{s}}(\boldsymbol{R})$.
- If $M_{\boldsymbol{s}}(\boldsymbol{R}) \neq M_{\boldsymbol{s}'}(\boldsymbol{R})$ for some $\boldsymbol{s},\boldsymbol{s}' \in \overline{\mathscr{S}}^*$, then we have scoring rules associated with scoring vectors of $\overline{\mathscr{S}}^*$ that provide different outcomes. In such cases it would be convenient to use an iterative process for breaking ties. A possibility is to fix a sequence of OWA operators and to proceed in a lexicographic manner. Suppose $A' : [0,1]^m \longrightarrow [0,1]$ is the next OWA operator of that sequence. We now consider the collective utility function $u_{A'} : \mathscr{S}^* \longrightarrow [0,1]$ and by solving the problem

  $$\max_{\boldsymbol{s} \in \overline{\mathscr{S}}^*} u_{A'}(\boldsymbol{s})$$

  we obtain the scoring rules of $\overline{\mathscr{S}_1}^* = \overline{\mathscr{S}}^*$ that maximize consensus according to A':

  $$\overline{\mathscr{S}_2}^* = \{\boldsymbol{s} \in \overline{\mathscr{S}_1}^* \mid u_{A'}(\boldsymbol{s}) \geq u_{A'}(\boldsymbol{s}') \ \forall\, \boldsymbol{s}' \in \overline{\mathscr{S}_1}^*\}.$$

  If, among the solutions of the previous problem, there are already scoring rules that supply different sets of winning alternatives, we consider the next OWA operator in the sequence for obtaining the corresponding set $\overline{\mathscr{S}_3}^* \subseteq \overline{\mathscr{S}_2}^*$. The process continues until a single outcome is obtained.

3.1 How to Define the (Indirect) Utility Functions?

It is well known that usually agents have difficulties in assigning exact numerical utility values to alternatives. Our proposal is to assign these utility values taking into account the following information provided by the voters:

1. The rankings included in the profile $\boldsymbol{R} = (R_1, \ldots, R_m) \in L(X)^m$.
2. The alternatives they approve of through the mapping $g : V \longrightarrow \{0, 1, \ldots, n\}$, where $g(i)$ shows that voter i declares the alternatives in the positions $j \leq g(i)$ are good and those in positions $j > g(i)$ are bad. Thus, we have the following cases:

   a. If $g(i) = 0$, then voter i thinks that all the alternatives in X are bad.
   b. If $0 < g(i) < n$, then $x^i_{(1)}, \ldots, x^i_{(g(i))}$ are good and $x^i_{(g(i)+1)}, \ldots, x^i_{(n)}$ are bad for voter i.
   c. If $g(i) = n$, then voter i thinks that all the alternatives in X are good.

3. An *attitudinal character* $\alpha_i \in [0, 1]$ for each voter $i \in V$ that represents the degree of optimism or riskiness.

Taking into account all these information, we would construct a *satisfaction function* $\varphi_i : \{1, \ldots, n\} \longrightarrow [0, 1]$ for each voter $i \in V$. The meaning of $\varphi_i(j)$ is the satisfaction that the alternative $x^i_{(j)}$ provides to voter i, whenever $x^i_{(j)} \in M_{\boldsymbol{s}}(\boldsymbol{R})$. Consequently, we assume that every φ_i is decreasing, i.e., $\varphi_i(1) = 1 \geq \varphi_i(2) \geq \cdots \geq \varphi_i(n)$.

Given a decreasing mapping $\varphi : \{1, \ldots, n\} \longrightarrow [0, 1]$, we define the satisfaction functions φ_i associated with φ and g by

$$\varphi_i(j) = \begin{cases} \varphi(j), & \text{if } j \leq g(i), \\ 0, & \text{if } j > g(i). \end{cases}$$

In Section 4 we use the mapping $\varphi : \{1, \ldots, n\} \longrightarrow [0, 1]$ defined by

$$\varphi(j) = \frac{n + 1 - j}{n}. \tag{1}$$

We use this mapping because, without additional information, we assume that the satisfaction of voters for having alternatives in the set of winners of a scoring rule decreases with the position of the alternatives in the individual ranking in a regular way. According to (1), these satisfactions decrease in arithmetic progression of difference $\frac{1}{n}$.

We also consider an OWA operator $A_i : \cup_{k=1}^{n} [0, 1]^k \longrightarrow [0, 1]$ with attitudinal character α_i for each voter $i \in V$.

We now introduce the utility functions associated with φ_i and A_i for each voter $i \in V$ as

$$u_i(\boldsymbol{s}) = A_i\big((\varphi_i(j))_{x^i_{(j)} \in M_{\boldsymbol{s}}(\boldsymbol{R})}\big), \quad i = 1, \ldots, m.$$

3.2 Special Cases

In some situations we may assume that individuals have the same attitudinal character. For example, if voters can choose within the set of winning alternatives or all of these winning alternatives would remain in the final outcome, they should consider the maximum satisfaction of their good alternatives in the winning set.

On the other hand, if a single alternative would be selected from the set of winning alternatives and voters do not know what is the procedure for that election, a prudent (or pessimistic) behaviour consists in considering the minimum satisfaction of their good alternatives in the winning set.

1. When the attitudinal character is 1 for all individuals, i.e., A_i is the maximum for every $i \in V$, we can simplify the iterative process. Given two scoring rules associated with the scoring vectors $\boldsymbol{s}$ and $\boldsymbol{s}'$, if $M_{\boldsymbol{s}'}(\boldsymbol{R}) \subseteq M_{\boldsymbol{s}}(\boldsymbol{R})$, then $u_i(\boldsymbol{s}') \leq u_i(\boldsymbol{s})$ for every $i \in V$. Since the collective utility is calculated through OWA operators, which are monotonic, we have $u(\boldsymbol{s}') \leq u(\boldsymbol{s})$. Therefore, in order to find scoring rules that maximize the collective utility, it is only necessary to calculate the collective utility for those scoring vectors $\boldsymbol{s}$ that generate maximal sets $M_{\boldsymbol{s}}(\boldsymbol{R})$ with respect to the inclusion. In other words, we only take into account scoring vectors $\boldsymbol{s}$ such that $M_{\boldsymbol{s}}(\boldsymbol{R})$ is not strictly included in $M_{\boldsymbol{s}'}(\boldsymbol{R})$ for any $\boldsymbol{s}'$ in the corresponding $\overline{\mathscr{S}_i}^*$.
2. When the attitudinal character is 0 for all individuals, i.e., A_i is the minimum for every $i \in V$, it is easy to check that given two scoring rules associated with the scoring vectors $\boldsymbol{s}$ and $\boldsymbol{s}'$, if $M_{\boldsymbol{s}}(\boldsymbol{R}) \subseteq M_{\boldsymbol{s}'}(\boldsymbol{R})$, then $u_i(\boldsymbol{s}') \leq u_i(\boldsymbol{s})$ for every $i \in V$. Since the collective utility is calculated through an OWA operator, which is monotonic, we have $u(\boldsymbol{s}') \leq u(\boldsymbol{s})$. Now for obtaining scoring rules that maximize the collective utility, it is only necessary to calculate the collective utility for those scoring vectors $\boldsymbol{s}$ that generate minimal sets $M_{\boldsymbol{s}}(\boldsymbol{R})$ with respect to the inclusion. In other words, we only take into account scoring vectors $\boldsymbol{s}$ such that $M_{\boldsymbol{s}'}(\boldsymbol{R})$ is not strictly included in $M_{\boldsymbol{s}}(\boldsymbol{R})$ for any $\boldsymbol{s}'$ in the corresponding $\overline{\mathscr{S}_i}^*$.

4 An Illustrative Example

In this section we provide an example to show the model implementation. Suppose five individuals that show their preferences on four alternatives $\{A, B, C, D\}$ through the profile $\boldsymbol{R}$ of linear orders included in Table 2.

Table 2 Voter's linear orders

1	2	3	4	5
A	B	D	C	A
B	C	C	D	D
C	A	B	A	C
D	D	A	B	B

Since the number of alternatives is four, we will use the following set of normalized scoring vectors:

$$\mathscr{S}^0 = \{\mathbf{s} = (1,s,t,0) \mid 0 \leq t \leq s \leq 1\}.$$

For each alternative, the collective assignment generated by an arbitrary scoring rule $\mathbf{s}$ is

$$r_{\mathbf{s}}(\mathrm{A}) = \sum_{i=1}^{5} r_{\mathbf{s}}^{i}(\mathrm{A}) = 2 + 2t,$$

$$r_{\mathbf{s}}(\mathrm{B}) = \sum_{i=1}^{5} r_{\mathbf{s}}^{i}(\mathrm{B}) = 1 + s + t,$$

$$r_{\mathbf{s}}(\mathrm{C}) = \sum_{i=1}^{5} r_{\mathbf{s}}^{i}(\mathrm{C}) = 1 + 2s + 2t,$$

$$r_{\mathbf{s}}(\mathrm{D}) = \sum_{i=1}^{5} r_{\mathbf{s}}^{i}(\mathrm{D}) = 1 + 2s.$$

It is easy to check that

$$\begin{array}{lll} \mathrm{A} \in M_{\mathbf{s}}(\boldsymbol{R}) & \Leftrightarrow & s \leq 0.5, \\ \mathrm{B} \notin M_{\mathbf{s}}(\boldsymbol{R}) & & \forall \mathbf{s} \in \mathscr{S}^0, \\ \mathrm{C} \in M_{\mathbf{s}}(\boldsymbol{R}) & \Leftrightarrow & s \geq 0.5, \\ \mathrm{D} \in M_{\mathbf{s}}(\boldsymbol{R}) & \Leftrightarrow & s \geq 0.5,\ t = 0. \end{array}$$

Therefore, the sets of winning alternatives that can be obtained with the different scoring rules are

$$\begin{array}{lll} M_{\mathbf{s}}(\boldsymbol{R}) = \{\mathrm{A}\} & \Leftrightarrow & s < 0.5, \\ M_{\mathbf{s}}(\boldsymbol{R}) = \{\mathrm{A},\mathrm{C}\} & \Leftrightarrow & s = 0.5, t > 0, \\ M_{\mathbf{s}}(\boldsymbol{R}) = \{\mathrm{A},\mathrm{C},\mathrm{D}\} & \Leftrightarrow & s = 0.5, t = 0, \\ M_{\mathbf{s}}(\boldsymbol{R}) = \{\mathrm{C}\} & \Leftrightarrow & s > 0.5, t > 0, \\ M_{\mathbf{s}}(\boldsymbol{R}) = \{\mathrm{C},\mathrm{D}\} & \Leftrightarrow & s > 0.5, t = 0. \end{array}$$

In Fig. 3 we show graphically the subsets of $\{(s,t) \in [0,1]^2 \mid t \leq s\}$ that generate the previous sets of winning alternatives.

Suppose now that individuals declare which alternatives are good and bad according to the function $g : V \longrightarrow \{0,1,2,3,4\}$ given by $g(1) = 2$, $g(2) = 3$, $g(3) = 4$, $g(4) = 3$ and $g(5) = 1$. The good alternatives for each individual are shown in Table 3.

In order to reach consensus on the scoring rule to use in this decision problem, we consider for all individuals the satisfaction function associated with (1); i.e.,

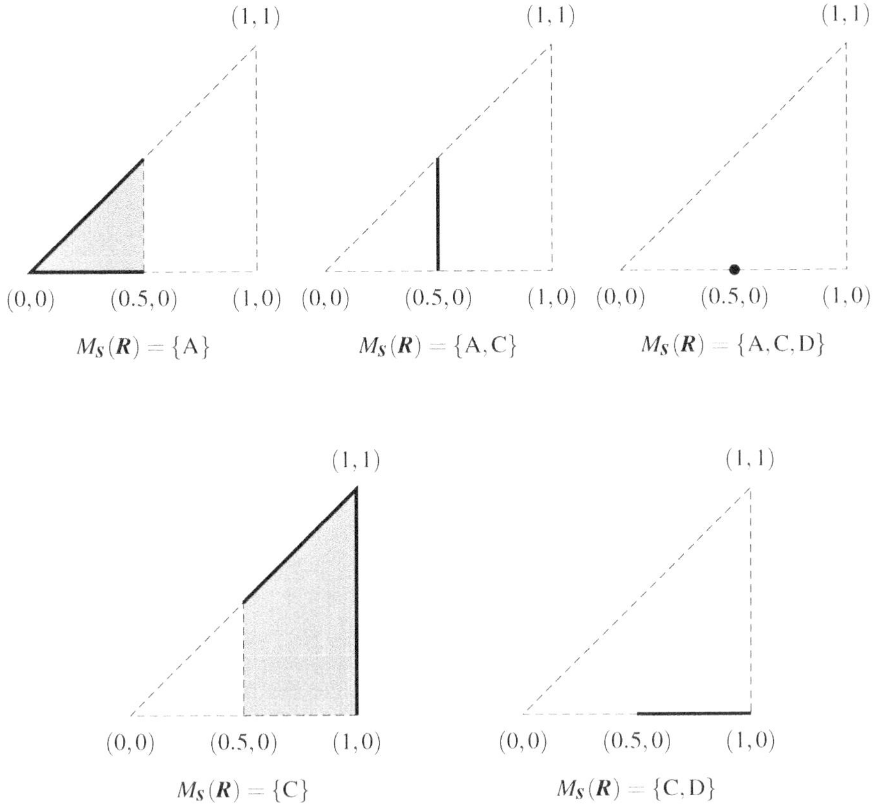

Fig. 3 Subsets of $\{(s,t) \in [0,1]^2 \mid t \leq s\}$ that generate the different $M_S(\boldsymbol{R})$

Table 3 Good alternatives for each voter

1	2	3	4	5
A	B	D	C	A
B	C	C	D	
	A	B	A	
		A		

$$\varphi_i(j) = \begin{cases} \frac{5-j}{4}, & \text{if } j \leq g(i), \\ 0, & \text{if } j > g(i), \end{cases}$$

for every $i \in \{1, \dots, 5\}$.

It is worth noting that, for each voter, bad alternatives give no satisfaction, and a good alternative ranked in the j-th position ($j \in \{1,\ldots,4\}$) provides a satisfaction of $\frac{5-j}{4}$. Table 4 summarizes the satisfaction obtained for the voters in each alternative.

Table 4 Voter's satisfaction with each alternative

	1	2	3	4	5
A	1	$\frac{1}{2}$	$\frac{1}{4}$	$\frac{1}{2}$	1
B	$\frac{3}{4}$	1	$\frac{1}{2}$	0	0
C	0	$\frac{3}{4}$	$\frac{3}{4}$	1	0
D	0	0	1	$\frac{3}{4}$	0

To generate voters' utilities, we have considered the same degree of optimism or riskiness for all agents, and we have aggregated the satisfaction associated with the good alternatives belonging to the winning set through three representative OWA operators: the maximum, the arithmetic mean and the minimum, whose attitudinal character are 1, 0.5 and 0, respectively.

On the other hand, we have used three OWA operators for aggregating individual utilities: the minimum, the arithmetic mean and the median –with the weighting vector $(0,0,1,0,0)$. The respective collective utilities are denoted by u_{min}, u_{arit} and u_{med}.

Maximum. According to the remarks of Subsection 3.2, and given that the sets $M_{\mathbf{s}}(\boldsymbol{R})$ are included in $\{\text{A},\text{C},\text{D}\}$ for every $\mathbf{s} \in \overline{\mathscr{S}}^*$, the winners are A, C, and D. These alternatives are obtained by using the scoring rule associated with the scoring vector $(1,0.5,0,0)$.

Arithmetic mean. In Table 5 we show the values of $u_i(\mathbf{s})$, $i = 1,\ldots,5$, when the arithmetic mean is used to obtain the individual utilities, and the corresponding values of u_{min}, u_{arit} and u_{med}.

Table 5 Individual utilities for the arithmetic mean and collective utilities

	$M_{\mathbf{s}}(\boldsymbol{R})$	$u_1(\mathbf{s})$	$u_2(\mathbf{s})$	$u_3(\mathbf{s})$	$u_4(\mathbf{s})$	$u_5(\mathbf{s})$	$u_{min}(\mathbf{s})$	$u_{arit}(\mathbf{s})$	$u_{med}(\mathbf{s})$
$s < 0.5$	{A}	1	$\frac{1}{2}$	$\frac{1}{4}$	$\frac{1}{2}$	1	$\frac{1}{4}$	0.65	$\frac{1}{2}$
$s = 0.5, t > 0$	{A, C}	$\frac{1}{2}$	$\frac{5}{8}$	$\frac{1}{2}$	$\frac{3}{4}$	$\frac{1}{2}$	$\frac{1}{2}$	0.575	$\frac{1}{2}$
$s = 0.5, t = 0$	{A, C, D}	$\frac{1}{3}$	$\frac{5}{12}$	$\frac{2}{3}$	$\frac{3}{4}$	$\frac{1}{3}$	$\frac{1}{3}$	0.5	$\frac{5}{12}$
$s > 0.5, t > 0$	{C}	0	$\frac{3}{4}$	$\frac{3}{4}$	1	0	0	0.5	$\frac{3}{4}$
$s > 0.5, t = 0$	{C, D}	0	$\frac{3}{8}$	$\frac{7}{8}$	$\frac{7}{8}$	0	0	0.425	$\frac{3}{8}$

As we can see in Table 5, we have the following winners:

1. For the minimum, the winners are A and C. These alternatives are obtained by using any scoring vector of the set $\{(1,0.5,t,0) \mid 0 < t \leq 0.5\}$.
2. For the arithmetic mean, the winner is A. This alternative is obtained by using any scoring vector of the set $\{(1,s,t,0) \mid 0 \leq s < 0.5, 0 \leq t \leq s\}$.
3. For the median, the winner is C. This alternative is obtained by using any scoring vector of the set $\{(1,s,t,0) \mid 0.5 < s \leq 1, 0 < t \leq s\}$.

Minimum. According to the remarks of Subsection 3.2, and given that $\{A\}$ and $\{C\}$ are the minimal sets with respect to the inclusion, it is only necessary to calculate the collective utilities for those scoring rules that generate these sets. Table 6 provides the values of $u_i(s)$, $i = 1,\ldots,5$, and the values of u_{min}, u_{arit} and u_{med}.

Table 6 Individual utilities for the minimum and collective utilities

	$M_S(R)$	$u_1(s)$	$u_2(s)$	$u_3(s)$	$u_4(s)$	$u_5(s)$	$u_{min}(s)$	$u_{arit}(s)$	$u_{med}(s)$
$s < 0.5$	$\{A\}$	1	$\frac{1}{2}$	$\frac{1}{4}$	$\frac{1}{2}$	1	$\frac{1}{4}$	0.65	$\frac{1}{2}$
$s > 0.5, t > 0$	$\{C\}$	0	$\frac{3}{4}$	$\frac{3}{4}$	1	0	0	0.5	$\frac{3}{4}$

As we can see in Table 6, we have the following winners:

1. For the minimum and the arithmetic mean, the winner is A. This alternative is obtained by using any scoring vector of the set $\{(1,s,t,0) \mid 0 \leq s < 0.5, 0 \leq t \leq s\}$.
2. For the median, the winner is C. This alternative is obtained by using any scoring vector of the set $\{(1,s,t,0) \mid 0.5 < s \leq 1, 0 < t \leq s\}$.

5 Concluding Remarks

There exist in the literature a wide variety of consensual processes. Some of them try to minimize the disagreement among decision makers with respect to the collective decision. We follow this approach in the framework of scoring rules: once decision makers show their preferences, we find a set of scoring rules that minimize the disagreement with respect to a specific consensual perspective where the information provided by decision makers is aggregated by means of appropriate OWA operators. It is worth mentioning that our proposal is very flexible and allows us to apply it to different scenarios.

Although our decision mechanism selects the set of scoring rules that maximizes consensus with respect to the devised procedure, other voting systems may reach higher consensus than the one generated by scoring rules. It is not a contradiction, because we are only interested in which scoring rules maximize consensus among decision makers.

Scoring rules are initially designed for linear orders. However, they can be extended to weak orders by averaging the scores of indifferent alternatives. For simplicity, we have only considered the standard case of linear orders, but our analysis may be extended in a natural way to weak orders.

Acknowledgements. The authors gratefully acknowledge the useful comments of Jorge Alcalde-Unzu. This work is partially supported by the Spanish Ministerio de Ciencia e Innovación (Project SEJ2006-04267/ECON), ERDF and Junta de Castilla y León (Consejería de Educación, Projects VA092A08 and VA002B08).

References

1. Arrow, K.J.: Social Choice and Individual Values, 2nd edn. [1st edn. (1951)]. Wiley, New York (1963)
2. Beliakov, G., Pradera, A., Calvo, T.: Aggregation Functions: A Guide for Practicioners. Springer, Heidelberg (2007)
3. Brams, S.J.: Mathematics and Democracy: Designing Better Voting and Fair-Division Procedures. Princeton University Press, Princeton (2008)
4. Brams, S.J., Fishburn, P.C.: Approval voting. American Political Science Review 72, 831–857 (1978)
5. Brams, S.J., Sanver, M.R.: Voting systems that combine approval and preference. The Mathematics of Preference, Choice and Order. In: Essays in Honor of Peter C. Fishburn. Studies in Choice and Welfare, pp. 215–237. Springer, Heidelberg (2009)
6. Calvo, T., Kolesárova, A., Komorníková, M., Mesiar, R.: Aggregation operators: Properties, classes and construction methods. In: Calvo, T., Mayor, G., Mesiar, R. (eds.) Aggregation Operators: New Trends and Applications, pp. 3–104. Physica-Verlag, Heidelberg (2002)
7. Chebotarev, P.Y., Shamis, E.: Characterizations of scoring methods for preference aggregation. Annals of Operations Research 80, 299–332 (1998)
8. Fodor, J., Roubens, M.: Fuzzy Preference Modelling and Multicriteria Decision Support. Kluwer Academic Publishers, Dordrecht (1994)
9. Gärdenfors, P.: Positionalist voting functions. Theory and Decision 4, 1–24 (1973)
10. Grabisch, M., Marichal, J.L., Mesiar, R., Pap, E.: Aggregation Functions. Cambridge University Press, Cambridge (2009)
11. Smith, J.H.: Aggregation of preferences with variable electorate. Econometrica 41, 1027–1041 (1973)
12. Yager, R.R.: On ordered weighted averaging operators in multicriteria decision making. IEEE Trans. Syst. Man. Cybern. 18, 183–190 (1988)
13. Young, H.P.: Social choice scoring functions. SIAM Journal on Applied Mathematics 28, 824–838 (1975)

Dominance-Based Rough Set Approach to Interactive Evolutionary Multiobjective Optimization

Salvatore Greco, Benedetto Matarazzo, and Roman Słowiński

Abstract. We present application of Dominance-based Rough Set Approach (DRSA) to interactive Evolutionary Multiobjective Optimization (EMO). In the proposed methodology, the preference information elicited by the decision maker in successive iterations consists in sorting some solutions of the current population as "good" or "bad", or in comparing some pairs of solutions. The "if ..., then ..." decision rules are then induced from this preference information using Dominance-based Rough Set Approach (DRSA). The rules are used within EMO in order to focus on populations of solutions satisfying the preferences of the decision maker. This allows to speed up convergence to the most preferred region of the Pareto-front. The resulting interactive schemes, corresponding to the two types of preference information, are called DRSA-EMO and DRSA-EMO-PCT, respectively. Within the same methodology, we propose DARWIN and DARWIN-PCT methods, which permit to take into account robustness concerns in multiobjective optimization.

1 Introduction

Real life decision problems usually involve consideration of multiple conflicting objectives. For example, product mix involves multiple objectives of the type: profit, time machine, sales, market share, net present value, resource consumption, and so on. As, in general, there does not exist a single solution which optimizes simultaneously all objectives, one has to search for Pareto-optimal

Salvatore Greco · Benedetto Matarazzo
Faculty of Economics, University of Catania, 95129 Catania, Italy
e-mail: {salgreco,matarazz}@unict.it

Roman Słowiński
Institute of Computing Science, Poznań University of Technology, Poznań, and Systems Research Institute, Polish Academy of Sciences, 00-441 Warsaw, Poland
e-mail: roman.slowinski@cs.put.poznan.pl

S. Greco et al. (Eds.): Preferences and Decisions, STUDFUZZ 257, pp. 225–260.
springerlink.com

solutions. A solution is Pareto-optimal (also called efficient or non-dominated) if there is no other feasible solution which would be at least as good on all objectives, while being strictly better on at least one objective. Finding the whole set of Pareto-optimal solutions (also called Pareto-set or Pareto-front) is usually computationally hard. This is why a lot of research has been devoted to heuristic search of an approximation of the Pareto-front. Among the heuristics proposed to this end, Evolutionary Multiobjective Optimization (EMO) procedures appeared to be particularly efficient (see, e.g., (7; 8)).

The underlying reasoning behind the EMO search of an approximation of the Pareto-front is that, in the absence of any preference information, all Pareto-optimal solutions have to be considered equivalent.

On the other hand, if the decision maker (DM) (alternatively called user) is involved in the multiobjective optimization process, then the preference information provided by the DM can be used to focus the search on the most preferred part of the Pareto-front. This idea stands behind Interactive Multiobjective Optimization (IMO) methods proposed long time before EMO has emerged (see, e.g., (28; 31; 39; 41)).

Recently, it became clear that merging the IMO and EMO methodologies should be beneficial for the multiobjective optimization process (4). Several approaches have been presented in this context:

- (13), (10) and (40) are based on various ways of guiding the search by an achievement scalarizing function taking into account a user-specified reference point,
- (6) proposes the guided MOEA in which the user is allowed to specify preferences in the form of maximally acceptable trade-offs like "one unit improvement in objective i is worth at most a_{ji} units in objective j",
- (11) proposes an interactive decision support system called I-MODE that allows the DM to interactively focus on interesting region(s) of the Pareto-front using several tools, such as weighted sum approach, utility function based approach, Chebycheff function approach or trade-off information,
- (26) suggests a procedure which asks the user to rank a few alternatives, and from this derives constraints for linear weighting of the objectives consistent with the given ordering, which are used within an EMO to check whether there is a feasible linear weighting such that solution x is preferable to solution y,
- (34) proposes an interactive evolutionary algorithm that allows the user to provide preference information about pairs of solutions during the run, computes the "most compatible" weighted sum of objectives by means of linear programming, and uses this as single substitute objective for some generations of the evolutionary algorithm,
- (27) uses preference information from pairwise comparisons of solutions for sampling sets of scalarizing functions by drawing a random weight vector for each single iteration, and using this for selection and local search,

– (5) proposes to apply robust ordinal regression (24; 12) to take into account the whole set of additive utility functions compatible with preference comparisons of some pairs of solutions by the DM, in order to guide exploration of the Pareto-front using a necessary preference relation which holds when a solution is at least as good as another solution for all compatible utility functions.

This paper presents another approach to merging IMO and EMO considering combination of EMO with Dominance-based Rough Set Approach (DRSA). DRSA is a methodology of reasoning about partially inconsistent and preference-ordered data (see (15), (17), (18), (36), (37), (38)), which has already been applied successfully to IMO (19).

In multiple criteria decision analysis, DRSA aims at obtaining a representation of the DM's preferences in terms of easily understandable "*if ..., then ...*" decision rules, on the basis of some exemplary decisions (past decisions or simulated decisions) made by the DM. The exemplary decisions can be:

1. assignments of selected alternatives to some ordered classes, such as "bad", "medium", "good", or
2. specifications of some holistic preferences on selected pairs of alternatives.

In case 1), the induced decision rules are of the form:
"*if on criterion i_1 alternative x has an evaluation at least α_1, and on criterion i_2 has an evaluation at least α_2 ..., and on criterion i_h has an evaluation at least α_h, then alternative x is at least medium*".

In case 2), decision rules are of the form:
"*if on criterion i_1 alternative x is at least strongly better than alternative y, and on criterion i_2 x is at least weakly better than y ..., and on criterion i_h x is at least indifferent to y, then x is comprehensively weakly preferred to y*".

DRSA can take into account uncertainty in decision problems (21; 23) inducing decision rules of the form:

"*if on criterion i_1 alternative x has an evaluation at least α_1 with probability at least p_{i1}, and on criterion i_2 x has an evaluation at least α_2 with probability at least p_{i2} ..., and on criterion i_h x has an evaluation at least α_h with probability at least p_{ih}, then alternative x is at least medium*"

or

"*if on criterion i_1 alternative x is at least strongly better than alternative y with probability at least p_{i1}, and on criterion i_2 x is at least weakly better than y with probability at least p_{i2} ..., and on criterion i_h x is at least indifferent with y with probability at least p_{ih}, then x is comprehensively weakly preferred to y*".

The methodology of interactive EMO based on DRSA (25) involves application of decision rules in EMO, which are induced from easily elicited preference information by DRSA, according to two general schemes, called DRSA-EMO and DRSA-EMO-PCT. This results in focusing the search of the

Pareto-front on the most preferred region. More specifically, DRSA is used for structuring preference information obtained through interaction with the user, and then a set of decision rules representing user's preferences is induced from this information; these rules are used to rank solutions in the current population of EMO, which has an impact on the selection and crossover. Within interactive EMO, DRSA for decision under uncertainty is important because it can take into account robustness concerns in the multiobjective optimization (for robustness in optimization see, e.g., (29; 3); for robustness and multiple criteria decision analysis see (35)). In fact, two methods of robust optimization methods combining DRSA and interactive EMO have been proposed: DARWIN (*D*ominance-based rough set *A*pproach to handling *R*obust *W*inning solutions in *IN*teractive multiobjective optimization) (20) and DARWIN-PCT (DARWIN using *P*airwise *C*omparison *T*ables) (22). DARWIN and DARWIN-PCT can be considered as two specific instances of DRSA-EMO and DRSA-EMO-PCT, respectively.

We believe that integration of DRSA and EMO is particularly promising for two reasons:

1. The preference information required by DRSA is very basic and easy to be elicited by the DM. All that the DM is asked for is to assign solutions to preference ordered classes, such as "good", "medium" and "bad", or compare pairs of non-dominated solutions from a current population in order to reveal whether one is preferred over the other. The preference information is provided every k iterations (k depends on the problem and the willingness of the user to interact with the system. In our studies, k ranges from 10 to 30).
2. The decision rules are transparent and easy to interpret for the DM. The preference model supplied by decision rules is a "glass box", while many other competitive multiple criteria decision methodologies involve preference models that are "black boxes" for the user. The "glass box" model improves the quality of the interaction and makes that the DM accepts well the resulting recommendation.

The paper is organized as follows. The next Section describes the DRSA methodology and, more precisely, DRSA applied to ordinal classification and DRSA applied to approximation of a preference relation represented through a pairwise comparison table. Then, Section 3 presents DRSA-EMO method resulting from application of DRSA for ordinal classification to interactive EMO. The following Section 4 presents DRSA-EMO-PCT method that applies DRSA for approximation of a preference relation to interactive EMO. Section 5 recall DRSA for decision under uncertainty. Section 6 presents DARWIN and DARWIN-PCT methods that apply DRSA for decision under uncertainty to robust interactive EMO. The last Section contains conclusions.

2 Dominance-Based Rough Set Approach (DRSA)

2.1 Dominance-Based Rough Set Approach for Ordinal Classification

DRSA is a methodology of multiple criteria decision analysis aiming at obtaining a representation of the DM's preferences in terms of easily understandable "*if ..., then ...*" decision rules, on the basis of some exemplary decisions (past decisions or simulated decisions) given by the DM. In this Section, we present the DRSA to sorting problems because in the dialogue stage of some interactive EMO methods we are preare senting (DRSA-EMO and DARWIN) this multiple criteria decision problem is considered. In this case, exemplary decisions are *sorting examples*, i.e. objects (solutions, alternatives, actions) described by a set of criteria and assigned to preference ordered classes. Criteria and the class assignment considered within DRSA correspond to the *condition attributes* and the *decision attribute*, respectively, in the classical Rough Set Approach (33). For example, in multiple criteria sorting of cars, an example of decision is an assignment of a particular car evaluated on such criteria as maximum speed, acceleration, price and fuel consumption to one of three classes of overall evaluation: "bad", "medium", "good".

Let us consider a set of criteria $F = \{f_1, \ldots, f_n\}$, the set of their indices $I = \{1, \ldots, n\}$, and a finite universe of objects (solutions, alternatives, actions) U such that, without loss of generality, $f_i : U \rightarrow \Re$ for each $i = 1, \ldots, n$, and, for all objects $x, y \in U$, $f_i(x) \geq f_i(y)$ means that "x is at least as good as y with respect to criterion i", which is denoted as $x \succeq_i y$. Therefore, we suppose that $\succeq_i$ is a complete preorder, i.e. a strongly complete and transitive binary relation, defined on U on the basis of evaluations $f_i(\cdot)$. Note that in the context of multiobjective optimization, $f_i(\cdot)$ corresponds to objective functions. Furthermore, we assume that there is a decision attribute d which makes a partition of U into a finite number of decision classes called sorting, $\boldsymbol{Cl}=\{Cl_1, \ldots, Cl_m\}$, such that each $x \in U$ belongs to one and only one class Cl_t, $t = 1, \ldots, m$. We suppose that the classes are preference ordered, i.e. for all $r, s = 1, \ldots, m$, such that $r > s$, the objects from Cl_r are preferred to the objects from Cl_s. More formally, if $\succeq$ is a *comprehensive weak preference relation* on U, i.e. if for all $x, y \in U$, $x \succeq y$ reads "x is at least as good as y", then we suppose

$$[x \in Cl_r,\ y \in Cl_s,\ r > s] \Rightarrow x \succ y,$$

where $x \succ y$ means $x \succeq y$ and *not* $y \succeq x$. The above assumptions are typical for consideration of a multiple criteria sorting problem (also called ordinal classification with monotonicity constraints).

In DRSA, the explanation of the assignment of objects to preference ordered decision classes is made on the base of their evaluation with respect to a subset of criteria $P \subseteq I$. This explanation is called *approximation* of decision classes with respect to P. Indeed, in order to take into account the order of decision classes, in DRSA the classes are not considered one by one but, instead, unions of classes are approximated: *upward union* from class Cl_t to class Cl_m denoted by $Cl_t^{\geq}$, and *downward union* from class Cl_t to class Cl_1, denoted by $Cl_t^{\leq}$, i.e.:

$$Cl_t^{\geq} = \bigcup_{s \geq t} Cl_s, \quad Cl_t^{\leq} = \bigcup_{s \leq t} Cl_s, \quad t = 1, ..., m.$$

The statement $x \in Cl_t^{\geq}$ reads "x belongs to at least class Cl_t", while $x \in Cl_t^{\leq}$ reads "x belongs to at most class Cl_t". Let us remark that $Cl_1^{\geq} = Cl_m^{\leq} = U$, $Cl_m^{\geq} = Cl_m$ and $Cl_1^{\leq} = Cl_1$. Furthermore, for t=2,...,m, we have:

$$Cl_{t-1}^{\leq} = U - Cl_t^{\geq} \quad \text{and} \quad Cl_t^{\geq} = U - Cl_{t-1}^{\leq}.$$

In the above example concerning multiple criteria sorting of cars, the upward unions are: $Cl_{medium}^{\geq}$, that is the set of all the cars classified at least "medium" (i.e. the set of cars classified "medium" or "good"), and $Cl_{good}^{\geq}$, that is the set of all the cars classified at least "good" (i.e. the set of cars classified "good"), while the downward unions are: $Cl_{medium}^{\leq}$, that is the set of all the cars classified at most "medium" (i.e. the set of cars classified "medium" or "bad"), and $Cl_{bad}^{\leq}$, that is the set of all the cars classified at most "bad" (i.e. the set of cars classified "bad"). Notice that, formally, also $Cl_{bad}^{\geq}$ is an upward union as well as $Cl_{good}^{\leq}$ is a downward union, however, as "bad" and "good" are extreme classes, these two unions boil down to the whole universe U.

The key idea of the rough set approach is explanation (approximation) of knowledge generated by the decision attributes, by *granules of knowledge* generated by condition attributes.

In DRSA, where condition attributes are criteria and decision classes are preference ordered, the knowledge to be explained is the assignments of objects to *upward* and *downward unions of classes* and the granules of knowledge are sets of objects contained in *dominance cones* defined in the space of evaluation criteria.

We say that x *dominates* y with respect to $P \subseteq I$ (shortly, x *P-dominates* y), denoted by xD_Py, if for every criterion $i \in P$, $f_i(x) \geq f_i(y)$. The relation of P-dominance is reflexive and transitive, that is it is a partial preorder.

Given a set of criteria $P \subseteq I$ and $x \in U$, the granules of knowledge used for approximation in DRSA are:

- a set of objects dominating x, called *P-dominating set*, $D_P^+(x) = \{y \in U: yD_Px\}$,
- a set of objects dominated by x, called *P-dominated set*, $D_P^-(x) = \{y \in U: xD_Py\}$.

Let us recall that the dominance principle requires that an object x dominating object y with respect to considered criteria (i.e. x having evaluations at least as good as y on all considered criteria) should also dominate y on the decision (i.e. x should be assigned to at least as good decision class as y).

The *P-lower approximation* of $Cl_t^{\geq}$, denoted by $\underline{P}(Cl_t^{\geq})$, and the *P-upper approximation* of $Cl_t^{\geq}$, denoted by $\overline{P}(Cl_t^{\geq})$, are defined as follows (t=1,...,m):

$$\underline{P}(Cl_t^{\geq}) = \{x \in U : D_P^+(x) \subseteq Cl_t^{\geq}\},$$
$$\overline{P}(Cl_t^{\geq}) = \{x \in U : D_P^-(x) \cap Cl_t^{\geq} \neq \emptyset\}.$$

Analogously, one can define the *P-lower approximation* and the *P-upper approximation* of $Cl_t^{\leq}$ as follows (t=1,...,m):

$$\underline{P}(Cl_t^{\leq}) = \{x \in U : D_P^-(x) \subseteq Cl_t^{\leq}\},$$
$$\overline{P}(Cl_t^{\leq}) = \{x \in U : D_P^+(x) \cap Cl_t^{\leq} \neq \emptyset\}.$$

The *P-boundaries* of $Cl_t^{\geq}$ and $Cl_t^{\leq}$, denoted by $Bn_P(Cl_t^{\geq})$ and $Bn_P(Cl_t^{\leq})$, respectively, are defined as follows (t=1,...,m):

$$Bn_P(Cl_t^{\geq}) = \overline{P}(Cl_t^{\geq}) - \underline{P}(Cl_t^{\geq}), \quad Bn_P(Cl_t^{\leq}) = \overline{P}(Cl_t^{\leq}) - \underline{P}(Cl_t^{\leq}).$$

The dominance-based rough approximations of upward and downward unions of decision classes can serve to induce a generalized description of sorting decisions in terms of "*if ..., then ...*" decision rules. For a given upward or downward union of classes, $Cl_t^{\geq}$ or $Cl_s^{\leq}$, the decision rules induced under a hypothesis that objects belonging to $\underline{P}(Cl_t^{\geq})$ or $\underline{P}(Cl_s^{\leq})$ are positive examples (that is objects that have to be matched by the induced decision rules), and all the others are negative (that is objects that have to be not matched by the induced decision rules), suggest a *certain* assignment to "class Cl_t or better", or to "class Cl_s or worse", respectively. On the other hand, the decision rules induced under a hypothesis that objects belonging to $\overline{P}(Cl_t^{\geq})$ or $\overline{P}(Cl_s^{\leq})$ are positive examples, and all the others are negative, suggest a *possible* assignment to "class Cl_t or better", or to "class Cl_s or worse", respectively. Finally, the decision rules induced under a hypothesis that objects belonging to the intersection $\overline{P}(Cl_s^{\leq}) \cap \overline{P}(Cl_t^{\geq})$ are positive examples, and all the others are negative, suggest an assignment to some classes between Cl_s and Cl_t ($s < t$). These rules are matching inconsistent objects $x \in U$, which cannot be assigned without doubts to classes Cl_r, $s < r < t$, because $x \notin \underline{P}(Cl_r^{\geq})$ and $x \notin \underline{P}(Cl_r^{\leq})$ for all r such that $s < r < t$.

Given the preference information in terms of sorting examples, it is meaningful to consider the following five types of decision rules:

1) *certain* $D_{\geq}$*-decision rules*, providing lower profiles (i.e. sets of minimal values for considered criteria) of objects belonging to $\underline{P}(Cl_t^{\geq})$, $P = \{i_1, \ldots, i_p\} \subseteq I$:
   if $f_{i_1}(x) \geq r_{i_1}$ *and* ... *and* $f_{i_p}(x) \geq r_{i_p}$,
   then $x \in Cl_t^{\geq}$,
   $t = 2, \ldots, m$, $r_{i_1}, \ldots, r_{i_p} \in \Re$;

2) *possible* $D_{\geq}$-*decision rules*, providing lower profiles of objects belonging to $\overline{P}(Cl_t^{\geq})$, $P = \{i_1, \ldots, i_p\} \subseteq I$:
$if\ f_{i_1}(x) \geq r_{i_1}\ and \ldots and\ f_{i_p}(x) \geq r_{i_p}$,
then x *possibly belongs to* $Cl_t^{\geq}$,
$t = 2, \ldots, m$, $r_{i_1}, \ldots, r_{i_p} \in \Re$;

3) *certain* $D_{\leq}$-*decision rules*, providing upper profiles (i.e. sets of maximal values for considered criteria) of objects belonging to $\underline{P}(Cl_t^{\leq})$, $P = \{i_1, \ldots, i_p\} \subseteq I$:
$if\ f_{i_1}(x) \leq r_{i_1}\ and \ldots and\ f_{i_p}(x) \leq r_{i_p}$,
then $x \in Cl_t^{\leq}$,
$t = 1, \ldots, m-1$, $r_{i_1}, \ldots, r_{i_p} \in \Re$;

4) *possible* $D_{\leq}$-*decision rules*, providing upper profiles of objects belonging to $\overline{P}(Cl_t^{\leq})$, $P = \{i_1, \ldots, i_p\} \subseteq I$:
$if\ f_{i_1}(x) \leq r_{i_1}\ and \ldots and\ f_{i_p}(x) \leq r_{i_p}$,
then x *possibly belongs to* $Cl_t^{\leq}$,
$t = 1, \ldots, m-1$, $r_{i_1}, \ldots, r_{i_p} \in \Re$;

5) *approximate* $D_{\geq\leq}$-*decision rules*, providing simultaneously lower and upper profiles of objects belonging to $Cl_s \cup Cl_{s+1} \cup \ldots \cup Cl_t$, without possibility of discerning to which class:
$if\ f_{i_1}(x) \geq r_{i_1}\ and \ldots and\ f_{i_k}(x) \geq r_{i_k}\ and$
$f_{i_{k+1}}(x) \leq r_{i_{k+1}}\ and \ldots and\ f_{i_p}(x) \leq r_{i_p}$,
then $x \in Cl_s \cup Cl_{s+1} \cup \ldots \cup Cl_t$,
$\{i_1, \ldots, i_p\} \subseteq I$ $s, t \in \{1, \ldots, m\}$, $s < t$,
$r_{i_1}, \ldots, r_{i_p} \in \Re$.

2.2 DRSA Applied to Pairwise Comparison Tables (PCT)

DRSA can also be used to approximate a preference relation (15; 16?). The rules induced on the basis of the corresponding rough approximations are used in some interactive EMO we are presenting (DRSA-PCT-EMO and DARWIN-PCT). We consider a set of reference objects A on which a DM can express his/her own preferences by pairwise comparisons. More precisely, we take into consideration a weak preference relation $\succeq$ on A and a negative weak preference relation $x \succeq^c y$ on A such that, for a pair of objects $(x,y) \in A \times A$, $x \succeq y$ means that x is at least as good as y and $x \succeq^c y$ means that it is not true that x is at least as good as y. The only assumptions with respect to (wrt) these relations are that $\succeq$ is reflexive and $\succeq^c$ is irreflexive, and they are incompatible in the sense that for all $x, y \in A$ it is not possible that $x \succeq y$ and $x \succeq^c y$.

For each pair of reference objects $(x,y) \in A \times A$, the DM can select one of the three following possibilities:

1. Object x is as good as y, i.e., $x \succeq y$.
2. Object x is not as good as y, i.e., $x \succeq^c y$.
3. The two objects are incomparable at the present stage, in the sense that neither $x \succeq y$ nor $x \succeq^c y$ can be asserted.

Let $\succeq \cup \succeq^c = B$, with $card(B) = m$. We also suppose that objects from A are described by a finite set of criteria $C = \{f_1, \ldots, f_n\}$. Without loss of generality, for each $f_i \in C$ we suppose that $f_i : A \to \Re$, such that, for each $x, y \in A$, $f_i(x) \geq f_i(y)$ means that x is at least as good as y wrt criterion f_i which is denoted by $x \succeq_i y$. For each criterion $f_i \in C$, we also suppose that there exists a quaternary relation $\succeq_i^*$ defined on A, such that, for each $x, y, w, z \in A$, $(x, y) \succeq_i^* (w, z)$ means that, wrt f_i, x is preferred to y at least as strongly as w is preferred to z. We assume that for each $f_i \in C$, the quaternary relation $\succeq_i^*$ is monotonic wrt to evaluations on criterion f_i, such that, for all $x, y, w, z \in A$,

$$f_i(x) \geq f_i(w) \text{ and } f_i(y) \leq f_i(z) \Rightarrow (x, y) \succeq_i^* (w, z).$$

We shall denote by $\succ_i^*$ and $\sim_i^*$ the asymmetric and the symmetric part of $\succeq_i^*$, respectively, i.e., $(x, y) \succ_i^* (w, z)$ if $(x, y) \succeq_i^* (w, z)$ and not $(w, z) \succeq_i^* (x, y)$, and $(x, y) \sim_i^* (w, z)$ if $(x, y) \succeq_i^* (w, z)$ and $(w, z) \succeq_i^* (x, y)$. The quaternary relation $\succeq_i^*, f_i \in C$, is supposed to be a complete preorder on $A \times A$. For each $(x, y) \in A \times A$, $C_i(x, y) = \{(w, z) \in A \times A : (w, z) \sim_i^* (w, z)\}$, is the equivalence class of (x, y) wrt $\sim_i^*$. Intuitively, for each $(x, y), (w, z) \in A \times A$, $(w, z) \in C_i(x, y)$ meas that w is preferred to z with the same strength as x is preferred to y. We suppose also that for each $x, y \in A$ and $f_i \in C$, $(x, x) \sim_i^* (y, y)$ and, consequently, $(y, y) \in C_i(x, x)$. Assuming that they are finite, we denote the equivalence classes of $\sim_i^*$ by $\succ_i^{\alpha_i}, \succ_i^{\alpha_i - 1}, \ldots, \succ_i^{-1}, \succ_i^0, \succ_i^1, \ldots, \succ_i^{\beta_i - 1}, \succ_i^{\beta_i}$, such that

- for all $x, y, w, z \in A$, $x \succ_i^h y$, $w \succ_i^k z$, and $h \geq k$ implies $(x, y) \succeq_i^* (w, z)$,
- for all $x \in A$, $x \succ_i^0 x$.

We call *strength of preference* of x over y the equivalence class of $\succeq_i^*$ to which pair (x, y) belongs. For each $f_i \in C$, we denote by H_i the indices of the equivalence classes of $\succeq_i^*$, i.e.

$$H_i = \{\alpha_i, \alpha_i - 1, \ldots, -1, 0, 1, \ldots, \beta_i - 1, \beta_i\}.$$

Therefore, there exists a function $g : A \times A \times C \to H_i$, such that, for all $x, y \in A, x \succ_i^{g(x,y,f_i)} y$, i.e., for all $x, y \in A$ and $f_i \in C$, function g gives the strength of preference of x over y wrt f_i. Taking into account the dependence of $\succeq_i^*$ on evaluations by criterion $f_i \in C$, there also exists a function $g^* : R \times R \times C \to H_i$, such that $g(x, y, f_i) = g^*(f_i(x), f_i(y), f_i)$ and, consequently, $x \succ_i^{g^*(f_i(x), f_i(y), f_i)} y$. Due to monotonicity of $\succeq_i^*$ wrt to evaluations on f_i, we have that $g^*(f_i(x), f_i(y), f_i)$ is non-decreasing wrt $f_i(x)$ and non-increasing wrt $f_i(y)$. Moreover, for each $x \in A$, $g^*(f_i(x), f_i(x), f_i) = 0$.

An $m \times (n+1)$ Pairwise Comparison Table (PCT) is then built up on the basis of this information. The first n columns correspond to the criteria from set C, while the m rows correspond to the pairs from B, such that, if the DM judges that two objects are incomparable, then the corresponding pair does not appear in PCT. The last, i.e. the $(n+1)$-th, column represents the comprehensive binary preference relation $\succeq$ or $\succeq^c$. For each pair $(x, y) \in B$, and for each criterion $f_i \in C$, the respective strength of preference $g^*(f_i(x), f_i(y), f_i))$ is put in the corresponding column.

In terms of rough set theory, the *pairwise comparison table* is defined as a data table $PCT = \langle B, C\cup\{d\}, H_C \cup \{\succeq, \succeq^c\}, g\rangle$, where $B \subseteq A \times A$ is a non-empty *set of exemplary pairwise comparisons of reference objects*, d is a decision corresponding to the comprehensive pairwise comparison resulting in $\succeq$ or $\succeq^c$, $H_C = \bigcup_{f_i \in C} H_i$, and $g{:}B \times (C\cup\{d\}) \rightarrow H_C \cup \{\succeq, \succeq^c\}$ is a total function, such that $g[(x,y),f_i] \in H_i$ for every $(x, y) \in B$ and for each $f_i \in C$, and $g[(x,y),d] \in \{\succeq, \succeq^c\}$ for every $(x,y) \in B$. Thus, binary relations $\succeq$ and $\succeq^c$ induce a partition of B. In fact, PCT can be seen as a decision table, since the set of considered criteria C and the decision d are distinguished.

On the basis of preference relations $\succ_i^h$, $h \in H_i$, $f_i \in C$, upward cumulated preference relations $\succ_i^{\geq h}$, and downward cumulated preference relations $\succ_i^{\leq h}$, can be defined as follows: for all $x, y \in A$,

$$x \succ_i^{\geq h} y \Leftrightarrow x \succ_i^k y \text{ with } k \geq h,$$

$$x \succ_i^{\leq h} y \Leftrightarrow x \succ_i^k y \text{ with } k \leq h.$$

Given $P \subseteq C$ ($P \neq \emptyset$), $(x,y),(w,z) \in A \times A$, the pair of objects (x,y) is said to dominate (w,z) wrt criteria from P (denoted by $(x,y)D_P(w,z)$), if x is preferred to y at least as strongly as w is preferred to z wrt each $f_i \in P$, i.e.,

$$xD_Py \Leftrightarrow (x, y) \succeq_i^* (w, z) \text{ for all } f_i \in P,$$

or, equivalently,

$$xD_Py \Leftrightarrow g(x, y, f_i) \geq g(w, z, f_i) \text{ for all } f_i \in P.$$

Since $\succeq_i^*$ is a complete preorder for each $f_i \in C$, the intersection of complete preorders is a partial preorder, and $D_P = \bigcap_{f_i \in P} \succeq_i^*$, $P \subseteq C$, then the dominance relation D_P is a partial preorder on $A \times A$.

Let $R \subseteq P \subseteq C$ and $(x,y),(w,z) \in A \times A$; then the following implication holds:

$$(x, y)D_P(w, z) \;\Rightarrow\; (x, y)D_R(w, z).$$

Given $P \subseteq C$ and $(x,y) \in B$, the *P-dominating set*, denoted by$D_P^+(x, y)$, and the *P-dominated set*, denoted by$D_P^-(x, y)$, are defined as follows:

$$D_P^+(x,y) = \{(w,z) \in B : (w,z)D_P(x,y)\},$$
$$D_P^-(x,y) = \{(w,z) \in B : (x,y)D_P(w,z)\}.$$

The P-dominating sets and the P-dominated sets are "granules of knowledge" that can be used to express P-lower and P-upper approximations of the comprehensive weak preference relations $\succeq$ and $\succeq^c$, respectively:

$$\underline{P}(\succeq) = \{(x,y) \in B : D_P^+(x,y) \subseteq \succeq\},$$
$$\overline{P}(\succeq) = \{(x,y) \in B : D_P^-(x,y) \cap \succeq \neq \emptyset\},$$
$$\underline{P}(\succeq^c) = \{(x,y) \in B : D_P^-(x,y) \subseteq \succeq^c\},$$
$$\overline{P}(\succeq^c) = \{(x,y) \in B : D_P^+(x,y) \cap \succeq^c \neq \emptyset\},$$

The P-boundaries (P-doubtful regions) of $\succeq$ and $\succeq^c$ are defined as

$$Bn_P(\succeq) = \overline{P}(\succeq) - \underline{P}(\succeq), \quad Bn_P(\succeq^c) = \overline{P}(\succeq^c) - \underline{P}(\succeq^c).$$

In fact we have that $Bn_P(\succeq)=Bn_P(\succeq^c)$.

Using the rough approximations of $\succeq$ and $\succeq^c$, it is possible to induce a generalized description of the preference information contained in PCT in terms of suitable decision rules, having the following syntax:

1. *certain* D$_{\geq}$*-decision rules*, supported by pairs of objects from the P-lower approximation of $\succeq$ only:

   $$\textit{If } x \succ_{i1}^{\geq h(i1)} y, \textit{ and } ..., \textit{ and } x \succ_{ip}^{\geq h(ip)} y, \textit{ then } x \succeq y,$$

   where $P=\{f_{i1},...,f_{ip}\}\subseteq C$
   and $(h(i1),...,h(ip))\in H_{i1} \times ... \times H_{ip}$.
2. *certain* D$_{\leq}$*-decision rules*, supported by pairs of objects from the P-lower approximation of $\succeq^c$ only:

   $$\textit{If } x \succ_{i1}^{\leq h(i1)} y, \textit{ and } ..., \textit{ and } x \succ_{ip}^{\leq h(ip)} y, \textit{ then } x \succeq^c y,$$

   where $P=\{f_{i1},...,f_{ip}\}\subseteq C$
   and $(h(i1),...,h(ip)) \in H_{i1} \times ... \times H_{ip}$.
3. *approximate* D$_{\geq\leq}$*-decision rules*, supported by pairs of objects from the P-boundary of $\succeq$ and $\succeq^c$ only:

   $$\textit{If } x \succ_{i1}^{\geq h(i1)} y, \textit{ and } ..., \textit{ and } x \succ_{ie}^{\geq h(ie)} y,$$
   $$\textit{and } x \succ_{ie+1}^{\leq h(ie+1)} y \textit{ and } ..., \textit{ and } x \succ_{ip}^{\leq h(ip)} y, \textit{ then } x \succeq y \textit{ or } x \succeq^c y,$$

   where $P=\{f_{i1},\ldots,f_{ie},f_{ie+1},\ldots,f_{ip}\} \subseteq C$
   and $(h(i1),\ldots,h_{ie},h_{ie+1},$
   $\ldots,h(ip)) \in H_{i1} \times \ldots, H_{ie}, H_{ie+1}, \ldots \times H_{ip}$.

3 DRSA-EMO: DRSA Applied to Interactive EMO

In this section, we propose a new interactive EMO scheme involving DRSA and called DRSA-EMO. The method consists of a sequence of steps alternating calculation and elicitation of DM's preferences. In the calculation stage, a population of feasible solutions is generated. The population of feasible solutions is evaluated by multiple objective functions. The DM indicates the solutions which, according to his/her preferences, are relatively good, and this information is then processed by DRSA producing a set of "*if*..., *then* ..." decision rules representing DM's preferences. Then, an EMO stage starts with generation of a new population of feasible solutions. The solutions from the new population are evaluated again in terms of objective function values. The "*if*..., *then* ..." decision rules induced in the previous stage are then matched to the new population. In result of this rule matching, the solutions from the new population are ranked from the best to the worst. This is a starting point for selection and crossover of parent solutions, followed by a possible mutation of the offspring solutions. A half of the population of parents and all the offsprings form then a new population of solutions for which a new iteration of EMO starts. The process is iterated until the termination condition of EMO is satisfied. Then, the DM evaluates again the solutions from the last population and either the method stops, because the most satisfactory solution was found, or a new EMO stage is launched with DRSA decision rules induced from DM's sorting of solutions from the last population into relatively good and others.

DRSA-EMO is composed of two embedded loops: the exterior interactive loop, and the interior evolutionary loop. These loops are described in the next subsections.

3.1 The Exterior Interactive Loop of DRSA-EMO

Consider the following MultiObjective Optimization (MOO) problem:

$$max \rightarrow [f_1(\mathbf{x}), \ldots, f_k(\mathbf{x})]$$

subject to:

$$g_1(\mathbf{x}) \geq b_1$$
$$\ldots\ldots\ldots\ldots$$
$$g_m(\mathbf{x}) \geq b_m,$$

where $\mathbf{x} = [x_1, \ldots, x_n]$ is a vector of decision variables, called *solution*, $f_j(\mathbf{x})$, $j = 1, \ldots, k$, are real-valued objective functions, $g_i(\mathbf{x})$, $i = 1, \ldots, m$, are real-valued functions of the constraints, and b_i, $i = 1, \ldots, m$, are right-hand sides of the constraints.

The exterior interactive loop of DRSA-EMO is composed of the following steps.

***Step* 1.** Generate a set of feasible solutions X to the MOO problem, using a Monte Carlo method.

***Step* 2.** Evaluate each solution $\mathbf{x} \in X$ in terms of considered objective functions $[f_1(\mathbf{x}), \ldots, f_k(\mathbf{x})]$.

***Step* 3.** Present to the DM the solutions from X in terms of values of objective functions.

***Step* 4.** If the DM finds a satisfactory solution in X, then STOP, otherwise go to *Step* 5.

***Step* 5.** Ask the DM to indicate a subset of relatively "good" solutions in set X

***Step* 6.** Apply DRSA to the current set X of solutions sorted into "good" and "others", in order to induce a set of decision rules with the following syntax:
"if $f_{j_1}(\mathbf{x}) \geq \alpha_{j_1}$ and ... and $f_{j_p}(\mathbf{x}) \geq \alpha_{j_p}$, then solution $\mathbf{x}$ is good". The decision rules represent DM's preferences on the set of solutions X.

***Step* 7.** An EMO procedure guided by DRSA decision rules is activated [*Steps* a to k of the interior loop].

3.2 The Interior Evolutionary Loop of DRSA-EMO

The interior loop of DRSA-EMO is an evolutionary search procedure guided by DRSA decision rules obtained in *Step* 5 of the exterior loop.

Step a. Generate a new set of feasible solutions X to the MOO problem, using a Monte Carlo method.

Step b. Evaluate each solution $\mathbf{x} \in X$ in terms of considered objective functions $[f_1(\mathbf{x}), \ldots, f_k(\mathbf{x})]$.

Step c. If termination condition is fulfilled, then show the solutions to the DM, otherwise go to *Step* e.

Step d. If the DM finds in the current set X a satisfactory solution, then STOP, otherwise, if the condition to ask DM new preferential information is verified (e.g., a fixed number of iterations is reached), go to *Step* 5 of the exterior loop, otherwise go to *Step* e of this loop.

Step e. Compute a primary score of each solution $\mathbf{x} \in X$, based on the number of DRSA rules matching $\mathbf{x}$.

Step f. Compute a secondary score of each solution $\mathbf{x} \in X$, based on the crowding distance of $\mathbf{x}$ from other solutions in X.

Step g. Rank solutions $\mathbf{x} \in X$ lexicographically, using the primary and the secondary score.

Step h. Make Monte Carlo selection of parents, taking into account the ranking of solutions obtained in *Step* g.

Step i. Recombine parents to get offsprings.

Step j. Mutate offsprings.

Step k. Update the set of solutions X by putting in it a half of best ranked parents and all offsprings. Go back to *Step* b.

In *Step* e, the primary score of each solution $\mathbf{x} \in X$ is calculated as follows. Let *Rule* be a set of DRSA rules obtained in *Step* 6 of the exterior loop. Then, $Rule(\mathbf{x})$ is a set of rules $rule_h \in Rule$ matched by solution $\mathbf{x}$ from set X:

$$Rule(\mathbf{x}) = \{rule_h \in Rule : rule_h \text{ is matched by solution } \mathbf{x}\}.$$

For each $rule_h \in Rule$, the set of solutions matching it is defined as:

$$X(rule_h) = \{\mathbf{x} \in X : \mathbf{x} \text{ is matching } rule_h\}.$$

Then, each $rule_h \in Rule$ gets a weight related to the number of times it is matched by a solution:

$$w(rule_h) = (1 - \delta)^{card(X(rule_h))},$$

where δ is a decay of rule weight, e.g., $\delta = 0.1$. The above formula gives higher weights to rules matching less solutions – this permits to maintain diversity with respect to rules.

The primary score of solution $\mathbf{x} \in X$ is then defined as:

$$Score(\mathbf{x}) = \sum_{rule_h \in Rule(\mathbf{x})} w(rule_h).$$

Observe that we are considering $D_{\geq}$-*decision rules*. However, we can consider also $D_{\leq}$-*decision rules* or both $D_{\geq}$-*decision rules* and $D_{\leq}$-*decision rules*. Of course, in case we consider $D_{\geq}$-*decision rules* the primary score has to be maximised, while in case we consider $D_{\leq}$-*decision rules* the primary score has to be minimised. If both $D_{\geq}$-*decision rules* and $D_{\leq}$-*decision rules* are considered together, then the score relative to $D_{\geq}$-*decision rules*, $Score^{\geq}(\mathbf{x})$, and the score relative to $D_{\leq}$-*decision rules*, $Score^{\leq}(\mathbf{x})$, must be aggregated using some function $\phi : \Re_+ \times \Re_+ \rightarrow \Re$ non decreasing in the first argument and non increasing in the second argument, such that the primary score of solution $\mathbf{x}$ is given by $\phi(Score^{\geq}(\mathbf{x}), Score^{\leq}(\mathbf{x}))$. For example, the primary score can be given by $Score^{\geq}(\mathbf{x}) - Score^{\leq}(\mathbf{x})$ or $Score^{\geq}(\mathbf{x})/Score^{\leq}(\mathbf{x})$.

In *Step* f, the secondary score of each solution $\mathbf{x} \in X$ is calculated in the same way as the crowding distance in the NSGA-II method (9), i.e. it is defined as the sum of distances between the solution's neighbors on either side in each dimension of the objective space. Individuals with a large crowding distance are preferred, as they are in a less crowded region of the objective space, and favoring them aims at preserving diversity in the population.

In *Step* h, the probability of selecting solution $\mathbf{x} \in X$ as a parent is:

$$Pr(\mathbf{x}) = \left(\frac{card(X) - rank(\mathbf{x}) + 1}{card(X)}\right)^{\gamma} - \left(\frac{card(X) - rank(\mathbf{x})}{card(X)}\right)^{\gamma},$$

where $rank(\mathbf{x})$ is a rank of solution $\mathbf{x}$ in the ranking, and $\gamma \geq 1$ is a coefficient of elitism, e.g., $\gamma = 2$. When γ is increasing, then the probability of choosing solutions with a higher rank is increasing.

If solutions $\mathbf{x} = [x_i], \mathbf{y} = [y_i] \in X$ are selected parents, then the offspring (child) solution $\mathbf{z}$ is obtained in *Step* i from recombination of $\mathbf{x}$ and $\mathbf{y}$, e.g., as:

$$z_i = \lambda_i x_i + (1 - \lambda_i) y_i,$$

where multipliers $\lambda_i \in [0,1]$ are chosen randomly.

The probability of mutation of an offspring $\mathbf{z}$ in *Step* j of iteration t is:

$$Pr(t) = \epsilon(1 - \omega)^{t-1},$$

where ϵ is the initial mutation probability, and ω is the decay rate of the mutation probability, e.g., $\epsilon = 0.5$, and $\omega = 0.1$. One can see, that this probability is decreasing in successive iterations. The mutation picks randomly one variable $x_i, i = 1, \ldots, n$, and replace it with another value randomly generated within the range of variation of x_i.

4 DRSA-EMO-PCT: DRSA Applied to Interactive EMO Using Pairwise Comparison Tables

In this section, we propose a new EMO method based on DRSA approximation of a preference relation called DRSA-EMO-PCT: Dominance-based Rough Set Approach to Evolutionary Multiobjective Optimization using Pairwise Comparison Tables. More precisely, comparing to DRSA-EMO, we propose to change the preference information required from the DM and, consequently, to change its use in the procedure. Instead of asking the DM which solutions are for him/her relatively good, we ask the DM to compare some solutions pairwise. From this preference information some decision rules are inferred using DRSA. They are of the form:

"*if* $f_{j_1}(\mathbf{x}) - f_{j_1}(\mathbf{y}) \geq \alpha_{j_1}$, *and* ..., *and* $f_{j_p}(\mathbf{x}) - f_{j_p}(\mathbf{y}) \geq \alpha_{j_p}$,
then solution $\mathbf{x}$ is preferred to solution $\mathbf{y}$",

or

"*if* $f_{j_1}(\mathbf{x})/f_{j_1}(\mathbf{y}) \geq \alpha_{j_1}$, *and* ..., *and* $f_{j_p}(\mathbf{x})/f_{j_p}(\mathbf{y}) \geq \alpha_{j_p}$,
then solution $\mathbf{x}$ is preferred to solution $\mathbf{y}$".

More in general these rules can be of the form

"*if* $\Delta_{j_1}(f_{j_1}(\mathbf{x}), f_{j_1}(\mathbf{y})) \geq \alpha_{j_1}$, *and* ..., *and* $\Delta_{j_p}(f_{j_p}(\mathbf{x}), f_{j_p}(\mathbf{y})) \geq \alpha_{j_p}$,
then solution $\mathbf{x}$ is preferred to solution $\mathbf{y}$",

where for all $j \in I$, $\Delta_j : V_j \times V_j \rightarrow \Re$ is a function such that

$$\Delta_j(f_j(\mathbf{x}), f_j(\mathbf{y})) \geq \Delta_j(f_j(\mathbf{w}), f_j(\mathbf{z})) \Leftrightarrow (x, y) \succeq_j^* (w, z).$$

These decision rules are then used to build a preference relation to be applied within NSGA-II (9), instead of the dominance ranking. Two examples of these preference relations are the following:

- solution $\mathbf{x}$ is preferred to solution $\mathbf{y}$ if $NFS(\mathbf{x}) > NFS(\mathbf{y})$, with $NFS(\mathbf{z}) = \sum_{\mathbf{w}}[rule(\mathbf{z},\mathbf{w}) - rule(\mathbf{w},\mathbf{z})]$ being the Net Flow Score of solution $\mathbf{z}$,
- solution $\mathbf{x}$ is preferred to solution $\mathbf{y}$ if $PO(\mathbf{x}) > PO(\mathbf{y})$, with $PO(\mathbf{z}) = min_{\mathbf{w}} rule(\mathbf{z},\mathbf{w})$ being the Prudent "Outranking" of solution $\mathbf{z}$ (1),

where $rule(\mathbf{x} \succ \mathbf{y})$ is the number of rules for which $\mathbf{x} \succ \mathbf{y}$.

Preference relations based on $NFS(\mathbf{z})$ or $PO(\mathbf{z})$, or some other preference index based on the applicaton of the decision rules to the current population, are used instead of the dominance relation to define consecutive fronts in the population as follows: individuals are ranked by iteratively determining the solutions in the population for which there is no other solution preferred to them, assigning those individuals the next best rank and removing them from the population. It is worthwhile to remark that the preference relation replacing the dominance relation in NSGA-II has to be transitive, because on the contrary it is not possible to obtain consecutive fronts in the population with the above procedure. This means, for example, that the two following preference relations cannot be used even if they seems quite reasonable:

- solution $\mathbf{x}$ is preferred to solution $\mathbf{y}$, $\mathbf{x} \succ \mathbf{y}$, if there is at least one decision rule supporting this preference and there is no decision rule supporting the preference of $\mathbf{y}$ over $\mathbf{x}$,
- solution $\mathbf{x}$ is preferred to solution $\mathbf{y}$ if $rule(\mathbf{x} \succ \mathbf{y}) > rule(\mathbf{y} \succ \mathbf{x})$.

Algorithm 1. DRSA-EMO-PCT

```
Generate initial population of solutions randomly
Elicit user's preferences {Present to the user some pairs of solutions from the
population and ask for a preference comparison}
Determine primary ranking taking into account preferences between solutions
obtained using decision rules {Will replace dominance ranking in NSGA-II}
Determine secondary ranking {Order solutions within a preference front, based
on the crowding distance measured by dist_rule(x, y)}
repeat
   Mating selection and offspring generation
   if Time to ask DM then
      Elicit user's preferences
   end if
   Determine primary ranking
   Determine secondary ranking
   Environmental selection
until Stopping criterion met
Return all preferred solutions according to primary ranking
```

With respect to the crowding distance used in NSGA-II, it is replaced by diversity measure which avoids the arbitrariness of the normalization of the values of objective functions. In fact, we measure the distance between solution $\mathbf{x}$ and solution $\mathbf{y}$ as

$$dist_{rule}(\mathbf{x}, \mathbf{y}) = rule(\mathbf{x} \succ \mathbf{y}) + rule(\mathbf{y} \succ \mathbf{x}).$$

The overall algorithm of DRSA-EMO-PCT is outlined in Algorithm 1.

5 DRSA for Decision under Uncertainty

5.1 DRSA for Ordinal Classification under Uncertainty

To apply rough set theory to decision under uncertainty, we consider the following basic elements:

- a set $S=\{s_1, s_2, \ldots, s_u\}$ of states of the world, or simply *states*, which are supposed to be mutually exclusive and collectively exhaustive,
- an a priori *probability distribution* P over the states of the world: more precisely,
  the probabilities of states $s_1, s_2, \ldots, s_u$ are $p_1, p_2, \ldots, p_u$, respectively ($p_1 + p_2 + \ldots + p_u = 1$, $p_i \geq 0$, $i = 1, \ldots, u$),
- a set $A=\{a_1, a_2, \ldots, a_n\}$ of *acts*,
- a set $X=\{x_1, x_2, \ldots, x_r\}$ *consequences* or *outcomes* that, for the sake of simplicity, we consider expressed in monetary terms ($X \subseteq \Re$),
- a function g: $A \times S \rightarrow X$ assigning to each pair act-state $(a_i,s_j) \in A \times S$ an outcome $x_k \in X$,
- a set of classes $\boldsymbol{Cl}=\{Cl_1, Cl_2, \ldots, Cl_m\}$, such that $Cl_1 \cup Cl_2 \cup \ldots \cup Cl_m = A$, $Cl_r \cap Cl_q = \emptyset$ for each $r,q \in \{1, \ldots, m\}$ with $r \neq q$; the classes from $\boldsymbol{Cl}$ are preference-ordered according to the increasing order of their indices,
- a function e: $A \rightarrow \boldsymbol{Cl}$ assigning each act $a_i \in A$ to a class $Cl_j \in \boldsymbol{Cl}$.

In this context, two different types of dominance can be considered:

1) (classical) *dominance*: given $a_p,a_q \in A$, a_p dominates a_q iff, for each possible state of the world, act a_p gives an outcome at least as good as act a_q; formally, $g(a_p,s_j) \geq g(a_q,s_j)$, for each $s_j \in S$,
2) *stochastic dominance* (see, e.g., (30)): given $a_p,a_q \in A$, a_p dominates a_q according to *first order stochastic dominance* iff, for each outcome $x \in X$, act a_p gives an outcome at least as good as x with a probability at least as great as the probability that act a_q gives the same outcome, i.e. for all $x \in X$,

$$P[S(a_p, x)] \geq P[S(a_q, x)]$$

where, for each $(a_i,x) \in A \times X$, $S(a_i,x) = \{s_j \in S\text{: } g(a_i,s_j) \geq x\}$.

Acts can be compared also considering second order stochastic dominance: given a_p,$a_q \in A$, a_p dominates a_q according to *second order stochastic dominance* iff, for each outcome $x \in X$, the expected values of the outcomes not greater than x given by act a_p is at least as good as the expected values of the outcomes not greater than x given by act a_q.

Of course, classical dominance implies stochastic dominance as well as first order stochastic dominance implies second order stochastic dominance. In this subsection, we consider the first order stochastic dominance. In subsection 5.3 we consider second order stochastic dominance.

On the basis of an a priori probability distribution P, we can assign to each subset of states of the world $W \subseteq S$ the probability $P(W)$ that one of the states in W is verified, i.e. $P(W) = \sum_{i:s_i \in W} p_i$, and then we can build up the set Π of all possible values $P(W)$, i.e.

$$\Pi = \{\pi \in [0,1]: \pi = P(W),\ W \subseteq S\}.$$

Let us define the following functions z: $A \times S \rightarrow \Pi$ and z': $A \times S \rightarrow \Pi$ assigning to each act-state pair $(a_i,s_j) \in A \times S$ a probability $\pi \in \Pi$, as follows:

$$z(a_i, s_j) = \sum_{r:g(a_i,s_r) \geq g(a_i,s_j)} p_r$$

and

$$z'(a_i, s_j) = \sum_{r:g(a_i,s_r) \leq g(a_i,s_j)} p_r .$$

Therefore, $z(a_i,s_j)$ represents the probability of obtaining an outcome whose value is *at least* $g(a_i,s_j)$ by act a_i. Analogously, $z'(a_i,s_j)$ represents the probability of obtaining an outcome whose value is *at most* $g(a_i,s_j)$ by act a_i.

On the basis of function $z(a_i,s_j)$, we can define function ρ: $A \times \Pi \rightarrow X$ as follows:

$$\rho(a_i, \pi) = \max_{j:z(a_i,s_j) \geq \pi} \{g(a_i, s_j)\} .$$

Thus, $\rho(a_i,\pi) = x$ means that the outcome got by act a_i is greater than or equal to x with a probability *at least* π (i.e. probability π or greater).

On the basis of function $z'(a_i,s_j)$ we can define function ρ': $A \times \Pi \rightarrow X$ as follows:

$$\rho'(a_i, \pi) = \min_{j:z'(a_i,s_j) \geq \pi} \{g(a_i, s_j)\} .$$

$\rho'(a_i,\pi) = x$ means that the outcome got by act a_i is smaller than or equal to x with a probability *at least* π.

Let us observe that information given by $\rho(a_i,\pi)$ and $\rho'(a_i,\pi)$ is related. In fact, if the elements of Π, $0=\pi_{(0)}, \pi_{(1)}, \pi_{(2)}, \ldots, \pi_{(d)}=1$ ($d=card(\Pi)$) are reordered in such a way that $0=\pi_{(0)} \leq \pi_{(1)} \leq \pi_{(2)} \leq \ldots \leq \pi_{(d)}=1$, then we have

$$\rho(a_i, \pi_{(j)}) = \rho'(a_i, 1 - \pi_{(j-1)}) \tag{1}$$

for all $a_i \in A$ and $\pi_{(j-1)}, \pi_{(j)} \in \Pi$.

This implies that the analysis of the possible decisions can be equivalently conducted using either $\rho(a_i,\pi)$ or $\rho'(a_i,\pi)$. However, from the viewpoint of representation of results, it is interesting to consider both values $\rho(a_i,\pi)$ and $\rho'(a_i,\pi)$. The reason is that, contrary to intuition, $\rho(a_i,\pi) \leq x$ is not equivalent to say that by act a_i the outcome is smaller than or equal to x with a probability *at least* π. The following example clarifies this point. Let us consider a game a with rolling a dice, in which if the result is 1, then the gain is €1, if the result is 2 then the gain is €2, and so on. Suppose, moreover, that the dice is fair and thus each result is equiprobable with probability 1/6. If we calculate the values of $\rho(a,\pi)$ for all possible values of probability, we have:

$$\rho(a, 1/6) = €6,\ \rho(a,\ 2/6) = €5,\ \rho(a,\ 3/6) = €4,$$
$$\rho(a, 4/6) = €3,\ \rho(a,\ 5/6) = €2,\ \rho(a,\ 6/6) = €1.$$

To explain the above values, take, for example, $\rho(a,\ 5/6) = €2$: it is related to result 2 or 3 or 4 or 5 or 6, i.e. the probability of gaining €2 or more is 5/6. Let us remark that $\rho(a,\ 5/6) \leq €3$ (indeed $\rho(a,\ 5/6) = €2$, and thus $\rho(a,5/6) \leq €3$ is true), however, this is not equivalent to say that by act a the outcome is smaller than or equal to 3 with a probability *at least* 5/6. In fact, this is false because this probability is 3/6, which follows from $\rho'(a,\ 3/6)$ $=€3$ (related to result 1 or 2 or 3).

Analogously, if we calculate the values of $\rho'(a,\ \pi)$ for all possible values of probability, we have:

$$\rho'(a,\ 1/6) = €1,\ \rho'(a,\ 2/6) = €2,\ \rho'(a,\ 3/6) = €3,$$
$$\rho'(a,\ 4/6) = €4,\ \rho'(a,\ 5/6) = €5,\ \rho'(a,\ 6/6) = €6.$$

To explain the above values, take, for example, $\rho'(a,\ 4/6) = €4$: it is related to result 1 or 2 or 3 or 4, i.e. the probability of gaining €4 or less is 4/6. Let us remark that $\rho'(a,\ 5/6) \geq €3$ (indeed $\rho'(a,\ 5/6) = €5$, and thus $\rho'(a,\ 5/6) \geq €3$ is true), however, this is not equivalent to say that by act a the outcome is greater than or equal to 3 with a probability *at least* 5/6. In fact, this is false because this probability is 4/6, which follows from $\rho(a,\ 4/6) =$ €3 (related to result 3 or 4 or 5 or 6).

Therefore, in the context of stochastic acts, if we need to express an outcome in positive terms, we refer to $\rho(a,\pi)$ giving a lower bound of an outcome ("for act a there is a probability π to gain at least $\rho(a,\pi)$"), while if we need to express an outcome in negative terms, we refer to $\rho'(a,\pi)$ giving an upper bound of an outcome ("for act a there is a probability π to gain at most $\rho'(a,\pi)$").

Given $a_p, a_q \in A$, a_p *stochastically dominates* a_q if and only if $\rho(a_p,\pi) \geq \rho(a_q,\pi)$ for each $\pi \in \Pi$. This is equivalent to say: given $a_p, a_q \in A$, a_p stochastically dominates a_q if and only if $\rho'(a_p,\pi) \geq \rho'(a_q,\pi)$ for each $\pi \in \Pi$.

For example, consider the game a^* with rolling a dice, in which if the result is 1, then the gain is €7, if the result is 2 then the gain is €6, and so on, until the case in which the result is 6 and the gain is €2. In this case, game a^* stochastically dominates the above game a, because $\rho(a^*, 1/6)$ = €7 is not smaller than $\rho(a, 1/6)$ = €6, $\rho(a^*, 2/6)$ = €6 is not smaller than $\rho(a, 2/6)$ = €5, and so on. Equivalently, we can say that game a^* stochastically dominates the above game a, because $\rho'(a^*, 1/6)$ =€2 is not smaller than $\rho'(a, 1/6)$ =€1, $\rho'(a^*, 2/6)$ =€3 is not smaller than $\rho'(a, 2/6)$ = €2, and so on.

We can apply the DRSA in this context, considering as set of objects U the set of acts A, as set of attributes (criteria) Q the set $\Pi \cup \{\boldsymbol{cl}\}$, where $\boldsymbol{cl}$ is a decision attribute representing the classification of acts from A into classes from $\boldsymbol{Cl}$, as set V the set $X \cup \boldsymbol{Cl}$, as information function f a function $\boldsymbol{f}$ such that $\boldsymbol{f}(a_i, \pi) = \rho(a_i, \pi)$ and $\boldsymbol{f}(a_i, \boldsymbol{cl})=e(a_i)$. With respect to the set of attributes Q, the set C of condition attributes corresponds to the set Π, and the set of decision attributes D corresponds to $\{\boldsymbol{cl}\}$. Let us observe that, due to equation (1), one can consider alternatively information function $\boldsymbol{f}'(a_i, \pi) = \rho'(a_i, \pi)$.

The aim of the rough set approach to decision under uncertainty is to explain the preferences of the decision maker represented by the assignments of the acts from A to the classes from $\boldsymbol{Cl}$ in terms of stochastic dominance, expressed by means of function ρ. The syntax of decision rules obtained from this rough set approach is as follows:

1) D$_{\geq}$-*decision rules*:
   if $\rho(a,p_{h1}) \geq x_{h1}$ *and, ..., and* $\rho(a,p_{hz}) \geq x_{hz}$, *then* $a \in Cl_r^{\geq}$
   (i.e. "if by act a the outcome is at least x_{h1} with probability at least p_{h1}, and, ..., and the outcome is at least x_{hz} with probability at least p_{hz}, then $a \in Cl_r^{\geq}$"),
   where $p_{h1}, \ldots, p_{hz} \in \Pi$, $x_{h1}, \ldots, x_{hz} \in X$, and $r \in \{2, \ldots, m\}$;
2) D$_{\leq}$-*decision rules*:
   if $\rho'(a,p_{h1}) \leq x_{h1}$ *and, ..., and* $\rho'(a,p_{hz}) \leq x_{hz}$, *then* $a \in Cl_r^{\leq}$
   (i.e. "if by act a the outcome is at most x_{h1} with probability at least p_{h1}, and, ..., and the outcome is at most x_{hz} with probability at least p_{hz}, then $a \in Cl_r^{\leq}$"),
   where $p_{h1}, \ldots, p_{hz} \in \Pi$, $x_{h1}, \ldots, x_{hz} \in X$, and $r \in \{1, \ldots, m-1\}$;
3) D$_{\geq\leq}$-*decision rules*:
   if $\rho(a,p_{h1}) \geq x_{h1}$ *and, ..., and* $\rho(a,p_{hw}) \geq x_{hw}$ *and* $\rho'(a,p_{hw+1}) \leq x_{hw+1}$ *and, ..., and* $\rho'(a,p_{hz}) \leq x_{hz}$, *then* $a \in Cl_s \cup Cl_{s+1} \cup \ldots \cup Cl_t$
   (i.e. "if by act a the outcome is at least x_{h1} with probability at least p_{h1}, and, ..., and the outcome is at least x_{hw} with probability at least p_{hw}, and the outcome is at most x_{hw+1} with probability at least p_{hw+1}, and, ..., and the outcome is at most x_{hz} with probability at least p_{hz}, then $a \in Cl_s \cup Cl_{s+1} \cup \ldots \cup Cl_t$),
   where $p_{h1}, \ldots, p_{hw}, p_{hw+1}, \ldots, p_{hz} \in \Pi$, $x_{h1}, \ldots, x_{hw}, x_{hw+1}, \ldots, x_{hz} \in X$,
   and $s,t \in \{1, \ldots, m\}$, such that $s < t$.

According to the meaning of $\rho(a_i, \pi)$ and $\rho'(a_i, \pi)$ discussed above, $D_{\geq}$-*decision rules* are expressed in terms of $\rho(a_i, \pi)$, $D_{\leq}$-*decision rules* are expressed in terms of $\rho'(a_i,\pi)$, and $D_{\geq\leq}$-*decision rules* are expressed in terms of both $\rho(a_i, \pi)$ and $\rho'(a_i, \pi)$. Let us observe that, due to equation (1), all above decision rules can be expressed equivalently in terms of values of $\rho(a_i,\pi)$ or $\rho'(a_i,\pi)$. For example, a $D_{\geq}$-*decision rule*

$r_{\geq}(\rho) \equiv$ *if* $\rho(a, p_{h1}) \geq x_{h1}$ *and, ..., and* $\rho(a, p_{hz}) \geq x_{hz}$, *then* $a \in Cl_r^{\geq}$

can be expressed in terms of $\rho'(a_i,\pi)$ as

$r_{\geq}(\rho') \equiv$ *if* $\rho'(a, p^*_{h1}) \geq x_{h1}$ *and, ..., and* $\rho'(a, p^*_{hz}) \geq x_{hz}$, *then* $a \in Cl_r^{\geq}$

where, if $p_{hr} = \pi_{(jr)}$, then $p^*_{hr}=1 - \pi_{(jr-1)}$,

with r=1, ..., z, $0=\pi_{(0)},\pi_{(1)},\pi_{(2)}, \ldots, \pi_{(d)}=1$ ($d=card(\Pi)$) reordered in such a way that $0=\pi_{(0)} \leq \pi_{(1)} \leq \pi_{(2)} \leq \ldots \leq \pi_{(d)}=1$.

Analogously, a $D_{\leq}$-*decision rule*

$r_{\leq}(\rho') \equiv$ *if* $\rho'(a, p_{h1}) \leq x_{h1}$ *and, ..., and* $\rho'(a, p_{hz}) \leq x_{hz}$, *then* $a \in Cl_r^{\leq}$

can be expressed in terms of $\rho(a_i, \pi)$ as

$r_{\leq}(\rho) \equiv$ *if* $\rho(a, p^*_{h1}) \leq x_{h1}$ *and, ..., and* $\rho(a, p^*_{hz}) \leq x_{hz}$, *then* $a \in Cl_r^{\leq}$

where if $p_{hr} = \pi_{(jr)}$, then $p^*_{hr}=1 - \pi_{(jr-1)}$, with r=1, ..., z.

Let us observe, however, that $r_{\geq}(\rho)$ is an expression much more natural and meaningful than $r_{\geq}(\rho')$, as well as $r_{\leq}(\rho')$ is an expression much more natural and meaningful than $r_{\leq}(\rho)$.

Another useful remark concerns minimality of rules, related to the specific intrinsic structure of the stochastic dominance. Let us consider the following two decision rules:

r1 ≡ "*if* by act *a* the outcome is at least 100 with probability at least 0.25, *then* *a* is at least good",
r2 ≡ "*if* by act *a* the outcome is at least 100 with probability at least 0.50, *then* *a* is at least good".

r1 and r2 can be induced from the analysis of the same information table, because they involve different condition attributes (criteria). In fact,

- r1 involves attribute $\rho(a, 0.25)$
  (it can be expressed as "*if* $\rho(a, 0.25)\geq 100$, *then* *a* is at least good"),
- r2 involves attribute $\rho(a, 0.50)$
  (it can be expressed as "*if* $\rho(a, 0.50)\geq 100$, *then* *a* is at least good").

Considering the structure of the stochastic dominance, we observe that the condition part of rule r1 is the weakest. In fact, rule r1 requires a cumulated outcome to be at least 100 with probability at least 0.25, while rule r2 requires the same outcome but with a greater probability, 0.5 against 0.25. Taking into account that the decision part of these two rules is the same, we can conclude that rule r1 is minimal among these two rules. From a practical viewpoint, this observation says that, if one induces decision rules using the same algorithms as used for DRSA, it is necessary to further filter the obtained results in order to remove rules which are not minimal in the specific context of the DRSA analysis using stochastic dominance.

Let us observe that the rough set approach we proposed takes into account only ordinal properties of the probability. Hence, this approach remains valid in case of non-additive probability (i.e. probability measure P such that one can have $\mathrm{P}(R)+\mathrm{P}(T) \neq \mathrm{P}(R \cup T)$ in case $R \cap T \neq \emptyset$), or even a qualitative probability (i.e. a probability with an ordinal qualitative scale such as {impossible, little probable, moderately probable, strongly probable, certain}). Moreover, the proposed approach takes into account only ordinal properties of the consequences. Thus, it can handle ordinal qualitative scales of consequences, such as {bad, medium, good}.

Once accepted by the decision maker, these rules represent his/her preference model as explained in Section 2. Also in this case, evaluations appearing in the condition part of decision rules answer the concerns of robustness analysis in a proper way. More precisely, the argument given by a decision rule remains valid for a new act b, even if the probability distribution of consequences changes within the subspace defined by the condition part of the considered decision rule. For example, the $D_{\geq}$-decision rule

"*if* by act a the outcome is at least x_{h1} with probability at least p_{h1}, *and*, ..., *and* the outcome is at least x_{hz} with probability at least p_{hz}, *then* $a \in Cl_r^{\geq}$"

matched by b, continues to be an argument in favor of an assignment of b to at least class r, even if the probability distribution of outcomes changes such that the outcome remains at least x_{h1} with probability at least p_{h1}, and, ..., and the outcome remains at least x_{hz} with probability at least p_{hz}.

5.2 Dominance-Based Rough Set Approach to Preference Learning from Pairwise Comparisons in Case of Decision under Uncertainty

To perform rough set analysis of *PCT* data in case of decision under uncertainty, we consider the following basic elements:

- set $S=\{s_1, s_2, \ldots, s_u\}$ of states of the world, or simply *states*, which are supposed to be mutually exclusive and collectively exhaustive,
- a priori *probability distribution* P over the states of the world; more precisely, the probabilities of states $s_1, s_2, \ldots, s_u$ are $p_1, p_2, \ldots, p_u$, respectively, $(p_1 + p_2 + \ldots + p_u = 1,\ p_i \geq 0,\ i=1,\ldots,u)$,
- set $A=\{a_1, a_2, \ldots, a_o\}$ of *acts*,
- set $X=\{x_1, x_2, \ldots, x_r\}$ of *consequences*,
- function g: $A \times S \rightarrow X$ assigning to each pair act-state $(a_i,s_j) \in A \times S$ an outcome $x_k \in X$,

- quaternary relation $\succeq^*$ on X being a complete preoder on $X \times X$ with $\succ^{*\alpha}, \succ^{\alpha-1}, \ldots, \succ^{0}, \ldots, \succ^{\beta-1}, \succ^{\beta}$ being the equivalence classes of $\sim^*$, $\mathcal{H} = \{\alpha, \alpha-1, \ldots, 0, \ldots, \beta-1, \beta\}$, such that for all $x \in A, x \succ^0 x$,
- function z: $X \times X \rightarrow \mathcal{H}$, such that, for any $(x_{i_1}, x_{i_2}) \in X \times X$, $x_{i_1} \succ^{z(x_{i_1}, x_{i_2})} x_{i_2}$, i.e. $z(x_{i_1}, x_{i_2})$ assigns to each pair (x_{i_1}, x_{i_2}) some strength of the preference relation of x_{i_1} over x_{i_2},
- weak preference relation $\succeq$ and a negative weak preference relation $\succeq^c$ on A, such that $\succeq \cap \succeq^c = \emptyset$ (i.e. $\succeq$ and $\succeq^c$ are incompatible because for any $a, b \in A$ it is not possible that $a \succeq b$ and $a \succeq^c b$) and $\succeq \cup \succeq^c = B \subseteq A \times A$ (i.e. $\succeq$ and $\succeq^c$ are not necessarily exhaustive, because we can have pairs of actions $(a, b) \in A \times A$ for which not $a \succeq b$ and not $a \succeq^c b$).

On the basis of preference relations $\succ^h$, $h \in \mathcal{H}$, upward cumulated preference relations $\succ^{\geq h}$ and downward cumulated preference relations $\succ^{\leq h}$ can be defined as follows: for all $x, y \in X$,

$$x \succ^{\geq h} y \Leftrightarrow x \succ^k y \text{ with } k \geq h, \text{ and } x \succ^{\leq h} y \Leftrightarrow x \succ^k y \text{ with } k \leq h.$$

On the basis of the quaternary relation $\succeq^*$ on X, for each $s \in S$ one can define a quaternary relation $\succeq^*_s$ on A as follows: for all $a, b, c, d \in A$,

$$(a, b) \succeq^*_s (c, d) \Leftrightarrow (g(a, s), g(b, s)) \succeq (g(c, s), g(d, s)).$$

Analogously, for each $s \in S$ and for each $a, b \in A$, the strength of preference of $g(a, s)$ over $g(b, s)$ can be extended to the strength of preference of a over b wrt state of nature s, i.e.,

$$a \succ^h_s b \Leftrightarrow g(a, s) \succ^h g(b, s).$$

In the same way, upward and downward cumulated preference relations above defined on X can be extended to A: for any $a, b \in A$, $s \in S$ and $h \in \mathcal{H}$,

$$a \succ^{\geq h}_s b \Leftrightarrow g(a, s) \succ^{\geq h} g(b, s),$$

$$a \succ^{\leq h}_s b \Leftrightarrow g(a, s) \succ^{\leq h} g(b, s).$$

For each $a, b \in A, h \in \mathcal{H}$ and $s \in S$, it is possible to calculate the probability $\rho^{\geq}(a, b, h)$ that a is preferred to b with a strength at least h, and the probability $\rho^{\leq}(a, b, h)$ that a is preferred to b with a strength at most h:

$$\rho^{\geq}(a, b, h) = \sum_{s \in S:\ a \succ^{\geq h}_s b} p_s, \qquad \rho^{\leq}(a, b, h) = \sum_{s \in S:\ a \succ^{\leq h}_s b} p_s.$$

Given $a, b, c, d \in A$, (a, b) *stochastically dominates* (c, d) if, for each $h \in \mathcal{H}$, the probability that a is preferred to b with a strength at least h is not smaller than the probability that c is preferred to d with a strength at least h, i.e., for all $h \in \mathcal{H}$, $\rho^{\geq}(a, b, h) \geq \rho^{\geq}(c, d, h)$.

The stochastic dominance of (a, b) over (c, d) can be equivalently expressed in terms downward cumulated preference $\rho^{\leq}(a, b, h)$ and $\rho^{\leq}(c, d, h)$ as follows: given $a, b, c, d \in A$, (a, b) stochastically dominates (c, d) if, for each $h \in \mathcal{H}$, the probability that a is preferred to b with a strength at most h is not greater than the probability that c is preferred to d with a strength at most h, i.e., for all $h \in \mathcal{H}$, $\rho^{\leq}(a, b, h) \leq \rho^{\leq}(c, d, h)$.

It is natural to expect that for any $a, b, c, d \in A$, if (a, b) stochastically dominates (c, d), then

- if $c \succeq d$, then also $a \succeq b$,
- if $a \succeq^c b$, then also $c \succeq^c d$.

Considering 2^S, the power set of the set of states of nature S, one can define the set

$$Prob = \left\{ \sum_{s \in T} p_s, \; T \subseteq S \right\}.$$

For any $q \in Prob$ and $a, b \in A$, let

$$f^+(a, b, q) = max\{h \in \mathcal{H} : \rho^{\geq}(a, b, h) \geq q\}$$

and

$$f^-(a, b, q) = min\{h \in \mathcal{H} : \rho^{\leq}(a, b, h) \geq q\}.$$

The above definitions can be interpreted as follows: for any $q \in Prob$ and $a, b \in A$,

- there is a probability at least q that a is preferred to b with a strength not smaller than $f^+(a, b, q)$,
- there is a probability at least q that a is preferred to b with a strength not greater than $f^-(a, b, q)$.

Observe that for any $a, b \in A$,

$$f^+(a, b, q_{\pi(i)}) = f^-(a, b, 1 - q_{\pi(i+1)}), \tag{1}$$

where π is a permutation of the probabilities from $Prob$, such that

$$0 = \pi(1) < \pi(2) < \ldots < \pi(k) = 1, \quad k = card(Prob).$$

Using values $f^+(a, b, q)$ and $f^-(a, b, q)$, we can give an equivalent definition of stochastic dominance of (a, b) over (c, d), for any $a, b, c, d \in A$: (a, b) stochastically dominates (c, d), if for any $q \in Prob$, $f^+(a, b, q) \geq f^+(c, d, q)$, or, equivalently, $f^-(a, b, q) \leq f^-(c, d, q)$.

In this context, setting $m = card(B)$ and $n = card(Prob)$, an $m \times (n{+}1)$, Pairwise Comparison Table (PCT) can be set up as follows. The first n columns correspond to the probabilities $q \in Prob$, while the m rows correspond to the pairs from B. The last $(n{+}1)$-th column represents the comprehensive binary preference relation $\succeq$ or $\succeq^c$. For each pair $(a, b) \in B$, and for each probability $q \in Prob$, the respective value $f^+(a,b,q)$ is put in the corresponding column.

In terms of rough set theory, the PCT is defined as a data table

$$PCT = \langle B, Prob \cup \{d\}, \mathcal{H} \cup \{\succeq, \succeq^c\}, f^+ \rangle,$$

i.e. we can apply the DRSA in this context, considering as set of reference objects the set of pairs of acts $B = \succeq \cup \succeq^c$, as set of attributes (criteria) the set $Prob \cup \{d\}$, where to each $q \in Prob$ corresponds a condition attribute assigning some strength of preference $h \in \mathcal{H}$ to each pair $(a,b) \in B$ through function $f^+(a,b,q)$, and d is a decision attribute representing the assignments of pairs of acts $(a,b) \in B$ to classes of weak preference ($a \succeq b$) or negative weak preference ($a \succeq^c b$), as set V the set $\mathcal{H} \cup \{\succeq, \succeq^c\}$, and as information function a function f, such that for all $q \in Prob$, $f(a,b,q) = f^+(a,b,q)$, and $f(a,b,d) = \succeq$ if $a \succeq b$, and $f(a,b,d) = \succeq^c$ if $a \succeq^c b$.

The aim of the rough set approach to decision under uncertainty is to explain the preferences of the DM on the pairs of acts from B in terms of stochastic dominance on values given by functions $f^+(a,b,q)$ and $f^-(a,b,q)$. The resulting preference model is a set of decision rules induced from rough set approximations of weak preference relations. The syntax of decision rules is as follows:

1. **D$_{\geq}$-*decision rules:***
   If $f^+(a,b,q_{\gamma_1}) \geq h_1$, *and,..., and* $f^+(a,b,q_{\gamma_z}) \geq h_z$, *then* $a \succeq b$,
   (i.e. "if with a probability at least q_{γ_1} act a is preferred to act b with a strength at least h_1, and,..., with a probability at least q_{γ_z} act a is preferred to act b with a strength at least h_z, then $a \succeq b$"),
   where $q_{\gamma_1}, \ldots, q_{\gamma_z} \in Prob$, $\quad h_{\gamma_1}, \ldots, h_{\gamma_z} \in \mathcal{H}$;
2. **D$_{\leq}$-*decision rules:***
   If $f^-(a,b,q_{\gamma_1}) \geq h_1$, *and,..., and* $f^-(a,b,q_{\gamma_z}) \geq h_z$, *then* $a \succeq b$,
   (i.e. "if with a probability at least q_{γ_1} act a is preferred to act b with a strength at most h_1, and,..., with a probability at least q_{γ_z} act a is preferred to act b with a strength at most h_z, then $a \succeq^c b$"),
   where $q_{\gamma_1}, \ldots, q_{\gamma_z} \in Prob$, $\quad h_{\gamma_1}, \ldots, h_{\gamma_z} \in \mathcal{H}$;
3. **D$_{\geq\leq}$-*decision rules:***
   If $f^+(a,b,q_{\gamma_1}) \geq h_1$, *and,..., and* $f^+(a,b,q_{\gamma_e}) \geq h_e$, *and*
   $f^-(a,b,q_{\gamma_{e+1}}) \geq h_{e+1}$, *and,..., and* $f^-(a,b,q_{\gamma_z}) \geq h_z$ *then* $a \succeq b$,
   (i.e. "if with a probability at least q_{γ_1} act a is preferred to act b with a strength at least h_1, and,..., with a probability at least q_{γ_e} act a is preferred to act b with a strength at least h_e, and if with a probability at least $q_{\gamma_{e+1}}$ act a is preferred to act b with a strength at most h_{e+1}, and,..., with a probability at least q_{γ_z} act a is preferred to act b with a strength at most h_z, then $a \succeq b$ or $a \succeq^c b$"),
   where $q_{\gamma_1}, \ldots, q_{\gamma_e}, q_{\gamma_{e+1}}, \ldots, q_{\gamma_z} \in Prob$, $h_{\gamma_1}, \ldots, h_{\gamma_e}, h_{\gamma_{e+1}}, \ldots, h_{\gamma_z} \in \mathcal{H}$.

5.3 DRSA for Decision under Uncertainty Using Tail Means

Representation of uncertainty by consideration of quantiles within DRSA can be based also on lower tail means, being the expected values of a random variable within the considered quantiles in case its values are ordered from the lower to the upper, or upper tail means, being the expected values of the random variable within the considered quantiles in case its values are ordered from the upper to the lower. For example if the 5% lower tail mean of profit is 100,000 \$, then, taking into account the worst 5% of cases, the expected value of the profit is 100,000\$. Analogously, if the 5% upper tail mean of profit is 1,000,000 \$, then, taking into account the best 5% of cases, the expected value of the profit is 1,000,000\$. Of course, if decision tables are presented in terms of lower or upper tail means instead of quantiles, also induced decision rules are expressed in terms of lower or upper tail means instead of quantiles. Therefore, the syntax of the decision rules becomes the following:

1) $D_{\geq}$-*decision rules*:
"if by act a the average outcome in the best $p_{h1} \times 100\%$ of the cases is at least x_{h1}, and, ..., and the average outcome in the best $p_{hw} \times 100\%$ of the cases is at least x_{hw}, and the average outcome in the worst $p_{hw+1} \times 100\%$ of the cases is at least x_{hw+1}, and, ..., and the average outcome in the worst $p_{hz} \times 100\%$ of the cases is at least x_{hz}, then $a \in Cl_r^{\geq}$",
where $p_{h1}, \ldots, p_{hw}, p_{hw+1}, \ldots, p_{hz} \in \Pi$, $x_{h1}, \ldots, x_{hw}, x_{hw+1}, \ldots, x_{hz} \in X$, and $r \in \{2, \ldots, m\}$;
2) $D_{\leq}$-*decision rules*:
"if by act a the average outcome in the best $p_{h1} \times 100\%$ of the cases is at most x_{h1}, and, ..., and the average outcome in the best $p_{hw} \times 100\%$ of the cases is at most x_{hw}, and the average outcome in the worst $p_{hw+1} \times 100\%$ of the cases is at most x_{hw+1}, and, ..., and the average outcome in the worst $p_{hz} \times 100\%$ of the cases is at most x_{hz}, then $a \in Cl_r^{\leq}$",
where $p_{h1}, \ldots, p_{hw}, p_{hw+1}, \ldots, p_{hz} \in \Pi$, $x_{h1}, \ldots, x_{hw}, x_{hw+1}, \ldots, x_{hz} \in X$, and $r \in \{1, \ldots, m-1\}$;
3) $D_{\geq\leq}$-*decision rules*:
"if by act a the average outcome in the best $p_{h1} \times 100\%$ of the cases is at least x_{h1}, and, ..., and the average outcome in the best $p_{hu} \times 100\%$ of the cases is at leatst x_{hu}, and the average outcome in the worst $p_{hu+1} \times 100\%$ of the cases is at least x_{hu+1}, and, ..., and the average outcome in the worst $p_{hv} \times 100\%$ of the cases is at least x_{hv}, the average outcome in the best $p_{hv+1} \times 100\%$ of the cases is at most x_{hv+1}, and, ..., and the average outcome in the best $p_{hw} \times 100\%$ of the cases is at most x_{hw}, and the average outcome in the worst $p_{hw+1} \times 100\%$ of the cases is at most x_{hw+1}, and, ..., and the average outcome in the worst $p_{hz} \times 100\%$ of the cases is at most x_{hz}, then $a \in Cl_s \cup Cl_{s+1} \cup \ldots \cup Cl_t$",

where $p_{h1}, \ldots, p_{hu}, p_{hu+1}, \ldots, p_{hv}, p_{hv+1}, \ldots, p_{hw}, p_{hw+1}, \ldots, p_{hz} \in \Pi$, $x_{h1}, \ldots, x_{hu}, x_{hu+1}, \ldots, x_{hv}, x_{hv+1}, \ldots, x_{hw}, x_{hw+1}, \ldots, x_{hz} \in X$, and $s,t \in \{1, \ldots, m\}$, such that $s < t$.

Observe that in the above decision rules both upper tail means and lower tail means are considered. However, one can consider decision rules with only upper tail means or lower tail means. The presence of lower or upper tail means, or both of them, in the decision rules depends on the modalities to represent uncertainty preferred by the DM. In general, a more risk adverse DM should prefer that the uncertainty is represented in terms of lower tail means, a more risk prone DM should prefer that the uncertainty is represented in terms of upper tail means, and a DM with a more equilibrated propension to the risk should prefer the presence of both upper and lower tail means.

Remark that left quantiles consider only the best value of objectives in the meaningful quantiles, while lower tail means represent expected values of all the values taken by the objective functions in the considered left quantiles. In other words, worse scenarios gain importance considering lower tail means instead of corresponding left quantiles. Therefore, considering lower tail means is more risk averse than considering left quantiles. Observe however, that the situation is exactly the opposite when comparing upper tail means with corresponding right quantiles. In fact, right quantiles consider only the worst value of objectives in the meaningful quantiles, while upper tail means represent the expected values of all the values taken by the objective functions in the considered right quantiles. In other words, better scenarios gain importance considering upper tail means instead of corresponding right quantiles. Therefore, considering upper tail means is more risk prone than considering right quantiles.

Observe also that the comparisons of random variables using dominance relation with respect to left or right quantiles is equivalent to first-order stochastic dominance, in case all quantiles are considered. Instead, comparisons of random variables using dominance relation with respect to lower or upper tail means is equivalent to second-order stochastic dominance, in case all quantiles are considered. Notice also that lower and upper tail means can be used only in case it is meaningful to compute expected values. Therefore, upper and lower tail means cannot be used in case the probability or the values taken by the objective functions are qualitative (e.g. "very probable", "probable", "normally probable", "few probable", "very few probable" with respect to probability or "very high", "high", "medium", "low" and "very low" with respect to values taken by the objective functions).

6 DRSA for Robust Interactive EMO: DARWIN and DARWIN-PCT

It is rare that all data needed to formulate the MultiObjective Optimization (MOO) problem are known as precise numbers. Rather the opposite, they are often not precisely known, and thus the coefficients of the multiobjective optimization problem are given as intervals of possible values. In this situation, instead of seeking for the best solution with respect to the considered objectives, one is rather interested in the best robust solution with respect to the considered objectives and uncertainties.

In this sense, two interactive EMO methods taking into account such robustness concerns have ben proposed as specific instances of DRSA-EMO and DRSA-EMO-PCT, respectively: DARWIN (*DARWIN*: *D*ominance-based rough set *A*pproach to handling *R*obust *W*inning solutions in *IN*teractive multiobjective optimization) (20) and DARWIN-PCT (DARWIN-Pairwise Comparison Table) (22).

In DARWIN and DARWIN-PCT, it is assumed that some coefficients in the objective functions and/or constraints of the MOO problem are not precisely known and given as interval values. In the calculation stage, a population of feasible solutions is generated together with a sample of vectors of possible values of the imprecise coefficients - each such vector is called scenario. The population of feasible solutions is evaluated by multiple objective functions for all scenarios from the considered sample. In this way, one obtains for each feasible solution a distribution of the values of objective functions over possible scenarios. Some representative quantiles of these distributions are presented to the DM in the preference elicitation stage. In DARWIN method, the DM indicates the solutions which, according to his/her preferences, are relatively good. In DARWIN-PCT method, the DM compare some solutions pairwise, indicating some pairwise preferences on the set of solutions of the type "solution $\mathbf{x}$ is preferred to solution $\mathbf{y}$". This information is then processed by DRSA, producing a set of "*if*..., *then* ..." decision rules representing DM's preferences. Then, an EMO stage starts with generation of a new population of feasible solutions and of a new sample of possible scenarios. The solutions from the new population are evaluated again in terms of representative quantiles of the distribution of objective function values. The "*if*..., *then* ..." decision rules induced in the previous stage are then matched to the new population. In result of this rule matching, the solutions from the new population are ranked from the best to the worst. This is a starting point for selection and crossover of parent solutions, followed by a possible mutation of the offspring solutions. A half of the population of parents and all the offsprings form then a new population of solutions for which a new iteration of EMO starts. The process is iterated until the termination condition of EMO is satisfied. Then, the DM evaluates again the solutions from the last population and either the method stops because the most satisfactory solution was found, or a new EMO stage is launched with DRSA decision rules induced from DM's classification of or preferences between solutions of the last population. The way in which DARWIN and DARWIN-PCT handle the robustness concerns is different from ajority of current approaches to robust optimization (see e.g. (29; 3)) that are focused on the worst case scenario (so called Wald criterion) or on the maximal regret (so called Savage criterion). Instead, DARWIN and DARWIN-PCT consider a uniform probability distribution on the set of scenarios, which corresponds to the so called Laplace criterion (for a comprehensive discussion about Wald criterion, Savage criterion and Laplace criterion see (32)). In general, Laplace criterion does not seem appropriate for robust optimization, because it takes into account an average situation, but this does not ensure that there can be very bad results in some specific scenario. However, in DARWIN and DARWIN-PCT, a new set of scenarios is generated within the domain of variations of considered variables in each iteration. Therefore, the final population of solutions given by DARWIN and DARWIN-PCT is the final product of "natural selection" of individuals able to "survive" performing well in a wide range of different situations. In this sense, we claim that the solutions proposed by DARWIN and DARWIN-PCT are robust. The introduction of this idea in the algorithms of DRSA-EMO and DRSA-EMO-PCT, resulting, respectively, in DARWIN and DARWIN-PCT methods, is explained in the following subsections.

6.1 *DARWIN*

DARWIN is composed of two embedded loops: the exterior interactive loop, and the interior evolutionary loop. These loops are described in the following.

The exterior interactive loop of *DARWIN*. Consider the following MultiObjective Optimization (MOO) problem:

$$max \rightarrow [f_1(\mathbf{x}), \ldots, f_k(\mathbf{x})]$$

subject to:

$$g_1(\mathbf{x}) \geq b_1$$

$$\ldots\ldots\ldots\ldots$$

$$g_m(\mathbf{x}) \geq b_m,$$

where $\mathbf{x} = [x_1, \ldots, x_n]$ is a vector of decision variables, called *solution*, $f_j(\mathbf{x})$, $j = 1, \ldots, k$, are real-valued objective functions, $g_i(\mathbf{x})$, $i = 1, \ldots, m$, are real-valued functions of the constraints, and b_i, $i = 1, \ldots, m$, are right-hand sides of the constraints.

We assume that some coefficients in the objective functions and in the constraints of the MOO problem are not precisely known and given as interval values. A vector of "imprecise" coefficients fixed on single values within the corresponding intervals is called a possible *scenario* of the imprecision.

The exterior interactive loop of *DARWIN* is composed of the following steps.

***Step* 1.** Generate a set of feasible solutions X to the MOO problem, using a Monte Carlo method.

***Step* 2.** Generate a set of possible scenarios S using a Monte Carlo method.

***Step* 3.** For each scenario $\mathbf{s} \in S$ evaluate each solution $\mathbf{x} \in X$ in terms of considered objective functions $[f_1(\mathbf{x}), \ldots, f_k(\mathbf{x})]$.

***Step* 4.** Present to the DM the solutions from X in terms of meaningful quantiles of the distribution of values of objective functions over scenarios from S, e.g., $f_1^{1\%}(\mathbf{x}), f_1^{25\%}(\mathbf{x}), f_1^{75\%}(\mathbf{x}), \ldots, f_k^{1\%}(\mathbf{x}), f_k^{25\%}(\mathbf{x}), f_k^{75\%}(\mathbf{x})$, for all $\mathbf{x} \in X$, where, in general, $f_j^{\beta_{j_p}}(\mathbf{x}) = \gamma$ means that there is a probability β_{j_p} that $f_j(\mathbf{x})$ takes a value at least equal to γ.

***Step* 5.** If the DM finds in X a satisfactory solution, then STOP, otherwise go to ***Step* 6**.

***Step* 6.** Ask the DM to indicate a subset of relatively "good" solutions in set X.

***Step* 7.** Apply DRSA to the current set X of solutions sorted into "good" and "others", in order to induce a set of decision rules with the following syntax "if $f_{j_1}^{\beta_{j_1}}(\mathbf{x}) \geq \alpha_{j_1}$ and ... and $f_{j_p}^{\beta_{j_p}}(\mathbf{x}) \geq \alpha_{j_p}$, then solution $\mathbf{x}$ is good", e.g. $\beta_{j_1}, \ldots, \beta_{j_p} \in \{1\%, 25\%, 75\%\}$, $\{j_1, \ldots, j_p\} \subseteq \{1, \ldots, k\}$. The decision rules represent DM's preferences on the set of solutions X.

Step* 8.** An EMO procedure guided by DRSA decision rules is activated [Steps* a** to **m** of the interior loop].

The interior evolutionary loop of *DARWIN*. The interior loop of DARWIN is an evolutionary search procedure guided by DRSA decision rules obtained in ***Step* 7** of the exterior loop.

***Step* a.** Generate a new set of feasible solutions X to the MOO problem, using a Monte Carlo method.

***Step* b.** Generate a new set of possible scenarios S using a Monte Carlo method.

***Step* c.** For each scenario $\mathbf{s} \in S$ evaluate each solution $\mathbf{x} \in X$ in terms of considered objective functions $[f_1(\mathbf{x}), \ldots, f_k(\mathbf{x})]$.

***Step* d.** For each solution $\mathbf{x} \in X$ calculate all meaningful quantiles of the distribution of values of objective functions over scenarios from S.

***Step* e.** If termination condition is fulfilled, then show the solutions to the DM, otherwise go to ***Step* g**.

Step* f.** If the DM finds in the current set X a satisfactory solution, then STOP, otherwise, if the condition to ask DM new preferential information is verified (e.g., a fixed number of iterations is reached), go to ***Step 6 of the exterior loop, otherwise go to ***Step* g** of this loop.

***Step* g.** Compute a primary score of each solution $\mathbf{x} \in X$, based on the number of DRSA rules matching $\mathbf{x}$.

***Step* h.** Compute a secondary score of each solution $\mathbf{x} \in X$, based on the crowding distance of $\mathbf{x}$ from other solutions in X.

***Step* i.** Rank solutions $\mathbf{x} \in X$ lexicographically, using the primary and the secondary score.

***Step* j.** Make Monte Carlo selection of parents, taking into account the ranking of solutions obtained in ***Step* i**.

***Step* k.** Recombine parents to get offsprings.

***Step* l.** Mutate offsprings.

***Step* m.** Update the set of solutions X by putting in it a half of best ranked parents and all offsprings. Go back to ***Step* b**.

Notice that the difference between DRSA-EMO and DARWIN is that in DARWIN the criteria are in fact quantiles computed under the hypothesis of a uniform probability distribution all over the possible scenarios. In fact, considering a uniform probability seems quite appropriate for taking into account robustness concerns. In this case it is supposed that, in general, the probability distribution is not known a priori and, in simple words, the final aim is to find some solutions which are generally good in all the possible scenarios. However, if some more precise information about probability distribution over possible scenarios is available, then DARWIN can take into account any specific probability distribution. Observe also that in ***Step* 7** of the exterior loop we are considering D$_{\geq}$-*decision rules*, but we can consider D$_{\leq}$-*decision rules* or both D$_{\geq}$-*decision rules* and D$_{\leq}$-*decision rules* as explained for DRSA-EMO procedure. Moreover, notice that decision rules can be also expressed in terms of upper or lower tail means if the DM prefers such a representation of the uncertainty.

6.2 DARWIN-PCT

DARWIN-PCT presents to the DM the solutions from the current population X in terms of meaningful quantiles of the distribution of values of objective functions over scenarios from S, e.g., $f_1^{1\%}(\mathbf{x}), f_1^{25\%}(\mathbf{x}), f_1^{50\%}(\mathbf{x}), \ldots, f_k^{1\%}(\mathbf{x}), f_k^{25\%}(\mathbf{x}), f_k^{50\%}(\mathbf{x})$, for all $\mathbf{x} \in X$. The DM is asked to indicate some pairwise preferences on the solutions from X of the type "solution $\mathbf{x}$ is preferred to solution $\mathbf{y}$". From this preference information some decision rules are inferred using DRSA. They are of the form:

$$\text{"}\mathit{if}\ f_{j_1}^{\beta_{j_1}}(\mathbf{x}) - f_{j_1}^{\beta_{j_1}}(\mathbf{y}) \geq \alpha_{j_1}\ \mathit{and}\ \ldots\ \mathit{and}\ f_{j_p}^{\beta_{j_p}}(\mathbf{x}) - f_{j_p}^{\beta_{j_p}}(\mathbf{y}) \geq \alpha_{j_p},\ \mathit{then}\ \text{solution } \mathbf{x} \text{ is preferred to solution } \mathbf{y}\text{"},$$

or

$$\text{"}\mathit{if}\ f_{j_1}^{\beta_{j_1}}(\mathbf{x})/f_{j_1}^{\beta_{j_1}}(\mathbf{y}) \geq \alpha_{j_1}\ \mathit{and}\ \ldots\ \mathit{and}\ f_{j_p}^{\beta_{j_p}}(\mathbf{x})/f_{j_p}^{\beta_{j_p}}(\mathbf{y}) \geq \alpha_{j_p},\ \mathit{then}\ \text{solution } \mathbf{x} \text{ is preferred to solution } \mathbf{y}\text{"},$$

where $\beta_{j_1}, \ldots, \beta_{j_p}$ are the meaningful quantiles considered (e.g., $\beta_{j_1}, \ldots, \beta_{j_p} \in \{1\%, 25\%, 50\%\}$), $\{j_1, \ldots, j_p\} \subseteq \{1, \ldots, k\}$.

More in general these rules can be of the form

$$\text{“}if\ \Delta_{j_1}(f_{j_1}^{\beta_{j1}}(\mathbf{x}), f_{j_1}^{\beta_{j1}}(\mathbf{y})) \geq \alpha_{j_1},\ and\ ...,\ and\ \Delta_{j_p}(f_{j_p}^{\beta_{jp}}(\mathbf{x}), f_{j_p}^{\beta_{jp}}(\mathbf{y})) \geq \alpha_{j_p},$$
$$then \text{ solution } \mathbf{x} \text{ is preferred to solution } \mathbf{y}\text{”},$$

where, for all $j \in \{1, \ldots, k\}$, denoted by V_j the set of values taken by objective function f_j, $\Delta_j : V_j \times V_j \rightarrow \Re$ is a function such that

$$\Delta_j(f_j^\beta(\mathbf{x}), f_j^\beta(\mathbf{y})) \geq \Delta_j(f_j^\beta(\mathbf{w}), f_j^\beta(\mathbf{z}))$$

means that with respect to quantile β of objective function f_j solution $\mathbf{x}$ is preferred to solution $\mathbf{y}$ at least as strongly as solution $\mathbf{w}$ is preferred to solution $\mathbf{z}$.

As in DRSA-EMO-PCT, the above decision rules are then used to build a preference relation to be applied within the popular EMO procedure, called NSGA-II(9), instead of the dominance ranking.

As in DRSA-EMO-PCT, also in DARWIN-PCT, the crowding distance used in NSGA-II is replaced by a diversity measure which avoids the arbitrariness of the normalization of the values of objective functions, i.e. we measure the distance between solution $\mathbf{x}$ and solution $\mathbf{y}$ as $dist_{rule}(\mathbf{x}, \mathbf{y}) = rule(\mathbf{x} \succ \mathbf{y}) + rule(\mathbf{y} \succ \mathbf{x})$.

Notice that the difference between DRSA-EMO-PCT and DARWIN-PCT is that in DARWIN the criteria are in fact quantiles computed considering a

Algorithm 2. DARWIN-PCT

```
Generate set of feasible scenarios randomly
Generate initial population of solutions randomly
Elicit user's preferences {Present to the user some pairs of solutions from the
population and ask for a preference comparison}
Determine primary ranking taking into account preferences between solutions
obtained using decision rules {Will replace dominance ranking in NSGA-II}
Determine secondary ranking {Order solutions within a preference front, based
on the crowding distance measured by dist_rule(x, y)}
repeat
   Mating selection and offspring generation
   if Time to ask DM then
      Elicit user's preferences
   end if
   Determine primary ranking
   Determine secondary ranking
   Environmental selection
until Stopping criterion met
Return all preferred solutions according to primary ranking
```

uniform probability distribution over all the possible scenarios. As in DARWIN, also in DARWIN-PCT the use of a uniform probability seems more appropriate for taking into account robustness concerns. In any case, also in DARWIN-PCT there is the possibility of considering any kind of probability ditribution. Moreover, even if we are considering $D_{\geq}$-*decision rules*, also $D_{\leq}$-*decision rules* or both $D_{\geq}$-*decision rules* and $D_{\leq}$-*decision rules* can be used.

The overall algorithm of DARWIN-PCT is outlined in Algorithm 2.

7 Conclusions

DRSA-EMO and DRSA-EMO-PCT are interactive EMO procedures involving preferences of the DM represented by "*if*..., *then* ..." decision rules induced from preference information by Dominance-based Rough Set Approach (DRSA). As proved in (18), (36), the set of "*if*..., *then* ..." decision rules is the most general and the most comprehensible preference (aggregation) model. The rules obtained using DRSA have a syntax adequate to multiobjective decision problems: the condition part of a rule compares a solution in the objective space to a dominance cone built on a subset of objectives; if the solution is within this cone, then the rule assigns the solution to either a class of "good" solutions (the case of a positive dominance cone) or to a class of "other" solutions (the case of a negative dominance cone).

DM gives preference information by answering easy questions, and obtains transparent feedback in a learning oriented perspective (see (2)).

Moreover, DRSA decision rules do not convert ordinal information into numeric one, which implies that: (*i*) from the point of view of multiobjective optimization, no scalarization is involved, and (*ii*) from the point of view of decision under uncertainty, no specific model, such as expected utility, Choquet integral, Max-min expected utility, cumulative prospect theory, etc., has been imposed, and only a very general principle of stochastic dominance in the space of meaningful quantiles is considered.

DARWIN and *DARWIN-PCT* methods extend DRSA-EMO and DRSA-EMO-PCT procedures by taking into account robustness concerns. The robustness of their solutions is ensured twofold, since: (*a*) DRSA decision rules are immune to inconsistencies in preference information, and (*b*) *DARWIN* and *DARWIN-PCT* take into account many possible scenarios, and involve preferences on distribution of values of objective functions over possible scenarios.

The first computational experiments carried out with DRSA-EMO, DRSA-EMO-PCT, *DARWIN* and *DARWIN-PCT* on some benchmark MOO problems confirm the intuition that decision rules induced from the DM's preference information drive well the EMO procedure towards the most interesting region of non-dominated solutions. Assuming a given value function of a hypothetical DM at the stage of sorting a population of solutions, DRSA-EMO,

DRSA-EMO-PCT, *DARWIN* and *DARWIN-PCT* (which ignore this value function) converge to the most preferred solutions in a very similar way to an evolutionay procedure maximizing just the same value function.

DRSA-EMO, DRSA-EMO-PCT, *DARWIN* and *DARWIN-PCT* provide very general interactive EMO schemes which can be customized to a large variety of Operational Research problems, from location and routing to scheduling and supply chain management.

Acknowledgment. The third author wishes to acknowledge financial support from the Polish Ministry of Education and Science, grant no. N N519 314435.

References

[1] Arrow, K.J., Raynaud, H.: Social Choice and Multicriterion Decision-Making. MIT Press, Cambridge (1986)

[2] Belton, V., Branke, J., Eskelinen, P., Greco, S., Molina, J., Ruiz, F., Słowiński, R.: Interactive Multiobjective Optimization from a Learning Perspective. In: Branke, J., Deb, K., Miettinen, K., Słowiński, R. (eds.) Multiobjective Optimization: Interactive and Evolutionary Approaches. LNCS, vol. 5252, pp. 405–433. Springer, Heidelberg (2008)

[3] Ben-Tal, A., Nemirovski, A.: Robust Optimization - Methodology and Applications. Mathematical Programming Series B 92, 453–480 (2002)

[4] Branke, J., Deb, K., Miettinen, K., Słowiński, R. (eds.): Multiobjective Optimization: Interactive and Evolutionary Approaches. LNCS, vol. 5252. Springer, Heidelberg (2008)

[5] Branke, J., Greco, S., Słowiński, R., Zielniewicz, P.: Interactive evolutionary multiobjective optimization using robust ordinal regression. In: Ehrgott, M., Fonseca, C.M., Gandibleux, X., Hao, J.-K., Sevaux, M. (eds.) EMO 2009. LNCS, vol. 5467, pp. 554–568. Springer, Heidelberg (2009)

[6] Branke, J., Kaußler, T., Schmeck, H.: Guidance in evolutionary multi-objective optimization. Advances in Engineering Software 32, 499–507 (2001)

[7] Coello Coello, C.A., Van Veldhuizen, D.A., Lamont, G.B.: Evolutionary Algorithms for Solving Multi-Objective Problems. Kluwer, Dordrecht (2002)

[8] Deb, K.: Multi-Objective Optimization using Evolutionary Algorithms. Wiley, Chichester (2001)

[9] Deb, K., Pratap, A., Agarwal, A., Meyarivan, T.: A fast and elitist multi-objective genetic algorithm: NSGA-II. IEEE Transactions on Evolutionary Computation 6, 182–197 (2002)

[10] Deb, K., Sundar, J., Udaya Bhaskara Rao, N., Chaudhuri, S.: Reference point based multi-objective optimization using evolutionary algorithms. International Journal of Computational Intelligence Research 2(3), 273–286 (2006)

[11] Deb, K., Chaudhuri, S.: I-MODE: An interactive multi-objective optimization and decision-making using evolutionary methods. Applied Soft Computing 10, 496–511 (2010)

[12] Figueira, J., Greco, S., Słowiński, R.: Building a set of additive value functions representing a reference preorder and intensities of preference: GRIP method. European Journal of Operational Research 195, 460–486 (2009)

[13] Fonseca, C.M., Fleming, P.J.: Genetic algorithms for multiobjective optimization: Formulation, discussion, and generalization. In: Proceedings of the Fifth International Conference on Genetic Algorithms, pp. 416–423 (1993)
[14] Fortemps, P., Greco, S., Słowiński, R.: Multicriteria decision support using rules that represent rough-graded preference relations. European Journal of Operational Research 188(1), 206–223 (2008)
[15] Greco, S., Matarazzo, B., Słowiński, R.: The use of rough sets and fuzzy sets in MCDM. In: Gal, T., Hanne, T., Stewart, T. (eds.) Multicriteria Decision Making: Advances in MCDM Models, Algorithms, Theory and Applications, pp. 14.1—14.59. Kluwer, Dordrecht (1999)
[16] Greco, S., Matarazzo, B., Słowiński, R.: Rough approximation of a preference relation by dominance relations. European Journal of Operational Research 117, 63–83 (1999)
[17] Greco, S., Matarazzo, B., Słowiński, R.: Rough sets theory for multicriteria decision analysis. European Journal of Operational Research 129, 1–47 (2001)
[18] Greco, S., Matarazzo, B., Słowiński, R.: Axiomatic characterization of a general utility function and its particular cases in terms of conjoint measurement and rough-set decision rules. European Journal of Operational Research 158, 271–292 (2004)
[19] Greco, S., Matarazzo, B., Słowiński, R.: Dominance-based Rough Set Approach to Interactive Multiobjective Optimization. In: Branke, J., Deb, K., Miettinen, K., Słowiński, R. (eds.) Multiobjective Optimization: Interactive and Evolutionary Approaches. LNCS, vol. 5252, pp. 121–155. Springer, Heidelberg (2008)
[20] Greco, S., Matarazzo, B., Słowiński, R.: DARWIN: Dominance-based rough set Approach to handling Robust Winning solutions in INteractive multiobjective optimization. In: Proc. 5th Intl Workshop on Preferences and Decisions, Trento, April 6-8, pp. 34–41 (2009)
[21] Greco, S., Matarazzo, B., Słowiński, R.: Dominance-Based Rough Set Approach to Decision Under Uncertainty and Time Preference. Annals of Operational Research 176, 41–75 (2010)
[22] Greco, S., Matarazzo, B., Słowiński, R.: DARWIN-PCT: Dominance-based rough set Approach to handling Robust Winning solutions in INteractive multiobjective optimization using Pairwise Comparison Tables. In: Proc. Intl Conference on Uncertainty and Robustness in Planning and Decision Making, Coimbra, Portugal, April 15-17 (2010)
[23] Greco, S., Matarazzo, B., Słowiński, R.: Dominance-Based Rough Set Approach to Decision Under Uncertainty using Pairwise Comparison Tables. In: Hüllermeier, E., Kruse, R., Hoffmann, F. (eds.) IPMU 2010. LNCS, vol. 6178, pp. 584–594. Springer, Heidelberg (2010)
[24] Greco, S., Mousseau, V., Słowiński, R.: Ordinal regression revisited: multiple criteria ranking with a set of additive value functions. European Journal of Operational Research 191, 415–435 (2008)
[25] Greco, S., Matarazzo, B., Słowiński, R.: Interactive Evolutionary Multiobjective Optimization using Dominance-based Rough Set Approach. In: Proc. of WCCI 2010 (2010) (forthcoming)

[26] Greenwood, G.W., Hu, X.S., D'Ambrosio, J.G.: Fitness functions for multiple objective optimization problems: combining preferences with Pareto rankings. In: Belew, R.K., Vose, M.D. (eds.) Foundations of Genetic Algorithms, pp. 437–455. Morgan Kaufmann, San Francisco (1997)
[27] Jaszkiewicz, A.: Interactive multiobjective optimization with the Pareto memetic algorithm. Foundations of Computing and Decision Sciences 32(1), 15–32 (2007)
[28] Jaszkiewicz, A., Słowiński, R.: The "Light Beam Search" approach - an overview of methodology and applications. European Journal of Operational Research 113(2), 300–314 (1999)
[29] Kouvelis, P., Yu, G.: Robust Discrete Optimization and its Applications. Kluwer Academic Publishers, Boston (1997)
[30] Levy, H.: Stochastic Dominance and Expected Utility: Survey and Analysis. Management Science 38(4), 555–593 (1992)
[31] Miettinen, K.: Nonlinear Multiobjective Optimization. Kluwer, Dordrecht (1999)
[32] Milnor, J.: Games against Nature. In: Thrall, R.M., Coombs, C.H., Davis, R.L. (eds.) Decision Processes, pp. 49–54. Wiley, New York (1954)
[33] Pawlak, Z.: Rough Sets. Theoretical Aspects of Reasoning about Data. Kluwer, Dordrecht (1991)
[34] Phelps, S., Köksalan, M.: An interactive evolutionary metaheuristic for multiobjective combinatorial optimization. Management Science 49(12), 1726–1738 (2003)
[35] Roy, B.: Robustness in operational research and decision aiding: A multifaceted issue. European Journal of Operational Research 200, 629–638 (2009)
[36] Słowiński, R., Greco, S., Matarazzo, B.: Axiomatization of utility, outranking and decision-rule preference models for multiple-criteria classification problems under partial inconsistency with the dominance principle. Control and Cybernetics 31, 1005–1035 (2002)
[37] Słowiński, R., Greco, S., Matarazzo, B.: Rough Set based Decision Support. In: Burke, E., Kendall, G. (eds.) Search Methodologies: Introductory Tutorials in Optimization and Decision Support Techniques, ch. 16, pp. 475–527. Springer, New York (2005)
[38] Słowiński, R., Greco, S., Matarazzo, B.: Rough Sets in Decision Making. In: Meyers, R.A. (ed.) Encyclopedia of Complexity and Systems Science, pp. 7753–7786. Springer, New York (2009)
[39] Steuer, R.: Multiple Criteria Optimization: Theory, Computation and Application. Wiley, New York (1986)
[40] Thiele, L., Miettinen, K., Korhonen, P.J., Molina, J.: A preference-based interactive evolutionary algorithm for multiobjective optimization. Evolutionary Computation Journal 17(3), 411–436 (2009)
[41] Vincke, P.: Multicriteria Decision Aid. Wiley, Chichester (1992)

Supporting Consensus Reaching Processes under Fuzzy Preferences and a Fuzzy Majority via Linguistic Summaries*

Janusz Kacprzyk and Sławomir Zadrożny

Abstract. We consider the classic approach to the evaluation of of degrees of consensus due to Kacprzyk and Fedrizzi [6], [7], [8] in which a soft degree of consensus has been introduced. Its idea is to find a degree to which, for instance, "most of the important individuals agree as to almost all of the relevant options". The fuzzy majority, expressed as fuzzy linguistic quantifiers (most, almost all, ...) is handled via Zadeh's [46] classic calculus of linguistically quantified propositions and Yager's [44] OWA (ordered weighted average) operators. The soft degree of consensus is used for supporting the running of a moderated consensus reaching process along the lines of Fedrizzi, Kacprzyk and Zadrożny [3], Fedrizzi, Kacprzyk, Owsiński and Zadrożny [2], Kacprzyk and Zadrożny [22], and [24].

Linguistic data summaries in the sense of Yager [43], Kacprzyk and Yager [13], Kacprzyk, Yager and Zadrożny [14], in particular in its protoform based version proposed by Kacprzyk and Zadrożny [23], [25] are employed. These linguistic summaries indicate in a human consistent way some interesting relations between individuals and options to help the moderator identify crucial (pairs of) individuals and/options with whom/which there are difficulties with respect to consensus. An extension using ontologies representing both knowledge on the consensus reaching process and domain of the decision problem is indicated.

Keywords: consensus, consensus reaching support, fuzzy preference, fuzzy majority, fuzzy logic, linguistic quantifier, OWA (ordered weighted averaging) operator.

1 Introduction

The paper is concerned with the problem of consensus reaching. Basically, we assume that there is a group of individuals and a set of options. The individuals express

Janusz Kacprzyk · Sławomir Zadrożny
Systems Research Institute, Polish Academy of Sciences
ul. Newelska 6, 01–447 Warsaw, Poland
e-mail: kacprzyk, zadrozny@ibspan.waw.pl

* This work was supported by the Ministry of Science and Higher Education in Poland under Grant N N519 404734.

S. Greco et al. (Eds.): Preferences and Decisions, STUDFUZZ 257, pp. 261–279.
springerlink.com

their testimonies concerning their preferences as to the particular pairs of options. It is assumed that the group's testimonies are different in the beginning, and in a step-wise process of consensus reaching, guided by a moderator, they gradually change, possibly in the direction of a fuller and fuller consensus.

First, we assume a novel approach to the definition of a soft degree of consensus introduced by Kacprzyk and Fedrizzi [6], [7], [8] in which a degree of consensus as a degree to which, for instance "*most* of the *relevant* (knowledgeable, expert, ...) individuals agree as to *almost all* of the *important* options". This degree is then used to evaluate the extent of consensus in the group.

Then, we use a general architecture for a group decision support system for supporting consensus reaching proposed by Fedrizzi, Kacprzyk and Zadrożny [3], and then further developed by Fedrizzi, Kacprzyk, Owsiński and Zadrożny [2], Kacprzyk and Zadrożny [22], [24] and Zadrożny and Kacprzyk [49] in which the role of a moderator is defined and it is shown how soft degrees of consensus mentioned above can be used to monitor the dynamics and progress of consensus reaching.

The main part of this paper is a novel use of linguistic summaries in the sense of Yager [43], or – maybe rather in its extended and implementable version – of Kacprzyk and Yager [13] or Kacprzyk, Yager and Zadrożny [14] to provide some further information as to what is "going wrong" in the consensus reaching process, what is to be paid attention to, which pairs of individuals/options may pose some problems, etc. In the end we will briefly mention how domain knowledge concerning the issues in question and the consensus reaching process as such, represented by ontologies, can be used as proposed by Kacprzyk and Zadrożny [27].

The process of group decision making, including that of the reaching of consensus, is centered on human beings, with their inherent subjectivity and imprecision in the articulation of opinions (e.g., preferences). To account for this, a predominant research direction is based on the introduction of *individual* and *social fuzzy preference relations* – cf., e.g., Nurmi [35]. Further, one can introduce a fuzzy majority to group decision making and consensus reaching as proposed first by Kacprzyk [4], [5]; for a comprehensive review, cf. Kacprzyk, Zadrożny, Fedrizzi and Nurmi [29], [30]. A fuzzy majority is meant as a soft aggregation tool and is assumed to be represented by Zadeh's [46] fuzzy linguistic quantifier, and then Yager's [44] OWA (ordered weighted average) operator.

The concept of a fuzzy majority, which is a considerable departure from the traditional non-fuzzy majority (e.g., a half, at least $\frac{2}{3}$, ...), is very relevant for our purposes, both for the new definitions of a degree of consensus and linguistic summaries to be employed to help monitor the consensus reaching process.

Fuzzy majority is commonly used by the humans, and not only in everyday discourse. A good example in a biological context may be found in Loewer and Laddaga [34]:

> ...It can correctly be said that there is a *consensus* among biologists that Darwinian natural selection is an important cause of evolution though there is currently *no consensus* concerning Gould's hypothesis of speciation. This means that there is a *widespread agreement* among biologists concerning the first matter but *disagreement* concerning the second ...

A rigid majority as, e.g., more than 75% would not evidently reflect the very essence of the above statement. It should be noted that there are naturally situations when a strict majority is necessary, for obvious reasons, as in, e.g., political elections. Anyway, the ability to accommodate a fuzzy majority in consensus formation models should help make them more human consistent hence easier implementable.

A natural manifestations of a fuzzy majority are the so-called *linguistic quantifiers* exemplified by *most, almost all, much more than a half, ...*. Though they cannot be handled by conventional logical calculi, fuzzy logic provides here simple and efficient tools, i.e. calculi of linguistically quantified propositions, notably due to Zadeh [46]. Since the fuzzy linguistic quantifiers serve in our context the purpose of an aggregation operator, Yager's [44] OWA (ordered weighted average) operators can also be used, and they provide a much needed generality and flexibility.

These fuzzy logic based calculi of linguistically quantified propositions have been applied by the authors to introduce a fuzzy majority for measuring (a degree of) consensus and deriving new solution concepts in group decision making (cf. Kacprzyk [4], [5], Kacprzyk and Fedrizzi [6], [7], [8]; cf. also works on generalized choice functions under fuzzy and non-fuzzy majorities by Kacprzyk and Zadrożny [?], [22], [25]. For a comprehensive review, see Kacprzyk, Zadrożny, Fedrizzi and Nurmi [29], [30].

The degrees of consensus proposed in those works have proved to have much conceptual and intuitive appeal. Moreover, they have been found useful and implementable in a decision support system for consensus reaching proposed by Fedrizzi, Kacprzyk and Zadrożny [3], and then further developed by Fedrizzi, Kacprzyk, Owsiński and Zadrożny [2] or Kacprzyk and Zadrożny [22], [24], and Zadrożny and Kacprzyk [49].

Basically, this degree of consensus is meant to overcome some "rigidness" of the conventional concept of consensus in which (full) consensus occurs only when "*all* the individuals agree as to *all* the issues". This may often be counterintuitive, and not consistent with a real human perception of the very essence of consensus (see, e.g., the citation from a biological context given in the beginning of this paper). The new degree of consensus can be therefore equal to 1, which stands for full consensus, when, for instance, "*most* of the (important) individuals agree as to *almost all* (of the relevant) options". This new degree of consensus has been proposed by Kacprzyk and Fedrizzi [6], [7], [7] using Zadeh's [46] calculus of linguistically quantified propositions. Then, Fedrizzi, Kacprzyk and Nurmi [1] have proposed to use the OWA (ordered weighted average) operators instead of Zadeh's calculus. It works well though some deeper works on the semantics of the OWA operators is relevant as shown by Zadrożny and Kacprzyk [50].

For clarity, and to provide a point of departure for our further discussion, we will first review basic elements of Zadeh's calculus of linguistically quantified propositions. Then, a relation between this calculus and the OWA operators is shown, and finally we proceed to the reformulation of degrees of consensus proposed by the authors in terms of the OWA operators.

2 Linguistic Quantifiers and the OWA (Ordered Weighted Averaging) Operators

In this section we will briefly present Zadeh's [46] calculus of linguistically quantified proposition which is employed to deal with fuzzy linguistic quantifiers, and Yager's [44] OWA (ordered weighted averaging) operators which are meant here to provide a flexible aggregation, notably a fuzzy linguistic quantifier driven.

2.1 Linguistic Quantifiers and a Fuzzy Logic Based Calculus of Linguistically Quantified Propositions

A *linguistically quantified proposition* may be exemplified by "most individuals are convinced", and may be generally written as

$$Qy\text{'s are } F \tag{1}$$

where Q is a *linguistic quantifier* (e.g., most), $\mathcal{Y} = \{y\}$ is a *set of objects* (e.g., individuals), and F is a *property* (e.g., convinced). Importance can be added leading to "QBY's are F", but this will not be considered here.

For our purposes, the main problem is how to find the truth of such a linguistically quantified proposition, i.e. truth(Qy's are F) knowing truth(y is F), $\forall y \in \mathcal{Y}$, which can be done by using mainly Zadeh's [46] calculus.

It is assumed that property F is a fuzzy set in Y, truth(y_i is F) $= \mu_F(y_i)$, $\forall y_i \in Y = \{y_1, \ldots, y_p\}$, and a linguistic quantifier Q is represented as a fuzzy set in $[0,1]$ as, e.g.,

$$\mu_{\text{"}most\text{"}}(x) = \begin{cases} 1 & \text{for } x \geq 0.8 \\ 2x - 0.6 & \text{for } 0.3 < x < 0.8 \\ 0 & \text{for } x \leq 0.3 \end{cases} \tag{2}$$

Then

$$\text{truth}(Qy\text{'s are } F) = \mu_Q(\frac{1}{p}\sum_{i=1}^{p} \mu_F(y_i)) \tag{3}$$

2.2 The OWA (Ordered Weighted Average) Operators

The OWA (ordered weighted average) operators (cf. Yager [44]) provide an alternative and attractive means for a linguistic quantifier driven aggregation.

An *OWA (ordered weighted average) operator* of dimension n is a mapping $H : [0,1]^n \to [0,1]$ if associated with H is a weighting vector $W = [w_i]^T$ such that: $w_i \in [0,1]$, $w_1 + \cdots + w_n = 1$, and

$$H(x_1, \ldots, x_n) = w_1 b_1 + \cdots + w_n b_n \tag{4}$$

where b_i is the i-th largest element among $\{x_1, \ldots, x_n\}$. B is called an ordered argument vector if each $b_i \in [0,1]$, and $j > i$ implies $b_i \geq b_j$, $i = 1, \ldots, n$.

Then

$$H(x_1, \ldots, x_n) = W^T B \tag{5}$$

Example 1. Let $W^T = [0.2\ 0.3\ 0.1\ 0.4]$, and calculate $H(0.6, 1.0, 0.3, 0.5)$. Thus, $B^T = [1.0\ 0.6\ 0.5\ 0.3]$, and $H(0.6, 1.0, 0.3, 0.5) = W^T B = 0.55$.

Some hints as to how to determine the w_i's are given in Yager [44]. For our purposes relations between the OWA operators and fuzzy linguistic quantifiers are relevant. Basically, under some mild assumptions (cf. Yager [44], Yager and Kacprzyk [45]), a linguistic quantifier Q has the same properties as the H aggregation function, so that it is our conjecture that the weighting vector W is a manifestation of a quantifier underlying the process of aggregation of pieces of evidence.

Then, as proposed by Yager [44],

$$w_k = \mu_Q\left(\frac{k}{n}\right) - \mu_Q\left(\frac{k-1}{n}\right), k = 1, \ldots, n \tag{6}$$

For instance:

1. if $w_n = 1$, and $w_i = 0, \forall i \neq n$, then this corresponds to Q = all;
2. if $w_1 = 1$, and $w_i = 0, \forall i \neq 1$, then this corresponds to Q = at least one.

The intermediate cases, which correspond to, e.g., *a half, most, much more than 75%, a few, almost all, ...* may be therefore obtained by a suitable choice of the w_i's between the above two extremes.

The OWA operators are therefore an interesting and promising class of aggregation operators that can provide a linguistic quantifier driven aggregation.

3 Degrees of Consensus under Fuzzy Preferences and a Fuzzy Majority

Suppose that we have a set of n options, $O = \{o_1, \ldots, o_n\}$, and a set of m individuals, $E = \{e_1, \ldots, e_m\}$. Each individual e_k provides his or her *individual fuzzy preference relation*, P_k, given by its membership function $\mu_{P_k} : O \times O \to [0, 1]$ which, if card O is small enough, may be represented by a matrix $[r^k_{ij}]$ such that $r^k_{ij} = \mu_{P_k}(o_i, o_j)$; $i, j = 1, \ldots, n$; $k = 1, \ldots, m$; $r^k_{ij} + r^k_{ij} = 1$.

The degree of consensus is now derived in three steps. First, for each pair of individuals we derive a degree of agreement as to their preferences between all the pair of options, next we aggregate these degrees to obtain a degree of agreement of each pair of individuals as to their preferences between $Q1$ (a fuzzy linguistic quantifier as, e.g., *most, almost all, much more than 50%, ldots*) pairs of options, and, finally, we aggregate these degrees to obtain a degree of agreement of $Q2$ (another fuzzy linguistic quantifier) pairs of individuals as to their preferences between $Q1$ pairs of options. This is meant to be the degree of consensus sought. For simplicity, we will use below the OWA operators to indicate a linguistic quantifier driven aggregation though they can readily be replaced by Zadeh's calculus.

We start with the degree of (strict) agreement between individuals e_{k1} and e_{k2} as to their preferences between options o_i and o_j

$$v_{ij}(k1,k2) = \begin{cases} 1 \text{ if } r_{ij}^{k1} = r_{ij}^{k2} \\ 0 \text{ otherwise} \end{cases} \quad (7)$$

where: $k1 = 1,\ldots,m-1$; $k2 = k1+1,\ldots,m$; $i = 1,\ldots,n-1$; and $j = i+1,\ldots,n$.

The degree of agreement between individuals $k1$ and $k2$ as to their preferences between all the pairs of options is

$$v(k1,k2) = \frac{2}{n(n-1)} \sum_{i=1}^{n-1} \sum_{j=i+1}^{n} v_{ij}(k1,k2) \quad (8)$$

The degree of agreement between individuals $k1$ and $k2$ as to their preferences between $Q1$ pairs of options is

$$v_{Q1}(k1,k2) = \text{OWA}_{Q1}(\{v_{ij}(k1,k2)\}_{1 \leq i < j \leq n}) \quad (9)$$

where $\text{OWA}_{Q1}(.)$ is the aggregation of $v_{ij}(k1,k2)$'s with respect to $Q1$ via the OWA operator of dimension $\frac{n(n-1)}{2}$ as shown in Section 2.2.

In turn, the degree of agreement of all the pairs of individuals as to their preferences between $Q1$ pairs of options is

$$v_{Q1} = \frac{2}{m(m-1)} \sum_{k1=1}^{m-1} \sum_{k2=k1+1}^{m} (v_{Q1}(k1,k2)) \quad (10)$$

and, finally, the degree of agreement of $Q2$ pairs of individuals as to their preferences between $Q1$ pairs of options, called the *degree of Q1/Q2 - consensus* is

$$\text{con}(Q1,Q2) = \text{OWA}_{Q2}(\{v_{Q1}(k1,k2)\}_{1 \leq k1 < k2 \leq m}) \quad (11)$$

where $OWA_{Q2}(.)$ is defined similarly as $OWA_{Q1}(.)$.

Since the strict agreement (7) may be viewed too rigid, we can use the degree of *sufficient agreement* (at least to degree $\alpha \in [0,1]$) of individuals e_{k1} and e_{k2} as to their preferences between options o_i and o_j , as well as the the degree of strong agreement of individuals $k1$ and $k2$ as to their preferences between options s_i and s_j, obtaining the degree of $\alpha/Q1/Q2$ - consensus and $s/Q1/Q2$ - consensus, respectively (cf. Kacprzyk and Fedrizzi [6], [7], [8]).

An important issue is the addition of the importance of individuals and the relevance of options. This is nontrivial a problem which requires a deeper analysis, in particular in the context of the OWA operators (cf. Zadrożny and Kacprzyk [50]).

We have therefore some human consistent means for the evaluation of a degree of consensus, and now we will provide some extra tools to support the consensus reaching process.

4 A Consensus Reaching Process

We consider the consensus reaching process in the setting proposed along the lines of Fedrizzi, Kacprzyk and Zadrożny [3], Fedrizzi, Kacprzyk, Owsiński and Zadrożny [2], Kacprzyk and Zadrożny [22], [24], and Zadrożny and Furlani [48].

Basically, there is a group of individuals and a moderator whose task is to effectively and efficiently run the consensus reaching session. The individuals and the moderator exchange information and opinions, provide argumentation, operating in a network as shown in Figure 1.

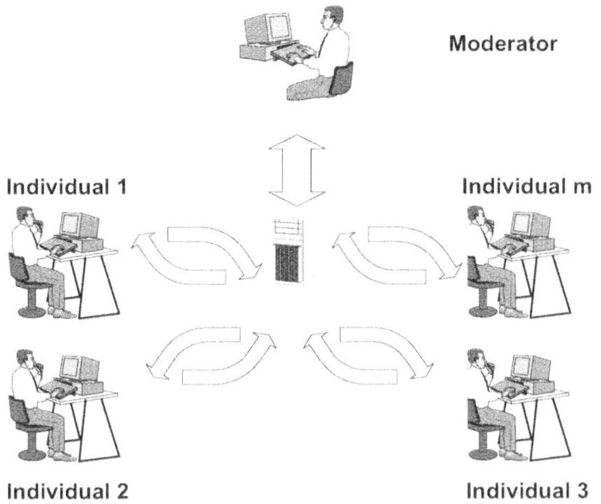

Fig. 1 Individuals and a moderator in a consensus reaching session

The consensus reaching has a dynamic character which can be depicted ans in Figure 2 to be meant as follows. In the beginning of the consensus reaching session, at $t = 0$, the individuals present their testimonies, i.e. their initial fuzzy preference relations, which may differ from each other to a large extent. The moderator tries to persuade them to change their preference relations using some argumentation. If the individuals are rationally committed to reaching consensus, they are willing to change their testimonies to get possibly closer to consensus.

To be more specific, let us repeat the basic setting. We assume that we have a set of m individuals $E = \{e_1, \ldots, e_m\}$ whose testimonies concerning a set of n options (alternatives) $O = \{o_1, \ldots, o_n\}$ are *individual fuzzy preference relations*. A moderator stimulates an exchange of information, rational argument, discussion, creative thinking, clarification of positions, etc. These should eventually lead to a change of the individual fuzzy preference relations. If the individuals are rationally committed to consensus, such a change usually occurs, and they get closer to consensus. Some individuals, even if willing to stick to their original preferences, can accept consensual preferences of the group provided their arguments has been heard and discussed. Thus, their acceptance of consensus may be viewed as a change of

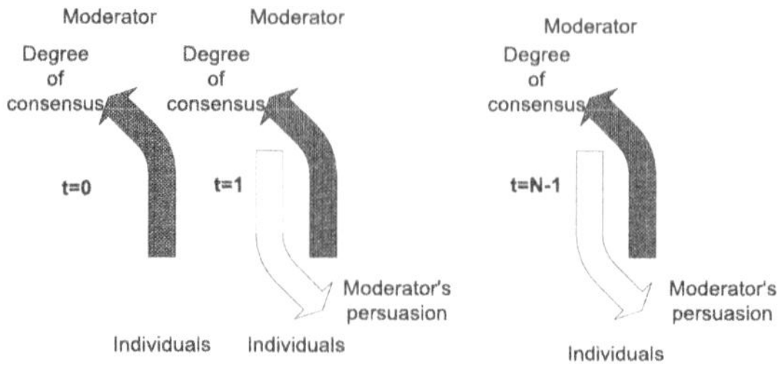

Fig. 2 Dynamics of the consensus reaching process

their preferences. This is repeated until the group gets sufficiently close to consensus, i.e. until the individual fuzzy preference relations become similar enough, or until some time limit is reached.

The moderator's job may be however difficult. First of all, in non-trivial situations the groups may include many individuals, and the number of options can be even higher. This all can make it difficult to grasp the very contents of all the individual fuzzy preference relations and dynamics of their possible changes.

Therefore, the moderator should be somehow supported to make his job easier, more efficient and faster. There may be many solutions adopted in this respect, like an effective and efficient human-computer interface, enhanced communication capabilities, advanced presentation tools for the visualization or verbalization of results obtained, etc.

In this work we propose to use a novel, natural language based support that is based on the verbalization of results obtained by using linguistic summaries of data in the sense of Yager [43], but in their implementable and extended version proposed by Kacprzyk and Yager [13] and Kacprzyk Yager and Zadrożny [14]. Even more so, we use here linguistic data summaries in the sense of the recent papers by Kacprzyk and Zadrożny [23], in which a protoform based analysis was presented, and in Kacprzyk and Zadrożny [28], in which an extremely powerful and far reaching relation to natural language generation (cf. Reiter and Dale [37]) was shown.

The rationale is that consensus reaching may be a lengthy process and its support may be greatly enhanced by various objective indicators summarizing various aspects of the process, notably preferences of individuals during a discussion, cf. [47]. One should however bear in mind that though consensus reaching needs, to be effective and efficient, a comprehensive decision support system, our aim is much more moderate. We focus on the reaching of consensus in a group with respect to directly expressed preferences on a set of alternatives.

Some guides for consensus reaching [51] list the following stages of this process:

1. brainstorming,
2. discussion,
3. clarification of interests,
4. criteria identification,
5. options reduction,
6. another round of discussion,

7. test for consensus (if negative, get back to Step 5), and
8. a formal adoption of a consensual solution.

In this paper we will deal with tools that can be of use for virtually all of the steps mentioned above.

5 A Concept of a Linguistic Data Summary

A linguistic data summary is meant as a natural language like sentence that subsumes the very essence (from a certain point of view) of a (numeric) set of data, too large to be comprehensible by humans. The original Yagers approach to the linguistic summaries (cf. Yager [43], Kacprzyk and Yager [13], Kacprzyk, Yager and Zadrożny [14] and Kacprzyk and Zadrożny [23]) may be expressed as follows: $Y = \{y_1, \dots, y_n\}$ is a set of objects, $A = \{A_1, \dots, A_m\}$ is a set of attributes characterizing objects from Y, and $A_j(y_i)$ denotes a value of attribute A_j for object y_i.

A linguistic summary of set Y consists of:

- a summarizer S, i.e. an attribute together with a linguistic value (label) defined on the domain of attribute A_j;
- a quantity in agreement Q, i.e. a linguistic quantifier (e.g. most);
- truth (validity) T of the summary, i.e. a number from the interval $[0, 1]$ assessing the truth (validity) of the summary (e.g., 0.7),

and, optionally, a qualifier R may occur, i.e. another attribute together with a linguistic value (label) defined on the domain of attribute A_k determining a (fuzzy subset) of Y.

In our context we may identify objects with individuals and their attributes with their preferences over various pairs of options. Then, the linguistic summary may be exemplified by

$$T(Most \text{ of individuals prefer option } o_1 \text{ to } o_2) = 0.7 \tag{12}$$

A richer form of the summary may include a qualifier as in, e.g.,

$$T(Most \text{ of } important \text{ individuals prefer option } o_1 \text{ to } o_2) = 0.7 \tag{13}$$

Thus, the core of a linguistic summary is a *linguistically quantified proposition* in the sense of Zadeh [46], briefly presented in Section 2.1. Thus, both linguistic summaries (12) and (13) may be written in a more general form as:

$$Qy\text{'s are } S \tag{14}$$

$$QRy\text{'s are } S \tag{15}$$

or more conveniently as

$$Qy\text{'s are } (R, S) \tag{16}$$

Then, T, i.e. its truth (validity), directly corresponds to the truth value of (14) or (15) which may be calculated by using mainly either Zadehs [46] calculus of linguistically quantified statements or the OWA operators (cf. Yager [44]).

Using Zadeh's [46] fuzzy logic based calculus of linguistically quantified propositions, a (proportional, nondecreasing) linguistic quantifier Q is assumed to be a fuzzy set in $[0, 1]$. Then, the truth values (from $[0, 1]$) of (14) and (15) are calculated, respectively, as:

$$\text{truth}(Qy\text{'s are } S) = \mu_Q[\frac{1}{n}\sum_{i=1}^{n} \mu_S(y_i)] \tag{17}$$

$$\text{truth}(QRy\text{'s are } S) = \mu_Q[\frac{\sum_{i=1}^{n}(\mu_R(y_i) \wedge \mu_S(y_i))}{\sum_{i=1}^{n} \mu_R(y_i)}] \tag{18}$$

where "$\wedge$" (minimum) can be replaced by, e.g., a t-norm.

The fuzzy predicates S and R are assumed here to be of a simplified, atomic form referring to just one attribute. They can be extended to cover more sophisticated summaries involving some confluence of various attribute values as, e.g, *young and well paid*. Clearly, the most interesting are non-trivial, human-consistent summarizers (concepts) as, e.g.: productive workers, difficult orders, etc. Their definition may require a complicated combination of attributes, a hierarchy (not all attributes are of the same importance for a concept in question), the attribute values are ANDed and/or ORed, k out of n, most, ... of them should be accounted for, etc.

Notice that the concept of a linguistic summary is closely related to the definitions of degrees of consensus discussed though it was more convenient to consider there the degrees of consensus in a "separate" setting, as linguistically quantified propositions. However, the setting of linguistic data summaries will be more convenient for our discussion of how some additional information (or knowledge) can be used for helping the moderator run a consensus reaching session.

6 Helping the Moderator Run a Consensus Reaching Session Using Linguistic Data Summaries

To reach consensus in our context means that most individuals are ready to change their original preference matrices in accordance with a consensual one. Thus, an important component of a consensus reaching support system is a set of indicators assessing how far the group is from consensus, what are the obstacles in reaching consensus, which preference matrix may be a candidate for a consensual one, etc. These indicators may be treated as some data summaries.

The original definition of a degree of consensus employed here is the degree to which "*Most* of the *important* individuals agree in their preferences as to *almost all* of the *important options*" which may be more formally expressed as follows:

$$Qh\text{'s are } (B', Qq\text{'s are } (I', \text{sim}(p_q^{h_1}, p_q^{h_2})) \tag{19}$$

where: $h \in E \times E$ is a pair of individuals, B' represents importance of a pair of individuals (related to B, an importance of particular individuals), $q \in O \times O$ is a pair of options, I' represents importance of a pair of options (related to I, an importance of particular options), $p_q^{h_i}$ is a preference degree of individual i of pair h for pair of options q, and sim$(\cdot, \cdot)$ is a measure of similarity between two preference degrees.

This definition is an example of a nested linguistic summary defined for the space of pairs of individuals and options. In what follows we propose the use of other linguistic summaries defined over various spaces. Anyway, summarizer S and qualifier R are composed of features of either individuals or options (depending on the perspective adopted; cf. subsections below) and fuzzy values (labels) expressing degree of preferences or importance weights of individuals/options.

6.1 Individuals as Objects

The objects of a linguistic summary may be identified with individuals and their attributes are preference degrees for particular pairs of options, as well as importance degrees of the individuals. Formally, referring to Section 5, we have:

$$Y = E \tag{20}$$

and

$$A = \{\mathcal{P}_{ij}\} \cup \{\mathcal{B}\} \tag{21}$$

where attributes $\mathcal{P}_{ij}$ correspond to preference degrees over pairs of options (o_i, o_j) and $\mathcal{B}$ represents the importance.

Then, the following types of summaries may be useful for consensus reaching session guidance.

Consensus indicating/building summary

It corresponds to a flexible definition of consensus previously proposed (cf. (19)) that states that *most of the individuals express similar preferences*, for instance "Most individuals definitely prefer o_{i1} to o_{i2}, moderately prefer o_{i3} to o_{i4}, ...", etc. formally written as

$$Qe_k(p^k_{i1,i2} = definite) \wedge (p^k_{i3,i4} = moderate) \wedge \ldots \tag{22}$$

If the list of conjuncts is long enough, then the truth of (22) means that there is a consensus among the individuals as to their preferences.

Thus, this type of summary may be used as another definition of consensus. Similarly to (19), it may be equipped with importance weights of individuals and/or options.

If the list of conjuncts is short, such a summary may be treated as a suggestion for building consensus. Namely, it indicates opinions that are shared by the group of individuals. Thus, they may be either further discussed to extend the common understanding in the group or assumed as agreed making the rest of the discussion focused on the remaining issues.

Discussion targeting summaries

They may be used to direct a further discussion in the group, for instance, may disclose some patterns of understanding, and may be exemplified by:

> Most individuals definitely preferring o_{i1} to o_{i2} also definitely prefer o_{i3} to o_{i4}

to be formally expressed as

$$Qe_k(p^k_{i1,i2} = definite, p^k_{i3,i4} = moderate) \tag{23}$$

The discovery of association expressed with such a summary may trigger a further discussion enabling a better understanding of the decision problem.

Option choice oriented summaries

So far we have not assumed much about the goal of discussion in the setting within a group. Thus, basically, the goal is to agree upon the content of the matrices of preferences. Usually, however, the aim of the discussion is to select either an option or a set of options preferred by the group. In such a case an agreement as to the preferences in respect to *all* or even *most* pairs of options may be unnecessary. In order to generate summaries taking that into account we have to assume a working definition of the concept of an option preferred by an individual as implied by his fuzzy preference relation. To this aim we can apply some *choice functions* as considered by the current authors, e.g., in Kacprzyk and Zadrożny [**?**], [22], [25]. They are based on the concept of the classical choice function, C, that may be defined in a slightly simplified general form as:

$$C(S, P) = S_0, \quad S_0 \subseteq S \tag{24}$$

that may be exemplified by

$$C(S, P) = \{o_i \in O : \forall_{i \neq j} P(o_i, o_j)\} \tag{25}$$

where P denotes a classical crisp preference relation.

In the case of a fuzzy preference relation, we assume C to be a fuzzy set of chosen options defined as:

$$\mu_C(o_i) = \min_j \mu_P(o_i, o_j) \tag{26}$$

which may lead to a more flexible formula by replacing the strict min operator with a linguistic quantifier Q (e.g., "most") yielding:

$$\mu_C(o_i) = T(Qo_j \; P(o_i, o_j)) \tag{27}$$

where, this time, P denotes a fuzzy preference relation.

For our further discussion the actual form of the choice function is not important, and we assume (27). In fact, it is possible that each individual adopts a different choice function, and hence a choice function assigned to each individual is denoted as C_k. For a related deep analysis of choice function in our setting, cf. Kacprzyk and Zadrożny [?], [22], [25].

Now, we can define a linguistic summary selecting a set of collectively preferred options, for instance as:

Most individuals choose options $o_{i1}, o_{i2}, \ldots$

to be formally expressed as, e.g.,

$$Qe_k\ (\mu_{C_k}(o_{i1}) = high) \wedge (\mu_{C_k}(o_{i2}) = very\ high) \wedge \ldots \tag{28}$$

where membership degrees to a choice set are discretized and expressed using linguistic labels.

The options referred to in such a summary qualify as a consensus solution if the goal of the group is to arrive at a subset of collectively preferred options. Then, such a summary is an alternative indicator of consensus.

On the other hand, a summary exemplified by

Most individuals reject options $o_{i1}, o_{i2}, \ldots$

to be formally expressed as, e.g.,

$$Qe_k\ (\mu_{C_k}(o_{i1}) = low) \wedge (\mu_{C_k}(o_{i2}) = very\ low) \wedge \ldots \tag{29}$$

make it possible to exclude these options from a further consideration, i.e., supports Step 5 of the consensus reaching procedure given in Section 4.

Therefore, by using the concept of a choice function we can get quite a practical definition of a consensus degree. Namely, both (19) and (22) refer to the preferences of the individuals over all pairs of options, possibly with importance weights. However, these importance weights are set independently of the current "standing" of the options implied by preference relations. A more practical definition should put more emphasis on preferences related to options preferred by individuals and less on those rejected by them. Thus, the importance weights of pairs of options in (19) may be assumed as:

$$\mu_{B'_{kl}}(o_i, o_j) = f(\mu_{C_k}(o_i), \mu_{C_l}(o_i), \mu_{C_k}(o_j), \mu_{C_l}(o_j)) \tag{30}$$

that is, importance weights of pairs of options are specific for each pair of individuals. Function f may be exemplified by a simple arithmetic average.

6.2 Options as Objects

Objects of linguistic summaries may also be equated with options and, then, their attributes are preference degrees over other options as expressed by particular individuals adding, possibly, importance degrees of the options. Formally, we have:

$$Y = O \tag{31}$$

and

$$A = \{\mathcal{P}_{ij}^k\} \cup \{\mathcal{I}\} \tag{32}$$

where attributes $\mathcal{P}_{ij}^k$ correspond to preference degrees over other options and $\mathcal{I}$ represents importance.

This perspective may give an additional insight into the structure of preferences of both the entire group and particular individuals. For example, a summary:

Most options are dominated by option o_i in opinion of individual k

formally expressed as, e.g.,

$$Qo_j \; p_{ij}^k = definite \tag{33}$$

directly corresponds to the choice function mentioned earlier. Namely, if such a summary is valid, then it means that option o_i belongs to the choice set of individual k. On the other hand, a summary like:

Most options are dominated by option o_i in opinion of individual $k1$, $k2$, ...

formally expressed as, e.g.,

$$Qo_j \; (p_{ij}^{k1} = definite) \wedge (p_{ij}^{k2} = definite) \wedge \ldots \tag{34}$$

indicates option o_i as a candidate for a consensual solution.

Interesting patterns in the group may be grasped via linguistic summaries exemplified by:

Most options dominating option o_i in opinion of individual $k1$ also dominate option o_i in opinion of individual $k2$

to be formally expressed as, e.g.,

$$Qo_j \; (p_{ji}^{k1} = definite, p_{ji}^{k2} = definite) \tag{35}$$

Such a summary indicates a similarity of preferences of individuals $k1$ and $k2$. This similarity is here limited to just a pair of options but may be much more convincing in case of the following summary:

$$Qo_j \; (p_{ji_1}^{k1} = definite \wedge p_{ji_2}^{k1} = definite \wedge \ldots \; ,$$
$$p_{ji_1}^{k2} = definite \wedge p_{ji_2}^{k2} = definite \wedge \ldots)$$

Another view may be obtained assuming a different set of attributes for options. Namely, we may again employ the concept of a choice set and characterize each option o_i by a vector:

$$[\mu_{C_1}(o_i), \mu_{C_2}(o_i), \ldots, \mu_{C_m}(o_i)] \tag{36}$$

Then, a summary like

Most options are preferred by individual e_k

formally expressed as, e.g.,

$$Qo_i \, \mu_{C_k}(o_i) = high \tag{37}$$

indicates individual e_k as being rather indifferent in his/her preferences, while a summary like

Most options are rejected by individual e_l

formally expressed as, e.g.,

$$Qo_i \, \mu_{C_l}(o_i) = low \tag{38}$$

suggests that individual e_l exposes a clear preference towards a limited subset of options.

The second representation of options as objects may be seen as a kind of compression of the first. Namely, for a given option o_i all p^k_{ij}'s related to individual e_k which represent o_i in (32) are compressed into one number $\mu_{C_k}(o_i)$ in (36), i.e.,

$$[p^k_{i1}, p^k_{i2}, \ldots, p^k_{in}] \longrightarrow \mu_{C_k}(o_i) \tag{39}$$

Another compression is possible by aggregating, for a given option o_i, all p^k_{ij}'s related to option o_j which represent o_i in (32) into one number, i.e.,

$$[p^1_{ij}, p^2_{ij}, \ldots, p^m_{ij}] \longrightarrow \text{aggr}(p^k_{ij})_{k=1,m} \tag{40}$$

The aggregation operator, denoted with aggr, may take various forms, including a linguistic quantifier guided aggregation. The representation of options as objects obtained thus far may be used to generate summaries with interpretations similar to (36), but with slightly different semantics. The difference is related to the *direct* and *indirect* approaches to group decision making as discussed in Kacprzyk [4], [5], and Zadrożny [47].

This subsumes some basic possible verbalized types of an additional information, which is based on linguistic summaries, that can be of a great help in supporting the moderator to effectively and efficiently run a consensus reaching session.

7 Concluding Remarks

The purpose of the paper was to present an extended, and a more unified and comprehensive approach to generate linguistic summaries of the "state of the matter" in the consensus reaching process run in a group of individuals by a moderator. We have presented first the concept of a soft degree of consensus in the setting of fuzzy preference relations and a fuzzy majority that is expressed as a linguistically quantified proposition, that is practically equivalent to some linguistic data summary.

Then, we have shown some other linguistically quantified propositions that are linguistic summaries of various relations between the individuals and their preferences over the set of options. These linguistic summaries may give an extraordinary insight into what is the present "state of the mind" of the group, and which paths (related to changes of testimonies of various individuals with respect to various options) may be promising for getting closer to consensus.

It should be noted that we have used our protoform based approach to linguistic data summaries as shown in Kacprzyk and Zadrożny [23] which provides a powerful general framework and also, as recently shown in Kacprzyk and Zadrożny [28] can make the use of tools and software developed in natural language generation (NLG) possible which may greatly simplify implementations.

Finally, one should mention that the inclusion of some additional representations of knowledge concerning both the very essence of a particular consensus reaching process and the domain in which it proceeds, as proposed in a different setting by Kacprzyk and Zadrożny [26], can be interesting for future works. Quite important for a deeper analysis of consensus reaching processes, and a more effective and efficient support for running them, may be an extension towards the use of linguistic summaries of trends in consensus reaching as proposed by Kacprzyk, Zadrożny and Wilbik [32], or towards more general representations of fuzzy preferences and majorities,exemplified by intuitionistic fuzzy sets as proposed by Szmidt and Kacprzyk [38], [39], [40], [41].

References

1. Fedrizzi, M., Kacprzyk, J., Nurmi, H.: Consensus degrees under fuzzy majorities and fuzzy preferences using OWA (ordered weighted average) operators. Control and Cybernetics 22, 71–80 (1993)
2. Fedrizzi, M., Kacprzyk, J., Owsińnski, J.W., Zadrożny, S.: Consensus reaching via a GDSS with fuzzy majority and clustering of preference profiles. Annals of Operations Research 51, 127–139 (1994)
3. Fedrizzi, M., Kacprzyk, J., Zadrożny, S.: An interactive multi - user decision support system for consensus reaching processes using fuzzy logic with linguistic quantifiers. Decision Support Systems 4, 313–327 (1988)
4. Kacprzyk, J.: Group decision - making with a fuzzy majority via linguistic quantifiers. Part I: A consensory - like pooling; Part II: A competitive - like pooling. Cybernetics and Systems: an Int. Journal 16, 119–129 (Part I), 131–144 (Part II) (1985)
5. Kacprzyk, J.: Group decision making with a fuzzy linguistic majority. Fuzzy Sets and Systems 18, 105–118 (1986)
6. Kacprzyk, J., Fedrizzi, M.: 'Soft' consensus measures for monitoring real consensus reaching processes under fuzzy preferences. Control and Cybernetics 15, 309–323 (1986)
7. Kacprzyk, J., Fedrizzi, M.: A 'soft' measure of consensus in the setting of partial (fuzzy) preferences. European Journal of Operational Research 34, 315–325 (1988)
8. Kacprzyk, J., Fedrizzi, M.: A 'human - consistent' degree of consensus based on fuzzy logic with linguistic quantifiers. Mathematical Social Sciences 18, 275–290 (1989)
9. Kacprzyk, J., Fedrizzi, M. (eds.): Multiperson Decision Making Models Using Fuzzy Sets and Possibility Theory. Kluwer, Dordrecht (1990)

10. Kacprzyk, J., Nurmi, H.: On fuzzy tournaments and their solution concepts in group decision making. European Journal of Operational Research 51, 223–232 (1991)
11. Kacprzyk, J., Fedrizzi, M., Nurmi, H.: Group decision making and consensus under fuzzy preferences and fuzzy majority. Fuzzy Sets and Systems 49, 21–31 (1992)
12. Kacprzyk, J., Wilbik, A., Zadrożny, S.: Linguistic summarization of time series using a fuzzy quantifier driven aggregation. Fuzzy Sets and Systems 159(12), 1485–1499 (2008)
13. Kacprzyk, J., Yager, R.R.: Linguistic summaries of data using fuzzy logic. International Journal of General Systems 30, 33–154 (2001)
14. Kacprzyk, J., Yager, R.R., Zadrożny, S.: A fuzzy logic based approach to linguistic summaries of databases. International Journal of Applied Mathematics and Computer Science 10, 813–834 (2000)
15. Kacprzyk, J., Zadrożny, S.: On a fuzzy querying and data mining interface. Kybernetika 36, 657–670 (2000)
16. Kacprzyk, J., Zadrożny, S.: Data mining via linguistic summaries of databases: an interactive approach. In: Ding, L. (ed.) A New Paradigm of Knowledge Engineering by Soft Computing, pp. 325–345. World Scientific, Singapore (2001)
17. Kacprzyk, J., Zadrożny, S.: Computing with words in decision making through individual and collective linguistic choice rules. International Journal of Uncertainty Fuzziness and Knowledge Based Systems 9, 89–102 (2001)
18. Kacprzyk, J., Zadrożny, S.: Computing with words in decision making through individual and collective linguistic choice rules. International Journal of Uncertainty, Fuzziness and Knowledge - Based Systems 9, 89–102
19. Kacprzyk, J., Zadrożny, S.: Collective choice rules in group decision making under fuzzy preferences and fuzzy majority: a unified OWA operator based approach. Control and Cybernetics 31, 937–948 (2002)
20. Kacprzyk, J., Zadrożny, S.: Collective choice rules in group decision making under fuzzy preferences and fuzzy majority: a unified OWA operator based approach. Control and Cybernetics 31, 937–948 (2002)
21. Kacprzyk, J., Zadrożny, S.: Dealing with imprecise knowledge on preferences and majority in group decision making: towards a unified characterization of individual and collective choice functions. Bulletin of the Polish Academy of Sciences (Tech. Sci.) 51, 279–302 (2003)
22. Kacprzyk, J., Zadrożny, S.: An Internet-based group decision support system. Management VII(28), 4–10 (2003)
23. Kacprzyk, J., Zadrożny, S.: Computing with words in intelligent database querying: standalone and Internet-based applications. Information Sciences 134, 71–109 (2005)
24. Kacprzyk, J., Zadrożny, S.: On a concept of a consensus reaching process support system based on the use of soft computing and Web techniques. In: Ruan, D., Montero, J., Lu, J., Martínez, L., D'hondt, P., Kerre, E.E. (eds.) Computational Intelligence in Decision and Control, pp. 859–864. World Scientific, Singapore (2008)
25. Kacprzyk, J., Zadrożny, S.: Towards a general and unified characterization of individual and collective choice functions under fuzzy and nonfuzzy preferences and majority via the ordered weighted average operators. International Journal of Intelligent Systems 24(1), 4–26 (2009)
26. Kacprzyk, J., Zadrożny, S.: Protoforms of linguistic database summaries as a human consistent tool for using natural language in data mining. International Journal of Software Science and Computational Intelligence 1(1), 100–111 (2009)
27. Kacprzyk, J., Zadrożny, S.: Soft computing and Web intelligence for supporting consensus reaching. Soft Computing (2010), doi:10.1007/s00500-009-0475-4

28. Kacprzyk, J., Zadrożny, S.: Computing with words is an implementable paradigm: fuzzy queries, linguistic data summaries and natural language generation. IEEE Transactions on Fuzzy Systems (in press, 2010a)
29. Kacprzyk, J., Zadrożny, S., Fedrizzi, M., Nurmi, H.: On group decision making, consensus reaching, voting and voting paradoxes under fuzzy preferences and a fuzzy majority: a survey and a granulation perspective. In: Pedrycz, W., Skowron, A., Kreinovich, V. (eds.) Handbook of Granular Computing, pp. 906–929. Wiley, Chichester (2008)
30. Kacprzyk, J., Zadrożny, S., Fedrizzi, M., Nurmi, H.: On group decision making, consensus reaching, voting and voting paradoxes under fuzzy preferences and a fuzzy majority: a survey and some perspectives. In: Bustince, H., Herrera, F., Montero, J. (eds.) Fuzzy Sets and Their Extensions: Representation, Aggregation and Models, pp. 263–295. Springer, Heidelberg (2008)
31. Kacprzyk, J., Zadrożny, S., Raś, Z.: Action rules in consensus reaching process support. In: Proceedings of ISDA 2009 - The 9th International Conference on Intelligent Systems Design and Applications, Pisa, pp. 827–832. IEEE Press, Los Alamitos (2009)
32. Kacprzyk, J., Zadrożny, S., Wilbik, A.: Linguistic summarization of some static and dynamic features of consensus reaching. In: Reusch, B. (ed.) Computational Intelligence, Theory and Applications, pp. 19–28. Springer, Heidelberg (2006)
33. Loewer, B. (Guest ed.): Special Issue on Consensus. Synthese 62(1) (1985)
34. Loewer, B., Laddaga, R.: Destroying the consensus. In: Loewer, B. (Guest ed.), pp. 79–96 (1985)
35. Nurmi, H.: Approaches to collective decision making with fuzzy preference relations. Fuzzy Sets and Systems 6, 187–198 (1981)
36. Nurmi, H., Kacprzyk, J., Fedrizzi, M.: Probabilistic, fuzzy and rough concepts in social choice. European Journal of Operational Research 95, 264–277 (1996)
37. Reiter, E., Dale, R.: Building Natural Language Generation Systems. Cambridge University Press, Cambridge (2000)
38. Szmidt, E., Kacprzyk, J.: Intuitionistic fuzzy sets in group decision making. Notes on Intuitionistic Fuzzy Sets 2, 15–32 (1996)
39. Szmidt, E., Kacprzyk, J.: Intuitionistic fuzzy relations and consensus formations. Notes on Intuitionistic Fuzzy Sets 6, 1–10 (2000)
40. Szmidt, E., Kacprzyk, J.: Using intuitionistic fuzzy sets in group decision making. Control and Cybernetics 31, 1055–1057 (2002)
41. Szmidt, E., Kacprzyk, J.: A consensus-reaching process under intuitionistic fuzzy preference relations. International Journal of Intelligent Systems 18(7), 837–852 (2003)
42. Szmidt, E., Kacprzyk, J.: A New Concept of a Similarity Measure for Intuitionistic Fuzzy Sets and its Use in Group Decision Making. In: Torra, V., Narukawa, Y., Miyamoto, S. (eds.) MDAI 2005. LNCS (LNAI), vol. 3558, pp. 272–282. Springer, Heidelberg (2005)
43. Yager, R.R.: A new approach to the summarization of data. Information Sciences (28), 69–86 (1982)
44. Yager, R.R.: On ordered weighted averaging operators in multicriteria decision making. IEEE Transactions on Systems, Man and Cybernetics SMC 18, 183–190 (1988)
45. Yager, R.R., Kacprzyk, J. (eds.): The Ordered Weighted Averaging Operators: Theory and Applications. Kluwer, Boston (1887)
46. Zadeh, L.A.: A computational approach to fuzzy quantifiers in natural languages. Computers and Mathematics with Applications (9), 149–184 (1983)
47. Zadrożny, S.: An approach to the consensus reaching support in fuzzy environment. In: Kacprzyk, J., Nurmi, H., Fedrizzi, M. (eds.) Consensus under Fuzziness, pp. 83–109. Kluwer, Boston (1997)

48. Zadrożny, S., Furlani, P.: Modelling and supporting of the consensus reaching process using fuzzy preference relations. Control and Cybernetics 20, 135–154 (1991)
49. Zadrożny, S., Kacprzyk, J.: An Internet-based group decision and consensus reaching support system. In: Yu, X., Kacprzyk, J. (eds.) Applied Decision Support with Soft Computing, pp. 263–275. Springer, Heidelberg (2003)
50. Zadrożny, S., Kacprzyk, J.: Issues in the practical use of the OWA operators in fuzzy querying. Journal of Intelligent Information Systems 33, 307–325 (2009)
51. Center for Excellence in Government, Facilitator's Toolbox (1996), http://www.employeesu.com/docs/EmployeeResources/Creating%20Consensus.pdf

Decision Making in Social Actions

Gabriella Marcarelli and Viviana Ventre

Abstract. Decision making in a social ambit involves both individual optimal choices and social choices. The theory of "perverse effects" by Boudon shows that the sum of rational individual choices can produce a very undesirable global effect. Then decision making in social action must keep into account the theory of cooperative games with many players, in order to obtain the optimal strategies. Because of the semantic uncertainty in the definition of social actions, it is preferable assume that the issues are represented by fuzzy numbers.

Keywords: Decision making, social choices, fuzzy numbers.

1 Introduction

As well as Natural Sciences, Social Sciences attempt to explain all phenomena through theories based on natural laws. We can assume that classical models for the Social Sciences are born as extensions of the models for the Natural Sciences. See e. g. the work by Auguste Comte (Comte, 1830-42, 1985).

Comte stressed that society must be viewed as an organism, drawing several parallels between biological organism and the social body. He drown the consequences that every social phenomenon has influences on the entire social system.

Social Sciences exhibit analogies with biology, that is based on the study of organic wholes; both sciences are holistic, concerned with physiological and environmental systems (Comte, 1830-42, 1985).

It is possible to explain social phenomena by simple assumptions about human nature (rationality and complete information) and by some ability to create and manipulate mathematical models (Boudon, 1967; Maturo et al., 2008).

But, as it results by many papers by Raymond Boudon (Boudon, 1967, 1969), and by some other authors (Sciarra, 2004; Maturo et al., 2008), the classical models can be used only for the study of the Social Structures, but they don't be useful to formalize *Values* or *Competition* of the Social Agents and so on.

Gabriella Marcarelli
University of Sannio, Via delle Puglie 82, Benevento, Italy
e-mail: marcarel@unisannio.it

Viviana Ventre
University of Sannio, Via delle Puglie 82, Benevento, Italy
e-mail: ventre@unisannio.it

S. Greco et al. (Eds.): Preferences and Decisions, STUDFUZZ 257, pp. 281–289.
springerlink.com

2 Some Characteristics of Social Organizations

Social Organization (or system) is a term usually used to indicate a group of social positions, connected by social relations, performing a social role. Common examples include education, governments, families, economic systems, religions. The environment of social systems includes other social systems (the environment of a family includes for example other families, political, economic and medical systems, and so on). No society exists without relations to an environment or without perceptions of environment. Social organizations can take many forms, depending on the social context.

(For example, for family context the corresponding social organization is the extended family; in the business context a social organization may be an enterprise, company, corporation; in the educational context, it may be a school, university; in the political context it may be a government, political party, etc.)

Social structure is a term rarely clearly conceptualized and it can refer to the relationship of definite entities or groups to each other, enduring patterns of behaviour by participants (agents) in a social system in relationship within a society and social norms and institutions or cognitive framework that becoming embedded into social systems in such a way that they shape the behaviour of actors within those social systems.

The study of social structures has informed the study of institutions, culture and agency, and social interactions as well as history. Weber investigated institutional arrangements of modern society and concluded that in the history of mankind, organizations evolved towards rationalization in the form of a rational – legal organization, like bureaucracy.

Social structure can be divided into microstructure and macrostructure, where microstructure is the pattern of relations between most basic elements of social life, that cannot be further divided and have no social structure of they own, and macrostructure is a pattern of relations between objects that have their own structure (Boudon, 1967, 1969; Sciarra, 2004; Maturo et al., 2008).

In social systems agents take their actions with a degree of freedom (autonomy) and these actions are mediated by existing institutional rules and expectations but, at the same time, they may influence institutional structure.

Social systems analysis is the study of social structure and its effects. Two key components define a social structure: actors, who represent different entities, such as groups, organizations, nations, as well as persons; and relationships, which represent flows of resources that can be related with aspects such as control, dependence, cooperation, information interchange, and competition.

We define the structure of social relations as the individuals with whom one has an interpersonal relationship and the linkages between these individuals. The structure has two dimensions: the formal relations, depending on positions and roles of individuals in society, and the informal social relations, i.e. the social network.

In a complex system many functional interactions take place simultaneously. Hierarchical control on the interacting agents both limit and give more freedom at the same time. The concepts of organization, control, self-regulation, equifinality and self-organization are as valid in the social and behavioral sciences as in the biological science.

Social systems can change their structures by evolution. Evolution changes the structural condition by differentiating mechanisms for variation, selection and stabilization. Intellectual evolution is the preponderant principle of social evolution.

What determines the evolution can be:

- the competition;
- the decision making in condition of uncertainty with multiple objectives;
- the cooperation by means of coalition.

It follows that many aspects of the evolution can be investigated in the ambit of multicriteria analysis or the game theory.

In effect both first and third issues lead to establish a connection with the game theory; in particular, the last point leads to consider cooperative games, in which players form coalitions.

3 Individual Decision Making and Social Effects

In a social organization single individuals create effective, coordinated, division of work groups at several levels of aggregation. At each level, there is not only competition between different groups at the same level, but also competition between the interests of the smaller incorporated units and the interests of the larger including unit.

Under the hypothesis of perfect competition individual interests are fundamental and social behaviour is an aggregation of individual behaviour; whereas when the assumption of perfect competition fails, (the concept of individual rationality becomes threatened) perceptions and rationality of others become part of one's own rationality (Arrow, 1951, 1986). So individual rationality is an inadequate model for synthesizing individual behaviour in a social system where a lot of actors are involved and cooperation is essential.

We can mark out two characteristics of a social organization: the division of work and the gradual belonging to a group. The division of the work encourages individuals to develop their talents (qualification and specialization) and contributes to the social bond by making each individual dependent on others. None could survive without the other.

With regard to the gradual belonging to a social group, if we consider a family with a given "head of family", the degree of belonging (dob) to the family is different for the various people. A "brother" as a dob different from a "nephew" or a "cousin" (Maturo et al., 2008).

The above characteristics of a social organization allow us to represent an organization by fuzzy sets and fuzzy relations.

Agreements on objectives and standards are based on networks of interpersonal ties, linking actors in different parts of the social structure, and on the flows of information and influence in these networks. Similarities in status, beliefs and behavior facilitate the formation of consensual relationships among individuals. The more dissimilar two individuals (social positions) are in status, beliefs and behavior, the "farther away" they are from one another in the system.

In order to formalize the social organization, we stress that it is very important the environment (natural, e.g. climatic conditions, or social, e.g. the civil or penal laws, the fiscal rules, etc...) in which the organisms (fuzzy systems) live. We can formalize this by assuming that the development depends on the axiomatic context considered.

Because of the interaction between actors into a group and in different groups of a social system, individual behaviour influences and, at the same time, is influenced by individual behaviours.

In the above axiomatic context an important role is played by the conditional probability. But unlike the Natural Sciences, in which the "axiomatic probability by Kolmogorov" is suitable, in the Social Ambit, the more general subjective conditional probability by de Finetti and Dubins is suitable (de Finetti, 1970; Dubins, 1975). A further generalization useful for the wideness of the applications is provided by the "generalized probability function" and the "fuzzy conditional probability" (Maturo, 2004, 2006).

We have to formalize the mechanisms of control, development, selection (how an individual enters, remains and go out of the system).

Social processes cannot be interpreted only as a sum of individual behaviours, but they have a proper life. We must consider also that in the ambit of the social action the "Perverse Effects" can occur (Boudon, 1967, 1969; Sciarra, 2004; Maturo et al., 2008).

The opposition between selfish individual behaviours makes some perverse effects since they are not provided but result by an arrangement of their rational actions. The aggregation of individual acts, each of these being rational individual choices, can have, as a consequence, a not desirable emergence.

The individual roles can be formalized as criteria for an individual belongs to a similarity class. These roles are in a continuous dynamical evolution which is influenced by the unforeseeable fluctuation of the set of the individual choices, in the ambit of the individual freedom granted by the institutional rules.

4 Fuzzy Set Modelling of Some Aspects of Social Behaviour

In social choice theory agent preferences on a set of alternatives are usually represented through binary relations.

However, human preferences are often vague; vagueness can be taken into account by means of fuzzy logic. The use of fuzzy relations in social choice theory for representing individual preferences has been justified by several authors (Basu, 1984; Dutta, 1987; Barrett et al., 1986; Yager, 2008).

The original concept of fuzzy set was introduced by Zadeh as an extension of crisp set, by enlarging the truth value set (or "grade of membership") from the two value set $\{0, 1\}$ to the unit interval $[0, 1]$ of real numbers (Zadeh, 1965; Bellman and Zadeh, 1970).

The basic arithmetic structure for fuzzy numbers was developed in (Dubois and Prade, 1978). The arithmetic operation was established either by the extension principle or observing the fuzzy numbers as a collection of α-levels. Alternative fuzzy operations for social applications are considered in (Maturo, 2009).

In fuzzy set theory, several triangular norms and conorms are used for defining the intersection and the union of fuzzy sets, respectively. Various factorizations of fuzzy weak preference relations have been given in the literature. See, for instance, (Ovchinnikov, 1981, 1991; Dutta, 1987; Richardson, 1998).

In order to deal with the problem of modeling in the above context we need:

- to assign numerical values (that are in general fuzzy values) to linguistic attributes;
- to give a way to aggregate information of perceptional or subjective nature;
- to formalize the interdependence between decision criteria;
- to have inference rules such that it is possible to manage imprecise information;
- to have a new way to a reconciliation between the abstract concepts of "competition", "control", and "selection", and the human way of thinking;
- to enlarge the possibility of formalization using probability functions and logical connectives;
- to have a defuzzification procedure (e.g., a fuzzy set is defuzzified to a crisp set by choosing the element of the fuzzy set with the highest degree of set membership).

Some of the above issues will be analyzed in the following section, utilizing the fuzzy set theory in the modeling decision making and game theory in the social context.

5 Competitive and Cooperative Situations in Social Sciences

Decision making in Social Sciences in a competitive or cooperative situation is modeled in Game Theory. Since the publication of "Theory of Games and Economic Behavior" (von Neumann and Morgenstern, 1944), game theory has been applied in different fields of research: economics, political science, management, philosophy. Game theory provides a normative rule to allow a rational person to do what is best for himself (principle of individual rationality). It take into account all possible actions and consequences for all participants, determining the best actions for all players.

Because of the different assumptions about the nature of game or about the character of rational human behavior, game theory has developed two great branches: cooperative and non-cooperative game theory. Competition within groups can have both benefits and costs for an individual.

We can observe that cooperative games and cooperative organizations have a common beginning in the concept of a group of agents that choose a common course of action for the mutual benefit (interest); in many social contexts, the presence of reciprocity motives allows to achieve a cooperative solution to the game. Cooperative game theory is applicable whenever the players in a game can form "coalitions", that is when groups choose a common strategy to improve the payoffs to the members of the group (McCain, 2008).

Let us recall the basic concepts of crisp n-person non-cooperative games in normal (strategic) form. See, e. g., (von Neumann and Morgenstern, 1944; Luce and Raiffa, 1957).

A crisp n-person (non-cooperative) game in normal form is a triplet G = (P, S, f) where:

- $P = \{P_1, P_2, \ldots, P_n\}$ is the set of players;
- $S = S_1 \times S_2 \times \ldots \times S_n$, with $S_j \neq \emptyset$ is the set of the pure strategies of the individual P_j;
- f is the vector of the payments $(f_1, f_2, \ldots, f_n)$, where $f_j: (s_1, s_2, \ldots, s_n) \in S \rightarrow f_j(s_1, s_2, \ldots, s_n) \in R$, is the utility of the player P_j if P_r chooses the strategy s_r, $r = 1, 2, \ldots, n$.

The n-tuple $(s_1^0, s_2^0, \ldots, s_n^0) \in S$ is said to be a *saddle point* or an *equilibrium point* of G if, for every $j \in \{1, 2, \ldots, n\}$, $s_j \in S_j$, we have:

$$f_j(s_1^0, s_2^0, \ldots, s_j, \ldots, s_n^0) \leq f_j(s_1^0, s_2^0, \ldots, s_j^0, \ldots, s_n^0).$$

A *mixed strategy* for the person P_j is a probability distribution on the set S_j of the pure strategies of P_j.

From now on we assume the set S is finite and n_j is the number of elements of S_j. In this case a *mixed strategy* for P_j is a vector $\sigma_j = (x_1, x_2, \ldots, x_{nj})$ of non negative real numbers such that their sum is 1.

Let $\Sigma = \Sigma_1 \times \Sigma_2 \times \ldots \times \Sigma_n$, where Σ_j is the set of mixed strategies of P_j. A function $F_j: (\sigma_1, \sigma_2, \ldots, \sigma_n) \in \Sigma \rightarrow F_j(\sigma_1, \sigma_2, \ldots, \sigma_n) \in R$ is defined as the linear extension of the function f_j, where $F_j(\sigma_1, \sigma_2, \ldots, \sigma_n)$ is the utility of the player P_j if every player P_r chooses the mixed strategy σ_r, $r = 1, 2, \ldots, n$.

It is well known that there are many games that haven't saddle points in pure strategies, but every finite game has at least an equilibrium point in mixed strategies (Nash, 1951).

Often, in modeling social situations by means of finite games, the players are not able to evaluate exactly some data of the game due to a lack of information or imprecision of the available information on the environment or on the behavior of other players. This drastic restriction made it difficult to apply the classical game theory to real problems.

In order to make the theory of games more applicable to real problems, fuzzy set theory has been introduced in non-cooperative game theory in (Butnariu, 1978).

Non-cooperative fuzzy games were studied also in (Tsurumi et al., 2001; Mares, 2001; Maturo et al., 2004; Kachera and Larbani, 2008), in which fuzzy utilities and strategies individuated by linguistic attributes are considered.

Fuzzy games are studied also by the cooperative point of view, in which fuzzy coalitions and fuzzy characteristic functions are introduced. The starting point is the concept of fuzzy coalition introduced in (Aubin, 1975).

Cooperative fuzzy games represent an extension of a fuzzy game with fuzzy coalitions and vague expectations together. Borkotokey has observed that the most of the properties satisfied by a crisp game hold good in the fuzzy sense in this extension (Borkotokey, 2008).

In Social Sciences the utilities are often given as values of a linguistic variable or, also in the case of a numerical variable, every utility has a degree of uncertainty.

Then we propose, in the applications of the game theory to competitive situations in Social Sciences, to assume that, the utilities are fuzzy numbers.

Moreover, by considering mixed strategies, practically we have a uncertainty of the probabilities assessed to the pure strategies.

Let Φ be the set of simple fuzzy numbers. We propose to consider fuzzy mixed strategies, using the following definition.

Definition A fuzzy mixed strategy for a player P_j is a vector $x = (x_1, x_2, \ldots, x_{nj}) \in \Phi^{nj}$ with the conditions:

- $\forall j \in \{1, 2, \ldots, nj\}$, the support of x_j is contained in $[0, 1]$;
- $C(x_1) + C(x_2) + \ldots + C(x_m) = 1$, where $C(x_i)$ is the core of the fuzzy number x_i.

Let us consider a simple example coming from a situation illustrated in (Crozier, 1963) and further dealt with in (Maturo et al., 2008). In the problem of Industrial Monopoly it is considered a precise system of roles of a great Enterprise. Let us limit our consideration to the functional roles and relations between the Director and the Budged Controller. We have:

- a hierarchical dependence of the Budged Controller by the Director, responsible of strategic and political trends;
- a joint responsibility for the important decisions with the obligation of a join signature.

The Director and the Budged Controller choose their strategies. In (Crozier, 1963) the system of relations between the Director and the Budged Controller is simplified by considering two possible strategies for everyone: *aggressive* (a) and *collaborative* (c). Moreover Crozier claimed that the *optimal join strategy* for the social organization of the Enterprise is (a) for the Director and (c) for the Budged Controller.

The remark that not always the Director is right and often he is "partially" right and not always it is convenient that the Budged Controller is remissive in defending own ideas, lead us to consider fuzzy mixed strategies, on the basis of the definition above.

References

Arrow, K.J.: Social Choice and Individual Values, 2nd edn. John Wiley, New York (1963)

Arrow, K.J.: Rationality of Self and Others. In: Hogarth, R.M., Reder, M.W. (eds.) Rational Choice. University of Chicago Press, Chicago (1986)

Aubin, J.P.: Fuzzy games, Report MRC-1480. Math. Res. Center Univ. of Wisconsin (1975)

Barrett, C.R., Pattanaik, P.K., Salles, M.: On the structure of fuzzy social welfare functions. Fuzzy Sets and Systems 19, 1–10 (1986)

Basu, K.: Fuzzy revealed preference theory. J. Econ. Theory 32, 212–227 (1984)

Bellman, R.E., Zadeh, L.A.: Decision Making in a fuzzy environment. Management Science 17B, 141–165 (1970)

Borkotokey: Cooperative games with fuzzy coalitions and fuzzy characteristic functions. Fuzzy Sets and Systems 159, 138–151 (2008)

Boudon, R.: L'analyse mathématique des faits socials, Paris, Plon (1967)

Boudon, R.: Les méthods in Sociologie, Paris, P.U.F. (1969)

Butnariu, D.: Fuzzy Games. A description of the concept. Fuzzy Sets and Systems 1, 182–192 (1978)

Comte, A.: Cours de Philosophie positive, Paris, vol. 4, 5, 6 (1830-1842)

Crozier, M.: Le Phénomène Bureaucratique, Paris, Seuil (1963)

de Finetti, B.: Teoria delle probabilità (1970) Einaudi, Torino English Translation: Theory of Probability. J. Wiley, New York (1975)

Dubins, L.E.: Finitely additive conditional probabilities, conglomerability, and disintegration. The Annals of Probability 3, 89–99 (1975)

Dubois, D., Prade, H.: Operations on Fuzzy Numbers. International Journal of Systems Science, 613–626 (1978)

Dutta, B.: Fuzzy preferences and social choice. Mathematical Social Sciences 13, 215–229 (1987)

Kachera, K., Larbani, M.: Existence of equilibrium solution for a non-cooperative game with fuzzy goals and parameters. Fuzzy Sets and Systems 159, 164–176 (2008)

Luce, R.D., Raiffa, H.: Games and Decisions. John Wiley and Sons, New York (1957)

Maturo, A.: Grandezze aleatorie fuzzy e loro previsioni per le decisioni in condizione di informazione parziale. In: Cortellini, Luchian (eds.) Current Topics in Computer Sciences, Panfilus, Iasi, pp. 15–24 (2004)

Maturo, A.: Fuzzy Conditional Probabilities by the Subjective Point of View. In: Tofan, I. (ed.) Advances in Mathematics of Uncertainty, pp. 99–108. Editura Performantica, Iasi (2006)

Maturo, A.: Alternative Fuzzy Operations and Applications to Social Sciences. International Journal of Intelligent Systems 24, 1243–1264 (2009)

Maturo, A., Sciarra, E., Tofan, I.: A formalization of some aspects of the Social Organization by means of the fuzzy set theory. Ratio Sociologica 1(1), 5–20 (2009)

Maturo, A., Tofan, I., Ventre, A.G.S.: Fuzzy Games and Coherent fuzzy previsions. Fuzzy Systems & A. I. X(3), 109–116 (2004)

McCain, R.A.: Cooperative games and cooperative organizations. The Journal of Socio-Economics 37, 2155–2167 (2008)

Nash, J.F.: Non-cooperative games. Annals of Mathematics 54, 286–295 (1951)

Ovchinnikov, S.: Structure of fuzzy binary relations. Fuzzy Sets Syst. 6, 169–195 (1981)

Ovchinnikov, S.: Social choice and Lukasiewicz logic. Fuzzy Sets Syst. 43, 275–289 (1991)

Richardson, G.: The structure of fuzzy preferences: Social choice implications. Soc. Choice Welfare 15, 359–369 (1998)

Sciarra, E.: Raymon Boudon e l'epistemologia dell'azione sociale. Edizioni Scientifiche Sigraf, Chieti (2004)
Tsurumi, M., Tanino, T., Inuiguchi, M.: A Shapley function on a class of cooperative fuzzy game. European Journal of Operational Research 129, 596–618 (2001)
von Neumann, J., Morgenstern, O.: Theory of Games and Economic Behavior. Princeton University Press, Princeton (1944)
Yager, R.R.: Intelligent Social Network Analysis Using Granular Computing. International Journal of Intelligent Systems 23, 1197–1220 (2008)
Zadeh, L.A.: Fuzzy Sets. Information and Control 8(3), 338–353 (1965)

Coherence for Fuzzy Measures and Applications to Decision Making

Antonio Maturo, Massimo Squillante, and Aldo G.S. Ventre

Abstract. Coherence is a central issue in probability (de Finetti, 1970). The studies on non-additive models in decision making, e. g., non-expected utility models (Fishburn, 1988), lead to an extension of the coherence principle in nonadditive settings, such as fuzzy or ambiguous contexts. We consider coherence in a class of measures that are decomposable with respect to Archimedean t-conorms (Weber, 1984), in order to interpret the lack of coherence in probability. Coherent fuzzy measures are utilized for the aggregations of scores in multiperson and multiobjective decision making. Furthermore, a geometrical representation of fuzzy and probabilistic uncertainty is considered here in the framework of join spaces (Prenowitz and Jantosciak, 1979) and, more generally, algebraic hyperstructures (Corsini and Leoreanu, 2003); indeed coherent probability assessments and fuzzy sets are join spaces (Corsini and Leoreanu, 2003; Maturo et al., 2006a, 2006b).

1 Aims and Topics

Theoretical and applied issues in decision making involve evaluations based on non-additive measures (Fishburn, 1988; Schmeidler, 1989; Dubois and Prade, 1992; Squillante and Ventre, 1988, 1992, 1998; Ventre, 1996; Diecidue and Maccheroni, 2003; Maturo, et al., 2006a, 2006b).

It is well known that the principle of coherence is a fundamental subject in probability (Coletti, 1994; Coletti et al. 1990, 2004; de Finetti, 1970). The coherence in a class of fuzzy measures, introduced in order to interpret the lack of coherence in probability (Squillante and Ventre, 1998; Maturo, et al., 2006a, 2006b), is here further studied.

Antonio Maturo
Department of Social Sciences, University "G. D'Annunzio" of Chieti-Pescara, Italy
e-mail: amaturo@unich.it

Massimo Squillante
Department of Economical and Social Sciences, University of Sannio at Benevento, Italy
e-mail: squillan@unisannio.it

Aldo G.S. Ventre
Department of Culture of the Project, and Benecon Research Center,
Second University of Napoli, Italy
e-mail: aldoventre@yahoo.it

S. Greco et al. (Eds.): Preferences and Decisions, STUDFUZZ 257, pp. 291–304.
springerlink.com

The coherence allows the construction of a measure that takes into account the evaluation systems of the decision makers.

Fuzzy measures allow to generalize the concept of consistency of a set of evaluations, and, in some cases, the initial assessment is interpolated by a plausibility or a belief function.

A geometrical representation of the fuzzy and probabilistic uncertainty is considered in the framework of the theory of the *algebraic hyperstructures* (Corsini and Leoreanu, 2003; Prenowitz and Jantosciak, 1979)). Indeed, coherent probability assessments and fuzzy sets are *join spaces* (Maturo, et al., 2006a, 2006 b).

The logical concepts of atoms and conditional events are framed as properties of particular hypergroupoids (Doria and Maturo, 1996).

A further goal will be obtained by setting in the same framework the assessments that are coherent with respect to decomposable measures of uncertainty.

2 Key Concepts and Preliminary Results

2.1 Belief and Plausibility Measures

A *belief function Bel* (see, e. g., (Banon, 1981)) is a function defined over an algebra $\mathfrak{I}$ of subsets of a set X s. t.

$$Bel(\emptyset) = 0,\ \ Bel(X) = 1, \tag{2.1}$$

$$Bel(A_1 \cup ... \cup A_n) \geq \sum Bel(A_i) - \sum Bel(A_i \cap A_j) + ... + (-1)^{n+1} Bel(A_1 \cap ... \cap A_n) \tag{2.2}$$

A *plausibility measure Pl* is characterized by $Pl(A)=1-Bel(A^c)$, with A^c the complement of A.

A measure (and integration) theory with respect to suitable decomposable set functions, that gives rise to another actual extension of Lebesgue measure theory, is due to (Weber, 1984).

2.2 Fuzzy Operations on [0, 1]

Let us recall, from (Weber, 1984), some definitions.

A *t-conorm* $\perp$ on the real unit interval [0, 1] is a binary operation

- non decreasing in each argument;
- associative;
- commutative;
- having 0 as neutral element.

A t-conorm is *Archimedean* if it is:

- continuous;
- $\perp(x, x) > x$, for every x in (0, 1).

An Archimedean t-conorm is *strict* if it is strictly increasing in the open square $(0, 1)^2$.

The following representation theorem holds:

Theorem 2.1 (Ling, 1965)**.** A binary operation $\perp$ on [0, 1] is an Archimedean t-conorm if and only if there exists a strictly increasing and continuous function

$$g: [0, 1] \to [0,+\infty], \text{ with } g(0) = 0,$$

such that

$$x \perp y = g^{(-1)}(g(x)+g(y)).$$

Function $g^{(-1)}$ denotes the pseudo-inverse of g, i.e.:

$$g^{(-1)}(x) = g^{-1}(\min(x, g(1))).$$

Moreover

$$\perp \text{ strict} \Leftrightarrow g(1) =+\infty.$$

The function g, called an *additive generator* of $\perp$, is unique up to a positive constant factor. The following identity holds:

$$g(g^{(-1)}(x)) = \min(x, g(1)).$$

The following is known as Sugeno t-conorm (Sugeno, 1974).

Example 2.1. For $\lambda > -1$, a, b$\in$[0, 1], let:

$$U_\lambda(a, b) = \min(a + b + \lambda\, ab, 1). \tag{2.3}$$

The function U_λ: (a, b)$\in[0, 1]^2 \to U_\lambda(a, b)$ is a non-strict Archimedean t-conorm with additive generator

$$g_\lambda(x) = (1/\lambda) \ln(1+\lambda x). \tag{2.4}$$

In particular:

- for $\lambda = 0$, we have the *bounded sum*:

$$U_0(a, b) = \min(a + b, 1), \tag{2.5}$$

with additive generator

$$g_0(x) = x.$$

- as $\lambda \to -1$ Sugeno t-conorm reduces to the following *algebraic sum*:

$$U_{-1}(a, b) = a + b - ab, \tag{2.6}$$

that is a strict Archimedean t-conorm with additive generator

$$g_{-1}(x) = -\ln(1-x).$$

3 Decomposable Measures

Definition 3.1. Let $(X, \Im)$ be a measurable space. A set function

$$m: \Im \to [0, 1], \text{ with } m(\emptyset) = 0 \text{ and } m(X) = 1,$$

is said to be:

(a) a $\perp$-*decomposable measure*, if

$$A \cap B = \emptyset \Rightarrow m(A \cup B) = m(A) \perp m(B); \tag{3.1}$$

(b) a σ-$\perp$-*decomposable measure*, if

$$A_i \cap A_j = \emptyset, \forall i \neq j \Rightarrow m(\cup_{n \geq 1} A_n) = \perp_{n \geq 1} m(A_n). \tag{3.2}$$

The following classification theorem holds:

Theorem 3.1 (Weber, 1984)**.** If the operation $\perp$ in $[0, 1]$ is a strict Archimedean t-conorm, then

- (S) $g \bullet m: \Im \to [0, +\infty]$ is an infinite (σ-)additive measure, whenever m is a (σ-)$\perp$-decomposable one.
- If $\perp$ is a non-strict Archimedean t-conorm, then $g \bullet m$ is finite and one of the following cases occurs:
  - (NSA) $g \bullet m: \Im \to [0, g(1) = (g \bullet m)(X)]$ is a finite (σ-)additive measure;
  - (NSP) $g \bullet m$ is a finite set function which is only pseudo (σ-)additive, i. e., if $A_i \cap A_j = \emptyset, \forall i \neq j$, then

$$(g \bullet m)(\cup_{n \geq 1} A_n) < g(1) \Rightarrow (g \bullet m)(\cup_{n \geq 1} A_n) = \sum_n (g \bullet m)(A_n); \tag{3.3}$$

$$(g \bullet m)(\cup_{n \geq 1} A_n) = g(1) \Rightarrow (g \bullet m)(\cup_{n \geq 1} A_n) \leq \sum_n (g \bullet m)(A_n). \tag{3.4}$$

The U_λ-decomposable measures are also called λ-*additive measures* or *Sugeno measures*.

Banon (1981) and Berres (1988) proved that λ-additive measures are *plausibility measures* if $-1 < \lambda < 0$, and *belief measures* if $\lambda > 0$.

For $\lambda = 0$ the λ-additive measures are *probability measures*.

4 Coherence for Decomposable Measures

Let $\boldsymbol{E}$ be a finite family of events A_i, $i = 1, \dots, n$. The *atoms* or *constituents* of $\boldsymbol{E}$ are defined as the non impossible events $\cap_i A'_i$, where $A'_i \in \{A_i, A_i^c\}$.

Let C_j, $j = 1, 2, \dots, s$ be the set of atoms.

The set of assessments $0 \leq m(A_i) = m_i < 1$, over the events A_i, is *coherent* w. r. to a $\perp$-decomposable measure m, with additive generator g, if there is a solution

$$\mathbf{w} = (w_1, \dots, w_s)$$

of the following system:

$$a_{i1}g(w_1) + a_{i2}g(w_2) + \ldots + a_{is}g(w_s) = g(m_i),\ i = 1,\ldots, n \quad (4.1)$$

$$g(w_1) + g(w_2) + \ldots + g(w_s) \geq g(1), \quad (4.2)$$

with $a_{ij} = 1$, if $C_j \subseteq A_i$ and $a_{ij} = 0$, if $C_j \subseteq A_i^c$, and $0 \leq w_j \leq 1$, $j = 1,\ldots, s$.
For Sugeno measures, system (4.1) – (4.2) reduces to:

$$a_{i1}\ln(1+\lambda w_1) + a_{i2}\ln(1+\lambda w_2) + \ldots + a_{is}\ln(1+\lambda w_s) = \ln(1+\lambda m_i),\ i=1,\ldots, n \quad (4.3)$$

$$\ln(1+\lambda w_1) + \ln(1+\lambda w_2) + \ldots + \ln(1+\lambda w_s) \geq \ln(1+\lambda),\ \text{for } \lambda>0, \quad (4.4)$$

$$\ln(1+\lambda w_1) + \ln(1+\lambda w_2) + \ldots + \ln(1+\lambda w_s) \leq \ln(1+\lambda),\ \text{for } \lambda<0, \quad (4.5)$$

$$0 \leq w_j,\ j = 1,\ldots, s. \quad (4.6)$$

Some examples show that an assessment of evaluations, inconsistent in the probabilistic framework, can actually be coherent w.r. to a suitable U_λ-decomposable measure.

Example 4.1. Let us consider the events A_1, A_2, A_3 with the evaluations

$$m(A_1) = 0.1,\ m(A_2) = 0.5,\ m(A_3) = 0.3,$$

and the relations

$$A_1 \cap A_2 \cap A_3 = \emptyset,\ A_3 = (A_1 \cap A_2) \cap (A_1 \cap A_2)^c.$$

The atoms are:

$$C_1 = A_1 \cap A_2^c \cap A_3,\ \ C_2 = A_1 \cap A_2 \cap A_3^c,\ \ C_3 = A_1^c \cap A_2 \cap A_3,\ \ C_4 = A_1^c \cap A_2^c \cap A_3^c.$$

In the ambit of the U_λ-decomposable measures, the system (4.3) –(4.6) gives coherence for $\lambda \geq 10/3$, i.e. for a belief measure that is not a probability measure.

Put $m_i = m(A_i)$. In the ambit of the U_λ-decomposable measures, if the system (4.3) –(4.6) gives coherence for some $\lambda \geq 0$, let us to define *a measure of inconsistency* w. r. to a probability as:

$\mu(m_1, \ldots, m_n) = \inf\{\lambda \geq 0$ such that there is a λ-additive measure interpolating the data$\}$.

5 An Application to Multiobjective and Multiperson Decision Making

We consider the problem of choosing an alternative in a set $\mathbf{A} = \{A_1, A_2, \ldots, A_m\}$ of *alternatives*, given:

- a set $D = \{d_1, d_2, \ldots, d_h\}$ of *decision makers*;
- a set $O = \{O_1, O_2, \ldots, O_n\}$ of *objectives*.

Each decision maker d_k assigns to any pair (alternative A_i, objective O_j) a real number $a_{ij}^k \in [0, 1]$ that measures to what extent A_i satisfies O_j.

We assume every O_j is a non trivial subset of a universal set U (or, in a probabilistic framework, we assume O_j is a non trivial event and U is the certain event).

Let m_i^k be the function defined on $O \cup \{\emptyset, U\}$ with values on [0, 1] such that:

$$m_i^k(\emptyset) = 0, \quad m_i^k(U) = 1, \quad m_i^k(O_j) = a_{ij}^k. \tag{5.1}$$

Assume m_i^k is a monotonic set function ("*finitely monotonic*" fuzzy measure).

Our aims are:

- to find Archimedean non strict t-conorms $\oplus_\lambda$, $\lambda \in \Lambda$, such that, for some λ, m_i^k is a restriction of a $\oplus_\lambda$-decomposable fuzzy measure;
- among these t-conorms, to individuate a suitable t-conorm $\oplus$ in order to aggregate scores of alternatives with respect to the objectives.

5.1 Decision Making with Disjoint Objectives

Let the set of objectives $O = \{O_1, O_2, \ldots, O_n\}$ be a family of disjoint subsets (resp. incompatible events) of U and let $\oplus$ be any non strict Archimedean t-conorm with additive generator g.

In this case

- the atoms are $C_1 = O_1$, $C_2 = O_2$, …, $C_n = O_n$ and (if it is not empty) C_{n+1} equal to the complement of the union of the O_j;
- then the system (4.1) – (4.2), with m_j replaced by a_{ij}^k and s = n+1, has solutions $(w_1, w_2, \ldots, w_n, w_{n+1})$, where $w_j = a_{ij}^k$ for j=1, 2, …n, and w_{n+1} is any value of a suitable close interval containing 1;
- the function m^{*k}_i that to every subset S of U obtained as union of atoms associates the number $\oplus\{w_r: C_r \in S\}$ is a $\oplus$-decomposable measure on the algebra $\mathfrak{I}$ generated by the objectives, extension of m_i^k;
- then, for every i, k, the assessment of the scores a_{ij}^k to the objectives is coherent w. r. to $\oplus$, and m^{*k}_i is its extension to the algebra $\mathfrak{I}$ of subsets of U.

Previous properties allows us to define the *global score* b_i^k of the alternative A_i w. r. to the decision maker d_k by the formula:

$$b_i^k = a_{i1}^k \oplus a_{i2}^k \oplus \ldots \oplus a_{in}^k, \quad i = 1, 2, \ldots, m \tag{5.1}$$

The global score of d_k is the point $P_k = (b_i^k, i = 1, 2, \ldots, m)$ of the Euclidean space R^m.

5.2 Decision Making with Non-disjoint Objectives

Let the set of objectives $O = \{O_1, O_2, \ldots, O_n\}$ be a family of non disjoint subsets (resp. non incompatible events) of U and let $\oplus$ be any non strict Archimedean t-conorm with additive generator g. Let $C = \{C_1, C_2, \ldots, C_s\}$ the set of atoms of O.

We say that $\oplus$ is *consistent* with the scores of the alternative A_i, with respect to the decision maker d_k, if the system (4.1) – (4.2), with m_i replaced by a_{ij}^k, has solutions $(w_{i1}, w_{i2}, \ldots, w_{is})$.

We say that $\oplus$ is an *aggregation criterion* for the decision maker d_k if it is consistent with the scores of all the alternatives A_i, with respect to d_k.

If a decision maker d_k has a own *aggregation criterion* $\oplus$ (consistent with the scores), then he can define the *global score* b_i^k of the alternative A_i by the formula:

$$b_i^k = \chi_1 w_{i1} \oplus \chi_2 w_{i2} \oplus \ldots \oplus \chi_s w_{is}, \tag{5.2}$$

where $i = 1, 2, \ldots, m$, and

- $(w_{i1}, w_{i2}, \ldots, w_{is})$ is a solution of the system (4.1) – (4.2) with m_i replaced by a_{ij}^k;
- $\chi_j = 1$ if the atom C_j is contained in at least an objective O_j and $\chi_j = 0$ otherwise.

As in the above case, the global score of d_k is the point $P_k = (b_i^k, i = 1, 2, \ldots, m)$ of R^m.

5.3 Some Further Problems in the Decision Making Process

- (*choice of solution*) if formula (5.2) is adopted, decision maker d_k must also provide for suitable criteria (e.g. maximize or minimize some "objective function") in order to choose the solution $(w_{i1}, w_{i2}, \ldots, w_{is})$ of the system (4.1) – (4.2), if it has more solutions;
- (*homogeneity*) in order to compare and activate a procedure of aggregation-consensus for a ranking of the alternatives that resumes the ranking of all the decision makers, it is necessary to believe in the "homogeneity" of the vectors b_i^k, $k = 1, \ldots, h$. This is important, in particular, if different decision makers use formula (5.2) with different t-conorms.
- (*normalization*) a criterion to obtain homogenous vectors of scores by different decision makers is the normalization, obtained by dividing scores by a suitable positive real number dependent by d_k. The most common normalization procedure is replacing the numbers b_{ik} with the numbers:

  $$\beta_i^k = b_i^k/(b_1^k + b_2^k + \ldots + b_m^k). \tag{5.3}$$

  After the normalization, the global score of d_k is the point

  $$Q_k = (\beta_i^k, i = 1, 2, \ldots, m) \tag{5.4}$$

  of the Euclidean space R^m belonging to the hyperplane:

  $$x_1 + x_2 + \ldots + x_m = 1. \tag{5.5}$$

5.4 Decision Making with Weighted Objectives

We can, in general, associate a weight ω_j to every objective O_j, that measures the importance of O_j. These weights can be obtained, e. g., by means of the Analytic

Hierarchy Process (AHP) (Saaty, 1980), by considering the pairwise comparisons of the objectives. These weights must satisfy the conditions:

$$\omega_j \geq 0, j = 1, 2, ..., n, \quad \omega_1 + \omega_2 + ... + \omega_n = 1. \tag{5.6}$$

If the objectives are disjoint, then the weights ω_j are consistent with a measure on the algebra containing the objectives.

In this case we replace the scores a_{ij}^k with the products $\omega_j\, a_{ij}^k$, and obtain the *weighted global score* b_i^k of the alternative A_i with respect to the decision maker d_k by the formula:

$$b_i^k = \omega_1\, a_{i1}^k \oplus \omega_2 a_{i2}^k \oplus \ldots \oplus \omega_n\, a_{in}^k, \; i = 1, 2, \ldots, m. \tag{5.7}$$

If the objectives are not disjoint the weights ω_j are consistent with respect to a $\oplus$-decomposable measure, with $\oplus$ non strict Archimedean t-conorm with generator g, if and only if the system (4.1) – (4.2), with m_i replaced by ω_j, has solutions $(c_1, c_2, \ldots, c_s)$.

In this case we obtain the *weighted global score* b_i^k by replacing the scores of the atoms w_{ij} with the products $c_j\, w_{ij}$ and by replacing the formula (5.2) with the following

$$b_i^k = \chi_1\, c_1\, w_{i1} \oplus \chi_2\, c_2\, w_{i2} \oplus \ldots \oplus \chi_s\, c_s\, w_{is}, \; i = 1, 2, \ldots, m. \tag{5.8}$$

6 Hyperstructures as a Tool for a Geometrical Representation of the Uncertainty

6.1 Main Concepts on Hyperstructures

Many new contributions to the representation of the uncertainty can be obtained as applications of the theory of the algebraic hyperstructures.

The theory of hyperstructures started with the paper (Marty, 1934), but its present development begins in 1978 with the First International Congress on Algebraic Hyperstructures and Applications (AHA), and the work (Prenowitz and Jantosciak, 1979). For further results and references see (Corsini and Leoreanu, 2003).

A *hypergroupoid*, or *hyperstructure with a hyperoperation*, is a pair (H, σ), where H is a non empty set and $\sigma\colon H\times H \to \wp^*(H) = \wp(H)-\{\varnothing\}$, called *hyperoperation* on H, is a function that associates to any ordered pair (a, b) of elements of H a non empty subset of H, denoted $a\sigma b$.

The elements of H are called *points* and any singleton {a}, $a\in H$, is identified with the point a. So, if $a\sigma b$ is a singleton for every a, $b\in H$, then the hyperoperation σ reduces to an operation on H.

For every A, $B\in \wp^*(H)$ we assume:

$$A\sigma B = \cup\{a\sigma b\colon a\in A, b\in B\}. \tag{6.1}$$

A hypergroupoid (H, σ) is said to be:

- *semihypergroup*, if, for every a, b, c∈H,

$$a\sigma(b\sigma c) = (a\sigma b)\sigma c \quad \textit{(associative property)};$$

- *weak semihypergroup*, if, for every a, b, c∈H,

$$a\sigma(b\sigma c)\cap(a\sigma b)\sigma c \neq \varnothing \quad \textit{(weak associative property)};$$

- *commutative* if, for every a, b∈H,

$$a\sigma b = b\sigma a \quad \textit{(commutative property)};$$

- *weak commutative hypergroupoid* if, for every a, b∈H,

$$a\sigma b \cap b\sigma a \neq \varnothing \quad \textit{(weak commutative property)};$$

- *quasihypergroup* if, for every a∈H,

$$a\sigma H = H = H\sigma a \quad \textit{(reproducibility property)}.$$

A hypergroupoid (H, σ) is said to be a *hypergroup* if it is both a semihypergroup and a quasihypergroup.

The commutative hypergroups are meaningful from a geometrical point of view.

By a geometrical point of view, the hyperproduct aσb is said to be the σ-segment (or simply the segment if there is no ambiguity) with endpoints a and b.

If A and B are two subsets of H, we say that A meets B, we write A ≈ B if A and B have at least a point in common.

In a commutative hypergroup, the concepts of *division* / and *half-line* are also introduced.

For every a, b∈H, the *division* a/b is the set:

$$a/b = \{x\in H: a\in x\sigma b\}, \tag{6.2}$$

called the σ-*half-line* (or simply the *half-line* if there is no ambiguity on the hyperoperation considered) with origin b and containing a.

6.2 Join Spaces

The commutative hypergroups that have the most meaningful geometric properties and that seem to be suitable for many extension of the concept of coherence are the join spaces.

A commutative hypergroup (H, σ) is said to be a *join space* if the following *incidence property* holds:

$$\forall a, b, c, d\in H,\ a/b\approx c/d \Rightarrow a\sigma d\approx b\sigma c. \tag{6.3}$$

In other words:

if the half-lines a/b and c/d have in common at least a point x, then also the segments aσd and bσc meet in at least a point y.

The following are particular join spaces (Prenowitz and Jantosciak, 1979):

- (*the Euclidean join space*) Every *Euclidean space* R^m, and every *convex set* S of R^m, with respect to the hyperoperation σ that to every pair (P, Q) of points associates the Euclidean segment with extremes P and Q. The incidence property is a different formulation of the Pasch axiom (see, e. g., Beutelspacher and Rosenbaum, 1998).
- (*the Cartesian join space*) The *Euclidean space* R^m, and every interval of R^m, with respect to the hyperoperation α that to every pair (P, Q) of points associates the Euclidean interval with extremes P and Q is a join space, called the *Cartesian join space*.
- (*the projective join space*) Let (S, L) be a projective space, with S the set of points and L the set of lines; S, with respect to the hyperoperation α that to every pair (P, Q) of points associates the line through the points P and Q, is a join space. The incidence axiom reduces to Veblen–Young axiom (see, e. g., Beutelspacher and Rosenbaum, 1998).

6.3 Hyperstructures Associated to a Fuzzy Set

(6.3.a) The Join Space Associated to a Fuzzy Set

Let α: U→[0, 1] be a fuzzy set with universe U. For every a, b in U, let:

$$a\sigma b = \{x \in U: \mathit{min}(\alpha(a), \alpha(b)) \le \alpha(x) \le \mathit{max}(\alpha(a), \alpha(b))\}. \quad (6.4)$$

The hyperoperation σ is commutative and we have {a, b} ⊆ aσb, then aσb ≠ ∅ and aσH =H = Hσb.

Then (U, σ) is a commutative quasihypergroup.

Because of the associativity of the operations min and max, the following equalities hold:

$$(a\sigma b)\sigma c=\{u\in U:\min(\alpha(a),\alpha(b),\alpha(c))\le\alpha(u)\le\max(\alpha(a),\alpha(b),\alpha(c))\}=a\sigma(b\sigma c), \quad (6.5)$$

and so (U, σ) is a commutative hypergroup.

As incidence property holds (Corsini and Leoreanu, 2003), then (U, σ) is a join space.

(6.3.b) The Hypergroup Associated to a Fuzzy Set of Type 2

Let (L, ∨, ∧) be a lattice. A function α: U→L is said to be a L-fuzzy set with universe U.

For every a, b in U, let:

$$a\tau b = \{x\in U: \alpha(a) \wedge \alpha(b)) \le \alpha(x) \le \alpha(a) \vee \alpha(b)). \quad (6.6)$$

The hyperoperation τ is commutative and {a, b} ⊆ aτb; then aτb ≠ ∅ and aσH=H = Hσb.

Hence (U, τ) is a commutative quasihypergroup, called the *hyperstructure associated* to α.

The following theorem holds:

Theorem 6.1 (Corsini and Leoreanu, 2003, p.178). If L is a distributive lattice then (U, τ) is associative and so it is a commutative hypergroup.

When, in particular, L equals the family of fuzzy sets with universe $J \subseteq [0,1]$ and $\vee$ and $\wedge$ are the usual operations of union and intersection between fuzzy sets with universe J, respectively, then the L-fuzzy sets are said to be *fuzzy sets of type* 2. Since the union is distributive with respect to the intersection, then the hyperstructure (U, τ) associated to a fuzzy set α of type 2 is a *commutative hypergroup*.

6.4 Join Spaces Associated to Coherent Probability or Prevision Assessments

(6.4.a) Join Space of the Coherent Probability Assessments

Let $\mathfrak{I} = \{E_1, E_2, \ldots, E_n\}$ be a finite set of events and let U be the Euclidean space having the events of $\mathfrak{I}$ as axes.

If K is an atom associated to $\mathfrak{I}$, we call *representative* of K the point P(K) such that, for every $E_i \in \mathfrak{I}$, its projection $P_i(K)$ on the axis E_i is 1 if $K \subseteq E_i$ and 0 if $K \subseteq E_i^c$.

Let Δ be the set of all the atoms associated to $\mathfrak{I}$, and P(Δ) the set of the points of R^n representative of the elements of Δ.

As shown by (de Finetti, 1970), a point of R^n

$$P = (p_1, p_2, \ldots, p_n), \text{ with } p_i = p(E_i)$$

is a coherent probability assessment on $\mathfrak{I}$, if and only if, for every $S = (S_1, S_2, \ldots, S_n) \in R^n$,

$$M_s = \max_{K \in \Delta} \Sigma_i S_i E_i^K \geq \Sigma_i S_i p_i, \qquad (6.7)$$

where

$$E_i^K = 1 \text{ if } K \subseteq E_i \text{ and } E_i^K = 0 \text{ if } K \subseteq E_i^c.$$

Let $\wp$ be the set of all the coherent probability assessments on $\mathfrak{I}$. For any P, $Q \in \wp$, let aσb be the closed Euclidean segment having P and Q as extreme points.

From (6.7) it follows that, if $A = (a_1, a_2, \ldots, a_n)$ and $B = (b_1, b_2, \ldots, b_n)$ are two elements of $\wp$, then any point

$$P = hA + (1-h)B, \ 0 \leq h \leq 1,$$

of aσb is also an element of $\wp$.

From suitable properties of convex sets of a Euclidean space (see, e. g., Prenowitz and Jantosciak, 1979), $(\wp, \sigma)$ is a commutative hypergroup, in particular a join space, that we call the *join space of the coherent probability assessments* on $\mathfrak{I}$.

By (6.7) it follows that $\wp$ contains all the points P(K), $K \in \Delta$ and it is the minimal convex subset of U containing all these points (de Finetti, 1970),

6.5 Logical Hypergroupoid

(6.5.a) Semihypergroup of Atoms

Let $\Im$ be an algebra of events. We define the hyperoperation σ on $\Im$ as,

$$\forall a, b \in \Im, a\sigma b \text{ is the set of the atoms generated by } \{a, b\}.$$

We have that $(\Im, \sigma)$ is a commutative semihypergroup, called the *semihypergroup of atoms*.

For further details, see (Corsini and Leoreanu, 2003; Doria and Maturo, 1996).

(6.5.b) Hypergroupoid of Conditional Events

Let $\Im$ be an algebra of events. We define on $\Im$ the hyperoperation σ such that, for every a, $b \in \Im$,

$$a\sigma b = \{a \cap b, b\}. \tag{6.8}$$

If $b \neq \emptyset$, then $a\sigma b$ can be interpreted as the conditional event a/b.

We have that $(\Im, \sigma)$ is a weak associative and weak commutative hypergroupoid, called the *hypergroupoid of conditional events*.

For further details see (Corsini and Leoreanu, 2003; Doria and Maturo, 1996).

6.6 Hyperstructures Associated to a Family of Fuzzy Sets

(6.6.a) Hypergroup Associated to a Lattice

Let $(L, \vee, \wedge)$ be a lattice. For every a, $b \in L$, we put:

$$a\sigma b = \{x \in L: a \wedge b \leq x \leq a \vee b\}. \tag{6.9}$$

Since $\{a, b\} \subseteq a\sigma b$, the pair (L, σ) is a quasi-hypergroup, called *hypergroupoid associated* to L.

The following theorem holds:

Theorem 6.1 (Varlet, 1975). The associativity holds if and only if L is distributive. In this case (L, σ) is a join space.

(6.6.b) Join Space of the Fuzzy Sets with the Same Universe U

Let Φ be the family of the fuzzy sets on the same universe U. For every a, $b \in \Phi$, we put:

$$a\sigma b = \{u \in \Phi: \forall x \in U, \min\{a(x), b(x)\} \leq u(x) \leq \max\{a(x), b(x)\}\}, \tag{6.10}$$

that is, if $\vee$ and $\wedge$ are the union and the intersection of fuzzy sets:

$$a\sigma b = \{u \in \Phi: a \wedge b \leq u \leq a \vee b\}. \tag{6.11}$$

Since $(\Phi, \vee, \wedge)$ is a distributive lattice, Varlet theorem implies (Φ, σ) is a *join space*.

(6.6.c) Hypergroup of the Generators of Non Strict Archimedean t-Conorms
Let G be the set of the generators of non strict Archimedean t-conorms such that $g(1) = 1$, for any $g \in G$.

We define on G the hyperoperation σ such that, for every a, $b \in G$,

$$a\sigma b=\{g\in T:\forall x\in [0, 1], \min\{a(x), b(x)\} \leq g(x) \leq \max\{a(x), b(x)\}. \qquad (6.12)$$

Since the elements of G are fuzzy sets with universe [0,1], by the previous theorem it follows that (G, σ) is a *join space.*

(6.6.d) Hypergroup of the Non Strict Archimedean t-Conorms
Let C be the set of the non strict Archimedean t-conorms. The function h: $C \rightarrow G$ that to every element $c\in C$ associates its generator $g_c = h(c)$ is a one-one and onto function.

Let us consider the hyperoperation α on C such that:

$$\forall a, b\in C, a\alpha b = h^{-1}(h(a)\sigma h(b)). \qquad (6.13)$$

Then h is an isomorphism between (G, σ) and (C, α) and therefore (C, α) is a join space.

6.7 Conclusions

Many other applications of hyperstructures can be considered in decision making and in probability.

If A_1 and A_2 are two possible alternatives, then we can define $A_1 \sigma A_2$ as the set of all the alternatives that are related to both A_1 and A_2 by a particular point of view, for instance they are not strictly preferable or they are not dominated.

The way to handle with the subjective conditional probability by dealing with a particular hyperstructure having two hyperoperations is supplied in (Maturo, 2000).

References

1. Banon, G.: Distinction between several subsets of fuzzy measures. Int. J. Fuzzy Sets and Systems 5, 291–305 (1981)
2. Berres, M.: Lambda additive measure spaces. Int. J. Fuzzy Sets and Systems 27, 159–169 (1988)
3. Beutelspacher, A., Rosenbaum, U.: Projective Geometry. Cambridge University Press, Cambridge (1998)
4. Coletti, G.: Comparative probabilities ruled by coherence conditions and its use in expert systems. Int. J. General Systems 22, 93–101 (1994)
5. Coletti, G., Gilio, A., Scozzafava, R.: Coherent qualitative probability and uncertainty in artificial intelligence. In: Manicopoulos, C.N. (ed.) Proc. 8th. Int. Conf. Cybernetics and Systems, vol. 1, pp. 132–138. NIJT Press, New York (1990)
6. Coletti, G., Scozzafava, R.: Conditional probability, fuzzy sets, and possibility: a unifying view. Int. J. Fuzzy sets and Systems 144, 227–249 (2004)

7. Corsini, P., Leoreanu, L.: Applications of the Hyperstructure Theory. Kluwer Academic Publishers, London (2003)
8. Diecidue, E., Maccheroni, F.: Coherence without additivity. Journal of Mathematical Psychology 47, 166–170 (2003)
9. Doria, S., Maturo, A.: Hyperstructures and geometric spaces associated to a family of events. Rivista di Matematica Pura e Applicata 19, 125–137 (1996)
10. Dubois, D., Prade, H.: Focusing and updating: two concepts of conditioning in the setting of non-additive uncertainty models. In: Proc. FUR 6th, Cachan, June 15-18 (1992) (session A7, 1-3)
11. de Finetti, B.: Teoria delle probabilità, Einaudi, Torino. Published in English as Theory of Probability. J. Wiley, New York (1970)
12. Fishburn, P.: Nonlinear Preference and Utility Theory. Wheatsheaf Books, Brighton (1988)
13. Ling, C.H.: Representation of associative functions. Publ. Math. Debrecen 12, 189–212 (1965)
14. Marty, F.: Sur une généralization de la notion de group. In: IV Congrès des Mathématiciens Scandinaves, pp. 45–49. Stockholm (1934)
15. Maturo, A.: On a non-standard algebraic hyperstructure and its application to the coherent probability assessments. Italian Journal of Pure and Applied Mathematics 7, 33–50 (2000)
16. Maturo, A., Squillante, M., Ventre, A.G.S.: Consistency for assessments of uncertainty evaluations in non-additive settings in: Amenta, D'Ambra, Squillante and Ventre. In: Metodi, modelli e tecnologie dell'informazione a supporto delle decisioni, pp. 75–88. Franco Angeli, Milano (2006a)
17. Maturo, A., Squillante, M., Ventre, A.G.S.: Consistency for nonadditive measures: analytical and algebraic methods. In: Reusch, B. (ed.) Computational Intelligence, Theory and Applications. Advances in Soft Computing. Springer, Heidelberg (2006b)
18. Prenowitz, W., Jantosciak, J.: Join geometries. Springer Verlag VTM, New York (1979)
19. Saaty, T.L.: The Analytic Hierarchy Process. McGraw-Hill, New York (1980)
20. Schmeidler, D.: Subjective probability and expected utility without additivity. Econometrica 57, 571–587 (1989)
21. Squillante, M., Ventre, A.G.S.: Generating fuzzy measures. J. Math. Anal. Appl. 165(2), 550–555 (1992)
22. Squillante, M., Ventre, A.G.S.: A Yoshida-Hewitt like theorem for decomposable measures. Ricerche di Matematica 37, 204–212 (1988)
23. Squillante, M., Ventre, A.G.S.: Consistency for uncertainty measures. International Journal of Intelligent Systems 13(5), 419–431 (1998)
24. Sugeno, M.: Theory of fuzzy integral and its applications, Ph. D. Thesis, Tokyo (1974)
25. Varlet, J.C.: Remarks on Distributive lattices. Bull. Dell'Acad. Polonnaise des Sciences, Série des Sciences Math. Astr. et Phys. XXII(11) (1975)
26. Ventre, A.G.S.: Decomposable expected utility. Rivista di Matematica della Università di Parma 5, 1–14 (1996)
27. Weber, S.: Decomposable measures and integrals for Archimedean t-conorms. J. Math. Anal. Appl. 101(1), 114–138 (1984)

Measures for Firms Value in Random Scenarios*

Paola Modesti

Abstract. The value of a firm cannot be totally independent of the financial context in which the firm operates. In this paper we propose a set of axioms in order to characterize appropriate measures of the (random) value of a company which provides a (sublinear) valuation functional consistent with the existence of a financial market. It allows to give an upper and a lower bound to the value of a firm.

Finally, in a random context, we consider some classical valuation methods and test them with respect to the axioms.

Keywords: Firms Valuation, Corporate Finance, Decision Theory.

1 Introduction

In this paper we propose a criterion to individuate appropriate measures for a company valuation in a random context. We will assume the existence of a financial market whose prices implicitly reveal (although partially) the probabilities of the states of the world underlying the traded assets. It will be used to give an upper and a lower bound to the value of a firm.

Our main result derives the representation for a valuation functional on the basis of a set of axioms, which appears to be quite natural from an economic point of view. The functional we will obtain is the maximum (or the minimum) of the expected values of the random variable describing the value of the firm in the different probabilistic scenarios consistent with a given market. Such a result is along the lines of the one of Arztner et al. [1] in Risk Measures.

Paola Modesti
Faculty of Economics, Parma University, via Kennedy 6/b, 43100 Parma, Italy
e-mail: paola.modesti@unipr.it

* This work was partially supported by MIUR.

S. Greco et al. (Eds.): Preferences and Decisions, STUDFUZZ 257, pp. 305–319.
springerlink.com

In the second part of the paper we will examine some among the most common valuation criteria for a company. In particular, methods based on the present value of future cashflows or returns and valuation through real options will be considered[1] and finally we will test such criteria with respect to our axioms.

To fix some idea let us refer to a very simplified example. Consider an oil company and assume that oil price will uniquely determine the value of the company. The current valuation is that the future oil price per barrel will vary in some interval. For the sake of simplicity, let us consider only two prices, say 50 and 100 dollars. The shareholders and a group of potential buyers ask to a statistician, a market analyst and a manager in the oil field to valuate the firm through the EVA method. The researchers attribute to the first scenario (50 dollars) probability 50%, 40% and $33,\bar{3}\%$ respectively. The values V of the company in the two cases are 150 or 210 million dollars. As a consequence, valuation gives the three expected values 180, 186 and 190. It seems natural that shareholders will take as ask price:

$$\max_p \mathrm{E}_p[V] = \max[180, 186, 190] = 190$$

whereas buyers will choose as bid price:

$$\min_p \mathrm{E}_p[V] = \min[180, 186, 190] = 180.$$

The usual methods for firms valuation are often based on investigations in sure contexts. But this seems to be rather unrealistic.

The events of Autumn 2008 show that the present crisis is, without doubt, the worst of the west world's Economics and that almost all valuation methods and control rules failed.

A long list of evidences supports our point (as Greeks used to say: Future is on Jupiter's lap!). Simply think, for instance:

(*i*) On July 2008 in three days the semipublic banks Fannie Mae and Freddie Mac, born with the aim to grant US middle classes loans for houses, suffered a 90% decrease in their stocks. At that moment their rating was triple A. Later on they became insolvent and the U. S. Government nationalized them in order to avoid their bankrupt.

(*ii*) The successive default of Lehman Brothers (September 2008) is giving rise to a chain reaction in firms whose long term effects nobody may foresee.

(*iii*) For months stocks values of quoted companies went through a frightful up and down showing that all the sure valuations turn out to be inconsistent. Only one instance: on December 2008 Morgan Stanley raised the objective-price of the Fastweb stock for the period 2008-2010 from 30 to 33 euros, whereas UBS lowered it from 24 to 17 euros. Both the motivations seem reasonable: the American bank values that the sector of telecommunications is underestimated and could

[1] See, among others, [6], [12], [13], [15] and [35].

produce surprises. On the contrary, the Swiss analysts fear the growing pressure of competition.

(*iv*) In sure processes of valuation catastrophic scenarios need to be totally ignored and in many cases this determines an irresponsible overestimate of the firm's value.

It is a matter of fact that the most important fields of Economics are actually investigated in a random context.

On a financial ground, dating from Fifties, markets are represented in random frameworks. Risk Measures Theory models risks (the same risks owned by firms) as random variables and the coherent risk measure proposed by Artzner et al.[2] in 1999, as well as some of common measure criteria (e.g. the S.P.A.N. method, the worst conditional expectation and the tail conditional expectation criteria), move from different states of the world. Of course every model in Decision Theory deals with random variables or random acts[3].

Hence, the possible absence of randomness in firms valuation methods appears deeply out of date. A more rational way to a realistic valuation would consists in considering the value of a firm as a random variable[4] to be synthesized through an appropriate functional.

Valuation and measure problems for economic or financial random variables are central in every economic field (Finance, Operational Research, Management Theory, Decision Theory, etc.). Although many valuations are ultimately expressed by a unique price, in the reality often two different prices are meaningful: an upper price and a lower price (in some sense, an ask and a bid price). In particular, a correct valuation of an economic phenomenon (think, for instance, of the frequent exigency to set some limitations to negotiations) makes sense if it is an "interval valuation".

In the last years, an innovative concept caught on. *By nature*, a coherent (in particular, leading to non-arbitrage in financial problems or to rational behaviours in Decision Theory) price system values a random variable through a maximum or a minimum of expected values with respect to a set of probabilities or, in particular, a single expectation. The idea is simple: often a market or an individual are not informed enough to express a unique probability measure for the states of the world, but acts as if it considers a set of probabilities which appears reasonable and consistent with the information available at the moment. Such family of probabilities leads to a collection of expected values among which the more adequate price has to be chosen. In the most conservative cases the choice will coincide with their maximum or their minimum.

In particular, such kind of characterization of prices and/or measures have been proposed first in Decision Theory by Gilboa and Schmeidler in 1989 [17], in Risk

[2] See [1].

[3] See [17].

[4] A good example of value measures in random scenarios is given by valuations in insurance contexts. See, among others, also for references, [14], [26], [30] and [31].

Theory by Arztner et al. in 1999 [1] and, a little bit later, in Insurance and Finance[5] (see, among others, [3], [8], [9], [10], [14], [18], [19], [20], [23], [24], [26], [30] and [31]).

As said above, our proposal consists in introducing a similar approach in the theory of firms valuation.

In Section 2 we will present a set of axioms for a value measure and the representation theorem. In Section 3, we will consider some class of valuation methods in a random framework and check if such criteria satisfy the axioms. In Section 4, some final consideration will conclude the paper.

2 Value Measures and Representation Theorem

In this section we propose axioms for a value measure of a firm and give a representation theorem. Such a measure is consistent with the existence of a financial market and leads to an upper and a lower value for the firm to be interpreted, roughly speaking, as the ask price and the bid price.

Let $[0,T]$ be one period of uncertainty. For now, we can think of T as a long period or as the period of life of the company[6]. Let $\Omega = \{\omega_1, \omega_2, \ldots, \omega_m\}$ be a finite set of m states of the world where Ω is large enough to include all the states of interest for the firm's valuation and for the reference market.

We assume the existence of a perfect (complete or incomplete) financial market without arbitrage opportunities. Let $M \subseteq \mathbb{R}^m$ be the linear subspace of the payoffs of the traded securities. Let the prices system for the assets be linear on M. Under these assumptions the market is frictionless and the *one price law* holds[7].

Furthermore, let the riskless bond paying 1 in each state of the world belong to M. Observe that such a requirement is always satisfied by the markets. This guarantees finite[8] prices for the quoted securities and that $\Phi = \Phi(T)$, the discount factor expressed by the market, is constant.

Because of no arbitrage hypothesis, the Fundamental Theorem of Asset Pricing guarantees the existence of one (at least) probability measure p such that, denoting by π the prices of the assets in M:

$$\pi(m) = \Phi E_p[m] \qquad \forall m \in M$$

---

[5] Even if we must observe that in these last contexts investigations are almost at the beginning, in spite of the first intuition that may be found in a paper of 1933 by de Finetti and Obry [16].

[6] About T see also Section 3.

[7] Prices also could be supposed to be sublinear. For a treatment of the Sublinear State Preference Model, see, for instance, [8] and, especially, [20].

[8] In fact, with this assumption the sufficient condition for prices finiteness:

$$M \cap \mathbb{R}^m_{++} \neq \emptyset$$

(that is, M contains at least one security with a strictly positive payoff) is satisfied.

Such probabilities are said to be risk neutral. Then, $\pi : M \to \mathbb{R}$ is a linear and increasing functional.

Let us consider the set of random variables X, where $X = X(\omega) \in \mathbb{R}^m$ represents the value of a firm at time T if state ω prevails.

Definition. A functional $\mu : \mathbb{R}^m \to \mathbb{R}$ is a value measure (or price) for a firm if satisfies the following axioms:

A1 Positive homogeneity:
$\mu(\alpha X) = \alpha\mu(X), \quad \forall\alpha \in \mathbb{R}_+, \quad \forall X \in \mathbb{R}^m.$

A2 Monotonicity:
If $Y(\omega) \geq X(\omega)$ for any $\omega \in \Omega$, then $\mu(Y) \geqslant \mu(X), \quad \forall X, Y \in \mathbb{R}^m.$

A3 Market consistency:
$\mu(X + m) = \mu(X) + \pi(m), \quad \forall X \in \mathbb{R}^m, \quad \forall m \in M.$

A4 Subadditivity:
$\mu(X + Y) \leqslant \mu(X) + \mu(Y), \quad \forall X, Y \in \mathbb{R}^m.$

Remarks. • A1 guarantees the invariance of μ in front of a change of currency;

- A2 is quite natural: if in each state of the world a firm has a higher value than another, the same must happen for their values;
- A3 states that traded marketed assets and liabilities are valued at their market prices;
- A4 says that the value of two (or more) companies cannot exceed the sum of their values. If not so, to buy them separately would be more convenient than a unique purchase. In many cases the two sides of inequality in A4 will be equal (the value of two independent companies is the sum of the values of each of them), but in case of a link between the two firms (they belong to the same group, they are partners in a same business, they are different divisions of a unique company, they cover an important share of the market, etc.), the strict inequality will hold. However, in some cases it may appear opportune to reverse the inequality, that is to ask for superadditivity, by introducing the axiom A4':

  A4' Superadditivity:
  $\mu(X + Y) \geqslant \mu(X) + \mu(Y), \quad \forall X, Y \in \mathbb{R}^m$

  This can be, for instance, the case of mergers between complementary firms where possible synergy effects[9] can create a future surplus of value for the new company.
- Finally, mathematically speaking, A1 and A4 guarantee the sublinearity of μ.

Representation Theorem. Let P be the set of risk neutral probabilities expressed by the market.

The functional $\mu : \mathbb{R}^m \to \mathbb{R}$ satisfies Axioms A1 - A4 and hence is a value measure (or price) for a firm if and only if there exists a set $Q \subseteq P$ of probabilities such that:

$$\mu(X) = \Phi \sup_{p \in Q} \mathrm{E}_p[X] \quad \forall X \in \mathbb{R}^m$$

[9] See, for instance, [25] and, also for a review, [11].

The set Q can be always taken compact and convex and in such a case it is unique and the supremum becomes a maximum.

Proof. See Appendix. □

The economic meaning of the theorem is clear: the valuation works as if different possible scenarios are considered. The upper value of every firm is the expected value of its future performances obtained through the scenario which is the most favourable for the firm. Furthermore, from the sublinearity of μ one has:

$$\mu(X) = \Phi \max_{p \in Q} \mathrm{E}_p[X] \geqslant \Phi \min_{p \in Q} \mathrm{E}_p[X] = -\mu(-X) \quad \forall X \in \mathbb{R}^m$$

which suggests to interpret $\mu(X)$ as the ask price and $-\mu(-X)$ as the bid price. The bid price is computed considering the scenario which is the most severe for the firm and, as usual, bid price is not greater than ask price.

Remarks. • An analogous representation holds with infimum instead of supremum if μ satisfies Axioms A1 - A4'.

- In the ideal case of a complete financial market, that is if each random variable representing the future value of a firm can be replicated, there is a unique risk neutral probability p and:

$$\mu(X) = \Phi \mathrm{E}_p[X] = \pi(X) \quad \forall X \in \mathbb{R}^m$$

  that is, μ is linear.
- Since a sublinear functional is increasing if and only if:

$$X \geqq 0 \quad \Longrightarrow \quad -\mu(-X) \geqslant 0 \quad \forall X \in \mathbb{R}^m$$

  whenever $X \geqq 0$ it will be:

$$\mu(X) \geqslant -\mu(-X) \geqslant 0$$

  which guarantees the positivity of both the prices.
- If, for some X it is $\mu(X) = -\mu(-X)$, then:

$$\mu(Y - X) = \mu(Y) - \mu(X) \quad \forall Y \in \mathbb{R}^m$$

- The inclusion $Q \subseteq P$ says that some scenarios considered by the financial market can be excluded. Furthermore, if we consider two different value measures μ_1 and μ_2 with the respective probability families Q_1 and Q_2 ($Q_1, Q_2 \subseteq P$), we can emphasize the role assumed by different level of information in the valuation process:

  (*i*) if (and only if) Q_1 is more informative than Q_2, both bid and ask Q_1-prices are more convenient than Q_2-prices:

$$Q_1 \subseteq Q_2$$

if and only if

$$\mu_1(X) = \Phi \max_{p \in Q_1} \mathrm{E}_p[X] \leqslant \Phi \max_{p \in Q_2} \mathrm{E}_p[X] = \mu_2(X) \quad \forall X \in \mathbb{R}^m$$

(*ii*) standard separation theorems allow to show the following results:
- if $Q_1 \cap Q_2 = \emptyset$, there exists a random variable $X \in \mathbb{R}^m$ such that:

$$-\mu_1(-X) \leqslant \mu_1(X) < -\mu_2(-X) \leqslant \mu_2(X)$$

that is the ask price obtained via Q_1 is smaller than the bid price obtained via Q_2;
- if $Q_1 \cap Q_2 \neq \emptyset$ and $Q_i \nsubseteq Q_j$, $i, j = 1, 2$, $i \neq j$, there exist at least two random variables X and Y in $\mathbb{R}^m$ such that:

$$\mu_1(X) > \mu_2(X) \qquad \text{and} \qquad \mu_1(Y) < \mu_2(Y)$$

that is some random variable is better valued by μ_1 and some other by μ_2.

Finally:

- positive homogeneity entails $\mu(0) = 0$;
- as every sublinear functional defined on a linear space is continuous, μ is continuous on $\mathbb{R}^m$.

3 Some Common Valuation Criteria

This section aims to consider some usual valuation methods. In particular, we will focus our attention on the large class of criteria based on future cashflows or returns and on the real options method.

3.1 DCF Methods

In the financial community mainly three classes of methods caught on in the last decades[10]: the Discounted Cashflows method (DCF) and some of its variants, like as the Free Cashflows (FCF) and the Dividend Discount Models (DDM), the market multiples criterion and the valuation through the Economic Value Added (EVA). Such methods are usually presented as deterministic, even if a simplified random framework can be recognized as starting point in several practical problems, i.e.:

(*i*) Every method is applied to more than one scenario. Usually at least three future possible situations are analyzed: a plausible (and/or forecast) performance for the company together with a better and a worse ones.
(*ii*) A good valuation is based on two or more criteria.
(*iii*) If the level of randomness is very high, a sensitivity analysis completes the investigation.

[10] See, among others, [2], [4], [6], [13], [33] and [34].

At the end of the process, the analyst expresses a valuation which represents a sort of compromise among the different results.

Therefore, although the context of valuation is deterministic, many scenarios are analyzed and, sometimes, stochastic tools contribute to the final valuation (think, for instance, of computing of expected earnings or expected dividends). Hence, also relying on the words of Modigliani and Miller who in their famous paper[11] of '61 suggested to approach the whole valuation problem in conditions of uncertainty, a random approach can at least partially mitigate subjectivity or factiousness.

Therefore, we will consider a random version of the FCF, DDM and EVA criteria[12].

The model we study is a unified generalized version of the mentioned criteria. We consider the capital structure of the company as given. Without entering upon the subject, the enterprise value measured through these methods has the form:

$$V = C + \sum_{t=1}^{T} \frac{a_t}{(1+\rho)^t} + \frac{F}{(1+\rho)^T} \qquad (1)$$

where usually C is not null only in the EVA model in which it is the capital committed to the business at the beginning of the period.

The valuation consists substantially of two addenda. The former is the sum of the present values of future free cashflows, or dividends or EVA's in the next T years. Usually in FCF and EVA models the discounting rate ρ is the weighted average cost of capital (WACC), whereas in DDM it is the opportunity cost of equity.

The latter is the "final value" of the company and represents the present value of all FCF's, dividends and EVA's respectively, after the year T. The quantity F is frequently assumed to be the present value in T of a perpetual annuity whose installments grow exponentially at rate[13] $g < \rho$ and has the form:

$$F = \left[C_T + \frac{a_{T+1}}{(\rho - g)} \right]$$

where usually C_T is null in FCF model and in DDM, whereas in EVA model it is the sum of C and all the next investments up to T (included). Furthermore, we suppose $a_{T+1} = a_T(1+g)$.

[11] See [29], p. 426.

In a previous paper (see [28], p. 267 and following) the Authors, moving from the formula $V = E + D$ where V is the value of the firm, E is the equity and D is the debt, obtain, under opportune assumptions, $V = \overline{X}/\rho$ where $\overline{X}$ is the expected return of a stream of earnings due to the assets of the firm and ρ may be seen as the expected rate of return of any share of the company or as the market rate of capitalization for the expected value of the uncertain cashflows generated by the firm.

[12] Because of their wide diffusion, we limit our analysis to these three indicators, but numerous other models are on the same lines (think, for instance, of the cashflows return on investment (CFROI) model).

[13] After T years the company is supposed to be *mature*, that is to have run out its skill to grow at elevated rhythms. Usually g is assumed to be near to zero.

The period $[0,T]$ is said Competitive Advantage Period (CAP) or explicit period and usually its length is of 5 -10 years. It is the period in which it is easier to express a precise forecast and/or the time in which the (*growth*) firm profits by a competitive advantage due, for instance, to recent investments. Often it depends on the economic (maybe sectorial) cycle as well.

It is well known[14] that, under opportune hypotheses, the three models are equivalent, that is, they attribute to a firm the same value.

Criterion (1) may be restated in random terms. Let us assume that information on the market is described in the usual form[15] through a filtration of algebras $\mathscr{A}_t$, $t=0,1,\ldots,T$. Therefore each a_t, since it is a risky amount[16] with maturity t, has to be intended as a random variable which results to be measurable with respect to $\mathscr{A}_t$.

It is immediate to verify the result:

Proposition. Let $\Omega=\{\omega_1,\omega_2,\ldots,\omega_m\}$ be the set of the states on $[0,T]$. If:

$$X=C(1+\rho)^T+\sum_{t=1}^{T}a_t(1+\rho)^{T-t}+F$$

is the random value of a firm in T, the quantity

$$\mu(X)=\frac{1}{(1+\rho)^T}\sup_{p\in Q}\mathrm{E}_p[X]=C+\sup_{p\in Q}\sum_{s=1}^{m}\left[\sum_{t=1}^{T}\frac{a_t}{(1+\rho)^t}+\frac{F}{(1+\rho)^T}\right]p_s$$

(where $p_s=p(\omega_s)$, $s=1,2,\ldots,m$, is the $s-$th component of $p\in Q$), is a value measure (or price) for the firm for any $Q\subseteq P$, where P is the set of risk neutral probabilities expressed by the market.

The following example illustrates the criterion. For the sake of simplicity, let $T=2$. Let $\Omega=\{\omega_1,\omega_2,\omega_3,\omega_4\}$ and:

$$p(\omega_1)=\frac{15-55\alpha}{60}$$
$$p(\omega_2)=\frac{24+20\alpha}{60}$$
$$p(\omega_3)=\alpha$$
$$p(\omega_4)=\frac{21-25\alpha}{60}$$

The condition $0<\alpha<3/11$ guarantees the absence of arbitrage opportunities.

Let WACC = 5% and the growth rate $g=1\%$. Let us assume that the future free cashflows of a company (in hundreds of thousands of euros) are:

[14] See, for instance, in a very wide literature, [27], [29], [33] and [34].

[15] Such data can be obtained on the basis of time series and/or of the opinions of the analysts.

[16] A more general version of the criterion may be obtained taking one or more of the considered rates as random.

$$a_1(\omega_1) = a_1(\omega_2) = 100$$
$$a_1(\omega_3) = a_1(\omega_4) = 200$$

and:

$$a_2(\omega_1) = 50$$
$$a_2(\omega_2) = 110$$
$$a_2(\omega_3) = 150$$
$$a_2(\omega_4) = 180$$

Therefore, the value of the firm in $T = 2$ is:

$$X = \begin{cases} X(\omega_1) = 100 \cdot 1.05 + 50 + \frac{50 \cdot 1.01}{0.04} = 1,522.5 \\ X(\omega_2) = 100 \cdot 1.05 + 110 + \frac{110 \cdot 1.01}{0.04} = 3,097.5 \\ X(\omega_3) = 200 \cdot 1.05 + 150 + \frac{150 \cdot 1.01}{0.04} = 4,042.5 \\ X(\omega_4) = 200 \cdot 1.05 + 180 + \frac{180 \cdot 1.01}{0.04} = 4,83 \end{cases}$$

The consequent ask and bid prices with the above probabilities are:

$$\mu(X) = \frac{1}{1.05^2} \sup_p \mathrm{E}_p[X] = \frac{1}{1.05^2} \sup_{0<\alpha<3/11} [3,012.22 + 1,516.86\alpha]$$
$$= 3,425.91$$

and

$$-\mu(-X) = \frac{1}{1.05^2} \inf_p \mathrm{E}_p[X] = \frac{1}{1.05^2} \inf_{0<\alpha<3/11} [3,012.22 + 1,516.86\alpha]$$
$$= 3,012.22$$

3.2 Real Options

In their celebrated paper [5], Black and Scholes, under simple assumptions, showed also that the future value of a company for its shareholders can be seen as the final payoff of an option which will pay the difference between the shares value and a possible debt with bondholders or nothing in case of default.

In the Eighties a complete theory for the valuation of a firm through options was developed[17], yielding to a random value at maturity for a firm and/or a project.

Often in the valuation of a firm or of a project, the usual methods turn out to be exaggeratedly static, that is, they can fail to capture the flexibility embedded in the project. Then, many different options should be considered: the option to defer

[17] See, for instance, [12], [15], [21], [32] and [35].

an investment, to alter the operating scale, to abandon the project, etc. Hence the theory of real options proposes to valuate the project as a financial option. The final valuation will give the price of the option or the sum of it, seen as the value of the potential future activities, and of the value of the existing ones. Many real options are similar to European or American calls or puts written on a stock giving or not dividends during the life of the option.

Generally speaking, in the cases in which the price of the option is the present value of the (maximum or minimum) expected value of its final payoff depending on risk neutral probabilities, the resulting value for the project is a measure satisfying the above axioms and, hence, a value measure.

Sometimes real options result to be very complex, but here we do not desire to enter in technicalities. We confine ourselves to examine a simple case of an *option to defer* we will present together with an example.

Let us consider the opportunity to invest now the amount I in a project which will pay in $T = 1$ the random amount V which will assume the value V^+ if a given event A will be true and V^- if A^{C}, the complement of A, will be true. Denote with p the probability of A in a risk neutral world and with r the riskless interest rate. Finally, in order to consider the most meaningful case, let $V^+ > I(1+r) > V^-$.

It is:

$$X = \begin{cases} V^+ - I(1+r) & \text{if } A \\ V^- - I(1+r) & \text{if } A^{\mathrm{C}} \end{cases}$$

and:

$$\mu(X) = \frac{1}{1+r} \max_p \mathrm{E}_p[X] = \frac{1}{1+r} \max_p \mathrm{E}_p[V] - I$$

is the current value of the project[18].

For instance, we could think of a company which can invest 100 millions of dollars to buy a TV network, knowing that, in one year, the Government will decide if it can broadcast on all the national territory without particular advertising constraints. If the decision will be favourable to the company (event A), the expected value from subsequent cashflows discounted back to an estimate of WACC is, in $T = 1$, 550 millions of dollars, whereas, in the opposite case, is 55. Let $r = 10\%$. It is:

$$\mu(X) = \max_p \,(450p - 50)$$

Let us suppose that the firm could defer undertaking the project in T, when the Government's decision will be known. In other words, the company owns an European call option[19] to wait on the project value with a strike price equal to $I(1+r)$. The value of the project in T is now the payoff of the call, that is the random variable:

---

[18] The quantity $\mu(X)$ is a value measure for the project because obtained with the above DCF method.

[19] For the sake of simplicity we ignore possible dividend-like effects.

$$Y = \begin{cases} \max\left[V^+ - I(1+r), 0\right] = V^+ - I(1+r) & \text{if } A \\ \max\left[V^- - I(1+r), 0\right] = 0 & \text{if } A^C \end{cases}$$

and the value measure for the project is:

$$\mu(Y) = \frac{1}{1+r}\max_p \mathrm{E}_p[Y] = \frac{1}{1+r}\max_p \left[V^+ - I(1+r)\right]p$$

It is easily to verify that also $\mu(Y)$ is a value measure for the firm.

Finally, the results provide also the value of the option to defer:

$$\mu(Y) - \mu(X) = \frac{1}{1+r}\max_p \left[I(1+r) - V^-\right](1-p) > 0$$

In the example, it is:

$$Y = \begin{cases} 440 & \text{if } A \\ 0 & \text{if } A^C \end{cases}$$

$$\mu(Y) = \max_p \left[400p\right]$$

and:

$$\mu(Y) - \mu(X) = \max_p 50(1-p)$$

4 Conclusions

The valuation of a firm is a deeply subjective process. To a great extent the final result depends on the beliefs of the analyst and on the inputs he considers. Surely a measure of value based on a set of axioms does not solve the problem (which, however, from a theoretical point of view, cannot be solved: probably a really objective value does not even exist), but may help at least to provide a rational starting point for the valuation and to avoid some excess.

In a globalization time, the price of a firm cannot be totally independent of the financial context in which the firm operates. Furthermore, this approach focuses the attention on the states of the world: often the interest of the valuation is not only in the random future flows and in the consequent value, but also in the events which can give the different earnings or losses. We start from a financial market providing linear prices for quoted assets and individuate a link between the market and the value of a firm. The presence of the market determines an upper and a lower bound for the value of the firm. Thus, there exists an interval of plausible values for the firm, $\left(\Phi \inf_{p\in Q} \mathrm{E}_p[X], \Phi \sup_{p\in Q} \mathrm{E}_p[X]\right)$, and, according to the different exigencies of the valuation (purchase, quotation on Stock Exchange, etc.), the economic context will determine the final price. As natural choice, we suggest to fix the ask and bid price to the maximum (from the best scenario of the firm) and the minimum (from the worst scenario) of the interval respectively. Mathematically, this means to extend the price functional for the financial market defined on M to a sublinear functional on the set of all the random variables defined on Ω.

The incompleteness of the market implies the existence of infinitely many risk neutral probabilities: it is as though the market would consider possible different probabilistic scenarios, the ones consistent with the prices of the quoted assets. Hence, the probabilistic ambiguity gives more that one price.

Further research may develop along two lines: the model can be stated in the continuous case and current criteria of valuation may be deeply investigated. In particular, real options, at the beginning seen as a too abstract tool, are now recognized as a valid instrument for the valuation especially in the cases of new lines of production in which an initial absence of earnings can make inadequate the traditional methods. Then, different types of real options may deserve to be considered. Finally, a good matter could be the analysis of the market multiples method.

References

1. Artzner, P., Delbaen, F., Eber, J.-M., Heath, D.: Coherent Measures of Risk. Math. Finance 9, 203–228 (1999)
2. Bacidore, J.M., Boquist, J.A., Milbourn, T.T., Thakor, A.V.: The Search for the Best Financial Performance Measure. Financial Analysts J. 53, 11–20 (1997)
3. Bensaid, B., Lesne, J.P., Pagës, H., Scheinkman, H.: Derivative Asset Pricing with Transaction Costs. Math. Finance 2, 63–86 (1992)
4. Berk, J.B., Green, R.C., Naik, V.: Optimal Investment, Growth Options and Security Returns. J. Finance 54, 1553–1607 (1999)
5. Black, F., Scholes, M.: The Pricing of Options and Corporate Liabilities. J. Political Econ. 81, 637–659 (1973)
6. Brealey, R., Myers, S.: Principle of Corporate Finance. McGraw-Hill, New York (2008)
7. Brézis, H.: Analyse Fonctionelle. Thé orie et Applications. Masson, Milan (1983)
8. Castagnoli, E., Favero, G.: Going Sublinear: the Fundamental Theorem of Asset Pricing in the State Preference Model with Proportional Frictions, Studi di Matematica finanziaria ed attuariale 37, Università Commerciale "L. Bocconi", Milan (2008)
9. Castagnoli, E., Favero, G.: On the Completeness of a Constrained Market. Int. J. Appl. Manag. Sci. 1, 90–96 (2008)
10. Castagnoli, E., Maccheroni, F., Marinacci, M.: Insurance Premia Consistent with the Market, Insur. Math. Econ. 31, 267–284 (2002)
11. Cigola, M., Modesti, P.: A Note on Mergers and Acquisitions, Manag. Finance 34, 221–238 (2008)
12. Copeland, T., Antikarov, V.: Real Options, a Practitioner's Guide, Texere, New York (2001)
13. Copeland, T.E., Weston, J.F.: Financial Theory and Corporate Policy. Addison Wesley, New York (2004)
14. Daykin, C.D., Pentikäinen, T., Pesonen, M.: Practical Risk Theory for Actuaries. Chapman & Hall, London (1995)
15. Dixit, A.K., Pindyck, R.S.: Investment Under Uncertainty. Princeton University Press, Princeton (1994)
16. de Finetti, B., Obry, S.: L'Optimum nella Misura del Riscatto. In: Atti del Secondo Congresso Nazionale di Scienza delle Assicurazioni, Rome, vol. II, pp. 99–123 (1933)

17. Gilboa, I., Schmeidler, D.: Maxmin Expected Utility with Non-Unique Prior. J. Math. Econ. 18, 141–153 (1989)
18. Harrison, J.M., Kreps, D.: Martingales and Arbitrages in Multiperiods Securities Markets. J. Econ. Theory 20, 381–408 (1979)
19. Jouini, E.: Price Functionals with Bid-Ask Spreads: an Axiomatic Approach. J. Math. Econ. 34, 547–558 (2000)
20. Jouini, E., Kallal, H.: Martingales and Arbitrages in Securities Markets with Transactions Costs. J. Econ. Theory 66, 178–197 (1995)
21. Kellogg, D., Charnes, J.M.: Real-Options Valuation for a Biotechnology Company. Financial Analysts J. 56, 76–84 (2000)
22. Kolmogorov, A.N., Fomin, S.V.: Introductory Real Analysis. Dover Publications, New York (1970)
23. Koehl, P.F., Pham, H.: Sublinear Price Under Portfolio Constraints. J. Math. Econ. 33, 339–351 (2000)
24. Kreps, D.: Arbitrage and Equilibrium Economics with Infinitely Many Commodities. J. Math. Econ. 8, 15–35 (1981)
25. Larson, K.D., Gonedes, N.J.: Business Combinations: an Exchange Ratio Determination Model. Account. Rev. 44, 720–728 (1969)
26. Longley-Cook, A.G.: Risk-Adjusted Economic Value Analysis. North Am. Actuar. J. 2, 87–100 (1998)
27. Modesti, P.: EVA and NPV: Some Comparative Remarks. Math. Methods Econ. Finance 2, 55–70 (2007)
28. Modigliani, F., Miller, M.H.: The Cost of Capital, Corporate Finance and the Theory of Investment. Am. Econ. Rev. 48, 261–297 (1958)
29. Modigliani, F., Miller, M.H.: Dividend Policy, Growth and the Valuation of Shares. J. Bus. 34, 411–433 (1961)
30. Pemberton, J., Allen, I., Brown, D., Hardwick, S., Shah, N., Stevens, A., Wilson, C.: Classifying DCF Models to Explicate EVA. The Staple Inn Actuarial Society, London (2000)
31. Rolski, T., Schmidli, H., Schmidt, V., Teugels, J.: Stochastic Processes for Insurance and Finance. Wiley & Sons, New York (2000)
32. Schwartz, E.S., Moon, M.: Rational Pricing of Internet Companies. Financial Analysts J. 56, 62–75 (2000)
33. Shrieves, R.E., Wachowicz Jr., J.M.: Economic Value Added and Net Present Value: a Reconciliation of Variations of Discounted Cashflow Valuation. Eng. Econ. 46, 33–51 (2001)
34. Stewart, G.B.: The Quest for Value: the EVA^{TM} Management Guide. Harper Business, New York (1991)
35. Trigeorgis, L. (ed.): Real Options in Capital Investment, Models, Strategies and Applications, Praeger, Westport (1995)

A Appendix: Representation Theorem: Proof and Comments

Definition. A functional $\mu : \mathbb{R}^m \to \mathbb{R}$ is:

(i) positive if whenever $x \geqq 0$, $\mu(x) \geqslant 0$;
(ii) increasing if whenever $y \geqq x$, $\mu(y) \geqslant \mu(x)$.

Let us remind that if M is a proper linear subspace of $\mathbb{R}^m$, a linear functional π on M may be linearly extended in infinite ways on $\mathbb{R}^m$. The maximum of such extensions is sublinear and finite.

The proof of the theorem is similar to the proof of the main result of Artzner et al. [1] for coherent risk measures. It is based on some well known characterizations of sublinear functionals which we recall briefly. In particular, the first result is a consequence of the Hahn-Banach theorem[20].

Lemma 1. *A functional $\mu : \mathbb{R}^m \to \mathbb{R}$ is sublinear if and only if it has the form:*

$$\mu(x) = \max_{\varphi \in Q} \varphi(x)$$

where Q is a set of linear functionals $\varphi : \mathbb{R}^m \to \mathbb{R}$ which may be taken compact and convex (and in such a case Q is unique).

Lemma 2. *A sublinear functional $\mu(x) = \max_{\varphi \in Q} \varphi(x)$ is increasing if and only if all the elements of Q are positive.*

Proof of the Representation Theorem. The first part of the implication (if) is obvious. For the second part (only if):

- A1 and A4 state the sublinearity of μ. Hence, for Lemma 1, there exists a unique compact and convex set Q of linear functionals $\varphi : \mathbb{R}^m \to \mathbb{R}$ such that:

$$\mu(x) = \max_{\varphi \in Q} \varphi(x)$$

- A2 asks for increasing monotonicity of μ. Hence, for Lemma 2, all the elements of Q are positive;
- A3 implies:

$$\mu(m) = \pi(m) \qquad \forall m \in M$$

Then, for the Hahn - Banach Theorem, μ is the maximum finite sublinear extension of π. Furthermore, $Q \subseteq P$ and $\varphi(x) = \Phi px = \Phi \mathrm{E}_p[X]$.
The thesis follows. □

[20] See, for instance, [7] and [22] and, for financial aspects, [3], [8], [18], [20], [23] and [24].

Thin Rationality and Representation of Preferences with Implications to Spatial Voting Models

Hannu Nurmi*

1 Introduction

Much of current micro economic theory and formal political science is based on the notion of thin rationality. This concept refers to the behavioral principle stating that rational people act according to their preferences. More precisely, a rational individual chooses A rather than B just in case he/she (hereafter he) prefers A to B. Provided that the individual's preference is a binary, connected and transitive relation over alternative courses of action, we can define a utility function that represents the individual's preferences so that when acting rationally – i.e. in accordance with his preferences – he acts as if he were maximizing his utility.[1] When considering risky alternatives, i.e. probability mixtures of certain outcomes, similar representation theorem states that the individual's preferences can be represented as a utility function with an expected utility property. These utility functions assign risky prospects utility values than coincide with weighted sums of the utility values of those outcomes that may materialize in the prospect. The weights, in turn, are identical with the probabilities of the corresponding outcomes.

In spatial models, the individuals are identified as their ideal points in a space. Similarly the decision alternatives are represented as points in the

Hannu Nurmi
Public Choice Research Centre
and
Department of Political Science
University of Turku
FI-20014 Turku
Finland
e-mail: hnurmi@utu.fi

* The author is grateful to Marlies Ahlert, John Roemer and Donald G. Saari for advice, encouragement and comments on earlier drafts of this article.

[1] In infinite alternative settings some additional technical assumptions are needed, see [12].

S. Greco et al. (Eds.): Preferences and Decisions, STUDFUZZ 257, pp. 321–337.
springerlink.com

space. In strong spatial models the individual i's evaluations of alternatives are assumed to be related to a distance measure d_i defined over the space. Moreover, each individual i is assumed to have an ideal point x_i in the space so that

$$x \succeq y \Leftrightarrow d_i(x, x_i) \leq d_i(y, x_i), \forall x, y \in W$$

In words, each individual is assumed to prefer those alternatives closer to his ideal point to those further away from it. This article focuses on the plausibility of this apparently obvious assumption.

Spatial models occupy an important position in modern social choice theory. From the early applications to party competition and electoral equilibrium they have spread to the study of inter-institutional power in the European Union (EU) and cabinet coalitions in multiparty systems ([10],[24],[25],[19]). They have also found new applications in expert systems advising the voters in making choices in elections. Work on spatial models has produced a wide variety of results ranging from the existence of stable outcomes (equilibria) of various various kinds ([23],[31]) to power distributions among voters ([39]) and suggestions for the design of institutions ([37]).

We approach the spatial voting games from the angle of aggregation paradoxes. These are surprises pertaining to inferring system properties from component properties or vice versa. In the social sciences typical instances of aggregation paradoxes are cross-level inferential fallacies, i.e. situations where one tries to determine individual-level properties from aggregate-level data. Suppose, for example, that we have voting data on electoral districts suggesting that in there is a positive correlation between the percentage of low-income voters and the support of the left-wing parties. To infer from these data that low-income voters are more likely to vote for left-wing parties is to commit an ecological (cross-level) fallacy.

The particular type of aggregation paradoxes we shall be mostly dealing with bears the name of Ostrogorski, a Russian diplomat and political theorist whose magnum opus [27] appeared in the opening years of the 20th century (see also [29] and [4]). It will be introduced and analyzed in the next section. Its variant – the exam paradox ([26]) – together with a majority-rule related paradox will be dealt with in the section that follows it. The next section introduces another aggregation paradox, viz. Simpson's paradox ([38]). The remaining sections discuss the conditions under which the paradoxes can be expected to occur.

The aggregation paradoxes have been the focus of scholarly attention for some time. The paradox bearing the name of Simpson was in fact mentioned nearly two decades before Simpson's important article by Cohen and Nagel ([8, 449]). In 1940's and 1950's several important contributions were made by Kenneth O. May ([20], [21]). Especially [22] is a pioneering work that shows i.a. that cyclical preferences make perfect sense when individual preference are formed on the basis of multiple criteria of performance. Our main point is related to May's: we aim to show that the basic tenet underlying many spatial models – i.e. that the voter preferences have a spatial representation – is far from innocuous. Indeed, there are situations where one should expect rational individuals to choose of two alternatives the one which is further away from the individuals' optimum point.

2 Ostrogorski's Paradox

Suppose that in an election there are 5 voters, 2 parties and 3 issues. Suppose, moreover, that each voter considers these issues to be of equal importance and that there are no other considerations in their mind that would determine their opinion on the parties. Consider two ways of determining the election result. (1) Each voter votes for the party that is closer to his/her (hereinafter his) opinion on more issues than the other party and whichever party gets more votes than the other is the winner. (2) For each issue the winner is the party that gets more votes than its competitor and the election winner is the party winning on more issues than the other. In a nutshell, Ostrogorski's paradox occurs when the election result differs in these two cases. Consider the following distribution of opinions on parties X and Y (Table 1).

Table 1 Ostrogorski's paradox

issue	*issue 1*	*issue 2*	*issue 3*	*the voter votes for*
voter A	X	X	Y	X
voter B	X	Y	X	X
voter C	Y	X	X	X
voter D	Y	Y	Y	Y
voter E	Y	Y	Y	Y
winner	Y	Y	Y	?

This is a rather strong version of the paradox since not only are the results different under procedures (1) and (2), but the winner under (2) is a unanimous one. Replacing any one Y with an X in the table would result in a weaker version of the paradox where "just" a majority winner is different under (1) and (2).

Replace now "voter" with "criterion" throughout in the preceding table and consider the procedure of forming an individual preference over two candidates X and Y. For example, in political competition the criteria could be relevant educational background, political experience, negotiation skills in the issue at hand, relevant political connections, etc. The issues might be e.g. education, economy and foreign policy. Each entry in the table then indicates which alternative is better on the criterion represented by the row when the issue is the one represented by the column. Suppose that the criterion-wise preference is formed on the basis of which alternative is better on more issues than the other. If all issues and criteria are deemed importance, the decision of which candidate the individual should vote is ambiguous: the row-column aggregation with the majority principle suggests X, but the column-row aggregation with the same principle yields Y.

Suppose now that the issues span a 3-dimensional Euclidean space where X and Y are located as two distinct points. The individual whose views are represented in the above table would then be located in this space so that on each dimension his ideal point (i.e. the point that represents him) is closer to Y than to X. However, it can not be inferred on this basis alone that in a pairwise comparison between X and Y he would vote for Y. In fact, if he resorts to the wholly reasonable principle of basing his choice on the criterion-wise performance of candidates, he will vote for X. After all, X outperforms Y on three criteria, while Y beats X on only two.

It is worth pointing out that the problem here cannot be resolved by assigning salience weights to issue dimensions, since Y is closer to the individual on each dimension. Strategic considerations – which of course may underly occasional votes against preferences – do not enter into the calculus dictating the choice of X rather than Y since the the agenda consists of only two alternatives and the ideal points of other voters are not known.

3 The Anscombe and Exam Paradoxes

Apparently closely related to Ostrogorski's is the paradox described by Anscombe [1]. In a nutshell, it says that it is possible that a majority of voters is in a minority (i.e. on the losing side) on a majority of issues involving dichotomous choices. Table 2 illustrates this paradox.

Table 2 Anscombe's paradox

issue	*issue 1*	*issue 2*	*issue 3*
voter A	X	X	Y
voter B	Y	Y	Y
voter C	Y	X	X
voter D	X	Y	X
voter E	X	Y	X
winner	X	Y	X

Voters A, B and C are on the losing side on a majority of issues: A on issues 2 and 3, B on issues 1 and 3 and C on issues 1 and 2. It is worth noticing that Table 2 does not exhibit Ostrogorski's paradox. Thus these two types are non-equivalent. Translated into the multi-criterion setting, Anscombe's paradox states that it is possible that a majority of criteria fails to coincide with the majority of criteria on a majority of issues. It is, thus, possible that more than 50% of the criteria fails to predict the "best" choice on more than 50% of the issues. In other words, a majority of criteria disagrees with the majority of criteria on more issues than those where they agree with the chosen alternative.

The exam paradox introduced and analyzed by Nermuth [26] has many similarities with Ostrogorski's paradox. In a way it is a generalization of the latter in a domain where the proximity of alternatives to ideal points takes on degrees instead of dichotomous values. The following is an adaptation of Nermuth's example. There are four issues and five criteria. One of two competitors, X, is located at the following distance from the voter's ideal point in a multi-dimensional space (Table 3). The score of X on each criterion is simply the arithmetic mean of its distances rounded to the nearest integer and in the case of a tie down to the nearest integer.

Table 3 X's distances from the voter's ideal point

criteria	issue 1	issue 2	issue 3	issue 4	average	score
criterion 1	1	1	2	2	1.5	1
criterion 2	1	1	2	2	1.5	1
criterion 3	1	1	2	2	1.5	1
criterion 4	2	2	3	3	2.5	2
criterion 5	2	2	3	3	2.5	2

X's competitor Y, in turn, is located in the space as indicated in Table 4.

Table 4 Y's distances from the voter's ideal point

criteria	issue 1	issue 2	issue 3	issue 4	average	score
criterion 1	1	1	1	1	1.0	1
criterion 2	1	1	1	1	1.0	1
criterion 3	1	1	2	3	1.75	2
criterion 4	1	1	2	3	1.75	2
criterion 5	1	2	1	2	1.75	2

4 Simpson's Paradox and the Sure-Thing Principle

When dealing with rates of improvement, recovery, growth etc. one may encounter Simpson's paradox. To quote Blyth ([6, 364])[2]:

> ... Savage's sure-thing principle ("if you would definitely prefer g to f, either knowing that event C obtained, or knowing that C did not obtain, then you definitely prefer g to f.") is not applicable to alternatives f an d g that involve sequential operations.

To consider a two-dimensional spatial setting, assume that the dimensions represent success rates of two policies in two separate settings. Suppose that

[2] For a general method for generating these paradoxes, see [30].

policy g is associated with a higher success rate than policy f in both settings. The sure-thing principle would then dictate that g be preferred to f. Yet, it may well be that the overall success rate associated f is much higher than that of g.

The following example illustrates (Table 5):

Table 5 Relative frequency data

	g	f
event C	1/3	1/4
event non-C	2/3	1/2

Suppose that f and g are some incentive schemes or experimental treatments and that the numbers indicate efficiency or quality (e.g. frequencies of exceeding some performance threshold). Then it would seem that g is, indeed, preferable to f. However, the following table (Table 6) is entirely consistent with the above data:

Table 6 Absolute frequencies

	g	f
event C	40	10
event non-C	10	45
total	50	55

Looking just at the "total" row one could easily be led to the opposite conclusion than previously, i.e. one could now suggest that f is preferable to g.

In typical spatial models, the sure-thing principle is implicitly assumed in virtue distance-based calculus. Simpson's paradox shows that there are limits in the plausibility of this assumption.

5 Core Conditions and Aggregation Paradoxes

Perhaps the best-known results on spatial models pertain to the conditions under which a core outcome exists ([2], [23], [31]). The core, it will be recalled, is the set of majority undominated outcomes:

$$x \in C \Leftrightarrow xMy, \forall y \in W$$

Here M is the weak majority preference relation so that xMy means that x either beats y with a majority of votes or there is a tie between the two. These results are based on the assumption that the ideal points as well as the distance measures and, consequently, the utility functions of the individuals are well-defined in the policy space. The results characterize in general terms the structural conditions under which stable outcomes exist and those under which they don't. Moreover, they tell us how the majority rule performs under circumstances when the core is empty and the preferences are smooth, i.e. have continuous utility representations.

What the preceding example of Ostrogorski's paradox amounts to, however, is that the core may exist in terms of a distance measure-based utility function, but the individual may plausibly vote for x even though y would be closer to his ideal point in all dimensions. So, the structurally stable core – i.e. stable in terms of the smooth preferences and distance measure – may not be stable after all.

Would this, then, imply that nonempty cores are even less common than the structural stability theorems suggest? Not at all, since a nonempty core may also exist in cases where none exists in terms of distance-based utility functions. Indeed, Humphreys and Laver argue that the smoothness of preferences runs counter the evidence on individual decision making experiments where other types of preferences have been found more common than smooth ones [14].

6 Indices of Voting Power

Spatial models are also resorted to in the study of voting power. Usually a distinction is made between spatial or preference-based and cooperative power indices. The latter typically refer to the Shapley-Shubik, Banzhaf and Holler-Packel indices (Banzhaf 1965; Shapley and Shubik 1954; Holler and Packel 1983). In fact, one of these, viz. the Shapley-Shubik one is based on a spatial intuition as well. To wit, it equates the voting power of a player with the relative number of times he is pivotal in all permutations of players. Since each permutation can be viewed as an attitudinal dimension, the Shapley-Shubik index has a spatial ring to it.

In typical spatial voting power measures, the crux is to regard a player more powerful than another just in case the the distance between the former's ideal point and the equilibrium outcome is smaller than the distance between the latter's ideal point and the equilibrium outcome (see e.g. [39]). Prima facie, this seems to be another situation where Ostrogorski's paradox renders the results of modeling somewhat questionable since it can again be argued that outcomes closer to player ideal points may be less preferred than those further away. On closer inspection this conclusion is, however, not warranted. The spatial power measures are based on hypothetical reasoning: if the players prefer outcomes closer to their ideal points to those at greater distance from them, then their power is inversely related to the distance between equilibrium outcomes and their ideal points. This interpretation makes these spatial models obviously weakly spatial. As such they are on par with the generic existence/nonexistence results on cores in spatial voting games. These are based on the assumption that the voters' utilities are simple functions of the relevant distances between alternatives and ideal points.

7 The Conditions of Paradoxes

The conditions under which Ostrogorski's paradox occurs have been studied by several authors (e.g. [9], [15], [16], [29], [36]). Daudt and Rae point out that there is a connection between Ostrogorski's and cyclic majorities. If candidates can be identified with different k-tuples of positions in the issue space – rather than just X or Y as in the preceding – an instance of Ostrogorski's paradox implies some intransitivities in majority preference relations over candidates. Kelly [15] shows that with three issues the occurrence of Ostrogorski's paradox implies that there can be no Condorcet winner. His conjecture is that this holds for all situations with an odd number of issues.

As was pointed out above, Ostrogorski's paradox may take on degrees according the "degree of contradiction" involved in computing election outcomes (i) over issues separately and then determining the overall winner, or (ii) over voters and then determining the election winner. In some occasions decision rules larger than simple majority are used. E.g. a voter may have a *status quo* favorite party that he votes for, unless its competitor is closer to his position on more than 2/3 of the issues. There may also be a *status quo* policy that is adopted unless more than, say, 3/5 of the electorate prefers its competitor. With k issues and n voters, assume that party 1's position is 1 on every issue and party 0's position 0 on every issue. An example is shown in Table 7. Clearly, if each voter votes for 0 unless 1 is closer to his position in 4/5 of the issues, no Ostrogorski's paradox emerges, while imposing the somewhat lower 3/5 requirement we would have an instance of the paradox.

Table 7 Ostrogorski's paradox: 0-1 version

issue	*issue 1*	*issue 2*	*issue 3*
criterion A	1	1	0
criterion B	1	0	1
criterion C	0	1	1
criterion D	0	0	0
criterion E	0	0	0

Deb and Kelsey (1987) show that the following condition expresses the necessary and sufficient condition for Ostrogorski's paradox:

$$kn - 2ny - 2kx - 12xy \geq 0. \tag{1}$$

Here $x = 1$ when n (the number of voters or criteria) is even, and $x = 1/2$ when it is odd. Similarly, $y = 1$ when the number k of issues is even and $y = 1/2$ when k is odd. Suppose now that 1 is chosen according to a criterion if the number of 1's in the corresponding row is at least K. Let the number of such rows (criteria) be G in number. Suppose, analogously, that for 0 to be chosen on any issue, there has to be at least S 0's in the corresponding column. Denote the number of such columns by I. Let $m = K/k$. Another result of Deb and Kelsey (1987) states that for any rational-valued $k_0 \in (1/2, 1)$ there exists a set of values for k, n, K, I, S and G so that Ostrogorski's paradox is possible.

Now, the inequality expressed in Equation 1 is true for nearly all values of k and n and would seem to be increasing with both of these variables. It would, however, be incorrect to relate Equation 1 to "frequency" of Ostrogorski's paradox so that the larger the positive value of the left-hand side, the more likely is the paradox. On the basis of the computer simulations performed by Kelly [15] under a version of the impartial culture assumption, the probability of the paradox increases with the increase of voters (or criteria), but decreases with the increase in issues. This is partially contradicted by the Deb and Kelsey result since the necessary and sufficient condition for the paradox is, for any value of n, fulfilled by increasing the value of k.

In analyzing Anscombe's paradox Wagner [40, 305-306] suggests a three-fourth's rule for avoiding it:

> If N individuals cast yes-or-no votes on K proposals then, whatever the decision method employed to determine the outcomes of these proposals, if the average fraction of voters, across all proposals, comprising the prevailing coalitions is at least three-fourths, then the set of voters who disagree with a majority of outcomes cannot comprise a majority.

8 Non-dichotomous Settings

The possibility and frequency of the paradox are two different things. What the above results suggest is that it is nearly always possible to encounter an instance of Ostrogorski's paradox unless one is restricted to a very small number of issues and criteria (voters). A step towards answering the question of how often one might encounter such paradoxes has been taken by Laffond and Lainé [17]). Their result states that for the avoidance of Ostrogorski's paradox it is sufficient that there exists a permutation of issues so that each row (criterion) exhibits a unique switch from 1's to 0's (single-switch condition). If the preferences satisfy another condition, viz. richness, the sufficient condition is also necessary for the avoidance of the paradox. A preference profile is rich if for every sequence of 0's and 1's representing an individual's stand on issues, a complementary sequence – i.e. a sequence where each 0 is replaced by 1 and every 1 by 0 – exists in the profile. In other words, in a rich profile each policy ideal is matched by its polar opposite.

The avoidance of Ostrogorski's paradox is, thus, possible only under very special profiles. The sufficiency of the single-switch condition for the avoidance of the paradox is due to the fact that under single switched profiles, there is a core winner platform, i.e. one that cannot be defeated by any other platform with a simple majority of votes in a pairwise comparisons assuming that the voters prefer the platform that is closest to their own ideal platform in the sense of implying the smallest number of switches from 0 to 1 or vice versa ([17, 57]). The core platform is one that consists of the majority winners on each issue. The relationships between Ostrogorski's and Condorcet's paradoxes have been explored e.g. by Bezembinder and Van Acker [4] and Lagerspetz [18]. The general conclusion from these studies is that each instance of Ostrogorski's paradox implies an underlying Condorcet's paradox in terms of platforms.

Thus far we have focused on $0-1$ matrices, but more generally one could ask under which conditions the same outcomes are reached when the entries are real numbers in the unit interval and the outcomes are determined in two different ways: 1) by first aggregating over rows and then aggregating the results (rows first method) and 2) by first aggregating the entries in each column and thereupon the results (columns first method)[28]. In the preceding we have considered the majority rule in determining the results of each aggregation phase. The more general $0-1$ matrix setting has been studied in the context of deprivation measures by Dutta et al. [11]. Specifically they study the properties of two functions $g : [0,1]^m \rightarrow [0,1]$ and $h : [0,1]^n \rightarrow [0,1]$. The former is interpreted as the method for aggregating individual deprivation degrees on various attributes (income, housing, education, etc) into an overall degree of deprivation of an individual. The latter, in turn, is the method for aggregating the individual deprivation degrees into an overall degree of deprivation characterizing the society. In the deprivation research setting the question is now under which conditions one can arrive at the same result concerning the overall degree of deprivation in a society following two paths: (i) form first an index of deprivation of each individual by aggregating her deprivation values in all m attributes and then compute the social degree of deprivation from those n index values, and (ii) compute for each attribute a deprivation index by aggregating the n individual deprivation values on that attribute, and aggregate then these indices to a social one. Let the matrix $A = [a_i j]$ be a $n \times m$ matrix of entries $a_i j \in [0,1]$. Each entry then gives the degree of individual i' deprivation on attribute j. Denote now by a_i the row vector representing individual i's deprivation degrees on all m attributes and by a^j the column vector giving each individual's deprivation degree on attribute j. The question thus becomes under which conditions the following holds:

$$h(g(a_1), \ldots, g(a_n)) = g(h(a^1), \ldots, h(a^m)) \tag{2}$$

Focusing on functions g and h that satisfy the following two conditions Dutta et al. (2003) [11] show that only a very restricted set of functions satisfy equation (2). The conditions are (i) that g be continuous, strictly increasing in all arguments and satisfies non-diminishing increments, and (ii) that h be continuous, symmetric and strictly increasing. Moreover, it is assumed that if all arguments of g are zero (unity, respectively) , so is the value of g. The same assumption is made regarding h. Let $g* : [0,1]^m \to [0,1]$ so that for any $x \in [0,1]^m$ there is a set of weights $w_1, \ldots w_m$ each between 0 and 1 with $\sum w_j = 1$ and $g*(x) = w_1x_1 + \cdots + w_mx_m$. Denote by $G*$ the set of all such functions $g*$. Moreover, let $h*(y) = (y_1 + \cdots + y_n)/n$ and denote by $H*$ the consisting of this function. Then all functions $g* \in G*$ together with the function $h*$ satisfy equation (2) and, conversely, any pair of functions (g, h) that satisfies (2) must be such that $g \in G*$ and $h = h*$.

The converse result is more important for our purposes. What it says is that in order to derive the same result using rows first and columns first methods, the indices attached to rows must be weighted averages of the row entries and the overall index value must be the arithmetic mean of the row indices.

9 Extreme Cases and Saari's Paradox Machine

Formally, Simpson's paradox can be expressed as follows [6]. Let A, B and C denote three distinct properties or predicates, such as being a victorious candidate, being a big campaign spender, supporting certain legislation, living in a given neighborhood etc. Furthermore, let A', B' and C' denote the absence of A, B and C, respectively. Let now

$$P(A|B) < P(A|B') \tag{3}$$

Simpson's paradox occurs whenever the following inequalities (3) and (4) hold as well.

$$P(A|BC) \geq P(A|B'C) \tag{4}$$

$$P(A|BC') \geq P(A|B'C') \tag{5}$$

The paradox is intuitively the more dramatic the larger the margins by which (2), on the one hand, and (3)-(4), on the other, hold. Blyth gives the conditions for extreme forms of Simpson's paradox. To outline those conditions, let us give the following interpretation to A, B and C.

A = the property of getting elected in a given election,
B = the property of spending less than X dollars in one's campaign,
C = campaigning in district 1 and
C' = campaigning in district 2.

Assuming that (2)-(4) hold, the paradox thus consists of finding that there seems to be an association between big spending and electoral success, while in both districts the big spenders are less likely to get elected than the small spenders. Now, the extreme case of the paradox is the following. Let $\gamma \geq 1$ and assume that (5) - (6) hold.

$$P(A|BC) \geq \gamma P(A|B'C) \tag{6}$$

$$P(A|BC') \geq \gamma P(A|B'C') \tag{7}$$

It is now possible that

$$P(A|B) \approx 0 \tag{8}$$

and

$$P(A|B') \approx 1/\gamma \tag{9}$$

With $\gamma = 1$, we may thus have $P(A|B) = 0$ and $P(A|B') = 1$ suggesting an extremely strong association between spending and electoral success. Yet, in both districts this association is reversed. The crux in constructing instances of Simpson's paradox is in the decomposition of the conditional probabilities $P(A|B)$ and $P(A|B')$ in terms of C and C':

$$P(A|B) = [P(C|B)]P(A|BC) + [P(C'|B)]P(A|BC') \tag{10}$$

and

$$P(A|B') = [P(C|B')]P(A|B'C) + [P(C'|B')]P(A|B'C') \tag{11}$$

It is evident that $P(A|B)$ is a weighted average of $P(A|BC)$ and $P(A|BC')$. Similarly $P(A|B')$ is a weighted average of $P(A|B'C)$ and $P(A|B'C')$. If the weights were the same or – expressed in another way – if B and C were independent, no paradox could ensue. It is precisely the association between B and C that explains the paradox. A person willing to construct an instance of Simpson's paradox should start from expressions (5) and (6) above. Since, $P(A|BC)$ by assumption is larger than $P(A|B'C)$ and $P(A|BC')$ is larger than $P(A|B'C')$, the paradox occurs because the expressions $P(C|B)$, $P(C'|B)$, $P(C|B')$ and $P(C'|B')$ are related so that $P(A|B) < P(A|B')$. It should be observed, though, that it is necessary for the paradox that either $P(A|B'C)$ or $P(A|B'C')$ has to be larger than the smaller of the pair $[P(A|BC), P(A|BC')]$. To fix our ideas, let us assume that the smaller of the pair is $P(A|BC)$, i.e. the probability of election of a small spender in district 1 is smaller than the probability of election of small spender in district 2. Assume, moreover, that it is $P(A|B'C)$ rather than $P(A|B'C')$ which is larger than $P(A|BC)$. In other words, the probability of winning the election is larger for a big spender in district 1 than a small spender in district 1 (or a big spender in district 2). The paradox can now be constructed by making the weight of $P(A|B'C)$ as large as possible and that of $P(A|B'C')$ as small as possible in (10). Similarly, the weight of $P(A|BC')$ should made as small as possible vis-à-vis that of $P(A|BC)$.

If the ease with which instances of Simpson's paradox can be constructed is an indicator of how often one may expect to encounter them in real world, then Saari's procedure suggests that they can be pretty common. To keep the interpretation simple, assume that we are looking at the possible effects of two different instructional modules, new and standard, on racial prejudice. Our data set consists of four-tuples (u_1, x, u_2, y) where u_1 indicates the number of individuals subjected to the new instruction and u_2 those subjected to the standard instruction. x, in turn, indicates the judged success rate (the decrease in racial prejudice) of the new instruction and x that of the standard education. Ignoring the sizes of the sub-populations, we can represent each four-tuple of empirical observations as a point in a two-dimensional space so that the x-value indicates its horizontal and x-value its vertical coordinate. Each such point thus represents the success rates in a locale with a group of subjects some of which have been given the new and the rest the standard instruction.

Since the x, y- values represent proportions, they range from 0 to 1. In other words, our possible observation points range from (0,0) to (1,1), the first (second, respectively) representing a locale where the success rate is zero (one) both for the students exposed to new and to standard material. To generate an instance of Simpson's paradox choose a pair of points (X, Y) with coordinates (X_x, X_y) and (Y_x, Y_y), respectively, in the two-dimensional space so that both points are located below the line connecting (0,0) and (1,1). Form now a rectangle by drawing lines parallel to the coordinate axes through X and Y as in Figure 1.

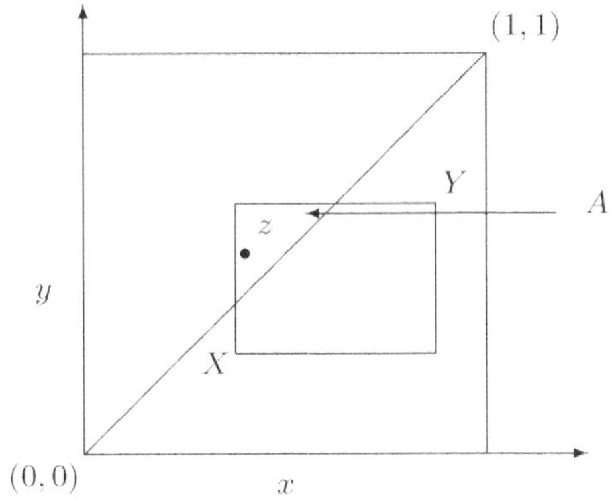

Fig. 1 Generating Simpson's paradoxes

A portion, denoted by A, of the area of the rectangle spanned by X and Y is located above the line connecting (0,0) and (1,1). Pick now any point z in A and find its coordinates (z_x, z_y). Let now

$$s = (z_x - X_x)/(Y_x - X_x)$$

and

$$t = (z_y - X_y)/(Y_y - X_y).$$

The values s and t thus determined can now be used to construct an instance of Simpson's paradox. A moment's reflection reveals that z_x, the horizontal coordinate value of the selected point, is a weighted average of the horizontal coordinate values of X and Y . Similarly, the vertical coordinate value z_y

of the selected point is a weighted average of the corresponding coordinate values of X and Y (the weights in those averages are not the same, though). Since the latter two points represent the success rates of two groups in two different locales, we can move point z closer to X (Y, respectively) along the horizontal dimension by including in the total population more individuals exposed to new (standard) instruction in locale 1 (locale 2). For example, the corner point of the rectangle above X represents a population in which all individuals exposed to the new material are drawn from locale 1 and all exposed to the standard material are from locale 2. By fiddling with the portions of individuals exposed to the two types of material and with the portion of individual selected from each locale, one can generate an instance of Simpson's paradox by pointing to the success rates represented by the selected point above the (0,0)-(1,1) line and to the success rates represented by points used in drawing the rectangle. Obviously, the first named point represents a distribution in which the success rate is higher for individuals exposed to the standard material. The latter two points, on the other hand, represent distributions where the individuals exposed to new material get higher success rates than those exposed to the standard material.

These are the basic outlines of Saari's procedure. It is based on geometrical properties of cones and, in particular, on the fact that cones can be used to represent a wide class of decision situations (for details, see [30]). Saari's paradox machine, thus, begins with two sub-population distributions located on the same side of the line connecting (0,0) and (1,1). The closer the points representing those sub-populations are to (0,0) and (1,1), respectively, the more freedom one has in working out instances of Simpson's paradox. Comparing Saari's procedure with what was said above in the context of Blyth's analysis of Simpson's paradox, we notice that moving the selected point y about is tantamount to manipulating the weights $P(C|B)$, $P(C|B')$, $P(C'|B)$ and $P(C'|B')$.

10 Conclusion

The basic question addressed in this paper is under which conditions one may expect that the assumption that individuals support the alternatives closer to their ideal points than competing alternatives can be expected to hold. It turns out that the validity conditions of this assumption are significant and should be taken seriously. The early observations on the connections between Condorcet's and Ostrogorski's paradoxes give us some indication as to what kinds of situations are riddled with the latter paradox. More pertinent, however, are the results of Laffond and Lainé suggesting that for the avoidance of Ostrogorski's paradox one needs to assume an underlying dimension of issues so that the criteria exhibit the single-switch property. This is a serious domain restriction on par with Black's single-peakedness condition [5]. In the dichotomous $0-1$ setting, the reason for Ostrogorski's paradox seems to be the non-commutativeness of the majority rule (Bezembinder and Van Acker 1985): the alternative resulting from first aggregating over rows (determining the majority alternative) and then over those majority alternatives may well differ from the alternative obtained by determining the majority alternatives in the opposite order. If – instead of the majority procedure – we would simply tally the number of 0's and 1's in each row and column and declare the alternative with the largest tally in the entire table the overall winner, then no paradox could ensue: tallying first in the rows first and columns first manner would make no difference.

In the more general setting of real-values distances, the row first and column first procedures as defined by Dutta et al. result in the same overall scores only under very restricted circumstances. To wit, if the row-wise distances are weighted averages of the issue-specific ones on each criterion and the overall score of an alternative is the arithmetic mean of those averages, then the two procedures can be expected to yield the same scores for alternatives. So, it is only in the linear world that Ostrogorski's paradox can be avoided.

References

1. Anscombe, G.E.M.: On Frustration of the Majority by Fulfillment of the Majority's Will. Analysis 36, 161–168 (1976)
2. Banks, J.: Singularity Theory and Core Existence in the Spatial Model. Journal of Mathematical Economics 24, 523–536 (1995)
3. Banzhaf, J.: Weighted Voting Doesn't Work: A Mathematical Analysis. Rutgers Law Review 13, 317–343 (1965)
4. Bezembinder, T., Van Acker, P.: The Ostrogorski Paradox and Its Relation to Nontransitive Choice. Journal of Mathematical Sociology 11, 131–158 (1985)
5. Black, D.: Theory of Committees and Elections. Cambridge University Press, Cambridge (1958)
6. Blyth, C.: On Simpson's Paradox and the Sure-Thing Principle. Journal of the American Statistical Association 67, 364–366 (1972)
7. Brams, S., Kilgour, M., Zwicker, W.: The Paradox of Multiple Elections. Social Choice and Welfare 15, 211–236 (1998)
8. Cohen, M., Nagel, E.: Introduction to Logic amd Scientific Method. Routledge and Kegan Paul, New York (1934)
9. Deb, R., Kelsey, D.: On Constructing a Generalized Ostrogorski Paradox: Necessary and Sufficient Conditions. Mathematical Social Sciences 14, 161–174 (1987)
10. Downs, A.: An Economic Theory of Democracy. Harper and Row, New York (1957)
11. Dutta, I., Pattanaik, P., Xu, Y.: On Measuring Deprivation and the Standard of Living in a Multidimensional Framework on the Basis of Aggregate Data. Economica 70, 197–221 (2003)
12. Harsanyi, J.C.: Rational Behavior and Bargaining Equilibrium in Games and Social Situations. Cambridge University Press, Cambridge (1977)
13. Holler, M., Packel, E.: Power, Luck and the Right Index. Zeitschrift für Nationalökonomie 43, 21–29 (1983)

14. Humphreys, M., Laver, M.: Spatial Models, Cognitive Metrics and Majority Voting Equilibria (2007), http://politics.as.nyu.edu/object/MichaelLaver.html
15. Kelly, J.: The Ostrogorski Paradox. Social Choice and Welfare 6, 71–76 (1989)
16. Laffond, G., Lainé, J.: Representation in Majority Tournaments. Mathematical Social Sciences 39, 35–53 (2000)
17. Laffond, G., Lainé, J.: Single-Switch Preferences and the Ostrogorski Paradox. Mathematical Social Sciences 52, 49–66 (2006)
18. Lagerspetz, E.: Paradoxes and Representation. Electoral Studies 15, 83–92 (1996)
19. Laver, M., Shepsle, K.: Making and Breaking of Governments. Cambridge University Press, New York (1996)
20. May, K.: The Aggregation Problem for a One-Industry Model. Econometrica 14, 285–298 (1946)
21. May, K.: Technological Change and Aggregation. Econometrica 15, 51–63 (1947)
22. May, K.: Intransitivity, Utility and the Aggregation of Preference Patterns. Econometrica 22, 1–13 (1954)
23. McKelvey, R., Schofield, N.: Generalized Symmetry Conditions at a Core Point. Econometrica 55, 923–933 (1987)
24. Napel, S., Widgrén, M.: Power Measurement as Sensitivity Analysis: A Unified Approach. Journal of Theoretical Politics 16, 517–538 (2004)
25. Napel, S., Widgrén, M.: The Inter-Institutional Distribution of Power in EU Codecision. Social Choice and Welfare 27, 129–154 (2006)
26. Nermuth, M.: Two-Stage Discrete Aggregation: The Ostrogorski Paradox and Related Phenomena. Social Choice and Welfare 9, 99–116 (1992)
27. Ostrogorski, M.: Democracy and the Organization of Political Parties, vol. I-II. Haskell House Publishers, New York (1970); Original: Ostrogorski, M.: La démocratie et l'organisation des partis politiques. Calmann-Levy, Paris (1902)
28. Pattanaik, P., Reddy, S., Xu, Y.: On Procedures for Measuring Deprivation and Living Standards of Societies in a Multi-Attribute Framework, Andrew Young School of Policy Studies, Research Paper Series, Working Paper 08-02 (2007)
29. Rae, D., Daudt, H.: The Ostrogorski Paradox: A Peculiarity of Compound Majority Decision. European Journal of Political Research 4, 391–398 (1976)
30. Saari, D.: Consistency of Decision Processes. Annals of Operations Research 23, 103–137 (1990)
31. Saari, D.: The Generic Existence of a Core for q-Rules. Economic Theory 9, 219–260 (1997)
32. Saari, D.: Decisions and Elections. Cambridge University Press, New York (2001)
33. Saari, D., Sieberg, K.: The Sum of the Parts Can Violate the Whole. American Political Science Review 95, 415–433 (2001)
34. Saari, D., Sieberg, K.: Are Partwise Comparisons Reliable? Research in Engineering Design 15, 62–71 (2004)
35. Shapley, L., Shubik, M.: A Method for Evaluating the Distribution of Power in a Committee System. American Political Science Review 48, 787–792 (1954)

36. Shelley, F.: Notes on Ostrogorski's Paradox. Theory and Decision 17, 267–273 (1984)
37. Shepsle, K., Weingast, B.: Why Are Congresssional Committees Powerful? American Political Science Review 81, 935–945 (1987)
38. Simpson, E.: The Interpretation of Interaction in Contingency Tables. Journal of Royal Statistical Society B 13, 238–241 (1951)
39. Steunenberg, B., Schmidtchen, D., Koboldt, C.: Strategic Power in the European Union. Journal of Theoretical Politics 11, 339–366 (1999)
40. Wagner, C.: Anscombe's Paradox and the Rule of Three-Fourths. Theory and Decision 15, 303–308 (1983)
41. Wagner, C.: Avoiding Anscombe's Paradox. Theory and Decision 16, 233–238 (1984)

Quantum Dynamics of Non Commutative Algebras: The SU(2) Case

C.M. Sarris and A.N. Proto

Abstract. Applying Maximum Entropy Formalism (MEP) the dynamics of Hamiltonians, associated to non commutative Lie algebras, can be found. For the $SU(2)$ case, it is easy to show that the Generalized Uncertainty Principle (GUP) is an invariant of motion. The temporal evolution of the system is confined to Bloch spheres whose radius lay on the interval $(0;1)$. The GUP, defines the fizziness of these spheres inside the $\hbar$ domain for the $SU(2)$ Lie algebra.

Keywords: Quantum Dynamics, $SU(2)$ algebras, Uncertainty Principle.

1 Introduction

The fuzzy nature of the $SU(2)$ Lie algebra was posed in quantum field theory some years ago, using differential calculus, and D-Branes approaches. In both cases clearly it could be seen that the fuzziness of the internal structure of a system could not be greater than the quantum uncertainty in the position of a particle. In the present contribution we present a general formalism to deal with quantum and semiquantum (time-dependent or not) Hamiltonian dynamics associated to non-commutative algebras, and expose the particular case of the $SU(2)$ Lie algebra.

The knowledge of the mean values of the operators and correlation functions, are of main interest in usual quantum mechanics (QM) applications. The maximum entropy formalism (MEP) allows us to describe quantum or semiquantum Hamiltonian systems in terms of those, and those only, quantum operators *relevant* to the problem at hand.

In order to make clear the fundamental features of our approach we begin summarizing the principal concepts of the (MEP) [3, 4]. Given the

C.M. Sarris
Lab. de Sistemas Complejos, Facultad de Ingenieria, Universidad de Buenos Aires

A.N. Proto
Comision de Investigaciones Cientificas PBA (CIC), Argentina

S. Greco et al. (Eds.): Preferences and Decisions, STUDFUZZ 257, pp. 339–351.
springerlink.com

expectation values $< \hat{O}_j >$ of the operators $\hat{O}_j$, the statistical operator $\hat{\rho}(t)$ is defined by [3]

$$\hat{\rho}(t) = \exp\left(-\lambda_0 \hat{I} - \sum_{j=1}^{q} \lambda_j \hat{O}_j\right), \tag{1}$$

where q is a non negative integer and the $q+1$ Lagrange multipliers λ_j, are determined to fulfill the set of constraints

$$< \hat{O}_j >= \mathrm{Tr}[\hat{\rho}(t)\hat{O}_j], \tag{2}$$

($\hat{O}_0 = \hat{I}$ is the identity operator) and the normalization condition, Tr $[\hat{\rho}(t)] = 1$, in order to maximize the entropy, defined by

$$S(\hat{\rho}) = -\mathrm{Tr}[\hat{\rho} \ln \hat{\rho}] \tag{3}$$

in terms of $q+1$ Lagrange multipliers λ (λ_0 is associated with the identity operator), so that, we can write

$$S = \langle \ln \hat{\rho}(t) \rangle = \lambda_0 \hat{I} + \sum_{j=1}^{q} \lambda_j(t) \left\langle \hat{O}_j \right\rangle_t \tag{4}$$

If the temporal evolution of the density operator follows the Liouville equation for all t, then we can write

$$i\hbar \frac{\partial \hat{\rho}(t)}{\partial t} = \left[\hat{H}(t), \hat{\rho}\right] \tag{5}$$

so, the entropy S *is a constant of motion*, and the relevant operators to be considered to construct the density matrix are those which fulfill the closure condition [3]

$$\left[\hat{H}(t), \hat{O}_j\right] = i\hbar \sum_{i=0}^{q} g_{ij}(t) \hat{O}_i \qquad j = 1, 2, \ldots, q \tag{6}$$

The $g_{ij}(t)$ are the elements of a $q \times q$ matrix $G(t)$ and it will help us to find invariants of the motion. Depending on what kind of operators compose the Hamiltonian $\hat{H}(t)$ of the system, the closure condition given by Eq.(6) may give rise to a complete set of noncommuting observables which constitute by themselves the generators of a Lie algebra as we will see in Sec. 3.2. The matrix $G(t)$ *entails the whole dynamics* of the system [3]. Besides, on account of Eq.(6) the temporal evolution of the expectation values is given by the generalized Ehrenfest relationships [4]

$$\frac{d\langle \hat{O}_i \rangle_t}{dt} = -Tr\left(\hat{\rho} \sum_{j=0}^{q} g_{ji} \hat{O}_j\right) = -\sum_{j=1}^{q} g_{ji} \langle \hat{O}_j \rangle_t \tag{7}$$

The generalized Ehrenfest theorem yields to a set of first-order differential equations for the temporal evolution of the expectation values which are our quantal relevant variables. of our. We do not need to appeal to any consideration about the wave function at the instant $t = 0$ due to it is replaced by the set of initial conditions imposed on Eqs.(7) without violating the uncertainty principle, making it easy to be compared with their classical counterpart if necessary. Besides, for the λ Lagrange multipliers it holds [3]

$$\frac{d\lambda_i(t)}{dt} = \sum_{j=0}^{q} g_{ij}\lambda_j(t), \qquad i = 1, 2, \ldots, q \tag{8}$$

Through Eqs.(6) and (7) we have connected the density operator (1) to the Lie algebras raising from Eq. (6), and we have *translated the quantum Hamiltonian dynamics into the language of dynamical systems*. This fact stresses the importance of having a method to construct invariants of motion, which is so relevant in the study of dynamical systems, and avoids any consideration about wave funtions. If commutators are replaced by Poisson brackets (see for instance, [3]) the method is straightfowardly extended to classical systems.

Once we are within the dynamical systems context, the search of invariants of motion plays an important role especially if the problem is the study of time-dependent Hamiltonians, or semiquantum systems with or without chaotic classical limit. In previous works, we have analyzed the genesis *of invariants of motion for quantum, semiquantum or classical time-dependent Hamiltonians*, through the MEP like in ref. [5] and references therein.

This contribution is devoted to derive the invariants of motion from the information entropy, and connect them with the Lie algebras associated to the Hamiltonian of the system, and exemplify the procedure for the $SU(2)$ case. It will be shown that for this particular Lie algebra one of the invariants is the Generalized Uncertainty Principle (GUP) itself and, we will also show that the system is confined to evolve on hypersurfaces defined on the generalized phase space spanned by the mean values of the generators of the $SU(2)$ Lie algebra $\mathbb{V} = span\{\langle\hat{\sigma}_x\rangle, \langle\hat{\sigma}_y\rangle, \langle\hat{\sigma}_z\rangle\}$. These hypersurfaces are defined by the different sets of initial conditions $\{\langle\hat{\sigma}_x\rangle_0, \langle\hat{\sigma}_y\rangle_0, \langle\hat{\sigma}_z\rangle_0\}$ or $\{\lambda_{x(0)}, \lambda_{y(0)}, \lambda_{y(0)}\}$ imposed on the system, through Eqs.(7), (8). The Hamiltonian dynamics is contained in these two equations, and the initial conditions (IC) should be imposed taking into account the GUP. The temporal evolution is confined to the Bloch spheres whose radius lay on the interval $(0; 1)$. The GUP defines the fuzziness of these spheres inside the $\hbar$ domain for the $SU(2)$ Lie algebra.

2 Invariants of Motion for Non-commutative Operators

The closure condition (6) and the generalized Ehrenfest theorem (7) play a fundamental rol. Eq. (6) defines the set of relevant operators which, generally, constitutes a complete set of non-commuting observables ($CSNCO$) and, as a particular case, they are a complete set of commuting observables. *As we have said in the previuos section*, the temporal evolution of the mean values of the relevant set is given by Eq. (7).

Let be $\{\hat{O}_1, ..., \hat{O}_q\}$ the set of relevant non-commuting observables defined by the closure condition (6) and let be the operator $\hat{L}_{ij} = \frac{1}{2}(\hat{O}_i\hat{O}_j + \hat{O}_j\hat{O}_i)$ where $\hat{O}_i$ and $\hat{O}_j$ are two belonging to the set. It is easy to see that $\hat{L}_{ij} = \hat{L}_{ji}$, and that $\hat{L}_{ii} = \hat{O}_i^2$. Besides, the closure condition allows us to obtain the following commutation relationships [6]

$$\left[\hat{H}(t), \hat{O}_j^2\right] = 2i\hbar \sum_{r=1}^{q} g_{rj}(t)\hat{L}_{rj}\,, \tag{9}$$

$$\left[\hat{H}(t), \hat{L}_{ij}\right] = i\hbar \sum_{r=1}^{q} \left[g_{rj}(t)\hat{L}_{ir} + g_{ri}(t)\hat{L}_{jr}\right], \tag{10}$$

where $g_{rj}(t)$ are the coefficients of the dynamical matrix $G(t)$ and they may depend upon time if the Hamiltonian of the system, $\hat{H}(t)$, is time-dependent. The Ehrenfest theorem (7) and Eqs. (9) and (10) give the temporal evolution of $\left\langle \hat{O}_j^2 \right\rangle$ and $\left\langle \hat{L}_{ij} \right\rangle$ in the following fashion

$$\frac{d\left\langle \hat{O}_j^2 \right\rangle}{dt} = -2\sum_{r=1}^{q} g_{rj}(t)\left\langle \hat{L}_{rj} \right\rangle, \tag{11}$$

$$\frac{d\left\langle \hat{L}_{ij} \right\rangle}{dt} = -\sum_{r=1}^{q} \left[g_{rj}(t)\left\langle \hat{L}_{ir} \right\rangle - g_{ri}(t)\left\langle \hat{L}_{jr} \right\rangle\right]. \tag{12}$$

Now, we are going to introduce the expression of the Generalized Uncertainty Principle, GUP: if we take into account two relevant operators belonging to the relevant set of the system, $\hat{O}_i$ and $\hat{O}_j$ we see that, generally, they do not commute and so they fulfill: $\left[\hat{O}_i, \hat{O}_j\right] = i\hat{C}$, where $\hat{C}$ is another observable which may or may not belong to the relevant set. The uncertainty relation between these two observables is [7]

$$\left(\Delta\hat{O}_i\right)^2 \left(\Delta\hat{O}_j\right)^2 - \left[\left\langle \hat{L}_{ij} \right\rangle - \left\langle \hat{O}_i \right\rangle \left\langle \hat{O}_j \right\rangle\right]^2 \geq -\frac{\left\langle \left[\hat{O}_i, \hat{O}_j\right] \right\rangle^2}{4}, \qquad i \neq j. \tag{13}$$

where $\left(\Delta\hat{O}_i\right)^2 = \left\langle \hat{O}_i^2 \right\rangle - \left\langle \hat{O}_i \right\rangle^2$ is the statistical fluctuation of $\hat{O}_i$. Eq. (13) is the basis on which we are going to build up the generalized version of the GUP given that it is possible to construct, in the same way, the uncertainty relation between all operators belonging to the relevant set. Now, we perform the summation all over the possible pairs of relevant observables which compose the relevant set and obtain the following expression [6]

$$\frac{1}{2}\sum_{i=1}^{q}\sum_{j=1}^{q}\left\{\left(\Delta\hat{O}_i\right)^2\left(\Delta\hat{O}_j\right)^2-\left[\left\langle\hat{L}_{ij}\right\rangle-\left\langle\hat{O}_i\right\rangle\left\langle\hat{O}_j\right\rangle\right]^2\right\}\geq-\frac{1}{4}\sum_{i=1}^{q}\sum_{j=1}^{q}\left\langle\left[\hat{O}_i,\hat{O}_j\right]\right\rangle^2, \tag{14}$$

which also can be expressed as

$$\sum_{\substack{i,j=1\\ i<j}}^{q}\left\{\left(\Delta\hat{O}_i\right)^2\left(\Delta\hat{O}_j\right)^2-\left[\left\langle\hat{L}_{ij}\right\rangle-\left\langle\hat{O}_i\right\rangle\left\langle\hat{O}_j\right\rangle\right]^2\right\}\geq-\frac{1}{4}\sum_{\substack{i,j=1\\ i<j}}^{q}\left\langle\left[\hat{O}_i,\hat{O}_j\right]\right\rangle^2. \tag{15}$$

Eq. (14) or Eq. (15) are the generalized GUP, considering these expressions take account the uncertainty relation between all the possible pairs of observables which compose the relevant set. Now, we focus on the left side of inequality (14)

$$I^H=\frac{1}{2}\sum_{i=1}^{q}\sum_{j=1}^{q}\left\{\left(\Delta\hat{O}_i\right)^2\left(\Delta\hat{O}_j\right)^2-\left[\left\langle\hat{L}_{ij}\right\rangle-\left\langle\hat{O}_i\right\rangle\left\langle\hat{O}_j\right\rangle\right]^2\right\}, \tag{16}$$

in order to find out under what condition it results an invariant of the motion. If we take the temporal evolution of Eq. (16) and take into account Eqs.(11) and (12) we obtain [6]

$$\begin{aligned}
\frac{dI^H}{dt}=&-\sum_{i=1}^{q}\sum_{j=1}^{q}(g_{ii}+g_{jj})\left\{\left(\Delta\hat{O}_i\right)^2\left(\Delta\hat{O}_j\right)^2-\left[\left\langle\hat{L}_{ij}\right\rangle-\left\langle\hat{O}_i\right\rangle\left\langle\hat{O}_j\right\rangle\right]^2\right\}-\\
&-\sum_{i=1}^{q}\sum_{j=1}^{q}(g_{ij}+g_{ji})\left\langle\hat{L}_{ij}\right\rangle\sum_{r=1}^{q}\left(\Delta\hat{O}_r\right)^2-\\
&-\sum_{i=1}^{q}\sum_{j=1}^{q}(g_{ij}+g_{ji})\left\langle\hat{O}_i\right\rangle\left\langle\hat{O}_j\right\rangle\sum_{r=1}^{q}\left\langle\hat{O}_r^2\right\rangle- &(17)\\
&-\sum_{i=1}^{q}\sum_{j=1}^{q}(g_{ij}+g_{ji})\left\langle\hat{O}_j\right\rangle\sum_{r=1}^{q}\left\langle\hat{L}_{ri}\right\rangle\left\langle\hat{O}_r\right\rangle-\\
&-\sum_{i=1}^{q}\sum_{j=1}^{q}(g_{ij}+g_{ji})\left\langle\hat{O}_i\right\rangle\sum_{r=1}^{q}\left\langle\hat{L}_{rj}\right\rangle\left\langle\hat{O}_r\right\rangle+ &(18)\\
&+\sum_{i=1}^{q}\sum_{j=1}^{q}(g_{ij}+g_{ji})\sum_{r=1}^{q}\left\langle\hat{L}_{ri}\right\rangle\left\langle\hat{L}_{rj}\right\rangle; \qquad i\neq j,\ r\neq i,\ r\neq j. &(19)
\end{aligned}$$

Eq.(19) implies that if the dynamical matrix $G(t)$ is an antisymmetric one $\left(G^T(t) = -G(t) \Rightarrow g_{ij}(t) = -g_{ji}(t)\right)$ then the *generalized GUP is a dynamical invariant* and, obviously, this constant, could not be minor than a certain value given by the condition $-\frac{1}{4}\sum_{\substack{i,j=1\\i<j}}^{q}\left\langle\left[\hat{O}_i,\hat{O}_j\right]\right\rangle^2$. This result is of general validity in a sense that the Hamiltonian of the system may have any temporal dependence or may be a semiquantum Hamiltonian in which coexist classical and quantaum degrees of freedom. The general validity of this last result was derived only by means of the closure condition (6) and the Eherenfest theorem (7). In other words: *the sufficient condition for the GUP be an invariant of the motion is that there exists a Lie algebra associated to the system which closes a commutation algebra with the Hamiltonian of the system which defines an antisymmetric dynamical matrix $G(t)$.*

Another dynamical invariant associated to the antisymmetry of the dynamical matrix $G(t)$ is the well-known Bloch vector. In fact, if we consider the expression [8]

$$B(t) = \sum_{j=1}^{q}\left\langle\hat{O}_j\right\rangle_t^2, \tag{20}$$

it defines the surface of a hypersphere on the space $\mathbb{V} = span\{\left\langle\hat{O}_1\right\rangle(t);...;\left\langle\hat{O}_q\right\rangle(t)\}$. If we consider its time derivative, it has the form

$$\frac{dB(t)}{dt} = -2\sum_{j=1}^{q} g_{jj}(t)\left\langle\hat{O}_j\right\rangle^2 - \sum_{j=1}^{q}\sum_{r=1}^{q}\left[g_{rj}(t) + g_{jr}(t)\right]\left\langle\hat{O}_r\right\rangle\left\langle\hat{O}_j\right\rangle \quad j \neq r. \tag{21}$$

and we see, once more, that an antisymmetric matrix $G(t)$ turns the Bloch vector into an invariant of the motion.

3 Generalized Metric Phase Space, Dynamic Evolution and the UP

The results obtained in the previous Section can be refined and extended as follows. Due to the $CSNCO$ (the set of relevant operators) obtained by means of Eq.(6) are linearly independent, and so are their mean values, it enables us to define the generalized phase space as the one spanned by the mean values of the set of relevant operators

$$\mathbb{V} = span\left\{\left\langle\hat{O}_1\right\rangle(t),...,\left\langle\hat{O}_q\right\rangle(t)\right\}. \tag{22}$$

As it can be seen from (22), this $\mathbb{V}-space$ is a $q - dimensional$ one.

With the help of Ehrenfest´s theorem (7) and the closure condition (6) it is possible to obtain the equations of motion of the mean values of the relevant operators whose solution (summation over repeated upper and lower indices in the same expression is assumed)

$$\left\langle \hat{O}_j \right\rangle (t) = F^l{}_j\,(t, t_0)\ \left\langle \hat{O}_l \right\rangle (t_0), \qquad j = 1, \ldots, q \tag{23}$$

defines the $q \times q$ non-singular matrix $F^T\,(t, t_0)$ which rules the evolution of the mean values at all times ($F\,(t_0, t_0) = \mathbb{I}$ in order to fulfill the initial conditions). If we focus our attention on this "evolution matrix" $F\,(t, t_0)$ we see: a) because of the linear independence of the mean values of the relevant operators, $F\,(t, t_0)$ is a non-singular matrix so, it belongs to the general linear group over the reals $GL(q, \mathbb{R})$, b) as it was demonstrated in ref. [3], it satisfies the equation of motion

$$\frac{\partial F(t, t_0)}{\partial t} = -F(t, t_0)\ G(t), \tag{24}$$

with $G(t)$ the matrix defined through Eq.(6); c) as it was said, due to the fact that (6) does not define univocally the $CSNCO$ for a given Hamiltonian, once a particular Lie algebra has been chosen to fulfill (6), the properties of $F\,(t, t_0)$ are completely established and this fact, in turn, is ruled by the properties of matrix $G(t)$; d) the evolution matrix $F\,(t, t_0)$ allows us to establish the covariant and contravariant nature of the mean values of the relevant set $\{\left\langle \hat{O}_1 \right\rangle (t), ..., \left\langle \hat{O}_q \right\rangle (t)\}$, and of the Lagrange multipliers, $\left\{\lambda^1(t), ..., \lambda^q(t)\right\}$, associated to them respectively. In fact, we notice that for a given instant t, the set

$$B(t) = \left\{\left\langle \hat{O}_1 \right\rangle (t), ..., \left\langle \hat{O}_q \right\rangle (t)\right\} \tag{25}$$

constitutes a basis for the generalized phase space $\mathbb{V}$, and that the evolution vector $u(t)$ of ref. [9] defined as

$$u(t) = S - \lambda_0 = \lambda^j(t) \left\langle \hat{O}_j \right\rangle (t), \qquad j = 1, ..., q \tag{26}$$

belongs to this $\mathbb{V} - space$. The Lagrange multipliers $\left\{\lambda^1(t), ..., \lambda^q(t)\right\}$ are the coordinates of $u(t)$ with respect to the basis (25) and follow the equations of motion [3]

$$\frac{d\lambda^i}{dt} = \sum_{r=1}^{q} g_{ir}\,(t)\,\lambda^r\,(t) \qquad i = 1, \ldots, q. \tag{27}$$

Taking into account Eq.(23), we can consider the temporal evolution of the set $\{\left\langle \hat{O}_1 \right\rangle (t), ..., \left\langle \hat{O}_q \right\rangle (t)\}$ as continuous changes of basis (beginning with the initial condition $B(t_0) = \left\{\left\langle \hat{O}_1 \right\rangle (t_0), ..., \left\langle \hat{O}_q \right\rangle (t_0)\right\}$) performed through the transformation matrix $F\,(t, t_0)$ which generates the successive time-dependent basis so, the representation of $u(t)$ with respect to any basis fulfills

$$u(t) = \lambda^j(t) \left\langle \hat{O}_j \right\rangle (t) = \lambda^j(t_0) \left\langle \hat{O}_j \right\rangle (t_0) = u(t_0). \tag{28}$$

Replacing Eq.(23) into Eq.(28) we can write

$$\left[\lambda^j(t)\ F^l{}_j\left(t,t_0\right)-\lambda^l(t_0)\right]\left\langle \hat{O}_l\right\rangle(t_0)=0, \tag{29}$$

and because of the $\left\langle \hat{O}_l\right\rangle(t_0)$'s are linearly independent [3, 10], we have

$$\lambda^l(t_0)=\lambda^j(t)\ F^l{}_j\left(t,t_0\right), \qquad j,l=1,...,q. \tag{30}$$

From Eqs.(23) and (30) we see that the Lagrange multipliers (the "coordinates") change in an opposite manner to that of the mean values (the "vectors") [3] so, the mean values and the Lagrange multipliers transform in covariant and contravariant way respectively, according to the transformation defined by $F\left(t,t_0\right)$. In the following we will derive some properties of the generalized phase space $\mathbb{V}$.

3.1 Metric on the q−Dimensional Phase Space and UP

Now, we are going to show that when it is possible to find a set of non-commutin observables through Eq. (6), the generalized phase space (22) is a metric space. Let's define a real scalar-valued function $\bullet$ of ordered pairs of elements belonging to the $\mathbb{V}-space$ such that [9]

$\bullet:\mathbb{V}\times\mathbb{V}\longrightarrow\mathbb{R}/$

$$\left\langle \hat{O}_i\right\rangle(t)\bullet\left\langle \hat{O}_j\right\rangle(t)=\frac{1}{2}Tr\left(\hat{\rho}\left[\hat{O}_i,\hat{O}_j\right]_+\right)-Tr\left(\hat{\rho}\hat{O}_i\right)Tr\left(\hat{\rho}\hat{O}_i\right), \tag{31}$$

where $[\hat{O}_i,\hat{O}_j]_+$ indicates anti-commutation and $\mathbb{R}$ is the set of real numbers (the field associated to the $\mathbb{V}-space$). Eq.(31) may be rewritten as

$$\left\langle \hat{O}_i\right\rangle(t)\bullet\left\langle \hat{O}_j\right\rangle(t)=\frac{1}{2}\left\langle \hat{O}_i\hat{O}_j+\hat{O}_j\hat{O}_i\right\rangle-\left\langle \hat{O}_i\right\rangle\ \left\langle \hat{O}_i\right\rangle=K_{ij}(t)=K_{ji}(t). \tag{32}$$

Certainly, Eq.(31) not only defines a positive definite metric on $\mathbb{V}-space$ [9] but also the components $K_{ij}(t)=K_{ji}(t)$

$$\begin{aligned}K_{ij}(t)&=F^l{}_i\left(t,t_0\right)\ \left\langle \hat{O}_l\right\rangle(t_0)\bullet F^k{}_j\left(t,t_0\right)\ \left\langle \hat{O}_k\right\rangle(t_0)=\\&=F^l{}_i\ \left(t,t_0\right)\ K_{lk}(t_0)\ F^k{}_j\left(t,t_0\right)\end{aligned} \tag{33}$$

of the second-rank covariant metric tensor $K(t)$. Notice that Eq.(33) not only exhibits the tensor nature [10] of $K(t)$ but also gives its temporal evolution. Eq.(33) can be put into matricial form as

$$K(t)=F^T\left(t,t_0\right)\ K(t_0)\ F\left(t,t_0\right). \tag{34}$$

This covariant metric tensor transforms the contravariant vector's components $\lambda^j(t)$ into covariant ones $\lambda_i(t)$ at any instant t

$$\lambda_i(t) = K_{ij}(t)\lambda^j(t), \qquad i = 1, ..., q. \tag{35}$$

Eq.(35) leads to a geometrical interpretation of the second order centered invariant $I^{(2)}$ considered in ref. [9]

$$I^{(2)} = \left\langle (\ln \hat{\rho})^2 \right\rangle - \langle \ln \hat{\rho} \rangle^2 = K_{ij}(t)\lambda^i(t)\lambda^j(t) \qquad i,j = 1, ..., q \tag{36}$$

this dynamical invariant is, indeed, the square of the norm (with respect to the inner product defined through Eq.(31)) of the evolution vector(26)

$$I^{(2)} = \|S - \lambda_0\|^2 = \|u(t)\|^2 = \lambda_i(t)\lambda^i(t). \tag{37}$$

Eq.(37) makes clear the fact that the *invariance of the entropy S gives rise to a metric $\mathbb{V}$ – space on which the norm $\|.\|$ induced by the metric, must be constructed with the entropy itself and* $\lambda_0 = \lambda_0\left(\lambda^1, ..., \lambda^q\right)$. It should be noticed that the components $K_{ij}(t)$ of the covariant metric tensor $K(t)$ are the quantum correlation coefficients [7] between the operators $\hat{O}_i$ and $\hat{O}_j$ belonging to the $CSNCO$ and, in virtue of this fact, we come to the following two conclusions

a) the metric defined by Eq.(31) is closely related to the GUP: indeed, the positive definiteness requirement of the metric makes possible the Schwarz inequality to hold

$$\left|\left\langle \hat{O}_i \right\rangle(t) \bullet \left\langle \hat{O}_j \right\rangle(t)\right|^2 \leq \left(\left\langle \hat{O}_i \right\rangle(t) \bullet \left\langle \hat{O}_i \right\rangle(t)\right)\left(\left\langle \hat{O}_j \right\rangle(t) \bullet \left\langle \hat{O}_j \right\rangle(t)\right), \tag{38}$$

or in an equivalent fashion

$$\left(\Delta\hat{O}_i\right)^2\left(\Delta\hat{O}_j\right)^2 \geq \left[\frac{1}{2}\left\langle \hat{O}_i\hat{O}_j + \hat{O}_j\hat{O}_i \right\rangle - \left\langle \hat{O}_i \right\rangle\left\langle \hat{O}_j \right\rangle\right]^2, \tag{39}$$

on the other hand, as the operators belonging to the $CSNCO$ are non-commuting ones they obey

$$\left[\hat{O}_j, \hat{O}_k\right] = i\hat{O}_l \ , \tag{40}$$

so, in virtue of the uncertainty relation between them we can write [7]

$$\left(\Delta\hat{O}_j\right)^2\left(\Delta\hat{O}_k\right)^2 - \left[\frac{1}{2}\left\langle \hat{O}_j\hat{O}_k + \hat{O}_k\hat{O}_j \right\rangle - \left\langle \hat{O}_j \right\rangle\left\langle \hat{O}_k \right\rangle\right]^2 \geq -\frac{1}{4}\left\langle\left[\hat{O}_j, \hat{O}_k\right]\right\rangle^2 \tag{41}$$

or

$$K_{jj}K_{kk} - K_{jk}^2 \geq -\frac{1}{4}\left\langle\left[\hat{O}_j, \hat{O}_k\right]\right\rangle^2 = -\frac{1}{4}\left\langle i\hat{O}_l \right\rangle^2. \tag{42}$$

From Eqs.(39) and (41) we see the connection between the metric (31) and the GUP as said previously, so

b) the second-rank covariant metric tensor $K(t)$ has the distinctive feature of being a non diagonal one (unless the density matrix $\hat{\rho}$ describes a minimum uncertainty state of the system [7] : $\frac{1}{2}\left\langle \hat{O}_j\hat{O}_k + \hat{O}_k\hat{O}_j\right\rangle - \left\langle \hat{O}_j\right\rangle\left\langle \hat{O}_k\right\rangle = 0$) and this is a direct consequence of the fact that the GUP must hold because of the quantum nature of the system (see Eq.(41)). Finally, we want to emphasize that the Generalized Uncertainty Principle, (15), can be recovered on the real linear generalized phase $\mathbb{V} - space$ by defining on it a proper metric (see Eq.(31)) provided that the left hand side of Eq.(15) can be obtained as the summation over the principal minors of second order of the covariant metric tensor [8].

Summarizing, here we have shown the connection between the Lie algebra defined ttrough the closure condition (6), the existence of the metric space $\mathbb{V}$ and the relation between this metric and the GUP.

3.2 The $SU(2)$ Lie Algebra Case

In order to show how MEP works, let's consider a very simple Hamiltonian [12]

$$\hat{H} = \frac{\hbar}{2}\omega_o\hat{\sigma}_z \tag{43}$$

which represents a 1/2 spin particle in in an external magnetic field B_0 (B_0 parallel to O_z direction), and $\hat{\sigma}_i$ are spin operators. It is well known that $\{\hat{\sigma}_x,\ \hat{\sigma}_y,\ \hat{\sigma}_z\}$ is a basis of the $SU(2)$ Lie algebra [11] and that they satisfy the commutation relatinships

$$[\hat{\sigma}_j, \hat{\sigma}_k] = 2i\varepsilon_{jkl}\hat{\sigma}_l \ , \tag{44}$$

so, this algebra closes algebra with quantum Hamiltonians of the form

$$\hat{H}(t) = \sum_{j=1}^{3} a_j(t)\ \hat{\sigma}_j \tag{45}$$

where $a_j(t)$ are functions that may depend on time explicitly or implicitly by means of another function that may be the clasical conjugate variables q and p as may be the case of semiquantum Hamiltonians.

Proposition

If a set of operators, which fulfills the commutation relation (44), closes a commutation algebra with a Hamiltonian of the type (45), then the dynamical matrix $G(t)$ of the system, defined by means of the closure condition (6), is an antisymmetric one.

Proof
Let be $\hat{\sigma}_k$ belonging to the relevant set $\Rightarrow \left[\hat{H}(t), \hat{\sigma}_k\right] = i\sum_{j=1}^{3} a_j(t)$ $[\hat{\sigma}_j, \hat{\sigma}_k] = i\sum_{j=1}^{3}\sum_{l=1}^{3} a_j(t)\ \varepsilon_{jkl}\ \hat{\sigma}_l$. Taking into account the closure condition (6): $\left[\hat{H}(t), \hat{\sigma}_j\right] = i\sum_{r=1}^{3} g_{rj}(t)\hat{\sigma}_r\,;\quad j = 1,\ldots,3, \Rightarrow \sum_{l=1}^{3} g_{lk}(t)$ $\hat{\sigma}_l = i\sum_{l=1}^{3}\sum_{j=1}^{3} a_j(t)\ \varepsilon_{jkl}\ \hat{\sigma}l \Rightarrow \sum_{l=1}^{3}\left[g_{lk}(t) - a_j(t)\varepsilon_{jkl}\right]\ \hat{\sigma}_l = 0$; as the operators $\hat{\sigma}_l$ are linearly independent $\Rightarrow$ any element belonging to $G(t)$ can be expressed as : $g_{lk}(t) = \sum_{j=1}^{3}\varepsilon_{jkl}\ a_j(t) \Rightarrow g_{ll}(t) = 0$ and $g_{lk}(t) = -g_{kl}(t)$ $\forall\ k, l \Rightarrow G(t)$ is antisymmetric.

Now, considering the $SU\,(2)$ Lie algebra $\{\hat{\sigma}_x, \hat{\sigma}_y, \hat{\sigma}_z\}$ (with $\hat{S}_i = \frac{\hbar}{2}\hat{\sigma}_i$) as the $CSNCO$, Eq.(6) leads to the following G antisymmetric matrix

$$G_{spin} = \begin{pmatrix} 0 & -\omega_0 & 0 \\ \omega_0 & 0 & 0 \\ 0 & 0 & 0 \end{pmatrix}. \tag{46}$$

The density matrix results

$$\hat{\rho} = \exp\left\{-\lambda_0 - \lambda_x\hat{\sigma}_x - \lambda_y\hat{\sigma}_y - \lambda_z\hat{\sigma}_z\right\}, \tag{47}$$

or equivalently [12]

$$\hat{\rho} = \frac{1}{2}\left(\hat{I} + \frac{\tanh|\boldsymbol{\alpha}|}{|\boldsymbol{\alpha}|}\ \boldsymbol{\alpha}\cdot\hat{\boldsymbol{\sigma}}\right), \tag{48}$$

with: $\hat{\sigma} = (\hat{\sigma}_x, \hat{\sigma}_y, \hat{\sigma}_z)$, $\quad |\boldsymbol{\alpha}| = \sqrt{\lambda_x^2 + \lambda_y^2 + {\lambda_z}^2}$, $\quad \lambda_0 = \ln\left[2\cosh|\boldsymbol{\alpha}|\right]$. This $\hat{\rho}$ matrix enables us to calculate the covariant metric tensor's components, $K_{xx}, K_{yy},\ K_{zz},\ K_{xy} = K_{yx},\ K_{xy} = K_{yx},\ and\ K_{yz} = K_{zy}$

So, the GUP is given by:

$$\begin{aligned} I_{spin}^{H} &= \left(\Delta\hat{\sigma}_x\right)^2\left(\Delta\hat{\sigma}_y\right)^2 - \left[\frac{1}{2}\left\langle\hat{\sigma}_x\hat{\sigma}_y + \hat{\sigma}_y\hat{\sigma}_x\right\rangle - \left\langle\hat{\sigma}_x\right\rangle\left\langle\hat{\sigma}_y\right\rangle\right]^2 + \\ &+ \left(\Delta\hat{\sigma}_x\right)^2\left(\Delta\hat{\sigma}_z\right)^2 - \left[\frac{1}{2}\left\langle\hat{\sigma}_x\hat{\sigma}_z + \hat{\sigma}_z\hat{\sigma}_x\right\rangle - \left\langle\hat{\sigma}_x\right\rangle\left\langle\hat{\sigma}_z\right\rangle\right]^2 + \\ &+ \left(\Delta\hat{\sigma}_y\right)^2\left(\Delta\hat{\sigma}_z\right)^2 - \left[\frac{1}{2}\left\langle\hat{\sigma}_y\hat{\sigma}_z + \hat{\sigma}_z\hat{\sigma}_y\right\rangle - \left\langle\hat{\sigma}_y\right\rangle\left\langle\hat{\sigma}_z\right\rangle\right]^2. \end{aligned} \tag{49}$$

Taking into account that [12]: $\hat{\sigma}_i\hat{\sigma}_j + \hat{\sigma}_j\hat{\sigma}_i = 0$ and $\left\langle\hat{\sigma}_i\right\rangle^2 = 1$ Eq.(50) can be written as [6]:

$$I_{spin}^{H} = 3 - 2\left[\left\langle\hat{\sigma}_x\right\rangle^2 + \left\langle\hat{\sigma}_y\right\rangle^2 + \left\langle\hat{\sigma}_z\right\rangle^2\right] = 3 - 2\left\langle\hat{\sigma}\right\rangle^2. \tag{50}$$

So, the GUP (50) is delimited by :

$$I_{spin}^{H} \geq -\frac{1}{4} \sum_{\substack{i,j=1 \\ i<j}}^{q} \langle [\hat{\sigma}_i, \hat{\sigma}_j] \rangle^2 \tag{51}$$

From Eqs. (50), (51) the condition $\langle \hat{\sigma} \rangle^2 \leq 1$ is easily recovered. Notice that Eq. (51) define the different sets of IC for the $\langle \hat{\sigma} \rangle$ value within the interval $0 < \langle \hat{\sigma} \rangle < 1$, exhibiting the non commutative quantum nature of the $\langle \hat{\sigma}_i \rangle$ operators. Notice that Eqs. (50) or (51) are a central point in our development, as they are the generalized version of the GUP, an invariant of motion, for the $SU(2)$ Lie algebra.

Now it is time to recover $\hbar$. Since $\hat{S} = \frac{\hbar}{2}\hat{\sigma}$, $\langle \hat{\sigma} \rangle^2 = \frac{4}{\hbar^2} \left\langle \hat{S} \right\rangle^2$ and as $\langle \hat{\sigma} \rangle^2 \leq 1$ then $0 < \left\langle \hat{S} \right\rangle^2 < \frac{\hbar^2}{4}$. This means that spin systems are confined to evolve on the Bloch sphere with a constant uncertainty $\left\langle \hat{S} \right\rangle < \frac{\hbar}{2}$. This uncertainty could be considered a sort of fuzziness as it was done in [1, 2]. Besides, as our approach deals with a set of differential equations, the different IC lead to different Bloch spheres, contained by Eq. (51), making clear the fuzzyness expressd in that equation.

4 Conclusions

The fuzzyness of the internal structure of a quantum system is analyzed, from the MEP perspective. We present a general formalism to deal with quantum and semiquantum (time-dependent or not) Hamiltonian dynamics associated to non-commutative algebras, and expose the particular case of the $SU(2)$ Lie algebra. MEP allows us to explore, as it was shown in Section 2, the connection between the closure condition (6), the existence of the metric space $\mathbb{V}$ and the relation between this metric and the GUP. All developed tools, applied to the non-commutative $SU(2)$ Lie algebra, show that the Generalized Uncertainty Principle (GUP) itself is an invariant of motion. The different sets of initial conditions $\langle \hat{\sigma}_x \rangle_0, \langle \hat{\sigma}_y \rangle_0, \langle \hat{\sigma}_z \rangle_0 \}$ that could be imposed on the system, through Eqs.(7), (8), should take into account the GUP, and for the $SU(2)$ algebra, the temporal evolution is confined on the Bloch spheres whose radius lay on the interval $(0; 1)$. So, the GUP,defines the fuzzyness of these spheres inside the $\hbar$ domain for this algebra.

References

1. Madore, J.: The fuzzy sphere. Class. Quantum Grav. 9, 69–87 (1992)
2. Alekseev, A.Y., Recknagel, A., Schomerus, V.: JHEP09, 0–23 (1999)
3. Alhassid, Y., Levine, R.D.: J. Chem. Phys. 67, 4321 (1977); Phys. Rev. A 18, 89 (1978)

4. Aliaga, J., Proto, A.N.: Phys. Lett. A 142 (1989), 63 and references therein
5. Gruver, J.L., Plastino, A., Proto, A.N.: Physics Letters A 246(1-2), 61–65 (1998)
6. Sarris, C.M., Caram, F., Proto, A.N.: Phys. Lett. A 324, 1–8 (2004)
7. Merzbaher, E.: Quantum Mechanics. Wiley, N.Y. (1963)
8. Sarris, C.M., Proto, A.N.: Phys. A 348, 97–109 (2005)
9. Sarris, C.M., Caram, F., Proto, A.N.: Phys. A 331, 125–139 (2004)
10. Gilmore, R.: Lie Groups, Lie Algebras and Some of Their Applications. Wiley, N. Y. (1974)
11. Tung, W.-K.: Group Theory in Physics. World Scientific Publishing, Singapore (1985)
12. Louisell, W.H.: Quantum Statistical Properties of Radiation. Wiley, Chichester (1973)

Benefits of Full-Reinforcement Operators for Spacecraft Target Landing

Rita A. Ribeiro, Tiago C. Pais, and Luis F. Simões

Abstract. In this paper we discuss the benefits of using full reinforcement operators for site selection in spacecraft landing on planets. Specifically we discuss a modified Uninorm operator for evaluating sites and a Fimica operator to aggregate pixels for constructing regions that will act as sites to be selected at lower spacecraft altitude. An illustrative case study of spacecraft target landing is presented to clarify the details and usefulness of the proposed operators.

1 Introduction

This paper discusses the suitability of full-reinforcement aggregation operators [1-3] for evaluating alternatives in multicriteria dynamic decision processes. Dynamic multicriteria decision making has been studied, from several different points of view [4-6], but here we focus on discrete spatio-temporal decision processes that involve feedback information for each step.

Moreover, this work extends a preliminary work by the authors [7], to highlight the benefits of using full reinforcement operators to aggregated past and current information in spatio-temporal decision making processes.

The choice of aggregation operator is extremely important in any decision process [8-9], particularly in dynamic decision processes, since they imply changes in input data, over time, as well as feedback from previous steps [10]. In this work, instead of using operators for aggregating criteria we focus on aggregation of alternative ratings at step *n* with respective feedback from past iteration *n-1* (i.e. discrete spatio-temporal decision process). Furthermore, we also discuss reinforcement operators to aggregate several alternatives into a single one to create regions, which will be evaluated as alternatives at lower altitudes (i.e. spacecraft's size at low altitude is bigger than a single site, hence regions become alternatives). To this aim we present two variations for well known classes of reinforcement operators, UNINORM and FIMICA [2-3]. We first present the minimal Uninorm operator and a

Rita A. Ribeiro · Tiago C. Pais · Luis F. Simões
Uninova
Campus UNL-FCT
2829-516 Caparica, Portugal
e-mail: {rar,tpp,lfs}@uninova.pt

S. Greco et al. (Eds.): Preferences and Decisions, STUDFUZZ 257, pp. 353–367.
springerlink.com

Uninorm extension called Hybrid-Reinforcement (HR). Afterwards, we introduce two functions for Sum and Product Fimica aggregation operators.

Most aggregation/rating methods are only either upward reinforcement methods (e.g. Hamacher and Dubois & Prade union operators) or downward methods (e.g. Hamacher and Dubois & Prade intersection operators). When we combine these two concepts we achieve what is called full reinforcement behavior [11]. In this work we highlight why full reinforcement operators are important for dynamic decision processes with feedback.

The case study goal, used to illustrate the suitability of reinforcement operators, is to recommend an adequate interplanetary spacecraft target-landing site [12]. The site adequacy is evaluated with respect to a set of requirements: (1) the site should be safe in terms of maximum local slope, light level and terrain roughness; (2) the site should be reachable with the available fuel; (3) the site should be visible from the camera during the final descent phase.

This chapter is organized as follows. Section 2 describes the case study and also presents the overall dynamic decision process. In Section 3 we briefly describe the UNINORM and FIMICA class of aggregation operators. Afterwards, in section 4, we present a detailed discussion and numerical examples regarding the proposed Uninorm and Fimica based aggregation operators. An assessment concerning the aforementioned operators within the case study is presented in Section 5. Section 6 contains the concluding remarks.

2 Spacecraft Landing Overview

The main objective of a descent phase, in spacecraft landing on planets, is to select the safest site for landing [12-14]. The goal of the case study was to provide an adequate target-landing site, evaluated with a set of requirements, as mentioned in the introduction. The case study was focused in the final descent phase (around 2 Km from surface), when hazard maps can be obtained [12, 14]. To achieve the objectives we had access to simulated hazard maps (images taken by onboard camara) of dimensions 512x512 pixels that provide assessments of terrain features and trajectory constraints on a landing scenario. Notice that some values used in the case study description are just indicative due to reasons of confidentiality.

The seven input criteria, which correspond to the hazard maps obtained during the descent phase, are [14]: slope, texture, fuel, reachability, distance, shadow and scientific interest. The alternatives are the pixels of the combined search space (derived from merging 7 matrices of 512x512, corresponding to the input images/hazard maps), resulting from a data preparation process requiring normalization and data fusion (out of scope here, some details can be seen in [12]) . The resulting set of alternatives is about 260.000 possible alternative sites per iteration (512*512= 262,144).

The dynamic decision process includes around 40-60 iterations, and, for each one, there is an evaluation process (called ranking process), which includes combining the k best alternatives from iteration n-1 (historic set feedback) with the current rated ones at the n iteration (for details about this process see [14]). We only consider the n-1 iteration, as historical information to aggregate with current rating, because we update the historic set per iteration and then pass this information to the next iteration (feedback of the dynamic process). When there are no more iterations the decision process stops and the best alternative is the one with the highest combined rate (after combining historic and current rating). For the dynamic evaluation process we used a Uninorm based operator, which we called Hybrid Reinforcement (HR) Operator.

Another important aspect to take in consideration in the dynamic evaluation process is the relative proximity of the spacecraft to planet surface. When the spacecraft altitude is high the current rating refers to pixels in the images (corresponding to site coordinates). However, in the final stages of landing on a planet, we have to consider that the spacecraft size is larger than the image pixels; hence instead of selecting single pixels (coordinates) we have to select regions (i.e. sets of single pixels) for landing. Details about the regions aggregation process are given in [13]. For the regions evaluation process we defined a Product FIMICA operator, which proved to be appropriate for aggregating the pixels into regions.

Figure 1 depicts the dynamic decision process of the case study.

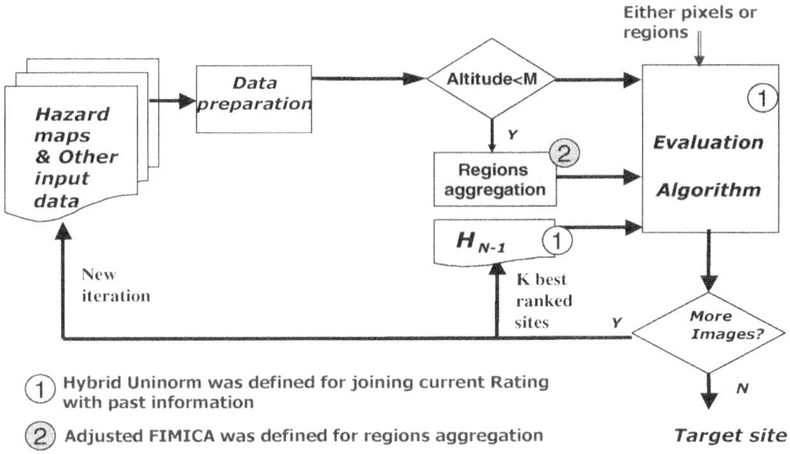

Fig. 1 Dynamic decision process of Spacecreaft Landing case study

In summary, the illustrative case study includes historical information from previous iterations (feedback) and uses suitable adaptations of full reinforcement operators in a dynamic decision process. Moreover, when spacecraft altitude is low a FIMICA reinforcement operator was devised to aggregate pixels into regions, the latter becoming alternatives of the multi-criteria dynamic decision system.

3 Background on Full Reinforcement Operators

Aggregation operators have been extensively studied in the literature and their usage in fuzzy multi-criteria problems is widely spread (see for example [8, 15-18]).

In this work we focus on aggregation operators with full-reinforcement behaviour, because this is an important quality for dynamic decision processes. Specifically, we discuss the Uninorm and Fimica classes [1-3] as beneficial for our type of problem. Full-reinforcement property means that: a) for a set of high scores to "positively" reinforce each other, it must obtain a higher score than any of the elements alone (upward reinforcement); b) for a set of low scores to "negatively" reinforce each other it must obtain a lower score than any of the elements alone (downward reinforcement). The main difference between the Fimica and Uninorm operators is that the former can be continuous.

3.1 UNINORM

The *UNINORM* class of aggregation operators was introduced by [1-2] as a generalization of T-norms and T-conorms. One of the main characteristics of this operator is the consideration of a neutral element, anywhere in the interval]0, 1[. A uninorm R is a mapping

$$R : [0,1]^2 \rightarrow [0,1] \tag{1}$$

having the following properties:

$$R(a,b) = R(b,a) \qquad \text{(commutativity);} \tag{2}$$

$$R(a,b) \geq R(c,d) \quad \text{if} \quad a \geq c \text{ and } b \geq d \qquad \text{(monotonocity);} \tag{3}$$

$$R(a, R(b,c)) = R(R(a,b),c) \qquad \text{(associativity);} \tag{4}$$

There exist some elements $e \in \left]0,1\right[$ called the neutral element such that for all $a \in [0,1]$, $R(a,e) = a$.

Moreover, uninorm operators also present a compensatory behaviour, i.e., any Uninorm R satisfies:

$$\begin{aligned} &\min(x,y) \leq U(x,y) \leq \max(x,y) \ , \ (x,y) \in [0,e[\times]e,1] \cup]e,1] \times [0,e[\\ &\text{where } e \text{ is a neutral element, i.e., } \exists e \in \left]0,1\right[\ \forall x \in [0,1] : U(x,e) = x \end{aligned} \tag{5}$$

It is known that any Uninorm operator have at least a discontinuity in $(1,0)$ and $(0,1)$. Therefore, uninorm operators may present asymptotic behaviour, i.e., a small change in the arguments implies a significant change in output value (cfr. Section 4.5). For example, the minimal uninorm operator [11] has discontinuities in $\{e\} \times [e,1] \wedge [e,1] \times \{e\}$, as shown in Figure 2.

3.2 FIMICA

The *FIMICA* class of aggregation operators [3] were derived from the MICA operators [15]. They are defined as a bag mapping:

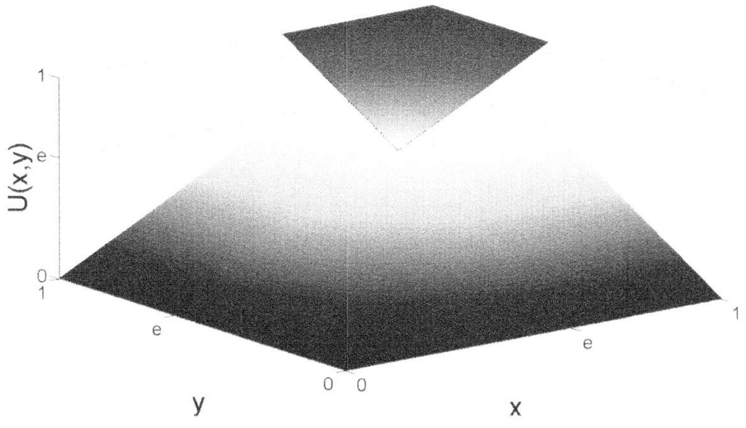

Fig. 2 Plot of minimal Uninorm (where $e = 0.65$ is the neutral element)

$$F : U^I \rightarrow I \tag{6}$$

having the following properties:

$$\text{If } A \geq B \text{ then } F(A) \geq F(B) \quad \text{(monotonocity);} \tag{7}$$

$$\exists g \in I \quad \forall A \in U^I : \; F(A \oplus \langle g \rangle) = F(A) \quad \text{(fixed identity);} \tag{8}$$

where $g \in]0,1[$ is called identity element, A and B are any bag, $A =< a_1,...,a_n >, a_i \in I = [0,1]$ and $B =< b_1,...,b_n >, b_i \in I = [0,1]$.

Fimica operators present full reinforcement behaviour, similar to Uninorm operator, and the F operator is also monotonic and commutative with respect to arguments in A. In addition, an important aspect of Fimica is the choice of an appropriate function, F, to control the operator behaviour, e.g., deciding if a small change in the arguments does (or not) imply a significant change in output value (cfr. Section 4.5).

4 Proposed Adjustments for Reinforcement Operators

In this section we discuss four variations for full reinforcement operators. First, we present the minimal Uninorm. Second we introduce one operator, called Hybrid Reinforcement (HR) operator, which is based on the minimal Uninorm [11] and includes a compensatory nature in all four quadrants of the image space (see Figure 2 and Figure 3). The third and fourth variations belong to the additive and product family of Fimica class aggregation operators, respectively, where particular functions were devised to better fit the illustrative example. All variations are discussed within the context of a small example and the illustrative case study. The authors presented a preliminary version of some of these operators in [7].

4.1 Minimal Uninorm

The proposed minimal Uninorm operator is formally defined as follows.

$$U_{\min}(x,y)=\begin{cases}\eta.T(\frac{x}{\eta},\frac{y}{\eta}) & , \quad \text{for } (x\le\eta \wedge y<\eta)\vee(x<\eta \wedge y\le\eta) \\ \eta+(1-\eta).S(\frac{x-\eta}{1-\eta},\frac{y-\eta}{1-\eta}), & \quad \text{for } x\ge\eta \text{ and } y\ge\eta \\ \mathrm{Min}(x,y) & , \quad \text{elsewhere}\end{cases} \tag{9}$$

where:
η is a neutral element;
T-norm is Hamacher intersection operator (T);
S- norm is Hamacher union operator (S);

The neutral element η is the parameter influencing the quantity of upward or downward reinforcement operations. In our case we use quantiles for neutral element because with a high quantile we ensure the majority of values fall before the bounded quantile value, hence more downward reinforcement operations. Using a lower quantile we ensure more upward reinforcement operations in the aggregation of alternative rating at iteration n with historic value from iteration *n-1*.

For S-norm and T-norm (S, T) we use the following Hamacher operator formulas [9]:

$$S_\alpha(x,y)=\frac{x+y-(2-\alpha)*x*y}{1-(1-\alpha)*x*y} \quad , \quad \text{where } \alpha\in[0;+\infty[\text{ and } x,y\in[0;1] \tag{10}$$

$$T_\alpha(x,y)=\frac{x*y}{\alpha+(1-\alpha)(x+y-y*x)} \quad , \quad \text{where } \alpha\in[0;+\infty[\text{ and } x,y\in[0;1] \tag{11}$$

In our case we use a low value for parameter α because we want to reward or penalize the rating values smoothly instead of using an aggressive aggregation behavior. The choice of Hamacher operators for the upper and lower reinforcement was based on its synergetic nature.

4.2 HR Operator

The motivation for this adapted operator was a need for an aggregation operator with full reinforcement characteristics, but flexible enough to include an averaging compensatory nature in the interval, $]0,\eta[\times]\eta,1[\cup]\eta,1[\times]0,\eta[$. Hence, we proposed an adaptation of the minimal Uninorm operator [14] which includes

Hamacher synergetic operators [9] for intersection and union and OWA [16], in the interval $]0,\eta[\times]\eta,1[\cup]\eta,1[\times]0,\eta[$. This latter interval is the one outside the image space of t-norms and s-norms, since these are bounded by the neutral element η (see Figure 3 a.). Formally, the proposed hybrid reinforcement operator, HR, is a mapping $[0,1]\times[0,1]\rightarrow[0,1]$, such that,

$$HR(x,y)=\begin{cases}\eta.T(\frac{x}{\eta},\frac{y}{\eta}) & \text{, for } (x\leq\eta\wedge y<\eta)\vee(x<\eta\wedge y\leq\eta)\\ \eta+(1-\eta).S(\frac{x-\eta}{1-\eta},\frac{y-\eta}{1-\eta}) & \text{, for } x\geq\eta \text{ and } y\geq\eta \\ \mathrm{OWA}(\mathrm{x},\mathrm{y}) & \text{, elsewhere}\end{cases} \quad (12)$$

where:
η is a neutral element;
T-norm represents Hamacher intersection operator (T) (cfr. Equation 10);
S- norm represents Hamacher union operator (S) (cfr. Equation 11);
OWA represents Yager´s OWA operator (cfr. Equation 13).

For OWA aggregation operator [16] we use the following formulation:

$$\begin{aligned}&OWA(x,y)=w_1\times\max(x,y)+w_2\times\min(x,y),\\&\text{where } w_1+w_2=1 \quad \text{and} \quad \mathrm{x},\mathrm{y}\in[0;1]\end{aligned} \quad (13)$$

The weights for the OWA aggregation operator (w_1 and w_2) will have in consideration that giving more weight to lower values will decrease the aggregation value; this is what we are looking to avoid selecting sites with lower rating values.

A similar argument as the one presented in section 4.1 can be made regarding the choice of Hamacher intersection and union functions for S and T-norm, respectively. Moreover, the same parameters were also used as in the minimal Uninorm operator.

The proposed *HR* operator is suitable for our case because our goal is to combine historical information, from previous iteration ($H_{n-1}\Rightarrow x$), with current rating value ($R_n\Rightarrow y$), until we reach a conclusion (stopping criterion). Combining current and feedback information is the feedback process in the dynamic decision process.

In summary there were four main requirements for the *HR* operator: to ensure full reinforcement capability; to include full compensatory nature in all quadrants; to take in consideration the order of elements; to ensure synergy between arguments by using Hamacher operators for T-norm and S-norm. Figure 3 (a) summarizes the combination of operators for each image space quadrant, determined by the neutral element; and plot (b) represents its behaviour for a neutral element of 0.5.

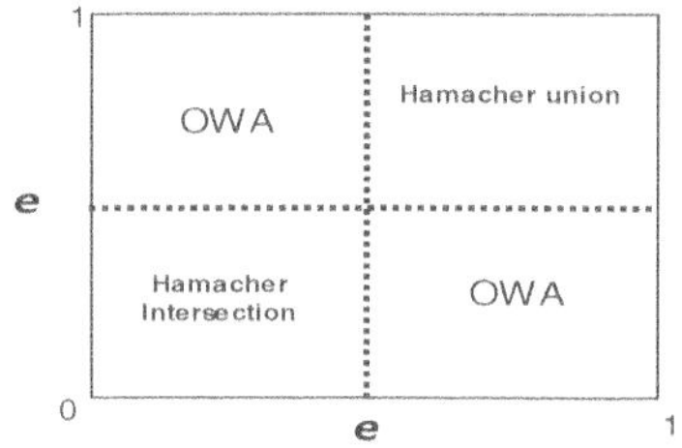

a) HR image space

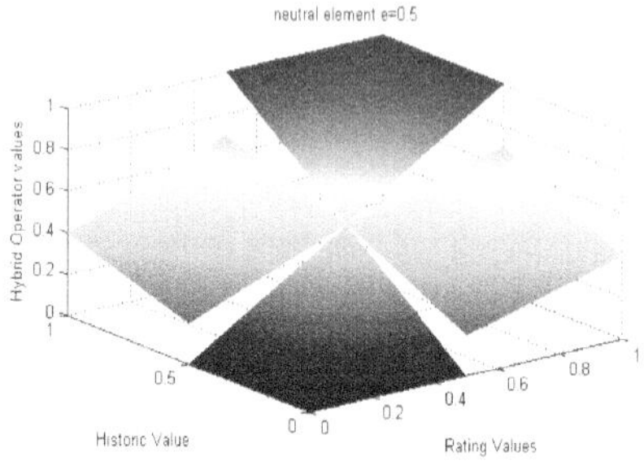

b) Plot example for e=0.5

Fig. 3 Search space of HR operator and a behavioural example.

The algorithm parameters were tuned to get a coherent behaviour for our case study. The neutral element η was set to a high quantile of R_n to obtain a small subset with high classifications. Then, we determined the α parameter for both Hamacher operators (S and T equations), and weights used for the OWA aggregation operator.

The last step of pixels evaluation (phase where spacecraft is at relative high altitude) is to rank the values decreasingly and, from this ordered list, we select a sub-set for the next iteration (historic set). At each iteration n we select the k best ranked sites and, depending on the altitude from the planet surface, the historical set size k varies. With this procedure, it can happen that the best choice of alternative is not the highest regarding its rating value, in the respective iteration. This situation is due to the use of historical feedback information and the behaviour of the hybrid reinforcement operator (HR, eq. 12) in the computation of the dynamic decision model. We want to select sites that proved to be good during a certain period of time, i.e. that provide some consensus about its suitability!

Finally, it is important to highlight that HR operator does not fulfil all properties of Uninorm operators, specifically the associativity condition. However, since for any iteration we just aggregate two values, historical and current rating values, there are no associative problems in our case. All other properties of Uninorm operators are satisfied on $[0,1]\times[0,1]$.

4.3 Additive FIMICA Operator

The additive family of the FIMICA class of aggregation operators is defined as follows [3].

$$\begin{aligned} &S(A) = f\Big(\sum_{i\in I} (a_i - g) \Big) \\ &\text{where}: \ f: D \subseteq \Re \to [0,1]; \\ &\qquad A = < a_1, \cdots, a_n > \text{ is a bag defined over the unit interval};\\ &\qquad g \in [0,1] \text{ is the identity.} \end{aligned} \tag{14}$$

Several f functions were tested, for our illustrative case study, and we chose the most suitable to ensure a smooth behaviour. Formally, the chosen function is,

$$S(<x,y>) = 0.5 - \frac{\arctan((x-g)+(y-g))}{\pi} \quad \text{, where } x,y \in [0,1] \tag{15}$$

In Figure 4 we show its respective plot.

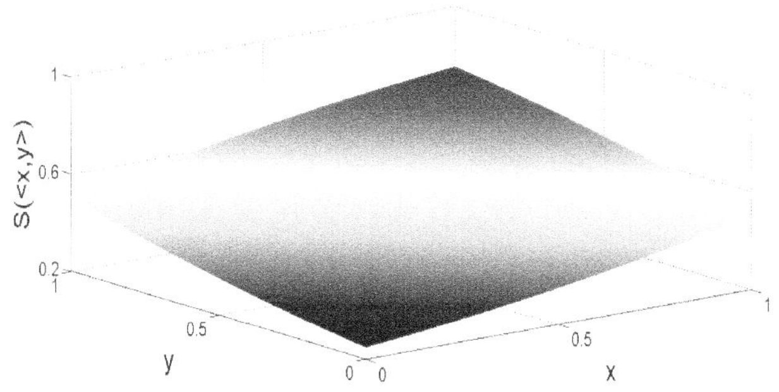

Fig. 4 Plot of additive Fimica operator for g= 0.5.

4.4 Product FIMICA Operator

The product Fimica aggregation operators are defined in a similar fashion as the previous additive Fimica operators. The product operator is defined as follows [3].

$$\begin{aligned} & V(A) = f\left(\prod_{i\in I}\left(\frac{a_i}{g}\right)\right) \\ & \text{where: } f : D \subseteq \Re_0^+ \rightarrow [0,1]; \\ & \qquad A = < a_1, \cdots, a_n > \text{ is a bag defined over the unit interval;} \\ & \qquad g \in]0,1] \text{ is the identity.} \end{aligned} \tag{16}$$

As in the previous section, several f functions were tested and the most suitable to ensure the necessary behaviour is defined as in Equation (17).

$$V(<x,y>) = 1 - \frac{1}{1 + \frac{x}{g} \times \frac{y}{g}} \quad \text{, where } x,y \in [0,1] \tag{17}$$

In Figure 5 we show its respective plot.

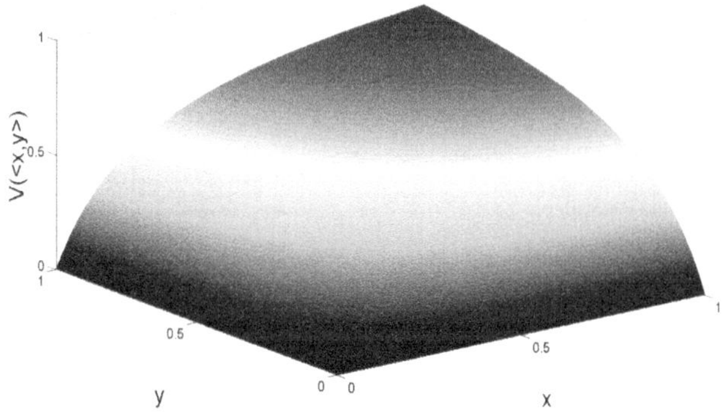

Fig. 5 Plot of product Fimica operator for g=0.5.

Comparing Figure 2 with Figure 5 it is obvious the latter continuous nature and the non-existence of intervals where smooth behaviour is not guaranteed.

4.5 Numerical Examples

This section illustrates the results of applying the proposed adjusted operators, to aggregate 2 criteria (columns) for 3 alternatives (rows), as depicted in the following matrix,

$$\begin{array}{c} \\ A_1 \\ A_2 \\ A_3 \end{array}\begin{array}{c} \begin{array}{cc} H_{n-1} & R_n \end{array} \\ \begin{bmatrix} 0.104 & 0.21 \\ 0.2 & 0.11 \\ 0.1039 & 0.21 \end{bmatrix} \end{array}$$

4.5.1 Example 1 - Umin Operator

Consider $e = 0.104$ (neutral element), Hamacher's parameters $\alpha = 0.8$. Applying U_{min}, we have:

$$U_{\min A_1}(x_{11}, x_{12}) = 0.2102$$

$$U_{\min A_2}(x_{21}, x_{22}) = 0.104 + (1 - 0.104) H_{\cup}\left(\frac{0.2 - 0.104}{1 - 0.104}, \frac{0.11 - 0.104}{1 - 0.104}\right) = 0.2052$$

$$U_{\min A_3}(x_{31}, x_{32}) = \min(0.1039\ , 0.21) = 0.1039$$

The results are: A1>A2>A3. Figure 6 a) depicts the plot for this operator behaviour.

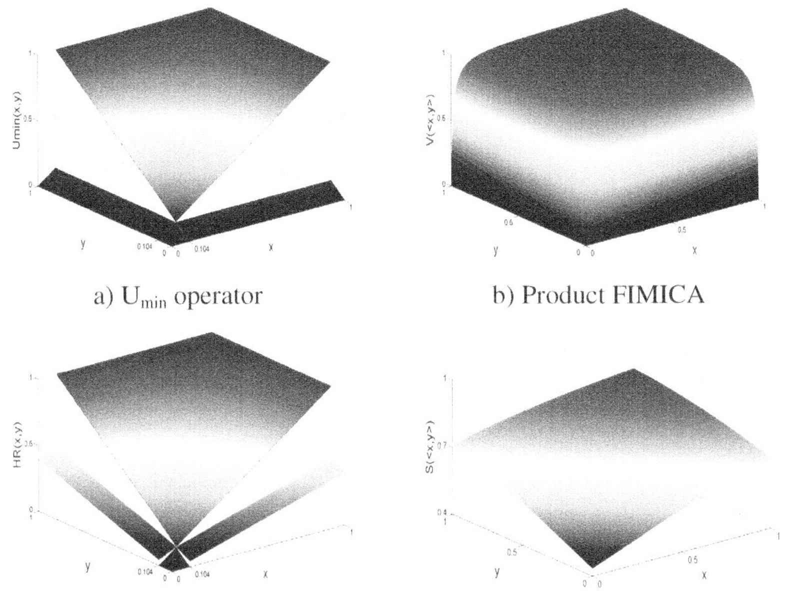

a) U_{min} operator b) Product FIMICA

c) HR operator d) Additive FIMICA

Fig. 6 Plots of operators used in numerical examples with e=g=0.104 and α =0.8.

4.5.2 Example 2 - HR Operator

Consider $e = 0.104$ (neutral element), Hamacher's parameters $\alpha = 0.8$ and OWA w_1 and w_2 equal to 0.4 and 0.6, respectively. Applying HR, we have:

$$HR_{A_1}(x_{11}, x_{12}) = 0.2102$$

$$HR_{A_2}(x_{21}, x_{22}) = 0.104 + (1 - 0.104)H_{\cup}\left(\frac{0.2 - 0.104}{1 - 0.104}, \frac{0.11 - 0.104}{1 - 0.104}\right) = 0.2052$$

$$HR_{A_3}(x_{31}, x_{32}) = OWA(0.1039, 0.21) = 0.1463$$

The results are: A1>A2>A3. Figure 6 c) shows the plot for this operator.

4.5.3 Example 3 - Product Fimica

Next we show the results obtained with the proposed product Fimica considering $g = 0.104$ for identity element:

$$V_{A_1}(< x_{11}, x_{12} >) = 1 - \frac{1}{1 + \left(\frac{0.104}{0.104} * \frac{0.21}{0.104}\right)} = 0.50476$$

$$V_{A_2}(< x_{21}, x_{22} >) = 0.50836$$

$$V_{A_3}(< x_{31}, x_{32} >) = 0.50429$$

The results are: A2>A1>A3. Figure 6 b) depicts the behaviour for this operator.

4.5.4 Example 4 - Additive Fimica

In this example is illustrated the results obtained with the proposed additive Fimica considering $g = 0.104$ for identity element:

$$V_{A_1}(<x_{11}, x_{12}>) = 0.5 - \frac{\arctan((0.21 - 0.104) + (0.104 - 0.104))}{\pi} = 0.53362$$

$$V_{A_2}(<x_{21}, x_{22}>) = 0.53236$$

$$V_{A_3}(<x_{31}, x_{32}>) = 0.53358$$

The results are: A1>A3>A2. Figure 6 d) depicts the plot of this operator´ behaviour.

4.5.5 Discussion of Examples Results

All reinforcement operators consider that alternative A3 is worse then A1 and this is a coherent result since A3 is dominated by A1.

Comparing Example 1 and Example 2 we can observe that the results are identical in terms of order of importance of alternatives but the Umin does not have a compensatory behaviour when one criteria is good (above the neutral element) and another is bad (below the threshold) hence it reduces the overall importance to its weakest criteria. In our case study we do not want this behaviour because we are aggregating historic information with current information and both should count to the overall result. Hence, for our dynamic model we selected the HR operator.

All operators consider that A1 is the best alternative except product FIMICA that considers A2 as the best alternative. This shows that for aggregating only two criteria this operator might not be the best choice but for aggregating several criteria it is another story. Comparing the results of Example 1 and 3, we can observe that product Fimica operator has a more coherent behaviour in the sense that similar alternatives have very close ratings (A1 and A3). In both operators, if we have one $a_i = 0$, the aggregated value will also be zero. This is a critical feature for our operator since their purpose is to aggregate a set of pixels into a single region. This means that if one region contains an unacceptable pixel the entire region is considered undesirable for landing.

In summary, it seems that HR is a good choice for aggregating two criteria that require synergy and compensatory behaviour (dynamic model), while product FIMICA seems good for aggregating several criteria when we want to ensure a smoother behaviour in the aggregation and the elimination of an alternative if there is one or more "bad" elements in that alternative. The choices of operators for the case study are further discussed in the next section.

In Figure 6 we show the four operators plots, the same neutral element and identity element value was used in all the examples. The differences are obvious in terms of behaviors' in the different quadrants of the search space.

5 Assessment of Reinforcement Operators within Case Study

In this work we described a dynamic evaluation process for rating sites, either for pixels or regions depending on the spacecraft altitude.

The dynamic decision process is done with the HR operator (12) and applies to either pixels or regions. To construct the regions (grouping of ratings of nearby pixels) we used the product Fimica (17). We used product FIMICA to ensure a smother aggregation behaviour and fulfilment of the associative property, when there are more than 2 elements to aggregate.

The main difference between the U_{min} and the product FIMICA operator is that the former is continuous. Further, choosing an appropriate function (17) guarantees a more smooth behaviour, i.e., a small change in the arguments does not imply a significant change in output values. This was illustrated in the small numerical examples above (Section 3.3), where a difference of only 0.01 (but it could be as small as we wanted) implies a change in the ranking order. When evaluating regions we observed that some had a poor ranking position, even though most of its pixels have high values and only a couple of them had smaller values than the neutral element (but very small differences).

Table 1 depicts results of grouping pixels into regions, for one iteration of the decision process. It can be observed that for the ten best regions, obtained with product FIMICA, the min Uninorm only ranks alternative 4 as the "best" and number 10 as the third ranked. It misses all other good regions that were obtained with product FIMICA. The results were also validated by Space experts and they concurred that FIMICA operator was more suitable to rate regions.

Table 1 Best landing site regions identified in one iteration, using the product FIMICA operator, and results for the same regions using the minimal Uninorm operator

Region 2D coordinates		FIMICA		Uninorm	
x	y	Rating	Ranking	Rating	Ranking
207	267	0,793156	1	0,983649	41
206	267	0,792631	2	0,983618	42
208	267	0,791215	3	0,983521	49
215	250	0,78965	4	0,984586	1
209	267	0,789257	5	0,983384	55
216	250	0,788848	6	0,984528	4
209	266	0,787445	7	0,983346	58
205	267	0,787431	8	0,983331	59
183	257	0,786907	9	0,984029	20
214	248	0,786865	10	0,984543	3

Now we will discuss the operators involved in the dynamic decision process of aggregating past and current information. The HR and additive Fimica operators have a completely different behaviour when compared with the two previous discussed operators. Also, as mentioned, in our case study they have a different purpose. These operators are must more adequate to be used in the dynamic aggregation phase due to their compensatory nature on the entire domain. When we observe the results presented in Section 4.5, again, the (additive) Fimica operator seems to have a more coherent behaviour. However, in this case, the function used in the Fimica operator is much more computational demanding and time consuming than the HR function even though it presents a more smooth behaviour. When we are dealing with dynamical decision models, where several decisions are made until a "consensus" is reached, a major constraint is the computational cost in terms of time. Hence, the HR operator is more suitable for dynamic aggregation. Moreover, for our case study computational time was an essential feature for the feasibility of the entire algorithm, so HR was chosen.

6 Concluding Remarks

In this work we discussed details about full reinforcement operators. Specifically, we focused on a hybrid operator (HR) used in a dynamic decision process for selecting alternatives and a product FIMICA operator used for aggregating (grouping) ratings of alternatives.

The suitability and flexibility of using full-reinforcement operators was assessed with an illustrative example and also with a case study of site selection for spacecraft landing on planets. Specifically the hybrid reinforcement operator was used to combine past and current ratings at each iteration of the dynamic decision process, while the proposed product Fimica was used for aggregating pixels into regions (when a spacecraft is close to the surface it is bigger than pixels in hazard maps, hence we need to select regions). Moreover, since the case study involves a dynamic decision process the usage of full-reinforcement operators proved quite successful for achieving a good decision after several iterations.

References

1. Fodor, J.C., Yager, R.R., Rybalov, A.: Structure of uninorms. International Journal of Uncertainty Fuzziness and Knowledge-Based Systems 5(4), 411–427 (1997)
2. Yager, R.R., Rybalov, A.: Uninorm aggregation operators. Fuzzy Sets and Systems 80(1), 111–120 (1996)
3. Yager, R.R., Rybalov, A.: Full Reinforcement Operators in Aggregation Techniques. IEEE Systems, Man and Cybernetics, Part B 28(6) (1998)
4. Diederich, A.: Dynamic stochastic models for decision making under time constraints. Journal of Mathematical Psychology 41(3), 260–274 (1997)
5. Khanmohammadi, S., Riberiro, R.A., Jassbi, J.: Multi criteria decision making using dynamics of criteria. In: Proceedings of the 11th Mediterranean Conference on Control and Automation, MED 2003 (2003)

6. Townsend, J.T., Busemeyer, J.: Dynamic representation of decision-making. In: Port, R., Gelder, T.V. (eds.) Mind in Motion: Dynamics, Behavior and Cognition, pp. 101–120 (1995)
7. Ribeiro, R.A., Pais, T.C.: Reinforcement Operators in Spacecraft Landing. In: Proceedings of the 5th Workshop on Preferences and Decisions, Trento, Italy (2009)
8. Beliakov, G., Warren, J.: Appropriate choice of aggregation operators in fuzzy decision support systems. IEEE Transactions on Fuzzy Systems 9(6), 773–784 (2001)
9. Zimmermann, H.J.: Fuzzy Set Theory - and its applications, 3rd edn. Kluwer Academic Publishers, Dordrecht (1996)
10. Richardson, G.P., Pugh, A.L.: Introduction to system dynamics modeling with DYNAMO. Journal of the Operational Research Society 48(11), 1146 (1997)
11. Detyniecki, M.: Fundamentals on Aggregation Operators. In: AGOP, Asturias (2001)
12. Devouassoux, Y., et al.: Hazard avoidance developments for planetary exploration. In: 7th International ESA Conference on Guidance, Navigation & Control Systems (2008)
13. Pais, T.C., et al.: Regions rating for selecting spacecraft landing sites. In: Ruan, D., Montero, J., Lu, J., Martinez, L., D'hondt, P., Kerre, E.E. (eds.) Computational Intelligence in Decision and Control - Proceedings of the 8th International FLINS Conference. World Scientific Proceeding Series on Computer Engineering and Information Science, vol. 1, pp. 1039–1044 (2008)
14. Pais, T.C., et al.: Dynamic ranking algorithm for landing site selection. In: Proceedings of the 12th International Conference on Information Processing and Management of Uncertainty in Knowledge-Base Systems, IPMU (2008)
15. Yager, R.R.: Aggregation operators and fuzzy systems modeling. Fuzzy Sets and Systems 67, 129–146 (1996)
16. Yager, R.R.: Ordered weighted averaging aggregation operators in multi-criteria decision making. IEEE Trans. on Systems, Man and Cybernetics 18(1), 636–643 (1988)
17. Calvo, T., Mayor, G., Mesiar, R. (eds.): Aggregation Operators: New Trends and Applications. Studies in Fuzziness and Soft Computing, vol. 97. Physica-Verlag, Heidelberg (2002)
18. Figueira, J., Greco, S., Ehrgott, M. (eds.): Multiple Criteria Decision Analysis: State of the Art Surveys. Kluwer Academic Publishers, Dordrecht (2005)

Neural Networks for Non-independent Lotteries

Giulia Rotundo

Abstract. The von Neuman-Morgenstern utility functions play a relevant role in the set of utility functions. This paper shows the density of the set von Neuman-Morgenstern utility functions on the set of utility utility function that can represent arbitrarily well a given continuous but not independent preference relation over monetary lotteries. The main result is that without independence it is possible to approximate utility functions over monetary lotteries by von Neuman-Morgenstern ones with arbitrary precision. The approach used is a constructive one. Neural networks are used for their approximation properties in order to get the result, and their functional form provides both the von Neumann-Morgenstern representation and the necessary change of variables over the set of lotteries.

1 Introduction and Basic Definitions

This paper focuses on von Neumann-Morgenstern (vNM) representation of utility functions [1, 2]. This representation is relevant also for computational issues leading to parallel calculus [3, 4]. Such a representation of utility functions holds under some hypotheses on the preference relation. Here I consider preferences over monetary lotteries, and I explore the role of vNM utility functions when the independence hypothesis is not holding any more. This generalization is most useful for research on consumers/customers, and for behavioral experiments when the functional form of the utility function is not given and concerns the individual preferences in some real world problem. In fact, in such cases the building of a functional form for the utility function needs to pass through an approximation problem, because the utility function must be extrapolated by the knowledge of some samples about the preferences of the individuals. Neural networks are particularly useful for their structure

Giulia Rotundo
Faculty of Economics, University of Tuscia, via del Paradiso 47, 01100 Viterbo, Italy
Tel.: +390761357727; Fax: +390761357707
e-mail: giulia.rotundo@uniroma1.it, giulia.rotundo@gmail.com

S. Greco et al. (Eds.): Preferences and Decisions, STUDFUZZ 257, pp. 369–375.
springerlink.com

when dealing with approximation tasks and with problems that require some separation in the dependence of a given function by the variables which the function depends on [6]. This particular form offers a way for finding a change of variables over the set of lotteries that leads to a representation of the preference relation by a vNM utility function. In order to improve the easy of read and the self-consistency of the paper, a few definitions and theorems are summed up here below.

About the theory of neural networks, a more general case of the following proposition is reported in [5].

Theorem 1. *Any continuous function $f : R^m \to R^q$ can be approximated uniformly on compacta by functions of the form*

$$\hat{f} = (\hat{f}^1, \cdots, \hat{f}^q) \tag{1}$$

where

$$\hat{f}^k(x) = \alpha \sum_{j=1}^{h} W^2_{kj} T(\sum_{i=1}^{m} W^1_{ji} x^i - \theta_j) - \vartheta_k \tag{2}$$

and $x = (x^1, \cdots, x^m) \in R^m$ is the input of the network, $W^1 \in R^{h \times m}$, $W^1 = (w^1_i)$ are the weights between the input and the hidden layer, $W^2 \in R^{q \times h}$ are the weights between the hidden and the output layer, $\theta \in R^h$ and $\theta \in R^q$ are the thresholds, $T : R^1 \to R^1$, $T(z) = (\frac{1}{1+exp(-z)})$, $\forall z \in R^1$ is the sigmoidal transfer function.

The above function (2) represents a particular feed forward neural network (known in literature also as multi layer perception) with only one hidden layer, sigmoidal transfer function and linear output units. The sigmoidal transfer function can be substituted by [6] threshold functions or linear truncated functions, and other ones. The thresholds can be merged in the sum by increasing both the input and the hidden layer of one unit, whose value is always equal to one, whose weights are the opposite of the corresponding thresholds and setting for these units some parameters that let the sigmoidal function behave like the identical function. Thus instead of (2) the following function can be used:

$$\hat{f}^k(x) = \sum_{j=1}^{h} W^2_{kj} T(\sum_{i=1}^{n} W^1_{ji} x^i) \tag{3}$$

And thus, from (1), writing the weights in a matrix form:

$$\hat{f}^k(x) = W^2 u(W^1 x) \tag{4}$$

where $u : R^h \to R^h$, $u(z) = (T(z^1), \cdots, T(z^h))$, $z \in R^h$ applies the transfer function to each component of the vector z.

Definition 1. *A sample for a neural network like (2) is a pair (x, y), x in the domain, y in the image set of the function (2).*

The set of multi layer perceptron is a universal function approximator. Neural network models are often used for economic problems, but, first of all, they must be "trained", i.e. weights need to be fixed such that the error of the approximation is minimized. From the approximation theory point of view the training procedure gets a set of samples for the network (x in the domain, y in the image set), it defines a function measuring the error of the approximation, and it determines through maximumn likekihood methods the values of the weights that minimize the error of the approximation, that is usually defined through a measure of distance between the values of the functions on x and the assigned value y. The target of learning of network (2) can be described by the minimization problem:

$$min_{W^1,W^2} \sum_t \| y^t - \hat{f}(x^t) \| \qquad (5)$$

We are going to weaken hypotheses. The following theorem [2]

Theorem 2. *Suppose that the rational preference relation $\succeq$ on a set A is continuous. Then there is a continuous utility function $u(x)$ that represents $\succeq$.*

shows that in order to work with continuous utility functions it is sufficient to deal with continuous preference relations. Other definitions are necessary to work with the function representation:

Definition 2. *The preference relation $\succeq$ has an extended expected utility representation if $\forall s \in S$, there is a function $u_s : R_+ \rightarrow R$ such that for any $(x_1, \cdots, x_s) \in R^s$ and $(x'_1, \cdots, x'_s) \in R^s$, $(x_1, \cdots, x_s) \succeq (x'_1, \cdots, x'_s)$ if and only if*

$$\sum_s \pi_s u_s(x_s) \geq \sum_s \pi_s u_s(x'_s) \qquad (6)$$

Remark 1. The above property (6) is verified under the hypotheses that $\pi_s \geq 0$, that is true when π_s are representing probabilities, and that $u_s : [0, \infty) \rightarrow R_+$ are strictly increasing $\forall s$.

The last property is verified, as example, by the sigmoidal function.

The following definitions introduce the target of lotteries [1]:

Definition 3. *A (monetary) lottery L over alternative levels of wealth w can be denoted by the collection of pairs $L = \{(p_i, w_i) : i = 1, ..., n\}$. The probability p_i is the probability of receiving the wealth w_i when the decision is made or the lottery is played.*

Definition 4. *An individual's choice over lotteries satisfies the continuity axiom whenever a sequence p_n of probabilities (i.e. $0 \leq p_n \leq 1$) converges to p, that is $p_n \rightarrow p$, and the lottery $p_n L_1 + (1 - p_n) L_2$ is preferred to a lottery L_3 $\forall n$, then $pL1 + (1 - p) L_2$ is preferred to L_3.*

Definition 5. *An individual's choice over lotteries satisfies the independence axiom whenever a lottery L_1 is preferred to another lottery L_2, then for each $0 < p < 1$ the compound lottery $pL_1 + (1 - p) L_3$ is preferred to the compound lottery $pL_2 + (1 - p) L_3$ for all lotteries L_3.*

This axiom is strongly connected with the linearity of the utility function that represents the preference relation.

If the continuity and the independence axioms hold it is possible to define a vNM utility function:

Theorem 3. *((Extended) Expected utility theorem) If an individual's utility function* $U : \mathscr{L} \to \mathscr{R}$ *over the set of lotteries satisfies independence and continuity, then there is a vNM utility function u over wealth such that* $U(L) = \sum_{i=1}^{n} p_1 u(w_i)$ *for every lottery* $L = \{(p_i, w_i) : i = 1, \cdots, n\}$.

The target is to weaken the independence hypothesis, and to show how to get a new space where the utility function is approximated:

Definition 6. *Given a transformation over a lottery set* $\mathscr{L}$*, let* $\succeq$ *be the preference relation on* $\mathscr{L}$ *and*

$$g : \mathscr{L} \subseteq R^{n\times n} \to \mathscr{L} \subseteq R^{h\times h} \tag{7}$$

$$(p,w) \to (p',w')$$

where $(p,w) \in \mathscr{L}$, $p,w \in R^n$, $(p',w') \in \mathscr{L}'$, $p',w' \in R^n$.

The induced preference relation $\succeq'$ over $\mathscr{L}'$ is such that if $(p_1,w_1) \succeq (p_2,w_2) \in \mathscr{L}$, then $(p'_1,w'_1) \succeq (p'_2,w'_2) \in \mathscr{L}'$, and $(p'_i,w'_i) = g((p_i,w_i)), i = 1,2$.

moreover in this paper are considered the transformations g that lead to the same utility function

Definition 7. *Given a transformation over a lottery set* $\mathscr{L}$,

$$g : \mathscr{L} \subseteq R^{n\times n} \to \mathscr{L} \subseteq R^{h\times h} \tag{8}$$

$$(p,w) \to (p',w')$$

and the utility function $U : \mathscr{L} \to R^1$ *that represents a preference relation over* $\mathscr{L}$*, the induced preference relation over* $\mathscr{L}'$, $U' : \mathscr{L}' \to \mathscr{R}^\infty$ *preserves the utility function if for each* $(p',w') = g(p,w))$ *it happens that* $U((p,w)) = U'((p',w'))$.

2 Approximation of Utility Function

Definitions reported in the previous section serve for a quick reference to the tools which the following theorem is built on:

Theorem 4. *For each continuous preference relation* $\succeq$ *over* $\mathscr{L} \in R^{2n}$ *there exists a change of variables* $l \in \mathscr{L} \to l' \in \mathscr{L}' \in R^h$ *such that the preference relation induced into the new lottery space is arbitrarily well approximated by a utility function that has a vNM representation. Moreover, this change of variables preserves the utility function.*

The theorem is willing to show how to build an approximating function and a transformation over the lottery space that leads to a vNM utility function.

Proof. Let $l = (p^1, \cdots, p^n, w^1, \cdots, w^n)^T \in \mathscr{L}$. Because of the continuity, $\succeq$ can be represented by a continuous utility function

$$U : \mathscr{L} \subseteq R^{n \times n} \rightarrow R^1 \tag{9}$$

$$l \rightarrow U(l)$$

that can be approximated by a neural network with $m = 2n$, $q = 1$, h to be fixed, and arbitrarily small error ε. In order to make the interpolation it is necessary to get a set of samples

$$\{(l_i, U(l_i))\}_{i \in I}$$

$l_i = (p_i^1, \cdots, p_i^n, w_i^1, \cdots, w_i^n)^T$, $p_i = (p_i^1, \cdots, p_i^n)^T$, $w_i = (w_i^1, \cdots, w_i^n)^T$ as a first step.

Once collected these samples it is also possible to use them for the interpolation of the function $l \rightarrow p(p^T p)^{-1} U(l)$, that is a continuous function of p and w ($p = (p^1, \cdots, p^n)^T, w = (w^1, \cdots, w^n)^T$).

Thus a learning procedure can be made for this network with $m = 2n$, $q = n$ in order to fix the weights W^1 and W^2 and the value of h that gives the desired error ε. Thus for each sample l_i

$$[p_i(p_i^T p_i)^{-1}] U(l_i) = W^2 u(W^1 l) \tag{10}$$

where $W^1 \in R^{h \times m}$ and $W^2 \in R^{q \times h}$, and u is the sigmoidal transfer function. Obviously the set of samples must be sufficiently large in order to ensure that the above correspondence (10) describes well the preference relation with an error lower than ε. Thus for each lottery $l = (p^1, \cdots, p^n, w^1, \cdots, w^n)^T$,

$$[p(p^T p)^{-1}] U(l) = W^2 u(W^1 l) \tag{11}$$

and then

$$U(l) = p^T [p(p^T p)^{-1}] U(l) = [p^T W^2 \parallel p^T W^2 \parallel^{-1}][\parallel p^T W^2 \parallel u(W^1 l)] \tag{12}$$

The factor $\parallel p^T W^2 \parallel$ is a normalization factor and $\parallel | \,|$ is a vector norm. Now it is possible to change variables in this way:

$$g : \mathscr{L} \subseteq R^{n \times n} \rightarrow \mathscr{L}' \subseteq R^{h \times h} \tag{13}$$

$$l \in \mathscr{L} \rightarrow l' \in \mathscr{L}$$

such that $l = (p^1, \cdots, p^n, w^1, \cdots, w^n)^T$, $l' = (p^{1'}, \cdots, p^{n'}, w^{1'}, \cdots, w^{n'})^T$, and

$$p' = p^T W^2 [\parallel p^T W^2 \parallel]^{-1} \tag{14}$$

and

$$w' = [\parallel p^T W^2 \parallel] U(W^1 l) \tag{15}$$

With this change of variables (12) becomes

$$U(l) = p'w' = \sum_{i=1,h} p'^i w'^i \tag{16}$$

This change of variables preserves the utility function and allows to write her in a vNM form. Because of the preservation of the utility function, it is possible to work into the new space for all the problems that involve the considered preference relation over $\mathscr{L}'$ and to use the reverse change of variables just in order to have the found solution in the original space $\mathscr{L}$, and the two solutions are ε-close [4].

Remark 2. The change of variables (14) and (15) can be inverted only under the hypotheses that $(W^2(W^2)^T)^{-1}$ and $(W^1(W^1)^T)^{-1}$ are invertible matrices.

Remark 3. The change of variables (14) and (15) must be restricted to the set of lotteries for which the images are in $R_+^{h\times h}$ in order to verify the necessary properties of the Remark 0.1

Remark 4. Each component of $\| p^T W^2 \| u(\cdot)$ is non - negative, continuously increasing function because $\| p^T W^2 \|$ is a positive number.

Remark 5. The change of variables for the new probabilities p' uses only the old probabilities p, while the change of variables for the wealth $w' = W^1 l = W^1(p,w)$ involves not only the wealth w, but also the probabilities p. Thus the new wealth w' can be seen as a particular weighted mean of the wealth w.

Remark 6. The normalization $[\| p^T W^{(2)} \|]$ is necessary in order to interpreter the vector p' as the probabilities into the new lottery space.

Remark 7. The utility function was built in order to approximate the utility function of a preference relation, and this is the interpretation that was evidenced, but she actually represents a preference relation over $\mathscr{L}'$ that is ε- close to the induced one. This can be expressed also by saying that the set of the vNM utility functions is dense in the set of the continuous utility functions.

Remark 8. The independence hypothesis of the extended expected utility theorem is a sufficient condition, and it is quite strong. However, it is possible to give a vNM utility function that does not verify the independence.

In fact it is possible, starting from (12), to continue the proof by using the following change of variables and utility function:

$$g : \mathscr{L} \subseteq R^{n\times n} \to \mathscr{L}' \subseteq R^{h\times h} \tag{17}$$

$$l \in \mathscr{L} \to l' \in \mathscr{L}$$

such that $l = (p^1, \cdots, p^n, w^1, \cdots, w^n)^T$, $l = (p'^1, \cdots, p'^n, w'^1, \cdots, w'^n)^T$, and

$$p' = p^T W^2 [\| p^T W^2 \|]^{-1}$$

and

$$w' = W^1 l$$

from which

$$p = [(W^{1^T} W^1)^{-1} (W^{1^T}) w']_I$$

and

$$w = [(W^{1^T} W^1)^{-1} (W^{1^T}) w']_{II}$$

where the operators $[\cdot]_I$ and $[\cdot]_{II}$ indicate, respectively, the components $1, \cdots, n$ and $n+1, \cdots, 2n$ of the given vector. The $u(\cdot)$ function must be charged in the following way:

$$u'(w')^j = [|| \ [(W^{1^T} W^1)^{-1} (W^{1^T}) w']_I]^T W^2 \ ||][1 + exp(w')^j]^{-1}], j = 1, \cdots, h \quad (18)$$

With this change of variables (12) becomes

$$U(l) = p' u'(w') = \sum_{i=1,h} p'^i (u'(w'))^i \quad (19)$$

than, in general, is not independent in the meaning of the Definition 4.

3 Conclusions

This paper shows a procedure that allows to get a useful representation of the utility functions for continuous preference relations. The possibility to merge the given lotteries into another lottery space allows to use an approximation to the utility function that represents the given preference relation by a vNM utility function. This result can be particularly useful if the analytic form of the utility function is not known. Moreover, the Theorem 4 can be generalized by introducing thresholds transferring functions and other non-continuous functions, and thus some further generalization to non-continuous preference relations would be possible.

References

1. Aliprantis, C.D., Chakrabarti, S.K.: Games and decision Making. Oxford Univ. Press, Oxford (1999)
2. Aliprantis, C.D., Chakrabarti, S.K.: Games and decision Making. Oxford Univ. Press, Oxford (1999)
3. Mas-Colell, A., Whinston, M.D., Green, J.R.: Microeconomic Theory. Oxford Univ. Press, Oxford (1995)
4. Rotundo, G.: Neural Networks for Partitionable Variational Inequalities. In: Proceedings of the APL 1998 (Array Parallel Processing) Conference, Rome (1998)
5. Rotundo, G.: Trasformazione di disequazioni variazionali in disequazioni variazionali partizionabili mediante reti neurali. In: XXIII A. M. A. S. E. S. Conference, Rende (1999)
6. Ripley, B.D.: Pattern recognition and Neural Netowrks. Cambridge University Press, Cambridge (1996)

Weak Implication and Fuzzy Inclusion

Romano Scozzafava

Abstract. The content of this talk is taken from joint papers with G. Coletti and B. Vantaggi.

We define weak implication $H \longmapsto_P E$ ("H weakly implies E under P") through the relation $P(E|H) = 1$, where P is a (coherent) conditional probability.

In particular (as a ... by-product) we get "inferential rules", that correspond to those of default logic. We discuss also connections between weak implication and fuzzy inclusion.

Keywords: conditional probability, weak implication, default logic, fuzzy inclusion.

1 Introduction

How can a seemingly loose concept like "weak implication" be embedded in a rigorous mathematical framework?

There is a huge relevant literature on this matter, but we will not undertake a discussion or a review on this.

Just to recall some "semantic" aspects, let us consider the trivial statement "if it rains on the spot x, then x is wet"; this is clearly a logical implication $R \subseteq W$ (with obvious meaning of the symbols concerning the *events* R and W). Conversely, assuming W (x is wet), we could conclude R (it rains on the spot x) if we are not made aware of possible water sources around x: shortly, we may say "W weakly implies R".

Now, given a conditional probability P, we can represent the above situation by the notation $W \longmapsto_P R$ and by assessing $P(R|W) = 1$, but for a

Romano Scozzafava
University "La Sapienza"
Dip. Metodi e Modelli Matematici,
Via Scarpa 16, 00161 Roma, Italy
e-mail: romscozz@dmmm.uniroma1.it

S. Greco et al. (Eds.): Preferences and Decisions, STUDFUZZ 257, pp. 377–382.
springerlink.com

rigorous formulation we need to say something more on events and conditional events.

2 Preliminaries

An *event* can be singled-out by a (nonambiguous) statement E, that is a (Boolean) *proposition* that can be either *true* or *false* (corresponding to the two "values" 1 or 0 of the *indicator* I_E of E).

The "logic of certainty" deals with *true* and *false* as final, and *not asserted*, answers concerning a *possible* event, while two particular cases are the *certain* event Ω (that is always true) and the *impossible* event $\emptyset$ (that is always false): notice that *only in these two particular cases the relevant propositions correspond to an assertion.* To make an assertion, we need to say something extra-logical, such as "we know that E is false" (so that $E = \emptyset$).

As far as conditional events are concerned, we generalize the idea of de Finetti of looking at a conditional event $E|H$, with $H \neq \emptyset$, as a 3-valued logical entity, which is *true* when both E and H are true, *false* when H is true and E is false, "undetermined" when H is false, by letting instead the third value *suitably depend on the given ordered pair* (E, H) and not being just an undetermined *common value* for all pairs: it turns out that this function is a measure of the degree of belief in the conditional event $E|H$, which under suitable – and natural – conditions is a conditional probability $P(E|H)$, in its most general sense related to the concept of *coherence*, satisfying the classic axioms as given by de Finetti [7].

Concerning coherence, we recall the following fundamental result:
Let $\mathcal{C}$ be any family of conditional events, and take an arbitrary *family $\mathcal{K} \supseteq \mathcal{C}$. Let P be an assessment on $\mathcal{C}$; then there exists a (possibly not unique) coherent extension of P to $\mathcal{K}$* if and only if *P is coherent on $\mathcal{C}$.*

3 Weak Implication

Given a conditional event $E|H$, notice that $P(E) = 1$ does not imply $P(E|H) = 1$ (as in the usual framework where it is necessary to assume $P(H) > 0$). We can take instead $P(H) = 0$ (the *conditioning* event H – which *must* be a *possible* one – may in fact have *zero probability*, since in the assignment of $P(E|H)$ we are driven only by coherence [3]). Then a probability equal to 1 *can be, in our framework, updated.*

Moreover, $P(E|H) = 1$ does not imply $H \subseteq E$: take in fact, e.g., an event E with $P(E) > 0$ and an event $H \supset E$ such that $P(H) = P(E)$ (that is $P(E^c \wedge H) = 0$). In particular, if we *assert* $H^c \vee E = \Omega$, then $H \subseteq E$ (H logically implies E), so we certainly have $P(E|H) = 1$.

It could appear that the most "natural" way to weaken inclusion should be the requirement $P(H^c \vee E) = 1$. But this weaker assumption in general is not enough (whenever $P(H) = 0$) to get $P(E|H) = 1$, while $P(E|H) = 1$ always entails $P(H^c \vee E) = 1$. The two latter statements easily follow from

$$P(H^c \vee E) = 1 - P(H \wedge E^c) = 1 - P(H)P(E^c|H) .$$

In other (semantic) words, we require that, even if a part of H is not inside E, this part can be considered, in a sense, as "ignorable" (with respect to H itself), while the probability of $H^c \vee E$ can be equal to 1, due to the circumstance that H may have probability equal to 0 even if a "large part" of it is not inside E.

The formal definition of weak implication follows:

Definition 1. *An event A weakly implies an event C under P (in symbols $A \longmapsto_P C$) iff $P(C|A) = 1$.*

We denote by Δ_P the set of given weak implications.

In most situations the base of knowledge is given by an arbitrary set $\mathcal{C}$ of (conditional) events, and the function P on them summarizes the state of information. Making inference means enlarging this assessment to new events, maintaining the rules required to the conditional probability P.

Notice that the single assessment $P(C|A) = 1$ is obviously coherent (not only for events $A \subseteq C$), except when A and C are incompatible. We remark that we can assign $P(C|A) = 1$ also in the case $P(C|\Omega) = 0$: then the only coherent value for $P(A|\Omega)$ will be 0.

Moreover, we recall that for any coherent assessment on $\mathcal{C}$, its enlargement to $\mathcal{K} \supseteq \mathcal{C}$ is not unique (in general). Nevertheless for some events we can have a unique coherent extension, so giving rise to the important concept of *entailment*.

Definition 2. *If an assessment on $\mathcal{C}$ contains a set of weak implications Δ_P and if for some event $E|H \in \mathcal{K} \supseteq \mathcal{C}$ every extension necessarily assumes value equal to 1, then we say that "Δ_P* entails *$H \longmapsto_P E$"*

We prove (see [6]) that entailment satisfies the following "inference" rules:

(i) Δ_P entails $A \longmapsto_P A$ for any $A \neq \emptyset$,

(ii) $(A = B)$, $(A \longmapsto_P C) \in \Delta_P$ entails $B \longmapsto_P C$,

(iii) $(A \subseteq B)$, $(C \longmapsto_P A) \in \Delta_P$ entails $C \longmapsto B$,

(iv) $(A \wedge B \longmapsto_P C)$, $(A \longmapsto_P B) \in \Delta_P$ entails $A \longmapsto_P C$,

(v) $(A \longmapsto_P B)$, $(A \longmapsto_P C) \in \Delta_P$ entails $A \wedge B \longmapsto_P C$,

(vi) $(A \longmapsto_P B)$, $(B \longmapsto_P A)$, $(A \longmapsto_P C) \in \Delta_P$ entails $B \longmapsto_P C$,

(vii) $(A \longmapsto_P B)$, $(A \longmapsto_P C) \in \Delta_P$ entails $A \longmapsto_P B \wedge C$,

(viii) $(A \longmapsto_P C)$, $(B \longmapsto_P C) \in \Delta_P$ entails $A \vee B \longmapsto_P C$.

Remark. Properties **(i)–(viii)** correspond to those that, in default logic (see, e.g., [8]), are called, respectively, *Reflexivity, Left Logical Equivalence, Right Weakening, Cut, Cautious Monotonicity, Equivalence, And, Or.*

The definition of weak implication has been extended to general conditional uncertainty measures. Different classes of such measures are singled–out on the basis of the family of conditional events where suitable properties of two operations $\oplus$ and $\odot$ are required to hold (see [2]).

For example, choosing ordinary sum and product, or *max* and any t-norm, we get, respectively, conditional probability or conditional possibility (for the latter, see [1]).

4 Fuzzy Inclusion

We recall that in our context a fuzzy set is defined through a coherent conditional probability in the following way (for details, see [3, 4]): if X is a variable and φ_X is a "property" of X, the membership function $\mu(x)$ is put equal to 1 for the elements x of X that certainly have the given property, while it is put equal to 0 for those elements that certainly do not have it; then it is given suitable values of the interval $(0, 1)$ for those elements of X for which we are doubtful (and then uncertain) on having or not the property φ_X.

Then the interest is in fact directed toward *conditional events* such as $E_\varphi|A_x$, where $A_x = \{X = x\}$, and

$$E_\varphi = \{\text{we claim (that } X \text{ has) the property } \varphi_X\}.$$

In other words, we identify the value (at x) of the membership function $\mu_\varphi(x)$ with a suitable conditional probability, according to the following

Definition 3. *Given a variable X with range $\mathcal{C}_X$ and a related property φ, a fuzzy subset E^*_φ of $\mathcal{C}_X$ is the pair*

$$E^*_\varphi = \{E_\varphi\,,\, \mu_\varphi\},$$

with $\mu_\varphi(x) = P(E_\varphi|A_x)$ for every $x \in \mathcal{C}_X$.

Consider any two fuzzy subsets $E^*_\varphi = (E_\varphi, \mu_\varphi)\,,\; E^*_\psi = (E_\psi, \mu_\psi)$ of $\mathcal{C}_X$, where $\mu_\varphi(\cdot) = P(E_\varphi|\cdot)$ is a coherent conditional probability (and analogously for ψ). Thus, since the assessment $P(\cdot|\cdot)$ defined on the following set of conditional events

$$\mathcal{C} = \{E_\varphi|A_x\,,\; E_\psi|A_x \;:\; A_x \in \mathcal{C}_X\}$$

is coherent, it can be extended (preserving coherence) to any set $\mathcal{D} \supset \mathcal{C}$.

Definition 4. *Consider the family $\mathcal{F}(\mathcal{C}_X)$ of fuzzy subsets of $\mathcal{C}_X$: the* degree $I(E^*_\varphi, E^*_\psi)$ of fuzzy inclusion *of the fuzzy subset $E^*_\varphi = (E_\varphi, \mu_\varphi)$ in the fuzzy subset $E^*_\psi = (E_\psi, \mu_\psi)$ is a function*

$$I : \mathcal{F}(\mathcal{C}_X) \times \mathcal{F}(\mathcal{C}_X) \to [0, 1]$$

with

$$I(E^*_\varphi, E^*_\psi) = P\big(E_\psi | E_\varphi\big)\,,$$

obtained as any coherent extension of $P(\cdot|\cdot)$ *from* $\mathcal{C}$ *to the conditional event* $E_\psi | E_\varphi$.

The existence of such a function is warranted by the fundamental extension theorem for coherent conditional probabilities.

The semantic behind this choice is the following: "the more" E^*_φ is included in E^*_ψ, "the more" if we claim the property φ we are willing to claim also the property ψ.

In the case of crisp sets we obtain that fuzzy inclusion holds with degree 1: in fact, if $A \subseteq B$, then $P(B|A) = P(B|B) = 1$. In the crisp case inclusion is reflexive, i.e. any set A is such that $A \subseteq A$. As far as fuzzy inclusion is concerned, we have that any fuzzy subset E^*_ψ is included in itself with maximum degree $I(E^*_\psi, E^*_\psi) = 1$, so also fuzzy inclusion can be seen as reflexive.

Clearly, the degree of inclusion of two fuzzy subsets has the lowest possible value 0 when they are "disjoint" (i.e., the corresponding membership functions have disjoint supports).

To compute the degree of fuzzy inclusion $I(E^*_\varphi, E^*_\psi)$, notice that, given the membership functions $\mu_\varphi(\cdot) = P(E_\varphi|\cdot)$ and $\mu_\psi(\cdot) = P(E_\psi|\cdot)$ defined on $\mathcal{C}_X$, we can find also the membership function $\mu_{\varphi\wedge\psi}(\cdot)$ of the fuzzy subset $E^*_\psi \cap E^*_\varphi$ (corresponding to a t-norm) as coherent extension of the assessment P given on $\{E_\psi|A_x\,,\ E_\varphi|A_x\ :\ A_x \in \mathcal{C}_X\}$.

We have the following result (see [9]):

If two fuzzy subsets $E^*_\varphi = (E_\varphi, \mu_\varphi)$, $E^*_\psi = (E_\psi, \mu_\psi)$ *of* $\mathcal{C}_X$ *are such that* $I(E^*_\varphi, E^*_\psi) = 1$ *for any probability distribution on* $\mathcal{C}_X$, *then they satisfy Zadeh's definition [10] of fuzzy inclusion, i.e.* $\mu_\varphi(x) \leq \mu_\psi(x)$ *for any* $x \in \mathcal{C}_X$.

Conversely, given two fuzzy subsets $E^*_\varphi = (E_\varphi, \mu_\varphi)$, $E^*_\psi = (E_\psi, \mu_\psi)$ *of* $\mathcal{C}_X$, *consider their intersection* $E^*_\varphi \cap E^*_\psi = \{E_{\varphi\wedge\psi}\,,\ \mu_{\varphi\wedge\psi}\}$ *and take* $\mu_{\varphi\wedge\psi}(x) = \min\{\mu_\varphi(x), \mu_\psi(x)\}$ *for any* x. *If* $\mu_\varphi(x) \leq \mu_\psi(x)$ *for any* $x \in \mathcal{C}_X$, *then* $I(E^*_\varphi, E^*_\psi) = 1$ *for any probability distribution on* $\mathcal{C}_X$.

In [9] we show also some connections between fuzzy inclusion and similarity.

In conclusion, coherent conditional probability *can act as a unifying tool* in dealing with these topics.

References

[1] Bouchon-Meunier, B., Coletti, G., Marsala, C.: Independence and Possibilistic Conditioning. Annals of Mathematics and Artificial Intelligence 35, 107–124 (2002)

[2] Coletti, G., Scozzafava, R.: From conditional events to conditional measures: a new axiomatic approach. Annals of Mathematics and Artificial Intelligence 32, 373–392 (2001)
[3] Coletti, G., Scozzafava, R.: Probabilistic Logic in a Coherent Setting. Trends in Logic, vol. 15. Kluwer, Dordrecht (2002)
[4] Coletti, G., Scozzafava, R.: Conditional Probability and Fuzzy Information. Computational Statistics & Data Analysis 51, 115–132 (2006)
[5] Coletti, G., Scozzafava, R., Vantaggi, B.: Coherent conditional probability as a tool for default reasoning. In: Bouchon-Meunier, B., Foulloy, L., Yager, R.R. (eds.) Intelligent Systems for Information Processing: from Representation to Applications, pp. 191–202. Elsevier, Amsterdam (2003)
[6] Coletti, G., Scozzafava, R., Vantaggi, B.: Weak implication in terms of conditional uncertainty measures. In: Mellouli, K. (ed.) ECSQARU 2007. LNCS (LNAI), vol. 4724, pp. 139–150. Springer, Heidelberg (2007)
[7] de Finetti, B.: Sull'impostazione assiomatica del calcolo delle probabilità. Annali Univ. Trieste 19, 3–55 (1949); Engl. transl. Probability, Induction, Statistics, ch. 5. Wiley, London (1972)
[8] Lehmann, D., Magidor, M.: What does a conditional knowledge base entail? Artificial Intelligence 55, 1–60 (1992)
[9] Scozzafava, R., Vantaggi, B.: Fuzzy Inclusion and Similarity through Coherent Conditional Probability. Fuzzy Sets and Systems 160, 292–305 (2009)
[10] Zadeh, L.: Fuzzy sets. Information and Control 8, 338–353 (1965)

The TOPSIS Method and Its Application to Linguistic Variables

M. Socorro García-Cascales, M. Teresa Lamata, and J. Luís Verdegay

Abstract. Most of the time, the input of the decision process is linguistic, but this is not the case for the output. For that reason, we have modified the TOPSIS model to make it so that the output of the process is the same as the input, that is to say linguistic. This proposal will be applied to the process of quality assessment and accreditation of the Industrial Engineering Schools within the Spanish university system.

1 Introduction

In decision making processes we are not always interested in knowing the ranking (e.g. it would be the case of finding the best individual for a specific position) but there are situations in which it is more interesting to know the final valuation given to an alternative (e.g. set of companies with good growth perspectives). This type of process is what we shall refer to in this work.

The Multiple Criteria Decision Making (MCDM) is a procedure that consists in finding the best alternative among a set of feasible alternatives. A MCDM problem with m alternatives and n criteria can be expressed in matrix format as follows[17,27]:

$$
\begin{array}{cc}
 & \begin{array}{cccc} w_1 & w_2 & \cdots & w_n \end{array} \\
 & \begin{array}{cccc} C_1 & C_2 & \cdots & C_n \end{array} \\
M = \begin{pmatrix} A_1 \\ A_2 \\ \vdots \\ A_m \end{pmatrix} & \begin{pmatrix} z_{11} & z_{12} & \cdots & z_{1n} \\ z_{21} & z_{22} & \cdots & z_{2n} \\ \vdots & \vdots & \cdots & \vdots \\ z_{m1} & z_{m2} & \cdots & z_{mn} \end{pmatrix}
\end{array} \qquad (1)
$$

M. Socorro García-Cascales
Dpto de Electrónica, Tecnología de Computadoras y Proyectos. Universidad de Politécnica de Cartagena. Murcia, Spain
e-mail: socorro.garcia@upct.es

M. Teresa Lamata · J. Luís Verdegay
Dpto. Ciencias de la Computación e Inteligencia Artificial. Universidad de Granada. 18071- Granada, Spain
e-mail: mtl@decsai.ugr.es, verdegay@decsai.ugr.es

S. Greco et al. (Eds.): Preferences and Decisions, STUDFUZZ 257, pp. 383–395.
springerlink.com

where $A_1, A_2, \cdots, A_m$ are feasible alternatives, $C_1, C_2, \cdots, C_n$ are evaluation criteria, z_{ij} is the performance rating of alternative A_i under criterion C_j, and w_j is the weight of criterion C_j.

Among the many compensatory approaches of MCDM, one of them is the Technique for Order Performance by Similarity to Ideal Solution (TOPSIS) method. This approach is employed for four main reasons [26]:

- TOPSIS logic is rational and understandable;
- The computation processes are straightforward;
- The concept permits the pursuit of best alternatives for each criterion depicted in a simple mathematical form; and
- The importance weights are incorporated into the comparison procedures.

But there are some disadvantages, such as:

- The existence of rank reversal, which is associated with the process of normalization [10].
- And that the linguistic labels may be, as an input, while the output is a number associated with the index of closeness [11].

With regard to this, we intend to make a modification in the TOPSIS method algorithm so that if the inputs are linguistic labels then the outputs are too. Within the literature there are a lot of works which develop the TOPSIS Method both for numerical and linguistic inputs, yet none of them is capable of facilitating a linguistic output. Thus, the TOPSIS approach was developed by [12], and improved by the same authors in 1987 and 1992. [20,26] and many others have also worked on this theme. Some examples using the fuzzy set theory can be seeing in [3,4,6,7,8,14].

In the last years, different papers have appeared in the literature in diverse applied fields. [5,7,8] give the extension for group decision, the first for solving supplier selection problems in a fuzzy environment and the second to location selection problems. [9] also apply the TOPSIS method for robot selection.

[7] proposes a fuzzy TOPSIS approach to resolve a problem in logistic information technology, also a comparison between fuzzy TOPSIS method is given in this paper. The total quality management consultant selection under fuzzy environment is viewed in [22] and the applications in aggregate planning in [25], whereas in [26], the application is related with Air Force Academy in Taiwan to evaluate the initial training aircraft. [23] develops a fuzzy TOPSIS method for evaluating the competitive advantages of shopping websites and mobile phone alternatives are studied in [13].

The process to obtain a linguistic output is simple, we only introduce two steps into the algorithm proposed by Hwang and Yoon [12]. The first will consist in introducing into the decision matrix not only the evaluation of the different alternatives but we also shall add as many alternatives as linguistic terms we have, in such a way that the evaluation vector is a repetition of the linguistic label in question and the second is solved by means of the definition of a distance. This will be seen by an example related to the accreditation of industrial engineering studies in the Spanish university system.

The paper is organized as follows: In section 2, we introduce some of the bases of linguistic variables and the associated fuzzy sets theory. In section 3, the framework for the TOPSIS method is defined. Section 4 describes the modification of the algorithm and an example is proposed. Finally we outline the most important conclusions.

2 Linguistic Variable and Fuzzy Sets

2.1 Linguistic Variable

Most of the times, the decision-maker is not able to define the importance of the criteria or the goodness of the alternatives with respect to each criterion in a strict way. In many situations, we use measures or quantities which are not exact but approximate.

Since Zadeh [28] introduced the concept of fuzzy set and subsequently went on to extend the notion via the concept of linguistic variables, the popularity and use of fuzzy sets has been extraordinary. We are particularly interested in the role of linguistic variables as an ordinal scale and their associated terms, in this case triangular fuzzy number, as used in the multi-criteria decision making.

By a linguistic variable, [29,30,31], we mean a variable whose values are words or sentences in a natural or artificial language. For example Age is a linguistic variable if its values are linguistic rather than numerical, i.e., young, not young, very young, quite young, old, not very old and not very young, etc., rather than numbers as 20, 21,22, 23,.... .

Definition 1. A linguistic variable is characterized by a quintuple

$$\{X;T(X);U;G;M\}$$

in which

1. X is the name of the variable,
2. $T(X)$ is the term set of X, that is, the collection of its linguistic values
3. U is a universe of discourse,
4. G is a syntactic rule for generating the elements of $T(X)$ and
5. M is a semantic rule for associating meaning with the linguistic values of X.

In general for the decision-maker it is easier when he/she evaluates their judgments by means of linguistic terms. In those cases, the concept of fuzzy number is more adequate than that of real number. In this paper, with the support of the fuzzy set theory, triangular fuzzy numbers, which are parameterized by triplet numbers, are used to represent the importance and alternative performance of linguistic evaluation of the criteria. Therefore, the relative importance contribution in the adjacent upper level can be described as gradual and not abrupt, and this gives a more exact representation of the relationship between candidate alternatives and the evaluation criteria. The basic theory of the triangular fuzzy number is described as follows.

2.2 Fuzzy Set Theory

We have identified the linguistic variable with a fuzzy set [2,15,18]. The fuzzy set theory, introduced by Zadeh [28] to deal with vague, imprecise and uncertain problems has been used as a modelling tool for complex systems that can be controlled by humans but are hard to define precisely. A collection of objects (universe of discourse) X has a fuzzy set A described by a membership function f_A with values in the interval $[0,1]$.

$$f_A : X \rightarrow [0,1]$$

Thus A can be represented as $A = \{f_A(x) | x \in X\}$. The degree that u belongs to A is the membership function $f_A(x)$.

The basic theory of the triangular fuzzy number is described in Klir [19].

With regard to the fuzzy numbers, we will show only the mathematical operations that will be used throughout the development of the paper.

Definition 2. If T_1 and T_2 are two triangular fuzzy numbers defined by the triplets (a_1, b_1, c_1) and (a_2, b_2, c_2), respectively. Then, for this case, the necessary arithmetic operations with positive fuzzy numbers are:

a) Addition

$$T_1 \oplus T_2 = [a_1 + a_2, b_1 + b_2, c_1 + c_2] \tag{1}$$

b) Subtraction

$T_1 \Theta T_2 = T_1 + (-T_2)$ when the opposite $-T_2 = (-c_2, -b_2, -a_2)$

$$\text{then} \quad T_1 \Theta T_2 = [a_1 - c_2, b_1 - b_2, c_1 - a_2] \tag{2}$$

c) Multiplication

$$T_1 \otimes T_2 = [a_1 \times a_2, b_1 \times b_2, c_1 \times c_2] \tag{3}$$

d) Division

$$T_1 \varnothing T_2 = [[a_1, b_1, c_1] \cdot [1/c_2, 1/b_2, 1/a_2]],\ 0 \neq [a_2, b_2, c_2] \tag{4}$$

e) Scalar Multiplication

$$k \circ T_1 = (k \circ a_1, k \circ b_1, k \circ c_1) \tag{5}$$

f) Root

$$T_1^{1/2} = \left[a_1^{1/2}, b_1^{1/2}, c_1^{1/2}\right] \tag{6}$$

3 Existing Framework for TOPSIS Evaluation

3.1 The TOPSIS Algorithm

A necessary step to be able to provide a linguistic output, consists in introducing into the decision matrix not only the alternatives themselves but also as many alternatives as terms that the linguistic variable may have. We shall call the latter fictitious alternatives. These alternatives are defined as those in which the valuation of each one of the criteria is the corresponding term of the linguistic variable.

Step 1: Identify the evaluation criteria and the appropriate linguistic variables for the importance weight of the criteria and determine the set of feasible alternatives with the linguistic score for alternatives in terms of each criterion. Once the decision matrix is formed, the normalized decision matrix (n_{ij} ; $i=1,2,..,m$ (number of alternatives); $j=1,2,..,n$ (number of criteria)) is constructed using equation (7)):

$$\overline{n}_{ij}^{1} = z_{ij} \Big/ \sqrt{\sum_{j=1}^{n} \left(z_{ij}\right)^{2}}, \; j = 1,\ldots,n, \; i = 1,\ldots,m. \tag{7}$$

where z_{ij} is the performance score of alternative i against criteria j.

Step 2: The weighted normalized decision matrix $\overline{v}_{ij}$ is calculated using equation (8).

$$\overline{v}_{ij} = w_j \otimes \overline{n}_{ij}, \; j = 1,\ldots,n, \; i = 1,\ldots,m, \tag{8}$$

where, w_j such that $1 = \sum_{j=1}^{n} w_j$ is the weight of the jth attribute or criterion.

Step 3: The positive ideal solution (PIS), $\overline{A}^{+}$ ($\overline{A}_i^{+}$; $i = 1,2,...,m$), is made of all the best performance scores

$$\overline{A}^{+} = \left\{\overline{v}_1^{+},\ldots,\overline{v}_n^{+}\right\} = \left\{\left(\max_i \overline{v}_{ij}, j \in J\right)\left(\min_i \overline{v}_{ij}, j \in J'\right)\right\} \quad i = 1,2,\ldots,m \tag{9}$$

and the negative ideal solution (NIS), $\overline{A}^{-}$ ($\overline{A}_i^{-}$; $j= 1,2,...,n$), is made of all the worst performance scores at the measures in the weighted normalized decision matrix.

$$\overline{A}^{-} = \left\{\overline{v}_1^{-},\ldots,\overline{v}_n^{-}\right\} = \left\{\left(\min_i \overline{v}_{ij}, j \in J\right)\left(\max_i \overline{v}_{ij}, j \in J'\right)\right\} \quad i = 1,2,\ldots,m \tag{10}$$

where J is associated with benefit criteria, and J' is associated with cost criteria. The idea of the TOPSIS method is represented in figure 1, where we would have to evaluate the five alternatives (A,B,C,D,E), and where for questions relating to the graphical representation we only consider two criteria.

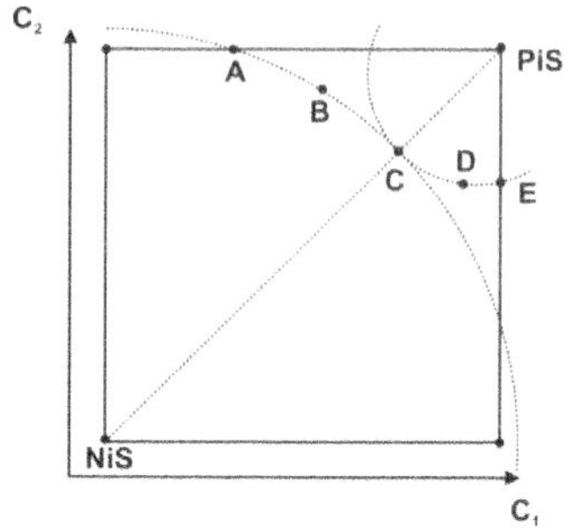

Fig. 1 PIS and NIS concept

If we only take into account the solution related with the PIS, Zeleny [27], the best alternatives would be C, D and E, since the three are at the same distance. If we add the NIS option to the decision process, then the best alternative would be exclusively alternative E, since it is the one which is closest to the PIS point and furthest from the NIS.

Thus, the solution is a compromise solution according to the decision-maker's preferences. It is based upon the concept that the chosen alternative should have the shortest distance from the positive ideal solution (PIS), in our case and the farthest from the negative ideal solution (NIS). Therefore, we need to define a distance.

Step 4: The distance of an alternative to the ideal solution $\bar{d}_i^+$,

$$\bar{d}_i^+ = \left\{ \sum_{j=1}^{n} \left(\bar{v}_{ij} - \bar{v}_j^+ \right)^2 \right\}^{\frac{1}{2}}, \quad i = 1,\ldots,m \tag{11}$$

and from the negative ideal solution $\bar{d}_i^-$,

$$\bar{d}_i^- = \left\{ \sum_{j=1}^{n} \left(\bar{v}_{ij} - \bar{v}_j^- \right)^2 \right\}^{\frac{1}{2}}, \quad i = 1,\ldots,m \tag{12}$$

In this case we use the 2-multidimensional Euclidean distance. But the result has not to be a new fuzzy number. Therefore, once the corresponding values obtained to $\left(d_1^+, d_2^+, d_3^+\right)$ we find the value of positive distance as

$$d^+ = \frac{1}{3}\left(d_1^+ + d_2^+ + d_3^+\right) \tag{13}$$

Similarly:

$$d^- = \frac{1}{3}\left(d_1^- + d_2^- + d_3^-\right) \tag{14}$$

Step 5: The ranking score R_i is calculated using equation (15). The obtained ranking scores represent the alternatives' performance achievement within their status. A higher score corresponds to a better performance.

$$R_i = \frac{d_i^-}{d_i^- + d_i^+}, \; i = 1, \ldots, m \tag{15}$$

This result is performed taking into account the operations defined in (3) and (6).

If $R_i = 1 \rightarrow A_i = A^+$

If $R_i = 0 \rightarrow A_i = A^-$

where the R_i value lies between 0 and 1. The closer the $R_i = 1$ value implies a higher priority of the ith alternative.

New Step 6: Rank the Linguistic values.

But it would be desirable to develop a decision method (not based on rules) in which output, in the case of linguistic inputs, was also linguistic. To obtain this linguistic output we calculated the distance of the alternatives to the labels, associating each alternative to that label whose distance is minimal.

Therefore, the distance between fuzzy numbers $A_1=(a_1,b_1,c_1)$ and $T_k=(a_2,b_2,c_2)$ is calculated as

$$d(A_i, T_k) = |A_i - T_k| \tag{16}$$

The output will be that linguistic term whose distance is minimal.

4 Example

One of the consequences derived from the European Space for Higher Education is the accreditation system for university qualifications. For this purpose the ANECA (the Spanish National Agency for Quality Assessment and Accreditation) was created in Spain. [1]

The scope and main activities of ANECA are:

- To enhance the improvement of university teaching, research and management activities.
- To foster Higher Education Institutions performance monitoring following objective and transparent processes.
- To provide public administrations with appropriate information for decision-making within the scope of their authority.
- To provide society with information about the achievement of the aims of universities.

The different university qualifications have been evaluated according to six criteria (1. Educational Programme, 2. Teaching Organization, 3. Human Resources, 4. Material Resources, 5. Educational Process and 6. Results) Of these, and as an example, we have taken the fourth of the criteria "Material Resources", which as can

Table 1 Material resources

4. Material Resources	4.1 Classrooms	4.1.1. Appropriateness for numbers of students
	4.2 Work Spaces	4.2.1. Appropriateness for numbers of students
		4.2.2. Appropriateness (Academic Staff and Administration and Services Staff)
		4.2.3. Infrastructures: practical
	4.3. Laboratories, workshops and experimental spaces	4.3.1. Appropriateness for number of students
	4.4. Library and document banks	4.4.1. Correctly furnished
		4.4.2. Quality, quantity, ...

be appreciated from Table 1, has four subcriteria and these in turn have different subcriteria. we have considered only five of all the qualifications evaluated.

To achieve this, we have developed a method which is simple to understand and implement, based in the TOPSIS method and with some modifications.

First of all we needed to construct the fuzzy numbers associated to the linguistic terms. These were obtained by means of a questionnaire sent to the experts of ANECA via e-mail. The question to obtain this was: "Define numerically from [0-10] the values of labels A, B, C and D." Then for each of them the mean and the deviation were calculated. Using expression (17), we obtained the relation between symmetric fuzzy numbers and linguistic terms (Table 2).

$$(a,b,c) = (\overline{X} - t\sigma_x, \overline{X}, \overline{X} + t\sigma_x) \qquad (17)$$

The insufficient evidence EI, is calculated as the media of the corresponding row.

Table 2 Linguistic labels and fuzzy numbers for t=2

Labels	Fuzzy number	Legend
A	(8.1354, 9.4054, 10.0000)	Excellent
B	(5.8108, 7.1081, 8.4054)	Good
C	(3.5090, 4.8108, 6.1126)	Average
D	(0.7355, 2.5135, 4.2916)	Deficient
EI		insufficient evidence

Supposing that the four criteria have the same weight *w*(4.1)=*w*(4.2)= *w*(4.3)= *w*(4.4)=1/4; as do the subcriteria, which have the same weight in each of the criteria: thus *w*(4.1.1)= 1/4, *w*(4.2.1)=*w*(4.2.2)=*w*(4.2.3)=1/12, *w*(4.3.1)=1/4 and *w*(4.4.1)= *w*(4.4.2)= 1/8.

We can see that *A*={UPCT, UCM, UNA, UPM, Unizar} and *T*=[A,B,C,D}. Once the data have been compiled the algorithm will be applied. So steps 1, 2 and 3 will be first, which give us the results in Table 4. See Appendix A.

Step 4. The positive and negative distances need to be calculated using expressions (11) and (12) as well as the real output associated with the fuzzy numbers taking into account the expressions (13) and (14) and these are shown in Table 5.

Table 3 Decisión Matrix

Criteria	4.1		4.2		4.3.	4.4	
Sub-criteria	4.1.1.	4.2.1.	4.2.2.	4.2.3.	4.3.1.	4.4.1.	4.4.2.
University							
UPCT	**B**	**D**	**B**	**D**	**C**	**D**	**B**
UCM	**B**	**C**	**B**	**EI**	**B**	**C**	**C**
UNA	**B**	**A**	**A**	**A**	**A**	**A**	**A**
UPM	**C**	**C**	**C**	**C**	**C**	**C**	**C**
Unizar	**C**	**C**	**C**	**C**	**C**	**C**	**B**
Excellent=A	**A**	**A**	**A**	**A**	**A**	**A**	**A**
B= Good	**B**	**B**	**B**	**B**	**B**	**B**	**B**
C= Average	**C**	**C**	**C**	**C**	**C**	**C**	**C**
D=Deficient	**D**	**D**	**D**	**D**	**D**	**D**	**D**

Table 5 Positive and negative ideal solution

d+UPCT	0.0888	**d-UPCT**	0.0745
d+UCM	0.0605	**d-UCM**	0.0906
d+UNA	0.0262	**d-UNA**	0.1313
d+UPM	0.0954	**d-UPM**	0.0494
d+Unizar	0.0728	**d-Unizar**	0.0853
d+A	0.0000	**d-A**	0.1448
d+B	0.0447	**d-B**	0.1001
d+C	0.0954	**d-C**	0.0494
d+D	0.1448	**d-D**	0.0000

Step 5. Using expression (13), we have obtained the closeness index (Table 6)

Table 6 Closeness index

RUPCT	0.4560
RUCM	0.5996
RUNA	0.8337
RUPM	0.3414
RUnizar	0.5397
RA	1.0000
RB	0.6914
RC	0.3414
RD	0.0000

With this, we could establish an order for the Universities, but as the input is linguistic then a linguistic output is sought.

Step 6. Output of linguistic labels.
From the data in Table 6, we calculated, using expression (16), the distance to the nearest label. Thus, for example:

$$d(UPCT, A) = |0.4560 - 1| = 0.5440$$
$$d(UPCT, B) = |0.4560 - 0.6914| = 0.2354$$
$$d(UPCT, C) = |0.4560 - 0.3414| = 0.1146$$
$$d(UPCT, D) = |0.4560 - 0| = 0.4560$$

Min (0.546, 0.2354, 0.1146, 0.4560) =0.1146⇒ Ouput Label is C=Average

Similarly, we calculated the distances for the rest of the alternatives, with the results shown in Table 7.

Table 7 Outputs of linguistic terms

RUPCT	Average
RUCM	Good
RUNA	Good
RUPM	Average
RUnizar	Average

5 Conclusions

When we deal with linguistic variables one of the points which has always been made is that the output was not a variable of the same type as the input.

The TOPSIS multicriteria decision making method has been applied by may authors when the evaluations are linguistic but the output has on all occasions been an index which provides a ranking.

In our case, by means of incorporating two steps into the algorithm proposed by Hwang and Yoon we have managed to obtain not only a ranking within the alternatives, but also a valuation for them can be established in the same terms as in the input.

Acknowledgements. This work is partially supported by the DGICYT under project TIN2008-06872-C04-04 and Junta de Andalucía (P07-TIC02970), respectively.

References

[1] Agencia Nacional de Evaluación de la Calidad y Acreditación (ANECA). Programa de Evaluación Institucional. Anualidad 2006/2007. Guía de evaluación externa 2007 (2007), `http://www.aneca.es/active/active_ense_pei0607.asp`
[2] Bellman, R.E., Zadeh, L.A.: Decision-making in a fuzzy environment. Management Sci. 17, 141–164 (1970)

[3] Braglia, M., Frosolini, M., Montanari, R.: Fuzzy TOPSIS approach for failure mode, effects and criticality analysis. Quality and Reliability Engineering International 19, 425–443 (2003)

[4] Chen, S.J., Hwang, C.L.: Fuzzy Multiple Attribute Decision Making: Methods and Applications. Springer, Berlin (1992)

[5] Chen, C.T.: Extensions of the TOPSIS for group decision-making under fuzzy environment. Fuzzy Sets and Systems 114, 1–9 (2000)

[6] Chen, M.F., Tzeng, G.H.: Combing grey relation and TOPSIS concepts for selecting an expatriate host country. Mathematical and Computer Modelling 40, 1473–1490 (2004)

[7] Chu, T.C.: Facility location selection using fuzzy TOPSIS under group decisions. International Journal of Uncertainty, Fuzziness and Knowledge-Based Systems 10, 687–701 (2002)

[8] Chu, T.C.: Selecting plant location via a fuzzy TOPSIS approach. The International Journal of Advanced Manufacturing Technology 20, 859–864 (2002)

[9] Chu, T.C., Lin, Y.C.: A fuzzy TOPSIS method for robot selection. International Journal of Advanced Manufacturing Technology 21, 284–290 (2003)

[10] Garcia-Cascales, M.S., Lamata, M.T.: On rank reversal and TOPSIS method. Under Review in Mathematical Computer Modelling

[11] Garcia-Cascales, M.S., Lamata, M.T.: Solving a decision problem with linguistic information. Pattern Recognition Letters 28(16), 2284–2294 (2007)

[12] Hwang, C.L., Yoon, K.: Multiple Attribute Decision Methods and Applications. Springer, Heidelberg (1981)

[13] Işıklar, G., Büyüközkan, G.: Using a multi-criteria decision making approach to evaluate mobile phone alternatives, vol. 29, pp. 265–274 (2007)

[14] Jahanshsloo, G.J., Hosseinzadeh Lofti, F., Izadikhah, M.: An algorithmic method to extend TOPSIS for decision-making problems with interval data. Applied Mathematics and Computation (2005)

[15] Kacprzyk, J., Yager, R.R.: Linguistic summaries of data using fuzzy logic. International Journal of General Systems 30(2), 133–154 (2001)

[16] Kahraman, C., Ates, N.Y., Cevik, S., Gulbay, M., Erdogan, S.A.: Hierarchical fuzzy TOPSIS model for selection among logistics information technologies. Journal of Enterprise Information Management 20(2), 143–168 (2007)

[17] Keeney, R., Raiffa, H.: Decisions with Multiple Objectives: Preferences and Value Tradeoffs. Wiley, New York (1976)

[18] Kerre, E.E.: The use of fuzzy set theory in electrocardiological diagnostics. In: Gupta, M.M., Sanchez, E. (eds.) Approximate Reasoning in Decision Analysis, pp. 277–282. North-Holland, Amsterdam (1982)

[19] Klir, G.J., Yuan, B.: Fuzzy sets and Fuzzy Logic. Prentice Hall PTR, New Jersey (1995)

[20] Lai, Y.L., Liu, T.Y., Hwang, C.L.: TOPSIS for MODM. European Journal of Operational Research 76(3), 486–500 (1994)

[21] Liou, T.S., Wang, M.J.: Ranking fuzzy numbers with integral value. Fuzzy Sets and Systems 50, 247–255 (1992)

[22] Saremi, M., Mousavi, S.F., Sanayei, A.: TQM consultant selection in SMEs with TOPSIS under fuzzy environment Expert Systems with Applications 36, 2742–2749 (2009)

[23] Sun, C.C., Lin, G.T.R.: Using fuzzy TOPSIS method for evaluating the competitive advantages of shopping websites. Expert Systems with Applications (in Press)
[24] Tsaur, S.H., Chang, T.Y., Yen, C.H.: The evaluation of airline service quality by fuzzy MCDM. Tourism Management 23, 107–115 (2002)
[25] Wang, R.C., Liang, T.F.: Application of fuzzy multi-objective linear programming to aggregate production planning. Computer & Industrial Engineering 46, 17–41 (2004)
[26] Wang, T.C., Chang, T.H.: Application of TOPSIS in evaluating initial training aircraft under a fuzzy environment. Expert Systems with Applications 33(4), 870 (2007)
[27] Zeleny, M.: Multiple Criteria Decision Making. McGraw-Hill, New York (1982)
[28] Zadeh, L.A.: Fuzzy sets. Information and Control 8, 338–353 (1965)
[29] Zadeh, L.A.: The concept of a linguistic variable and its application to approximate reasoning: Part 1. Information Sciences 8, 199–249 (1975)
[30] Zadeh, L.A., Kacprzyk, J. (eds.): Computing with Words in Information / Intelligent Systems 1. Foundations. Studies in Fuzziness and Soft Computing, vol. 34. Physica-Verlag/Springer, Heidelberg (1999)
[31] Zadeh, L.A., Kacprzyk, J. (eds.): Computing with Words in Information / Intelligent Systems 2. Applications. Studies in Fuzziness and Soft Computing, vol. 33. Physica-Verlag/Springer, Heidelberg (1999)

Appendix A

Table 4 The weighed normalized matrix

UPCT	0.0621	0.0896	0.1303	0.0029	0.0115	0.0243	0.0208	0.0295	0.0425	0.0028	0.0113	0.0238	0.0389	0.0620	0.0967	0.0040	0.0157	0.0326	0.0312	0.0442	0.5105
UCM	0.0621	0.0896	0.1303	0.0137	0.0220	0.0346	0.0208	0.0295	0.0425	0.0179	0.0267	0.0402	0.0644	0.0915	0.1330	0.0192	0.0301	0.0465	0.0188	0.0299	0.3712
UNA	0.0621	0.0896	0.1303	0.0317	0.0430	0.0566	0.0291	0.0390	0.0506	0.0312	0.0422	0.0554	0.0901	0.1211	0.1582	0.0446	0.0589	0.0760	0.0437	0.0585	0.6073
UPM	0.0375	0.0606	0.0947	0.0137	0.0220	0.0346	0.0126	0.0199	0.0309	0.0135	0.0216	0.0339	0.0389	0.0620	0.0967	0.0192	0.0301	0.0465	0.0188	0.0299	0.3712
Unizar	0.0621	0.0896	0.1303	0.0137	0.0220	0.0346	0.0126	0.0199	0.0309	0.0135	0.0216	0.0339	0.0389	0.0620	0.0967	0.0446	0.0589	0.0760	0.0312	0.0442	0.5105
A	0.0870	0.1185	0.1550	0.0317	0.0430	0.0566	0.0291	0.0390	0.0506	0.0312	0.0422	0.0554	0.0901	0.1211	0.1582	0.0446	0.0589	0.0760	0.0437	0.0585	0.6073
B	0.0621	0.0896	0.1303	0.0227	0.0325	0.0476	0.0208	0.0295	0.0425	0.0223	0.0319	0.0466	0.0644	0.0915	0.1330	0.0319	0.0445	0.0639	0.0312	0.0442	0.5105
C	0.0375	0.0606	0.0947	0.0137	0.0220	0.0346	0.0126	0.0199	0.0309	0.0135	0.0216	0.0339	0.0389	0.0620	0.0967	0.0192	0.0301	0.0465	0.0188	0.0299	0.3712
D	0.0079	0.0317	0.0665	0.0029	0.0115	0.0243	0.0026	0.0104	0.0217	0.0028	0.0113	0.0238	0.0081	0.0324	0.0679	0.0040	0.0157	0.0326	0.0039	0.0156	0.2606
A+	0.0870	0.1185	0.1550	0.0317	0.0430	0.0566	0.0291	0.0390	0.0506	0.0312	0.0422	0.0554	0.0901	0.1211	0.1582	0.0446	0.0589	0.0760	0.0437	0.0585	0.0759
A-	0.0079	0.0317	0.0665	0.0029	0.0115	0.0243	0.0026	0.0104	0.0217	0.0028	0.0113	0.0238	0.0081	0.0324	0.0679	0.0040	0.0157	0.0326	0.0039	0.0156	0.0326

Information Fusion with the Power Average Operator

Ronald R. Yager

Abstract. The power average provides an aggregation operator that allows argument values to support each other in the aggregation process. The properties of this operator are described. We see this mixes some of the properties of the mode with mean. Some formulations for the support function used in the power average are described. We extend this facility of empowerment to a wider class of mean operators such as the OWA and generalized mean.

Keywords: information fusion, aggregation operator, averaging, data mining.

1 Introduction

Aggregating information using techniques such as the average is a task common in many information fusion processes. Here we provide a tool to aid and provide more versatility in this process. In this work we introduce the concept of the power average [1]. With the aid of the power average we are able to allow values being aggregate to support each other. The power average is provides a kind of empowerment as it allows groups of values close to each other to reinforce each other. This operator is particularly useful in group decision making [2].

2 Power Average

In the following we describe an aggregation type operator called the **P**ower **A**verage (P–A), this operator takes a collection of values and provides a single value [1]. We define this operator as follows:

$$\text{P-A}(a_1, \ldots, a_n) = \frac{\sum_{i=1}^{n} (1 + T(a_i))\, a_i}{\sum_{i=1}^{n} (1 + T(a_i))}$$

Ronald R. Yager
Machine Intelligence Institute
Iona College
New Rochelle, NY 10801

S. Greco et al. (Eds.): Preferences and Decisions, STUDFUZZ 257, pp. 397–414.
springerlink.com

where $T(a_i) = \sum_{\substack{j=1 \\ j \neq i}}^{n} Sup(a_i, a_j)$ and is denoted the support for a from b.

Typically we assume that Sup(a, b) satisfies the following three properties:

1. $Sup(a, b) \in [0, 1]$
2. $Sup(a, b) = Sup(b, a)$
3. $Sup(a, b) \geq Sup(x, y)$ if $|a - b| \leq |x - y|$

In condition three we see the more similar, closer, two values the more they support each other.

We shall find it convenient to denote $V_i = 1 + T(a_i)$ and $w_i = \frac{V_i}{\sum_{i=1}^{n} V_i}$. Here the w_i are a proper set of weights, $w_i \geq 0$ and $\Sigma_i w_i = 1$. Using this notation we have

$$P\text{-}A(a_1, ..., a_n) = \Sigma_i w_i a_i,$$

it is a weighted average of the a_i. However, this is a non-linear weighted average as the w_i depend upon the arguments.

Let us look at some properties of the power average aggregation operator. First we see that this operator provides a generalization of the simple average, if $Sup(a_i, a_j) = k$ for all a_i and a_j then $T(a_i) = k (n - 1)$ for all i and hence $P\text{-}A(a_1, ..., a_n) = \frac{1}{n} \Sigma_i a_i$. Thus when all the supports are the same the power average reduces to the simple average.

We see that the power average is commutative, it doesn't depend on the indexing of the arguments. Any permutation of the arguments has the same power average.

The fact that $P\text{-}A(a_1, ..., a_n) = \Sigma_i w_i a_i$ where $w_i \geq 0$ and $\Sigma_i w_i = 1$ implies that the operator is bounded, $Min[a_i] \leq P\text{-}A(a_1, a_2, ..., a_n) \leq Max_i[a_i]$. This in turn implies that it is idempotent, if $a_i = a$ for all i then $P\text{-}A(a_1, ..., a_n) = a$.

As a result of the fact that the w_i depend upon the arguments, one property typically associated with averaging operator that is not generally satisfied by the power average is monotonicity. We recall that monotonicity requires that if $a_i \geq b_i$ for all i then $P\text{-}A(a_1, ..., a_n) \geq P\text{-}A(b_1, ..., b_n)$. As the following example illustrates, the increase in one of the arguments can result in a decrease in the power average.

Example: Assume the support function Sup is such that

$Sup(2, 4) = 0.5$	$Sup(2, 10) = 0.3$	$Sup(2, 11) = 0$
	$Sup(4, 10) = 0.4$	$Sup(4, 11) = 0$

the required symmetry means $S(a, b) = S(b, a)$ for these values.

Consider first P-A(2, 4, 10), in this case

$$T(2) = Sup(2, 4) + Sup(2, 10) = 0.8$$
$$T(4) = Sup(4, 2) + Sup(4, 10) = 0.9$$
$$T(10) = Sup(10, 2) + Sup(10, 4) = 0.7$$

and therefore $P\text{-}A(2, 4, 10) = \dfrac{(1 + 0.8)\,2 + (1 + 0.9)\,4 + (1 + 0.7)\,10}{(1 + 0.8) + (1 + 0.9) + (1 + 0.7)} =$ 5.22.

Consider now P-A(2, 4, 11), in this case

$$T(2) = Sup(2, 4) + Sup(2, 11) = 0.5$$
$$T(4) = Sup(4, 2) + Sup(4, 11) = 0.5$$
$$T(11) = Sup(11, 2) + Sup(11, 2) = 0$$

and therefore

$$P\text{-}A(2, 4, 11) = \frac{(1.5)(2) + (1.5)\,4 + (1)(1.1)}{1.5 + 1.5 + 1} = 5$$

Thus we see that P-A(2, 4, 10) > P(2, 4, 11).

As we shall subsequently see, this ability to display non-monotonic behavior provides one of the useful features of this operator that distinguishes it from the usual average. For example the behavior displayed in the example is a manifestation of the ability of this operator to discount outliers. For as we shall see in the subsequent discussion, as an argument moves away from the main body of arguments it will be accommodated, by having the average move in its direction, this will happen up to point then when it gets too far away it is discounted by having its effective weighting factor diminished.

To some degree this power average can be seen to have some of the characteristics of the mode operator. We recall that the mode of a collection of arguments is equal to the value that appears most in the argument. We note that the mode is bounded by the arguments and commutative, however as the following example illustrates it is not monotonic.

Example: Mode(1, 1, 3, 3, 3) = 3. Consider now Mode(1, 1, 4, 7, 8) = 1, here we increased all the threes and obtain a value less than the original.

As we shall subsequently see, while both the power average and mode in some sense are trying to find the most supported value, a fundamental difference exists between these operators. We note that in the case of the mode we are not aggregating, blending, the values we are counting how many of each, the mode must be one of the arguments. In the case of power average we are allowing blending of values.

It is interesting, however, to note a formal relationship between the mode and the power average. To understand this we introduce an operator we call a **P**ower **M**ode. In the case of the power mode we define a support function $Sup_m(a, b)$, indicating the support for a from b, such that

1) $Sup_m(a, b) \in [0, 1]$
2) $Sup_m(a, b) = Sup_m(b, a)$

3) $Sup_m(a, b) \geq Sup_m(x, y)$ if $|a - b| \leq |x - y|$

4). $Sup_m(a, a) = 1$.

We then calculate $Vote(i) = \sum_{j=1}^{n} Sup_m(a_i, a_j)$ and define

$$\text{Power Mode}(a_1, ..., a_n) = a_{i^*}$$

where i* is such that $Vote(i^*) = Max_i[Vote(i)]$, it is the argument with the largest vote.

If $Sup_m(a, b) = 0$ for $b \neq a$ ($Sup_m(a, a) = 1$ by definition) then we get the usual mode. Here we are allowing some support for a value by neighboring values). It is also interesting to note the close relationship to the mountain clustering method introduced by Yager and Filev [3] and particularly with the special case of mountain clustering called the subtractive method suggested by Chu [4]. Some connection also seems to exist between the power mode and the idea of fuzzy typical value introduced in [5].

3 Power Average with Binary Support Functions

In order to obtain some intuition for the power average aggregation operator we shall consider first a binary support function. Here we assume

$$Sup(a, b) = K \quad \text{if } |a - b| \leq d$$
$$Sup(a, b) = 0 \quad \text{if } |a - b| > d.$$

Thus two values support each if they are less than or equal d away, otherwise they supply no support. Here K is the value of support. In the following discussion we say a and b are neighbors if $|a - b| \leq d$. The set of points that are neighbors of x will be denoted N_x. We shall call a set of points such that all points are neighbors and no other points are neighbors to those points a cluster. We note if x and y are in the same cluster then the subset $\{x\} \cup N_x = \{y\} \cup N_y$ defines the cluster.

Let us first assume that we have two disjointed clusters of values $A = \{a_1, ..., a_{n_1}\}$ and $B = \{b_1, ..., b_{n_2}\}$. Here all points in A support each other but support none in B while the opposite holds for B. In this case for all i and j, $|a_i - a_j| \leq d$, $|b_i - b_j| \leq d$ and $|a_i - b_j| > d$. Here for each a_i in A, $T(a_i) = K(n_1 - 1)$ and for each b_j in B, $T(b_j) = K(n_2 - 1)$. From this we get $1 + T(a_i) = (1 - K) + n_1 K$ and $1 + T(b_j) = (1 - K) + n_2K$. Using this we have

$$P\text{-}A(a_1, ..., a_{n_1}, b_1, ..., b_{n_2}) = \frac{\sum_{i=1}^{n_1} ((1 - K) + n_1K)a_i + \sum_{j=1}^{n_2} ((1 - K) + n_2K))b_j}{n_1(1 - K + n_1K) + n_2(1 - K + n_2K)}$$

Letting $\bar{a} = \frac{1}{n_1} \sum_{i=1}^{n_1} a_i$ and $\bar{b} = \frac{1}{n_2} \sum_{j=1}^{n_2} b_j$ we have

$$PA(a_1, ..., a_{n_1}, b_1, ..., b_{n_2}) = \frac{((1 - K) + n_1K)n_1\bar{a} + ((1 - K) + n_2K))n_2\bar{b}}{n_1(1 - K + n_1K) + n_2(1 - K + n_2K)}$$

We get a weighted average of the cluster averages. If we let

$$w_a = \frac{(1 - K + n_1K)n_1}{n_1(1 - K + n_1) + n_2(1 - K + n_2K)} \text{ and } w_b = \frac{(1 - K + n_2K)n_2}{n_1(1 - K + n_1) + n_2(1 - K + n_2K)}$$

then $PA(a_1, ..., a_{n_1}, b_1, ..., b_{n_2}) = w_a\,\bar{a} + w_b\,\bar{b}$. We note $w_a + w_b = 1$ and $\frac{w_a}{w_b} = \frac{(1 - K + n_1K)\,n_1}{(1 - K + n_2K)\,n_2}$

We see that if k = 1, then $\frac{w_a}{w_b} = (\frac{n_1}{n_2})^2$, the weights proportional to the square of the number of elements in the clusters. Thus in this case $w_a = \frac{n_1^2}{n_1^2 + n_2^2}$ and $w_b = \frac{n_2^2}{n_1^2 + n_2^2}$. On the other hand if we allow no support, K = 0, then $\frac{w_a}{w_b} = \frac{n_1}{n_2}$, the weights are just proportional to the number of elements in each cluster. In this case $w_a = \frac{n_1}{n_1 + n_2}$ and $w_b = \frac{n_2}{n_1 + n_2}$. Thus we see as we move from K = 0 to K = 1 we move from being proportional to number of elements in each cluster to being proportional to the square of the number of elements in each cluster. We now begin to see the effect of this power average. If we allow support then elements that are close gain power. This becomes a reflection of the adage that there is power in sticking together. We also observe that if n_1K and n_2K >> (1 - K), there are a large number of arguments, then again $\frac{w_a}{w_b} = (\frac{n_1}{n_2})^2$. Furthermore we note if $n_1 = n_2$ then we always have $\frac{w_a}{w_b} = 1$, here we take the simple average.

Consider now the case when we have q disjoint clusters, each only supporting elements in its neighborhood. Let a_{ji} for i = 1 to n_j be the elements in the jth cluster. In this case

$$P\text{-}A = \frac{\sum_{j=1}^{q} \left(\sum_{i=1}^{n_j} (1 - K + n_jK)a_{ji} \right)}{\sum_{j=1}^{q} n_j(1 - K + n_jK)}$$

Letting $\frac{1}{n_j}\sum_{i=1}^{n_j} a_{ji} = \bar{a}_j$, the individual cluster averages, we can express this power average as

$$\text{P-A} = \frac{\sum_{j=1}^{q} ((1 - K + n_j K)\, n_j\, \bar{a}_j)}{\sum_{j=1}^{q} (1 - K + n_j K)\, n_j}$$

Again we get a weighted average of the individual cluster averages, $\text{P-A} = \sum_{j=1}^{q} w_j\, \bar{a}_j$. In this case $w_i = \frac{(1 - K + n_i K)\, n_i}{\sum_{j=1}^{q} (1 - K + n_j K)\, n_j}$ and $\frac{w_i}{w_j} = \frac{(1 - K + n_i K)\, n_i}{(1 - K + n_j K)\, n_j}$.

Again we see if K= 1, then $\frac{w_i}{w_j} = \frac{n_i^2}{n_j^2}$, the proportionality factor is the square of the number of elements. Here then $w_i = \frac{n_i^2}{\sum_{j=1}^{q} n_j^2}$. If we allow no support, K = 0, then $\frac{w_i}{w_j} = \frac{n_j}{n_i}$, here we get the usual average. We note that K is the value of support.

Consider a case with small value of support, $1 - K \approx 1$. Furthermore assume n_i is a considerable number of elements while n_j is a very small number. Here $l(1 - K) + n_j K \approx 1$ while $(1 - K) + n_i K \approx n_1 K$ then $\frac{w_i}{w_j} = \frac{n_i^2 K}{(1 - K) n_j} \approx \frac{n_i^2 K}{n_j}$.

On the other hand if n_i and n_j are large, $n_i K$ and $n_j K >>> 1$ then $\frac{w_i}{w_j} = \frac{n_i^2}{n_j^2}$.

We that if $(1 - K) << n_j K$ for all j then $\text{P-A} = \frac{\sum_{j=1}^{q} n_j^2\, \bar{a}_j}{\sum_{j=1}^{q} n_j^2}$, the weights in proportion to the square of the number of elements.

Let us observe another interesting property of this P-A. To most clearly illustrate the property we shall assign K = 1. Assume we have two clusters then with K = 1 we have

$$\text{P-A} = \frac{n_1^2\,\bar{a}_1 + n_2^2\,\bar{a}_2}{n_1^2 + n_2^2}$$

If $n_1 \approx n_2 = \frac{1}{2}n$, they have the same number of elements then P-A = $\frac{1}{2}\bar{a}_1 + \frac{1}{2}\bar{a}_2$. Assume now that the second cluster is broken into two equal disjoint clusters. Then $P(A) = \frac{n_1^2\,\bar{a}_1 + n_2^2\,\bar{a}_2 + n_3^2\,\bar{a}_3}{n_1^2 + n_2^2 + n_3^2}$ with $n_1 = \frac{1}{2}$ n, $n_2 = \frac{1}{4}$ n and $n_3 = \frac{1}{4}$ n. From this we see that

$$P(A) = \frac{\frac{1}{4}\bar{a}_1 + \frac{1}{16}\bar{a}_2 + \frac{1}{16}\bar{a}_3}{\frac{1}{4} + \frac{1}{16} + \frac{1}{16}} = \frac{4\,\bar{a}_1 + \bar{a}_2 + \bar{a}_3}{6}$$

We see cluster one's influence (power) has greatly increased because of the fragmentation of cluster two..

We now consider a situation in which we have three sets of elements, A = $\{a_1, ..., a_{n_1}\}$, B = $\{b_1, ..., b_{n_2}\}$ and C = $\{c_1, ..., c_{n_3}\}$. We assume all the elements in A are a neighbors with each other as well as with those in B. Those in B are neighbors with each other and also with those in both A and C. The elements in C are neighbors with themselves and B. Thus B is seen to be between A and C. Here we see that for all a_i we have $T(a_i) = K(n_1 + n_2 - 1)$, for all b_i $T(b_i) = K(n_1 + n_2 + n_3 - 1)$ and for all $c_i T(c_i) = K(n_2 + n_3 - 1)$. Let $\bar{a} = \frac{1}{n_j}\Sigma\, a_j$, $\bar{b} = \frac{1}{n_2}\Sigma\, b_j$ and $\bar{c} = \frac{1}{n_3}\Sigma\, b_j$. Using this we have

$$\text{P-A} = \frac{(1 - K + K(n_1 + n_2))n_1\bar{a} + (1 - K + K(n_1 + n_2 + n_3)n_2\bar{a} + (1 - K + K(n_2 + n_3)n_3\bar{a}}{(1 - K + K(n_1 + n_2))n_1 + (1 - K + K(n_1 + n_2 + n_3)n_2 + (1 - K + K(n_2 + n_3)n_3}$$

Again for illustrative purposes we assume K =1 hence

$$\text{P-A} = \frac{(n_1 + n_2)n_1\bar{a} + n\,n_2\bar{b} + (n_2 + n_3)n_3\bar{c}}{(n_1 + n_2)n_1 + n\,n_2 + (n_2 + n_3)n_3}$$

$$\text{P-A} = \frac{(n - n_2)n_1\bar{a} + n\,n_2\bar{b} + (n - n_1)n_3\bar{c}}{n^2 - 2n_1n_3}$$

We see that relationship between the weights associated A and C is

$$\frac{w_a}{w_c} = \frac{(n - n_3)n_1}{(n - n_1)n_3} = \frac{(n_2 + n_1)n_1}{(n_2 + n_3)n_3}$$

If n_2 is large compared with both n_1 and n_3 then $\frac{w_a}{w_c} = \frac{n_1}{n_3}$, their relationship is proportion to the number of elements in A and C. If n_2 is small compared with both n_1 and n_3 then $\frac{w_a}{w_c} = \frac{n_1^2}{n_3^2}$. Consider the relationship between A and B, which is analogous to B and C, $\frac{w_a}{w_b} = \frac{n_1(n_1 + n_2)}{(n)(n_2)}$. If n_2 is large compared with n_1 and n_3 then $\frac{w_a}{w_b} \approx \frac{n_1 n_2}{(n)(n_2)} \approx \frac{n_1}{n}$

We consider now another situation that exemplifies the possibility for non-monotonicity. Let $\{a_1, ..., a_n, a_{n+1}\}$ be a collection of points in the same cluster, for all a_i and a_j, $|a_i - a_j| \le d$. In this case $\text{P-A}\{a, ..., a_{n+1}\} = \frac{1}{n+1} \sum_{j=1}^{n+1} a_i = \bar{a}$.

Assume now that we replace a_{n+1} by $\hat{a}_{n+1}$ where $\hat{a}_{n+1} \ge a_{n+1}$ and $|a_{n+1} - a_j| > d$ for all other a_j. That is we have moved the n+1th observation all the way to the right. In this case we can view the situation having two disjoint clusters one being $\{a_1, ..., a_n\}$ and the other $\{\hat{a}_{n+1}\}$. As we already established the power average of this situation is

$$\text{P-A}(a_1, a_2, ..., a_n, \hat{a}_{n+1}\} = w_1 \tilde{a} + w_2 \hat{a}_{n+1}$$

here $\tilde{a} = \frac{1}{n} \sum_{i=1}^{n} a_i$ and $\hat{a}_{n+1} = a_{n+1} + \Delta$. We also note that $\bar{a} = \frac{1}{n+1} a_{n+1} + \frac{n}{n+1} \tilde{a}$ hence

$$\tilde{a} = \frac{(n+1)\,\bar{a} - a_{n+1}}{n}$$

In the situation where K = 1 we have $\frac{w_1}{w_2} = \frac{n_1^2}{n_2^2} = \frac{n^2}{1}$. This gives us $w_1 = \frac{n^2}{n^2 + 1}$ and $w_1 = \frac{1}{n^2 + 1}$ and hence

$$\text{P-A}(a_1, ..., \hat{a}_{n+1}) = \frac{n^2}{n^2 + 1} \tilde{a} + \frac{1}{n^2 + 1} \hat{a}) = a + \frac{\Delta - (n-1)\ (a_{n+1} - \bar{a})}{n^2 + 1}$$

Thus we see that if a_{n+1} was the right most element then we get a non-monotonicity as long as Δ is not too big.

4 Forms for the Support Function

The support function is a crucial part of the power average method. The form of the support function is context dependent. Here we describe some useful parameterized formulations for expressing the Sup function. The determination of the values of the parameters may require the use of some learning techniques. We recall if **R** is the range of the values to be aggregated then Sup:$\mathbf{R} \times \mathbf{R} \to [0, 1]$ such that Sup(a, b) = S(b, a), and Sup(a, b) ≥ Sup(x, y) if $|a - b| \leq |x - y|$.

In the preceding we assumed a binary Sup function, Sup(a, b) = K if $|a - b| \leq d$ and Sup(a, b) = 0 if $|a - b| > d$. A natural extension of this is to consider a partitioned type support function. Let K_i for i = 1 to p be a collection of values such that $K_i \in [0, 1]$ and where $K_i > K_j$ if $i < j$. Let d_i be a collection of values such that $d_i \geq 0$ and where $d_i < d_j$ if $i < j$. We now can define a support function as

If $|a - b| \leq d_1$ then Sup(a, b) = K_1

If $d_{j-1} < |a - b| \leq d_j$ then Sup(a, b) = K_j for j = 2 to p - 1

If $d_{p-1} < |a - b|$ then Sup(a, b) = K_p

Inherent in the above type of support function is a discontinuity as we move between the different ranges.

One form of the Sup function with a continuous transition is Sup(a, b) = $K\, e^{-\alpha(a - b)^2}$ where $K \in [0,1]$ and $\alpha \geq 0$. We easily see that this function is symmetric and lies in the unit interval. We see K is the maximal allowable support and α is acting as a attenuator of the distance. The larger the α the more meaningful differences in distance. We note here that a = b gives us Sup(a, b) = K and as the distance between a and b gets larger, Sup(a, b) →0.

Using this form for support function we have

$$P\text{-}A(a_1, \ldots, a_n) = \frac{\sum_{i=1}^{n} (1 + T(a_i))a_i}{\sum_{i=1}^{n} (1 + T(a_i))}$$

where $T(a_i) = \sum_{\substack{j=1 \\ j \neq i}}^{n} K e^{-\alpha(a_i - a_j)^2}$. Denoting $V_i = 1 + T(a_i)$ we express $P\text{-}A(a_1, \ldots, a_n) = \Sigma_i\, w_i\, a_i$ where $w_i = \frac{V_i}{\sum_{j=1}^{n} V_j}$. Since $e^{-\alpha(a_i - a_i)^2} = 1$ we can express $V_i = 1 - K + K\, M_i$ where $M_i = \sum_{j=1}^{n} e^{-\alpha(a_i - a_j)^2}$. Noting the similarly of M_i to the mountain function used in mountain clustering [3] we call M_i the support mountain at i. It's clear that if $a_p = a_q$ then $M_q = M_p$ and hence $V_q = V_p$. It is also noted that $M_i \geq 1$ for all i.

We see here that

$$\text{P-A}(a, \ldots, a_n) = \frac{\sum_{i=1}^{n} (1 - K)\, a_i + K \sum_{i=1}^{n} M_i a_i}{n(1 - K) + K \sum_{i=1}^{n} M_i}$$

In the special case where K = 1 then $V_i = M_i$ and hence

$$\text{P-A}(a_1, \ldots, a_n) = \frac{\sum_{i=1}^{n} M_i a_i}{\sum_{i=1}^{n} M_i}$$

A simple algorithm approach somewhat is in spirit of the mountain method is as follows:

1. For each argument value a_i, i = 1 to n, initialize $M_i = 0$
2. For each data point a_j j = 1 to n augment M_i, $M_i = M_i + e^{-\alpha(a_i - a_j)^2}$

This builds the support mountain.

3. Calculate $V_i = (1 - K) + K\, M_i$ - linear transformation of mountain values
4. Calculate $w_i = \dfrac{V_i}{\sum_{j=1}^{n} V_j}$
5. $\text{P-A} = \Sigma_i\, w_i\, a_i$

As we have noted an important characteristic of this power average is its possibility for displaying non-monotonicity, a feature that can provide one of the benefits of this method. The following example illustrates the occurrence of non-monotonicity.

Example: Consider the Power average of twenty elements, 10 of which are ten's and 10 of which are five's. In this case the ordinary average evaluates to 7.5 and for any choice of K and α the power average also evaluates to 7.5. The following table shows what happens as we change one of the values originally equal to 10. For illustrative purposes we used K = 1 and $\alpha = 0.3$

Value	**AVE**	**P-A**
10	7.5	7.5
9	7.45	7.398
8	7.4	7.278
7	7.35	7.193
6	7.3	7.083

5	7.25	6.982
11	7.55	7.4727
12	7.6	7.346
13	7.65	7.259
14	7.7	7.232
15	7.75	7.22856
16	7.8	7.22828
17	7.85	7.22829
18	7.9	7.22831
19	7.9	7.22832
20	8	7.22834

We see that as we decrease the value and move it towards the cluster of fives our P-A decrease, although more dramatically than the average. Essentially the variable value is beginning to join the cluster of fives and increase its power. In the case of increasing the value, initially the power average instead of increasing as does the average begins to decrease, exhibiting non–monotonicity. This decrease is a reflection of the fragmentation of the cluster at 10, it is losing its power because it lost a member and the cluster at five has gained in power more than compensating for the increase in value. This decreasing in the P-A continues as we increase the element until it reaches eighteen at which time we see a reversal and now the P-A starts increasing At this point the increase in value begins overcoming the loss of power. But still we are favoring the cluster of fives.

We describe another approach to obtaining the support function that combines the partitioning of the first method with the continuity displayed by the exponential function. This approach motivated by Zadeh's idea of computing with words [6] makes use of fuzzy systems modeling technology [7]. We shall briefly describe the possibilities for this approach. Using this approach we can express our support function by a description of its performance in terms of a set of rules using linguistic values. For example.

If difference is *very small* then support is K_1

If difference is *small* then support is K_2

If difference is *moderate* the support is K_3

If difference is *large* the support is K_4

If difference is *very large* the support is K_5

Representing the italic terms as fuzzy sets, VS, S, M, L, and VL respectively and denoting the difference between a and b as Δ than we have a collection of fuzzy if-then rules, a fuzzy systems model:

If Δ is VS then $S(a, b) = K_1$

If Δ is S then $S(a, b) = K_2$

If Δ is M then $S(a, b) = K_3$

If Δ is L then $S(a, b) = K_4$

If Δ is VL then $S(a, b) = K_4$

here $K_i < K_j$ if $i > j$.

To obtain the Sup(a, b) we use the inference mechanism of fuzzy systems modeling. Letting $\Delta = |a - b|$ then the analytic formulation of our support function is

$$Sup(a, b) = \frac{K_1 VS(\Delta) + K_2 S(\Delta) + K_3 M(\Delta) + K_4 L(\Delta) + VL(\Delta)}{VS(\Delta) + S(\Delta) + M(\Delta) + L(\Delta) + VL(\Delta)}$$

here $VS(\Delta)$ indicates the membership of Δ in the fuzzy subset VS.

We now look at the power average in the special situation in which the arguments that are being aggregated, the a_i, always be in the unit interval [0, 1]. This is a situation that occurs in many environments when the arguments are degrees of belief. We note a particular important situation is in the aggregation of fuzzy subsets.

In the case when the arguments lie in the unit interval a very natural definition for the Sup function is

$$Sup(a, b) = K(1 - |a - b|^{\alpha})$$

for $\alpha \geq 0$. Here we see that the term $|a - b|$ is a measure of distance between the arguments. We note since a and b are assumed to lie in the unit interval then $|a - b|$ must also lie in the unit interval as well as $|a - b|^{\alpha}$. We see $|a - b| \rightarrow 0$ indicates the elements are close and $|a - b| \rightarrow 1$ indicates the elements are far. We see that is Sup is related to the negation of the distance.

We notice that because a and b always lie in the unit interval, $|a - b| = 1$ if and only if one of the arguments equal zero and the other equals one. Furthermore we note that α modifies the effects of distance. Since $(a - b) < 1$ then $\alpha > 1$ reduces the effect of distance while $\alpha < 1$ increase the effects of distance. We note Sup(a, b) = K when a = b.

As in the preceding $P\text{-}A(a_1, ..., a_n) = \frac{\sum_{i=1}^{n} V_i a_i}{\sum_{i=1}^{n} V_i}$. Let us consider the case when $\alpha = 2$, $Sup(a, b) = K(1 - (a - b)^2)$. Here $V_i = 1 + T(a_i)$ with $T(a_i) = K \sum_{\substack{j=1 \\ i \neq j}}^{n} (1 - (a_i - a_j)^2)$. Realizing $1 - (a_i - a_i)^2 = 1$ then $V_i = (1 - K) + K \sum_{j=1}^{n} (1 - (a_i - a_j)^2)$. Letting $Q_i = \sum_{j=1}^{n} (a_i - a_j)^2$ we have

$$V_i = 1 - K + Kn - Kq_i$$

Let us carefully look at the term Q_i. We shall denote $\bar{a} = \frac{1}{n}\sum_{j=1}^{n} a_j$, it is the average, and denote $Var(a) = \frac{1}{n}\sum_{j=1}^{n} (a_j - \bar{a})^2$. Using these notations we can express

$$Q_i = \sum_{j=1}^{n} (a_i - a_j)^2 = \sum_{j=1}^{n} [(a_i - \bar{a}) - (a_j - \bar{a})]^2$$

$$Q_i = \sum_{j=1}^{n} (a_i - \bar{a})^2 + \sum_{j=1}^{n} (a_j - \bar{a})^2 - 2\sum_{j=1}^{n} (a_i - \bar{a})(a_j - \bar{a})$$

Realizing that $\sum_{j=1}^{n} (a_i - \bar{a})(a_j - \bar{a}) = (a_i - \bar{a})\sum_{j=1}^{n} (a_j - \bar{a}) = 0$ we have

$$Q_i = \sum_{j=1}^{n} (a_i - \bar{a})^2 + \sum_{j=1}^{n} (a_j - \bar{a})^2$$

Letting $\Delta_i = |a_i - \bar{a}|$. we have $Q_i = n\ \Delta_i^2 + n\ Var(a)$.

From this we have $V_i = (1 - K) + Kn - nK(\Delta_i^2 + Var(a))$. Using this we get that

$$\sum_{i=1}^{n} V_i = n(1 - K) + Kn^2 - n^2 K\, Var(a) - n K \sum_{i=1}^{n} \Delta_i^2$$

Since $\frac{1}{n}\sum_{i-} \Delta_i^2 = Var(a)$ then $\sum_{i=1}^{n} V_i = n(1 - K) + Kn^2 - 2n^2 K Var(a)$

Let us consider the special case where $K = 1$, here $V_i = n\,(1 - Var(a) - \Delta_i^2)$ and $\sum_{i=1}^{n} V_i = n^2\,(1 - 2\,Var(a))$. Using this

$$P\text{-}A(a_1, \ldots, a_n) = \frac{\sum_{i=1}^{n} V_i\, a_i}{n^2(1 - 2Var(a))} = \bar{a} + \frac{\bar{a}\sum_{i=1}^{n} \Delta_i^2 - \sum_{i=1}^{n} \Delta_i^2 a_i}{n(1 - 2Var(a))}$$

We see that if the arguments are such that there are a few large values far away from the the rest of the values mean then the power average tends to pull $\bar{a}$ downwards.

Another interesting case of $Sup(a, b) = K(1 - |a - b|^{\alpha})$ occurs when $\alpha = 1$, here $Sup(a\ b) = K(1 - |a - b|)$. We note that $|a - b| = Max(a, b) - Min(a, b) = (a \vee b) - (a \wedge b)$. Here again $P\text{-}A(a_1, ..., a_n) = \frac{\sum_{i=1}^{n} V_i a_i}{\sum_{i=1}^{n} V_i}$. In this case

$$V_i = 1 + (v - 1) K - K \sum_{j=1}^{n} [(a_i \vee a_j) - (a_j \wedge a_i)]$$

Without loss of generality let us assume that the a_i have been indexed in descending order, thus a_i is the ith largest of the arguments. In this case

$a_i = Min[a_i, a_j]$ and $a_j = Max[a_i, a_j]$ for $j = 1$ to $i - 1$
$a_j = Min[a_i, a_j]$ and $a_i = Max[a_i, a_j]$ for $j = i + 1$ to n
$a_i = Min[a_i, a_j] = Max[a_i, a_j]$ for $j = 1$

If we denote $Q_i = \sum_{j=1}^{n} |a_i - a_j|$ then

$$Q_i = \sum_{j=1}^{n} (a_i \vee a_j) - (a_i \wedge a_j) = \sum_{j=1}^{i-1} a_j + \sum_{j=i+1}^{n} a_i - (\sum_{j=1}^{i-1} a_i + \sum_{j=i+1}^{n} a_j)$$

$$Q_i = \sum_{j=1}^{i-1} a_j - \sum_{j=i+1}^{n} a_j - (\sum_{j=1}^{i-1} a_i - \sum_{j=i+1}^{n} a_i) = \sum_{j=1}^{i-1} a_j - \sum_{j=i+1}^{n} a_j + (n - 2i)a_i$$

Denoting $SL(i) = \sum_{j=1}^{i} a_j$ and $SU(i) = \sum_{j=i+1}^{n} a_j$ then $Q_i = SL(i) - SU(i) + (n - 2i)\, a_i$ and

$$V_i = 1 + (n - 1) K - K (SL(i) - SU(i) + (n - 2i)\, a_i)$$

and $\sum_{i=1}^{n} V_i = n + n (n - 1) K - K \sum_{j=1}^{n} (SL(i) - SU(i) + (n - 2i)\, a_i)$.

Let us consider the special case where $K = 1$, hence

$$V_i = n - (SL(i) - SU(i) + (n - i)a_i)$$

$$\sum_{i=1}^{n} V_i = n^2 - \sum_{i=1}^{n} SL(i) - S(u)i + (n - 2i)a_i$$

Since a_i appears in n - i + 1 of the SL and in i of SU, $\sum_{i=1}^{n} SL(i) - SU(i) = \sum_{i=1}^{n} (n-2i+1)a_i$ then $\sum_{i=1}^{n} V_i = n^2 - \sum_{i=1}^{n} (2n - 4i + 1)\, a_i = n^2 - n(2n-1)\bar{a} + 4\sum_{i=1}^{n} i\, a_i$

5 Empowering Alternative Mean Operators

The average operator, $\frac{1}{n}\sum_{i=1}^{n} a_i$, provides one example of mean type aggregation operators [8]. We recall that mean type operators are characterized by boundedness, commutativity and monotonicity. Other examples of mean type operators are the Max, Min, and Median. In the preceding with the power average we extended the average operator by introducing the idea of support. That is with the P-A operator we allowed arguments in the aggregation to support each. This effectively result is a weights associated with the different arguments depending upon the support they obtained from other elements being aggregated. In this section we want to generalize the idea of supported aggregation to a wider class of mean operators.

We first look at the OWA operator [9] and introduce the **Power-OWA** operator. An OWA operator can be defined in terms of function $g:[0, 1] \to [0, 1]$, called a BUM function, having the properties: **1**. $g(0) = 0$, **2**. $g(1) = 1$ and **3**. $g(x) \geq g(y)$ if $x > y$. Using this BUM function the OWA aggregation $OWA_g(a_1, ..., a_n)$ can be expressed as $OWA_g(a_1, ..., a_n) = \sum_{i=1}^{n} w_i b_i$ where b_i is the ith largest of a_j and the w_i are a collection of weights such that $w_i = g(\frac{i}{n}) - g(\frac{i-1}{n})$. It can be easily shown these weights are proper, $w_i \in [0, 1]$ and $\sum_{i=1}^{n} w_i = 1$.

By appropriately selecting g we can implement different types of aggregation imperative. For example if $g(x) = x$ then the OWA operator becomes the ordinary average with $w_j = \frac{1}{n}$ for all j. If g is such that $g(x) = 1$ for all $x > 0$ then we get the maximal aggregation, $OWA_g(a_1, ..., a_n) = Max_i[a_i]$. If g is such that $g(x) = 0$ for all $x < 1$ the we get the minimal aggregation, $OWA_g(a_1, ..., a_n) = Min_i[a_i]$. A median type operator can be implemented if $g(x) = 0$ for $x < 0.5$ and $g(x) = 1$ for $x \geq 0.5$. A class of OWA operators can be obtained if $g(x) = x^{\alpha}$ with $\alpha \geq 0$.

Before preceding we shall find it convenient to use a slightly different notation for the OWA operator. We shall let *index* be an indexing function such that index(i) is the index of the ith largest of the a_j. Thus we order the argument in descending order and then index(i) is the index of ith element in this list. Since b_i is the ith largest of the a_j using this index function we see that $b_i = a_{index(i)}$. Using this we can express the OWA aggregation as

$$OWA_g(a_i, ..., a_n) = \sum_{i=1}^{n} w_i\, a_{index(i)},$$

where the w_i as before are $w_i = g(\frac{i}{n}) - g(\frac{i-1}{n})$.

As in the preceding we shall let Sup(a, b) indicate the support for a from b. We note that using the index operator $Sup(a_{index(i)}, a_{index(j)})$ still represents the support of the second argument for the first. Because of the nature of the Sup function, $Sup(a, b) \geq Sup(x, y)$ when $|a - b| < |x - y|$, and the ordering captured by the index function we note that if $i < j < k$ then $Sup(a_{index(i)}, a_{index(j)}) \geq Sup(a_{index(i)}, a_{index(k)})$ and $Sup(a_{index(j)}, a_{index(k)}) \geq Sup(a_{index(i)}, a_{index(k)})$. We let $T(a_{index(i)})$ denote the support of the ith largest argument by all the other arguments, hence

$$T(a_{index(i)}) = \sum_{\substack{j=1 \\ j \neq i}}^{n} Sup(a_{index(i)}\ a_{index(j)}).$$

In addition we shall let $V_{index(i)} = 1 + T(a_{index(i)})$ and denote $TV = \sum_{i=1}^{n} V_{index(i)}$. We now can define the Power OWA operator as

$$POWA_g(a_1, ..., a_n) = \sum_{i=1}^{n} u_i\, a_{index(i)}$$

where $u_i = g(\frac{R_i}{TV}) - g(\frac{R_{i-1}}{TV})$ with $R_i = \sum_{j=1}^{i} V_{index(j)}$, by definition $R_{i-1} = 0$.

We note that $TV = R_n$. We also observe that $R_i = R_{i-1} + V_{index(i)}$.

We can show in the special case where $g(x) = x$ that this reduces to the Power Average. In this case

Another class of mean operators, called generalized means [8], are defined by

$$GM_\alpha(a_1, a_2, ..., a_n) = \left(\frac{1}{n}\sum_{j=1}^{n} a_j^\alpha\right)^{1/\alpha}$$

where $\alpha \in [-\infty, \infty]$. It is required when using these operators that $a_j \geq 0$. The inclusion of support in this class of mean operators can be accomplished in the following manner. Again let $T(a_i) = \sum_{\substack{j=1 \\ j \neq i}}^{n} Sup(a_i, a_j)$, $V_i = 1 + T(a_i)$ and $TV = \sum_{i=1}^{n} V_i$. the power generalized mean is defined as.

$$PGM_\alpha(a_1, ..., a_n) = \left(\frac{1}{TV}\sum_{i=1}^{n} V_i\, a_i^{\alpha}\right)^{\frac{1}{\alpha}}$$

We shall not further look at the properties of the Power OWA or the power generalized mean only to indicate that they act with respect to their mother operations in a manner similar to the way the power average acts with respect to the average.

In the preceding we assumed that all of the objects being aggregated were of equal importance. Here we shall consider the effect on the power operations of having differing importances associated with the objects being aggregated. We assume that each being aggregated has a weight $\omega_i \in [0, 1]$ indicating its importance. The procedure for including this importance involves a simple modification of the value V_i which we recall is defined as $V_i = 1 + T(a_i)$ where $T(a_i) = \sum_{\substack{j=1 \\ j \neq i}}^{n} Sup(a_i, a_j)$. In order to include the weights we suggest redefining V_i as

$$V_i = \omega_i \left(1 + \sum_{\substack{j=1 \\ j \neq i}}^{n} \omega_j\, Sup(a_i, a_j)\right)$$

and then continuing as described in preceding.

6 Conclusion

We introduced the power average operator to provide an aggregation operator which allows argument values to support each other in the aggregation process. The properties of this operator were described. We discussed the idea of a power median. We introduced some formulations for the support function used in the power average. We extended the idea of empowerment, supported aggregation, to a wider class of mean operators such as the OWA and generalized mean. Interesting applications of this approach to aggregation can be seen in data mining, group decision making and information fusion.

References

[1] Yager, R.R.: The power average operator. IEEE Transaction on Systems, Man and Cybernetics Part A 31, 724–730 (2001)
[2] Xu, Z., Yager, R.R.: Power geometric operators and their use in group decision making. IEEE Transactions on Fuzzy Sets and Systems (to appear)
[3] Yager, R.R., Filev, D.P.: Approximate clustering via the mountain method. IEEE Transactions on Systems, Man and Cybernetics 24, 1279–1284 (1994)
[4] Chiu, S.L.: Fuzzy model identification based on cluster estimation. Journal of Fuzzy and Intelligent Systems 2, 267–278 (1994)
[5] Yager, R.R.: A note on a fuzzy measure of typicality. International Journal of Intelligent Systems 12, 233–249 (1997)
[6] Zadeh, L.A.: Generalized theory of uncertainty (GTU)-principal concepts and ideas. Computational Statistics and Data Analysis 51, 15–46 (2006)
[7] Yager, R.R., Filev, D.P.: Essentials of Fuzzy Modeling and Control. John Wiley, New York (1994)
[8] Beliakov, G., Pradera, A., Calvo, T.: Aggregation Functions: A Guide for Practitioners. Springer, Heidelberg (2007)
[9] Yager, R.R.: On ordered weighted averaging aggregation operators in multi-criteria decision making. IEEE Transactions on Systems, Man and Cybernetics 18, 183–190 (1988)

Author Index

Baione, Fabio 123
Bertini, Cesarino 195
Bosi, Gianni 1
Brunelli, Matteo 159

Carlsson, Christer 11
Castagnoli, Erio 31
Cavallo, Bice 111
Cerqueti, Roy 53
Coletti, Giulianella 69

D'Apuzzo, Livia 111
De Angelis, Paolo 123
Dubois, Didier 139

Fargier, Hélène 139
Favero, Gino 31
Fedrizzi, Mario 159
Fedrizzi, Michele 159
Fodor, János 183
Fullér, Robert 11

Gambarelli, Gianfranco 195
García-Cascales, M. Socorro 383
García-Lapresta, José Luis 211
Greco, Salvatore 225

Isler, Romano 1

Kacprzyk, Janusz 261

Lamata, M. Teresa 383
Llamazares, Bonifacio 211

Marcarelli, Gabriella 111, 281
Marques Pereira, Ricardo Alberto 159
Matarazzo, Benedetto 225
Maturo, Antonio 291
Modesti, Paola 305

Nurmi, Hannu 321

Ottaviani, Riccardo 123

Pais, Tiago C. 353
Peña, Teresa 211
Proto, A.N. 339

Rambaud, Salvador Cruz 85
Ribeiro, Rita A. 353
Rotundo, Giulia 53, 369

Sarris, C.M. 339
Scozzafava, Romano 377
Simões, Luis F. 353
Słowiński, Roman 225
Squillante, Massimo 291

Torrecillas, María José Muñoz 85

Uristani, Angelo 195

Vantaggi, Barbara 69
Ventre, Aldo G.S. 291
Ventre, Viviana 281
Verdegay, J. Luís 383

Yager, Ronald R. 397

Zadrożny, Sławomir 261

GPSR Compliance

The European Union's (EU) General Product Safety Regulation (GPSR) is a set of rules that requires consumer products to be safe and our obligations to ensure this.

If you have any concerns about our products, you can contact us on ProductSafety@springernature.com

In case Publisher is established outside the EU, the EU authorized representative is:

Springer Nature Customer Service Center GmbH
Europaplatz 3
69115 Heidelberg, Germany

Batch number: 10755461

Printed by Printforce, the Netherlands